W0193184

Progress in Mathematics

Volume 299

Series Editors
Hyman Bass
Joseph Oesterlé
Yuri Tschinkel
Alan Weinstein

Folkert Müller-Hoissen
Jean Marcel Pallo
Jim Stasheff
Editors

Associahedra, Tamari Lattices and Related Structures

Tamari Memorial Festschrift

 Birkhäuser

Editors
Folkert Müller-Hoissen
Max-Planck-Institute for
 Dynamics and Self-Organization
Göttingen
Germany

Jean Marcel Pallo
Département d'Informatique, LE2I
Université de Bourgogne
Dijon
France

Jim Stasheff
Department of Mathematics
University of North Carolina
Chapel Hill, NC
USA

ISBN 978-3-0348-0404-2   ISBN 978-3-0348-0405-9 (eBook)
DOI 10.1007/978-3-0348-0405-9
Springer Basel Heidelberg New York Dordrecht London

Library of Congress Control Number: 2012942603

© Springer Basel 2012
This work is subject to copyright. All rights are reserved by the Publisher, whether the whole or part of the material is concerned, specifically the rights of translation, reprinting, reuse of illustrations, recitation, broadcasting, reproduction on microfilms or in any other physical way, and transmission or information storage and retrieval, electronic adaptation, computer software, or by similar or dissimilar methodology now known or hereafter developed. Exempted from this legal reservation are brief excerpts in connection with reviews or scholarly analysis or material supplied specifically for the purpose of being entered and executed on a computer system, for exclusive use by the purchaser of the work. Duplication of this publication or parts thereof is permitted only under the provisions of the Copyright Law of the Publisher's location, in its current version, and permission for use must always be obtained from Springer. Permissions for use may be obtained through RightsLink at the Copyright Clearance Center. Violations are liable to prosecution under the respective Copyright Law.
The use of general descriptive names, registered names, trademarks, service marks, etc. in this publication does not imply, even in the absence of a specific statement, that such names are exempt from the relevant protective laws and regulations and therefore free for general use.
While the advice and information in this book are believed to be true and accurate at the date of publication, neither the authors nor the editors nor the publisher can accept any legal responsibility for any errors or omissions that may be made. The publisher makes no warranty, express or implied, with respect to the material contained herein.

Printed on acid-free paper

Springer Basel AG is part of Springer Science+Business Media (www.birkhauser-science.com)

Preface

On the occasion of the *centennial birthday* of the mathematician *Dov Tamari* (1911–2006), born as the German *Bernhard Teitler*, this book commemorates his ground breaking work resulting in an *associativity theory*, with important contributions to the "word (decision) problem", as well as lattice theory and geometric combinatorics. The editors of this book invited designated researchers to present modern areas of mathematics that are related to Tamari's work.

To a monomial (word) formed from a set (of letters), one can assign different meanings by properly distributing brackets. If the bracketing expresses a binary operation on the set, associativity becomes an issue. Interpreting associativity as a (left- or rightward) substitution rule leads to what is known as a *Tamari lattice*. This partial order on a Catalan set (i.e., the number of its elements is a Catalan number) first appeared in 1951 in Dov Tamari's thesis at the Sorbonne in Paris. It turned out that these Tamari lattices possess realizations on special polytopes, called *associahedra*, which appeared in a different context in Jim Stasheff's thesis in 1961. In fact, associahedra already appeared in Tamari's thesis, but not in the part that was published. Since then these beautiful structures, and quite a number of important generalizations, have made their appearance in many publications in different areas of pure and applied mathematics, such as Algebra, Combinatorics, Computer Science, Category Theory, Geometry, Topology, and more recently also in Physics. It is this interdisciplinary nature of these structures that provides much of their fascination and value.

In the first chapter of this book, Folkert Müller-Hoissen and Hans-Otto Walther describe Tamari's extremely troubled life. When the Nazis came to power in Germany of 1933, he saw himself forced to leave Germany, losing the possibility of a smooth academic career he could have had in a less cruel political and social environment. All the obstacles along his further way luckily could not break his dedication and passion for mathematics. His uncompromising demand for honesty and fairness on all levels, including politics, surely did not make his life easier. The chapter about Dov Tamari also offers an elementary introduction to some aspects of his mathematical work. It is supplemented by Carl Maxson's reminiscences as a student of Tamari.

Jim Stasheff, whose name is firmly connected with associahedra, traces the latter back to Tamari's 1951 thesis, reviews their history, and leads the reader to modern developments. Jean-Louis Loday develops an arithmetic of (planar rooted binary) trees, a framework in which Tamari lattices find a natural place. He also reviews realizations of the latter as polytopes, the associahedra.

The further chapters in this book are of a somewhat more advanced nature. Susan Gensemer reviews the problem of extending a partially defined binary operation on a set (partial groupoid) to a completely defined associative binary operation (semigroup). This problem has been at the very roots of Tamari's mathematical research.

We grouped together articles that deal primarily with associahedra and related families of polytopes, and then those that center more around *Tamari lattices* and related families of posets.

Satyan Devadoss, Benjamin Fehrman, Timothy Heath and Aditi Vashist treat geometric and combinatorial aspects of the moduli space of "particles" on the Poincaré disk. In this framework, a generalization of associahedra shows up, the cyclohedra. Cesar Ceballos and Günter Ziegler summarize some mysteries and questions concerning realizations of the associahedra. A well-known class of polytopes, the permutahedra (also called permutohedra), and moreover polytopes obtained from Cambrian lattices, which generalize Tamari lattices, are the subject of Christophe Hohlweg's article, highlighting the role of finite reflection groups in their realizations. Further classes of polytopes, like flag nestohedra, graph-associahedra and graph-cubeahedra, arise from truncations of cubes, and their properties are described in the article by Victor Buchstaber and Vadim Volodin. Stefan Forcey reports on an extension of the Tamari order to families of polytopes called multiplihedra and composihedra, and explores the interplay between lattice structures and Hopf algebra structures.

Patrick Dehornoy presents an exhaustive study of the connection between Tamari lattices and the Thompson group F, which consists of a special class of piecewise linear homeomorphisms of the unit interval onto itself. Ross Street looks at the Tamari lattice as an example of an operad and dives into monoidal categories. Frédéric Chapoton considers the category of modules over the incidence algebra of a Tamari lattice, and also the derived category of the former. Hugh Thomas explains how the Tamari lattice arises in the context of the representation theory of quivers. Nathan Reading traces the way from Tamari lattices to Cambrian lattices, in the context of finite Coxeter groups, reviews the construction of Cambrian fans, and moreover makes contact with the important concept of cluster algebras. Filippo Disanto, Luca Ferrari, Renzo Pinzani and Simone Rinaldi present a unified setting for Dyck and Tamari lattices. Dyck lattices are a refinement of Tamari lattices. The restriction of the weak (strong) Bruhat order on permutations of a fixed length to "312-avoiding permutations" leads to the Tamari (Dyck) order. A generalization of the Tamari order to a partial order on the set of "tubings" of a simple graph is described in María Ronco's work. Jörg Rambau and Victor Reiner present a survey of higher Stasheff-Tamari orders (which first appeared in the work of Mikhail Kapranov and

Vladimir Voevodsky). These are posets defined on triangulations of cyclic polytopes and there is a relation with higher Bruhat orders (first introduced by Yuri Manin and Vadim Schechtman).

A physical realization of maximal chains of Tamari lattices in terms of tree-shaped soliton solutions of the Kadomtsev-Petviashvili (KP) equation, describing, e.g., shallow water waves, is the subject of the article by Aristophanes Dimakis and Folkert Müller-Hoissen. The analysis of KP solitons naturally leads to a reduction of higher Bruhat orders to higher Tamari orders, which is different from the relation described by Rambau and Reiner.

We hope that this book will convey to the reader a bit of the fascination that was experienced by those who contributed to the foundations and modern developments described in it.

Finally, we would like to thank Dr. Thomas Hempfling for his efficient help during the publishing process.

Göttingen, Dijon, Chapel Hill   *Folkert Müller-Hoissen*
October 2011   *Jean Pallo*
                                                                               Jim Stasheff

Contents

List of Contributors

Buchstaber, Victor
Geometry and Topology Department, Steklov Institute of Mathematics, Gubkina str. 8, 119991 Moscow, Russia
e-mail: *buchstab@mi.ras.ru*

Ceballos, Cesar
Institut für Mathematik, Freie Universität Berlin, Arnimallee 2, 14195 Berlin, Germany
e-mail: *ceballos@math.fu-berlin.de*

Chapoton, Frédéric
Institut Camille Jordan, Université Claude Bernard Lyon 1, Bâtiment Braconnier, 21 Avenue Claude Bernard, 69622 Villeurbanne Cedex, France
e-mail: *chapoton@math.univ-lyon1.fr*

Dehornoy, Patrick
Laboratoire de Mathématiques Nicolas Oresme, Université de Caen, 14032 Caen, France
e-mail: *patrick.dehornoy@gmail.com*

Devadoss, Satyan
Department of Mathematics and Statistics, Williams College, Williamstown, MA 01267, USA
e-mail: *satyan.devadoss@williams.edu*

Dimakis, Aristophanes
Department of Financial and Management Engineering, University of the Aegean, 41 Kountourioti Str., GR-82100 Chios, Greece
e-mail: *dimakis@aegean.gr*

Disanto, Filippo
Institut für Genetik, Universität Köln, Zülpicher Str. 47a, 50674 Köln, Germany
e-mail: *disafili@yahoo.it*

Fehrman, Benjamin
Department of Mathematics, University of Chicago, Chicago, IL 60637, USA
e-mail: *bfehrman@math.uchicago.edu*

Ferrari, Luca
Dipartimento di Sistemi e Informatica, Università degli Studi di Firenze, Viale G.B.
Morgagni 65, 50134 Firenze, Italy
e-mail: *ferrari@dsi.unifi.it*

Forcey, Stefan
Department of Mathematics, Buchtel College of Arts and Sciences, The University
of Akron, Akron, OH 44325-4002, USA
e-mail: *sforcey@gmail.com*

Gensemer, Susan Helene
Department of Economics, Maxwell School of Citizenship and Public Affairs,
Syracuse University, Syracuse, NY 13244-1020, USA
e-mail: *gensemer@maxwell.syr.edu*

Heath, Timothy
Department of Mathematics, Columbia University, New York, NY 10027, USA
e-mail: *timheath@math.columbia.edu*

Hohlweg, Christophe
Département de Mathématiques – LaCIM, Université du Québec à Montréal, CP
8888 Succ. Centre-Ville, Montréal, Québec, H3C 3P8 Canada
e-mail: *hohlweg.christophe@uqam.ca*

Loday, Jean-Louis
Institut de Recherche Mathématique Avancée, CNRS et Université de Strasbourg, 7
rue René-Descartes, 67084 Strasbourg, France
e-mail: *loday@math.unistra.fr*

Maxson, Carl
Mathematics Department, Texas A&M University, College Station, TX 77843-3368,
USA
e-mail: *cjmaxson@math.tamu.edu*

Müller-Hoissen, Folkert
Max-Planck-Institute for Dynamics and Self-Organization, Bunsenstrasse 10,
D-37073 Göttingen, Germany
e-mail: *folkert.mueller-hoissen@ds.mpg.de*

Pallo, Jean Marcel
Département d'Informatique, LE2I, Université de Bourgogne, B.P. 47870, 21078
Dijon Cedex, France
e-mail: *pallo@u-bourgogne.fr*

Pinzani, Renzo
Dipartimento di Sistemi e Informatica, Università degli Studi di Firenze, Viale G.B. Morgagni 65, 50134 Firenze, Italy
e-mail: *pinzani@dsi.unifi.it*

Rambau, Jörg
Lehrstuhl für Wirtschaftsmathematik, Universität Bayreuth, D-95440 Bayreuth, Germany
e-mail: *joerg.rambau@uni-bayreuth.de*

Reading, Nathan
Department of Mathematics, North Carolina State University, SAS Hall 4118, Box 8205, Raleigh, NC 27695, USA
e-mail: *nathan_reading@ncsu.edu*

Reiner, Victor
School of Mathematics, University of Minnesota, Minneapolis, MN 55455, USA
e-mail: *reiner@math.umn.edu*

Rinaldi, Simone
Dipartimento di Scienze Matematiche e Informatiche, Pian dei Mantellini, 44, 53100, Siena, Italy
e-mail: *rinaldi@unisi.it*

Ronco, Maria
Instituto de Matemática y Física, Universidad de Talca, 2 norte 685 Talca, Chile
e-mail: *mariaronco@inst-mat.utalca.cl*

Stasheff, Jim
Mathematics Department, University of North Carolina at Chapel Hill, CB #3250, Phillips Hall Chapel Hill, NC 27599, USA
e-mail: *jds@math.upenn.edu*

Street, Ross
Mathematics Department, Macquarie University, New South Wales 2109, Australia
e-mail: *ross.street@mq.edu.au*

Thomas, Hugh
Department of Mathematics and Statistics, University of New Brunswick, Fredericton, NB, E3B 5A3, Canada
e-mail: *hugh@math.unb.ca*

Vashist, Aditi
Department of Mathematics, University of Michigan, Ann Arbor, MI 48109, USA
e-mail: *avashist@umich.edu*

Volodin, Vadim
Steklov Institute of Mathematics, Gubkina str. 8, 119991 Moscow, Russia
e-mail: *volodinvadim@gmail.com*

Walther, Hans-Otto
Mathematisches Institut, University of Gießen, Arndtstraße 2, D-35392 Gießen,
Germany
e-mail: *hans-otto.walther@math.uni-giessen.de*

Ziegler, Günter M.
Institut für Mathematik, Freie Universität Berlin, Arnimallee 2, 14195 Berlin,
Germany
e-mail: *ziegler@math.fu-berlin.de*

Dov Tamari (formerly Bernhard Teitler)

Folkert Müller-Hoissen and Hans-Otto Walther

Abstract The life of Dov Tamari is described, including a brief introduction to his mathematical work.

1 Germany

Bernhard Teitler was born in Fulda, Germany, on April 29, 1911, as a child of Frieda and Levi Yitzchak Teitler, who was reputed to be a descendant of the famous rabbi Levi Yitzchak of Berdichev and originally came from Bistritz in Transilvania.[1] They had moved from Vienna, Austria, to Fulda, where Levi Yitzchak Teitler was registered as a "rubberstamp maker". He also served as a rabbi in Fulda and small communities nearby [2].[2] When Bernhard was one or two years old, the family moved to Giessen. Already at the age of three, Bernhard was able to read from newspapers. His early youth was overshadowed by World War I, when his father had to serve in the Austrian army, as an officer in the rank of Lieutenant, and the small family business, a printing

Folkert Müller-Hoissen
Max-Planck-Institute for Dynamics and Self-Organization
Bunsenstrasse 10, 37073 Göttingen, Germany
e-mail: *folkert.mueller-hoissen@ds.mpg.de*

Hans-Otto Walther
Mathematisches Institut, University of Giessen, Arndtstrasse 2, 35392 Giessen, Germany
e-mail: *hans-otto.walther@math.uni-giessen.de*

[1] A short description of Dov Tamari's life already appeared as an appendix "About the author", written by his son Doram Tamari, in Tamari's book about the mathematician Moritz Pasch ([1], pp. 332–334). The present essay partly draws upon the latter. Biographical data of mathematicians that played a role in Tamari's life were taken from various public sources, notably Wikipedia. Throughout this work, in quotes we replaced underlining by *italics*.

[2] Dov Tamari's daughter Tal remembers that Dov described his father as a Hebraist and a Kabbalist. Levi Yitzchak Teitler also performed as a cantor, in particular on Saturday mornings in Wieseck (Giessen) [2].

shop, declined. In the 1920s, he joined Betar, the Jewish youth organization of the Revisionist Zionist movement. Because of his overly enthusiastic Zionist activities, the Jewish community in Giessen finally decided to ban him from speaking.[3]

In the last years of his school education at the Landgraf-Ludwigs-Gymnasium in Giessen, from autumn 1926 to spring 1928, Bernhard Teitler voluntarily worked two hours a week as a secretary for the almost blind Professor Emeritus Moritz Pasch[4]. In his last years, Pasch tried to enrich an updated version of his *Ursprung des Zahlbegriffs* (origin of the notion of numbers) by a succinct system of axioms, but all his attempts remained unsuccessful. One day Teitler said to Pasch quite unthoughtfully "Perhaps this cannot be done." and Pasch reacted "Thank you very much, I already thought this too, but I still wanted to try it again." ([1], pp. 129–130, translation from the German). He did not live long enough to experience the revolution that Kurt Gödel[5] brought about. The new version of Pasch's work appeared in 1930 (Springer, Berlin) without a corresponding extension, but he commended Bernhard Teitler in the preface.

In 1929 Teitler passed the final school examination (Abitur)[6] and began to study mathematics, physics and philosophy in Vienna, where he stayed with an uncle ([1], p. 130). After the first semester he returned to Giessen for two semesters at the Hessische Ludwigs-Universität, followed by two semesters at the Universität Frankfurt am Main. The winter semester 1931–32 he spent again at the University of Giessen and then returned to Johann-Wolfgang-Goethe Universität Frankfurt, where he was matriculated until May 4, 1933. In 1932 he was accepted as a doctoral

---

[3] In a biographical note, probably written around 1952–53, Tamari mentioned, "I remember that Pasch and Dehn blamed me for this passionate and consuming activity 'which will finally prove incompatible with that of a mathematician.' I replied that Zionism is not a hobby and not a luxury, but a question of life and death. Unhappily, we both were only too right." [3]

[4] Moritz Pasch (1843–1930) was born in Breslau and received his doctoral degree in mathematics in 1865. In 1870 he moved to Giessen, where he was first a Privatdozent and later became Ordinarius at the university. After that, apparently, he never left the town any more, except for very short trips for personal reasons. Pasch was the first who rigorously axiomatized projective geometry. "... Pasch's ideas ... opened a new world of infinitely many finite and infinite elegant projective geometries of unlimited dimensions." ([1], p. 50, translation from the German)

[5] Kurt Friedrich Gödel (1906–1978) earned his doctorate and habilitation from the University of Vienna, Austria. At the end of 1939 he left for Princeton and became a member of the Institute for Advanced Study (IAS). He is famous for his incompleteness theorems, published in 1931. In particular, he showed that, given any system of axioms, a true proposition about the natural numbers exists that cannot be formally deduced from the axioms. Moreover, he is also well known in the field of Einstein's General Relativity, where the "Gödel space-time" is a fundamental example of a cosmological model allowing time travel.

[6] His "Zeugnis der Reife zum Besuche der Universität", Landgraf-Ludwigs-Gymnasium, Giessen, Febr. 15, 1929, shows best grades ("sehr gut") in German, mathematics and religion, second best grades ("gut") in biology, chemistry and physics. He did less well in French ("genügend"), Greek ("im ganzen gut") and Latin ("genügend"). There were slight deficits in behavior ("Betragen: im ganzen gut") and attentiveness ("Aufmerksamkeit: genügend"). The Hebrew University Central Archives preserve a copy of this document.

candidate by Carl Ludwig Siegel[7] with a topic about Diophantine approximation and hypergeometric functions.[8] In the same year he joined the Deutsche Mathematiker-Vereinigung (DMV).

Among Teitler's teachers at the University of Frankfurt was Max Dehn[9] (winter semester 1930–31: proseminar on determinants; mathematical historical seminar: Leonhard Euler (organized jointly with Paul Epstein, Ernst Hellinger and Siegel); summer semester 1931: seminar on foundations of geometry; winter semester 1932–33: higher algebra).

In March 1933 Bernhard Teitler got into a brawl with Nazis at the corner of Frankfurter Strasse and Alicenstrasse in Giessen [5]. He accompanied a traditionally clothed Jewish student to the train station. The aggression of the young Nazis was at first only directed against this student. But Teitler defended him, the police came and arrested him. He was taken into protective custody ("Schutzhaft"), a means by which the Nazis tried to get troublesome individuals under control, a pre-stage of a concentration camp. According to the mathematician Spiros P. Zervos[10], Teitler had a close friend (with main interest in philosophy) who, for unknown reasons, joined Röhm's organization (SA).

> "Now, the good luck of Dov was that the commander of this camp was F [the friend]. As soon as he discovers that Dov is there, he calls him in his bureau, alone, gives to him sufficient money to buy a railway ticket to Paris and tells Dov: 'I shall give you a permission for going to

---

[7] Siegel (1896–1981) was born in Berlin and studied mathematics and physics at the Humboldt University in Berlin. He received his doctoral degree from the University of Göttingen, under the supervision of Edmund Landau. In 1922 he became a professor at the University of Frankfurt am Main. As an opponent of the Nazi regime, in 1940 he emigrated via Norway to the USA, where he became a member of the Institute for Advanced Study in Princeton. He returned to Göttingen in 1951 and held a professor position until his retirement in 1959. Siegel is famous for his work in number theory and celestial mechanics and is one of the most important mathematicians of the last century.

[8] In summer semester 1932 Teitler took part in Siegel's seminar on Diophantine approximations (Teitler's record of study at the University of Frankfurt). The Hebrew University Central Archives possess copies of Teitler's records of study (Studienbücher) in Austria and Germany.

[9] Max Dehn (1878–1952) was born in Hamburg. He studied in Freiburg and Göttingen, and wrote a dissertation in 1900 under the guidance of David Hilbert. In his habilitation thesis (Münster, 1901) he solved *Hilbert's third problem*. In 1922 he succeeded Ludwig Bieberbach at the University of Frankfurt. After retirement in 1935, in 1939 he fled from Nazi Germany via Norway to the USA where he had to accept minor positions in order to make his living. Besides several important contributions to geometry, topology ("Dehn twist") and group theory, he is well known for the "Dehn problem", the *word problem* for groups. Also see [4].

[10] Spiros P. Zervos, born on March 17, 1930, in Athens, Greece, is a son of the mathematician Panayotis Zervos (1878–1952). Zervos had long conversations with Tamari and still has a very good memory. Tamari described Zervos as "an exceptionally fine human" [6]. They probably met for the first time in 1977 in Athens and were introduced by Zervos' "beloved Paris-Master Marc Krasner" [7]. Marc Krasner (1912–1985) was born in Odessa. He received his doctorate in 1935 from Université de Paris (advisor: Jacques Hadamard). After a position at CNRS, in 1960 he became professor at Clermont-Ferrand and in 1965 at the Université Pierre et Marie Curie in Paris. He worked on algebraic number theory and introduced the notion 'ultrametric space' [8].

a dentist. But, from the moment that you leave my bureau you go directly to the railway station and you do not leave your place in the train before you cross the border with France.' " [7]

In a curriculum vitae that Teitler (then Tamari) wrote much later, he only mentioned that after recovery from protective custody he crossed the border from the Saargebiet (Germany) to France at night ("in Nacht und Nebel") and spent about three months in Paris as a refugee. Then he took a ship to Palestine.

2 Palestine

When Teitler came to Palestine, there was no chance to continue academic studies. The first two years in Palestine he made his living as an agricultural laborer, then as an auxiliary surveyor and draftsman for the Public Works Department (PWD) in Jerusalem [9]. Together with his younger brother Joachim, in 1934 he arranged to get his parents and his sister Ruth out of Germany to Palestine.[11] In 1935 he enrolled at Hebrew University of Jerusalem as a student of mathematics, with philosophy and physics as subsidiary subjects.

Fig. 1 Excerpt from a record of study (No. 592, created on November 17, 1935), listing in Hebrew the courses Teitler took at Hebrew University of Jerusalem. Courtesy of the Hebrew University Central Archives.

---

[11] Joachim Teitler was born in Giessen on July 21, 1913. After he had finished school, he passed a Hechalutz training and emigrated to Palestine half a year before Bernhard. Joachim later changed his name to Elhanan Tamari and became an agronomist. In 1987 he was the victim of a robbery and died three months later. Teitler's sister Ruth Broza worked as a nurse. She lived in Motza and died in 2009.

In the Arab revolt in 1936 Teitler was seriously injured while on duty as a supernumerary constable [10]. A bullet shattered his right thigh and he was in a hospital for surgical and medical treatment for about three months, followed by several months in a sanatorium. The doctors had actually recommended amputation of his right leg, but Teitler refused. With such a disability, he would never have had a chance again to make his living with work like that for the PWD.

> "He was then told – by doctors – that it was observed that wounded who had their dressings changed and antiseptics applied regularly, died more frequently than patients not treated with antiseptics. He refused to have his bandages changed – with the result that the stench was so great that he had to be put in a separate room." (Doram Tamari, private communication)

According to Zervos, Teitler had read in a newspaper that during the Spanish Civil War the observation had been made that

> "... statistically, the recovering of wounded left to the care of peasants was better than that for who had been found a place in hospitals. The peasants used to apply a piece of mould on the wound. So Teitler asked the doctor to apply to him the method of Spanish peasants, and his leg was saved thanks to this primitive 'penicillin'." [7]

Teitler had to use crutches for many years and could not continue his previous work right away. But this led him to resume his studies. He took part in a Master of Science program at Hebrew University of Jerusalem with a thesis[12] about "Linear algebra in general rings", where "general" stands for non-commutative. It was supervised by Jacob Levitzki.[13]

Teitler had been active in the militant underground organization Irgun Zva'i Le'umi (IZL, National Military Organization, in Israel mostly called Etzel) that operated against the British and against the Arabs during the British Mandate in Palestine (see, e.g., [14]). On June 11, 1939, explosives (potassium chlorate mixed with sugar, six half-sticks of gelignite) were found in Teitler's room in a house in the area Zichron Moshe of Jerusalem [15]. On July 14, 1939, the Jerusalem Military Court sentenced him to seven years of imprisonment. About the trial we read:

> "Professor A. Fraenkel, Rector of Hebrew University, gave evidence as to the good character of the accused, whom he described as an excellent student of mathematics. For the Defence, Mr. E.D. Goitein argued that the Prosecution had failed to prove its case against the accused, who was 'a scatterbrain, like all mathematicians,' and therefore might have done something foolish." [16]

---

[12] In the "list of papers" attached to his "application for research leave" from SUNY at Buffalo in 1967 [11], he mentioned that the manuscript of about 200 pages (written in Hebrew) was "confiscated" and lost. In [12] and [13] he recalled a bit of its contents.

[13] Jacob Levitzki (Yaakov Levitski) was born in 1904 in the Ukraine. In 1913 his family emigrated to Palestine. In 1922 Levitzki began to study mathematics in Göttingen, Germany, and received his doctorate in 1929 under the supervision of Emmy Noether. He left Germany in 1931. After two years at Yale University (New Haven, USA), he became a professor at Hebrew University of Jerusalem. Together with his student Shimshon Amitsur (1921–1994), Levitzki was the first recipient (1953) of the Israel Prize in exact sciences (for their work on noncommutative rings). Levitzki died in 1956 in Jerusalem.

At first, Teitler was placed in the central prison of Jerusalem in the Russian Compound (Migraš ha-Rusim), now the Museum of Underground Prisoners (Asirei Hamachtarot, also see [17]). Later he spent some time in a prison in Akko near Haifa, which is now the Akko Museum of the Underground Prisoners. With a special permission of the authorities, Teitler was allowed to take part in the Hebrew University's M.Sc. examinations, while in prison. The graduation ceremony took place on April 30, 1940. Teitler received his Magister Scientiarum diploma in prison personally from the first president of Hebrew University, Judah Leon Magnes [18].[14]

Attempts by Leib Altman[15] [19, 20] and Teitler's mother [10] to achieve an early release from prison in 1941 failed. But Teitler was released from prison ahead of schedule in summer 1942. The army needed him. It was the height of Rommel's North African offensive, and he was sent to an engineering army unit in Egypt to help in the construction of fortifications at the Suez canal.

In 1942 Bernhard Teitler officially changed his name to Dov Tamari [21]. Unofficially Teitler had already used the new name earlier [22]. 'Bernhard' has the meaning 'strong as a bear', and 'Dov' is Hebrew for 'bear'. The name 'Teitler' may have its origin in the Yiddish word 'teytl' for 'date', the fruit, while 'tamar' may designate the fruit or the palm tree bearing it.

After his return from Egypt, Tamari became a regular employee of the PWD and worked as a civil engineer (although he did not have a corresponding diploma), in particular for the construction of roads, including the Jerusalem-Ramla roadway and the Hebron-Beit-Jubrin road. We know little about Tamari's further activities between 1943 and 1948. He was certainly still active in the underground.

In 1943 Tamari was one of ten recipients at Hebrew University of a (one year) Bialik scholarship, named after the poet Chaim Nachman Bialik. The scholarship was funded by Sir Montague Burton [23]. Though strongly supported in particular by Jacob Levitzki, he was not allowed to work on a doctoral thesis at Hebrew University. The reasons were obviously not of a scientific but rather of a political nature. The British had some influence on the university and perhaps tried to prevent that Tamari had access to laboratories. Moreover, Tamari was at odds with a highly influential person who had visited him in prison in 1942 and whom he exposed as an informer for the British.[16] This may indeed have caused problems, also for his later academic career.

In February 1947 Tamari finished his first scientific publication, "On a certain classification of rings and semigroups", which appeared in print in 1948 with Hebrew Uni-

---

[14] Magnes (1877–1948) was well known for his dedication to an Arab-Jewish reconciliation and for his view of a common state for Arabs and Jews.

[15] Aryeh (Arie) Leib Altman (1902–1982) followed Ze'ev Jabotinsky after his death in 1940 as the head of the Political Department of the Revisionist Zionist Movement in Jerusalem. In 1951 he became a Knesset member.

[16] Tamari wrote more about this in personal notes and some letters.

versity as his address [24].[17] It is based on publications by Oystein Ore[18], and it may well be that Tamari had already learned about them during his Master's thesis work.

A domain (a ring without zero divisors) is *right (left) regular in the sense of Ore* if, for any given $a \neq 0, b \neq 0$, there are elements c, d such that $ac = bd$ (respectively $ca = db$). This requirement of the existence of "common right (left) multiplum" has later been referred to as the *right (left) Ore condition* (see, e.g., [26, 27]). Ore showed that any regular domain is isomorphic to a subring of a division ring [28]. If a domain is *not* right regular, then at least two elements A, B exist such that $Aa + Bb = 0$ with elements a, b implies $a = b = 0$. The maximal number n of elements $A_1, \ldots, A_n$ such that $A_1 a_1 + \cdots + A_n a_n = 0$ implies $a_1 = \cdots = a_n = 0$ has been called (right) "order of irregularity" by Ore. Tamari proved in [24] that no rings of finite (left/right) order $n > 1$ exist and moreover showed that the remaining nine possibilities of orders (pairs built from $0, 1, \infty$) can all be realized. In fact, a generalization of Ore's result to rings appeared 1946 in Dubreil's book [29] (Théorème 3 on page 147, also see footnote 4 in [24]). Later, in [30] Tamari proved that the enveloping algebra of a finite-dimensional Lie algebra satisfies the right Ore condition (also see [27, 31]). Based on methods from the latter work, at the International Congress of Mathematicians in Amsterdam, 1954, he presented a refined analysis of the classification in his first paper [32].

Ore invited Tamari for studies at Yale University, but Tamari was unable to obtain a visa. Surely the British had blacklisted him as a terrorist and subversive subject and had supplied the American Consulate with corresponding records. Finally, Tamari decided to try his luck in Europe. In January 1948 he left for Paris.

3 France

In France, Tamari was involved in the well-known Altalena affair.[19] The Altalena was a ship that transported weapons and troops to Israel intended to support the Irgun (IZL) during the war of independence against the Arabs. The name was a pseudonym of Ze'ev (Vladimir) Jabotinsky, the founder of Betar. In May 1948, the State of Israel was declared. David Ben Gurion, head of the provisional government, meanwhile

---

[17] A footnote refers to related work by Paul Dubreil (1904–1994), who later became Tamari's doctoral thesis advisor. Dubreil studied in Paris and received his doctoral degree in 1930. A Rockefeller fellowship allowed him to study with Emil Artin (1898–1962) in Hamburg, Germany, and with Emmy Noether (1882–1935) in Göttingen, where he also worked with Bartel Leendert van der Waerden (1903–1996). After studies in Rome, with Guido Castelnuovo (1865–1952), Federigo Enriques (1871–1946) and Francesco Severi (1879–1961), he returned to France and held positions at the universities of Lille and Nancy. In 1954 he became a professor at the Sorbonne in Paris. He worked in algebra (semigroups), algebraic geometry and number theory. In 1930 he had married the mathematician Marie-Louise Jacotin-Dubreil (1905–1972).

[18] Oystein Ore (1899–1968) was born in Norway, received his doctorate in 1924 from the University of Kristiania (later renamed to Oslo) and became a professor at Yale University, USA, in 1927. His main aim was to set up general, formal foundations of algebra [25]. Besides Garrett Birkhoff (1911–1996), he was the leading proponent of lattice theory in the 1930s.

[19] Doram Tamari, private communication, and [7].

regarded the Irgun as a threat and gave order to the Israel Defense Forces (IDF)[20] to confiscate the weapons when the ship arrived in June. On the Irgun side, Menachem Begin negotiated with Ben Gurion, but refused to turn the weapons over to the IDF. Finally the Altalena was shelled and sank.

In the same year 1948 Tamari met Sara Slutzkai, a Jerusalem-born painter who studied at the École des Beaux-Arts in Paris. They were introduced to each other by Eri Jabotinsky and married in September 1948.[21]

From 1949 to 1953 Tamari was an Attaché Scientifique (research fellow) of CNRS (Centre National de la Recherche Scientifique) in Paris. Tamari's stay in Paris had been made possible by Szolem Mandelbrojt [7, 3].[22]

Tamari surely found an "oasis of scientific freedom and human atmosphere" in the "Paris mathematical paradise" of the nineteen fifties [7]. In 1951 he finished his thesis "Monoides préordonnés et chaînes de Malcev" at the Sorbonne and became a Docteur ès Sciences Mathématiques. The examination committee consisted of Henri Cartan[23], Albert Châtelet[24] (head of the committee) and Paul Dubreil.[25]

---

[20] The IDF originated from the Haganah (Hebrew for defense), which was the largest militant Jewish organization in Palestine.

[21] Doram Tamari, private communication. Eri Jabotinsky (1910–1969) was the son of Ze'ev Jabotinsky. Eri followed in his father's footsteps and became an Irgun activist. He was involved in saving thousands of Jews from Nazi Germany. Their immigration to Palestine violated the British *White Book* restrictions. Eri Jabotinsky was imprisoned by the British authorities in 1940 and moved to the USA after release, where he became one of the founders of the *Emergency Committee to Save the Jewish People of Europe*. Later he became a mathematician and in 1955 he submitted his Ph.D. thesis "Iteration" to Hebrew University of Jerusalem. He also worked at the Department of Mathematics of the Technion in Haifa, where in October 1960 he submitted a paper "On analytic iteration" jointly with the famous number theorist Paul Erdös, published in *J. d'Analyse Math.* **8** (1960–1961) 361–376.

[22] Szolem Mandelbrojt (1899–1983) was Jacques Hadamard's successor as a professor at the Collège de France in Paris. He was among the first members of the BOURBAKI group. During the German occupation of France, Mandelbrojt had to leave, but returned to France after World War II. Tamari first met him when he gave lectures in Palestine.

[23] Henri Cartan (1904–2008) was the son of Élie Cartan (1869–1951), both outstanding mathematicians. Henri studied at the École Normale Supérieure and received his doctoral degree in 1928. His supervisor was Paul Montel (1876–1975). After a position in Strasbourg, in 1940 he became a professor at the Sorbonne. His main contributions are in homological algebra, algebraic topology and complex analysis. Quite a number of important mathematicians were his students. Henri Cartan was among the founders of the BOURBAKI group.

[24] Albert Châtelet (1883–1960) studied mathematics at the École Normale Supérieure in Paris and received his doctorate in 1911 ("Sur certains ensembles de tableaux et leur application à la théorie des nombres", *Ann. Sci. É.N.S.* **28** (1911) 105–202). After positions at the Université de Lille, from 1940 he was professor in Paris, from 1945 he held the chair for arithmetic and number theory at the Faculté de Science, between 1949 and 1954 he was its dean. From 1937 to 1940 he also worked for the French ministry of education and in 1958 he was candidate for President of France (against Charles de Gaulle), as a member of *Union des forces démocratiques*. Châtelet's mathematical work concerned number theory and group theory.

[25] According to Zervos [7], who stayed in Paris from November 1955 to July 1959 and participated in the *Seminaire d'Algèbre et de Théorie des Nombres*, Dubreil was the "uncontested 'Great Master'; uncontested, because of the deep respect that, intuitively everybody felt for him." Together with

Most of Tamari's later publications can be regarded as extensions of material in his thesis.

In 1950 Tamari joined the American Mathematical Society (AMS) and also the Société Mathématique de France (SMF) [34].

4 Tamari's thesis and beyond

The thesis builds on results of Anatoly Maltsev[26], who in 1937 found the first example of a semigroup (a set with an associative binary operation) that cannot be embedded in a group. Maltsev also established necessary and sufficient conditions for a semigroup to admit such an embedding. Tamari achieved more general results by considering more generally

- a *partial* binary operation ("bin"), thus dropping the usual assumption that the binary operation is completely defined. Examples of *partial* binary operations are provided by *categories*.
- substitution rules instead of equality relations, hence preordered sets. This makes contact with what are nowadays called *term rewriting systems*, which in particular underlie programming languages.

In particular, *associativity* becomes a much more complicated property.

For a completely defined binary operation on a set, replacing the familar associativity relation by a substitution rule for a monomial of fixed length results in a poset (partially ordered set). Tamari conjectured in his thesis that the resulting posets are *lattices*.[27] They nowadays carry Tamari's name.

Expressing the binary operation in terms of brackets, the set of nodes corresponds to the set of all parenthesizations of a monomial formed with $n + 1$ different symbols.

Charles Pisot (1910–1984), Dubreil organized a renowned seminar on algebra and number theory to which also Tamari contributed.

[26] Anatoly Ivanovich Maltsev (1909–1967), sometimes written Malcev or Mal'cev, studied mathematics at the Moscow State University, graduated in 1931, but also worked as a teacher at a secondary school in Moscow. In 1932, he became an assistant at the Ivanovo Pedagogical Institute, north-east of Moscow, where in 1944 he was promoted to professor. His thesis in 1941 was about the structure of isomorphic representable infinite algebras and groups. In 1946 he received the (State) Stalin Prize (of 2nd degree) for his work on Lie groups. In 1960, Maltsev became chairman of the Mathematics Institute in Novosibirsk. He had studied with Andrey Nikolaevich Kolmogorov (1903–1987), who posed a question that led Maltsev to study the embeddability problem around 1937 [35, 36]. In 1962 Tamari gave a short talk at the ICM in Stockholm with the title "The associativity problem for finite monoids is unsolvable and equivalent to the word problem for finitely presented groups" [33]. At this congress he met Maltsev personally (see the acknowledgments in [65]).

[27] A lattice is a poset with the property that any two elements possess a least upper bound, called *join*, and a greatest lower bound, called *meet*. For example, the natural numbers with *gcd* (greatest common divisor) as meet and *lcm* (least common multiple) as join operation determine a (complete distributive) lattice. The existence of these operations follows from Euclid's algorithm, "the Ancestor of all Lattice theory" ([38], footnote 10). Also see [39].

The number of its elements is the *Catalan number*[28]

$$C_n = \frac{1}{n+1} \binom{2n}{n}.$$

For $n = 2$, there are only two elements, and the lattice $((ab)c) \mapsto (a(bc))$ expresses a rightward application of the usual associativity law. For $n = 3$, one obtains the pentagon lattice in Figure 2. In his thesis, Tamari also considered representations of these lattices as *polytopes*, which later became known as *associahedra* (also see [42]).

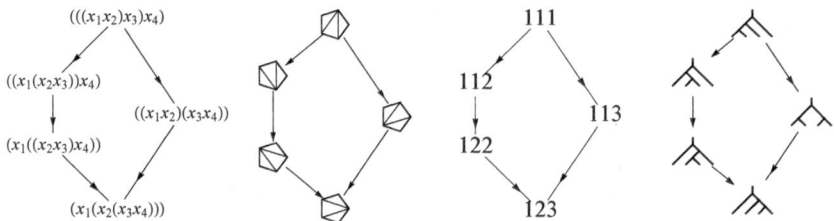

Fig. 2 The left figure shows the Tamari lattice $\mathbf{T}_3$ associated with a monomial of length 4, composed of different symbols $x_1, \ldots, x_4$. The outer brackets are usually dropped. The second figure shows an equivalent representation in terms of triangulations of a 5-gon, related by (left/right) *diagonal flips*. In the third figure the vertices are non-decreasing sequences of natural numbers n_i (i the position) with $n_i \leq i$ (Tamari called such sequences "normal lists" [43]). The last figure shows a representation in terms of planar rooted binary trees. Here the rightward associativity rule corresponds to *right rotation* in a tree. Many other representations of Tamari lattices are known by now.

The first proof of the lattice property of the associativity posets appeared in 1964 [44] and in a revised form in Tamari's publication with Haya Friedman in 1967 [45] (which, surprisingly, has no reference to the first version). In 1969 he achieved a much simpler and totally different proof, jointly with a student, Samuel Huang [46]. But the manuscript was rejected by two journals, with reference to the lesser interest of 2nd proofs [47]. In 1970 George Grätzer, an expert of lattice theory, invited Tamari to formulate the proof as a number of exercises for his forthcoming book [37]

---

[28] This was shown in 1838 by Eugène Charles Catalan (1814–1894). Tamari's work [40] contains a proof based on the simple identity $C_n = \binom{2n}{n} - \binom{2n}{n-1}$, where $\binom{2n}{n}$ is the number of sequences with n appearances of a symbol (here the left bracket) and n appearances of another symbol (here the right bracket). The problem is to extract the *correct* sequences. For $n = 2$, there are only two, namely $(())$ and $()()$, and this matches $C_2 = 2$. Now the following translation (change of bracketing convention) establishes contact with the original problem: $(()) \mapsto x_1(x_2(x_3)) \mapsto ((x_1x_2)x_3)$, $()() \mapsto x_1(x_2)(x_3) \mapsto (x_1(x_2x_3))$. The Catalan numbers already appeared in 1751 in work of Leonhard Euler on the number of triangulations of a planar convex $(n+2)$-gon. By now a lot of counting problems are known which are solved by the Catalan numbers [41].

(Problems 26–36). In 1971 Tamari contacted Gian-Carlo Rota[29], whom he had met some years before at a combinatorics meeting in Waterloo, Canada, and Rota finally arranged publication of the paper with Huang [47]. A third proof was published by Danièle Huguet in 1975 [48]. A fourth proof appeared in an unpublished manuscript in 1996 [38].

A completely defined binary operation on a set $\mathcal{M}$ is *n-associative* if any monomial composed of $n+1$ elements is independent of how it is (properly) parenthesized (also see [49]). As minor results, Tamari showed in his thesis that[30]

1. n-associativity implies $(n+1)$-associativity,
2. if $\mathcal{M} \subset \mathcal{M}^2$ (where $\mathcal{M}^2 = \{ab \,|\, a, b \in \mathcal{M}\}$), then n-associativity implies m-associativity for $2 \leq m \leq n$.

An example of a 4-associative binary operation that is not 3-associative is given by the following multiplication table:

$$
\begin{array}{c|cccc}
\cdot & a & b & c & d \\
\hline
a & b & c & d & d \\
b & d & d & d & d \\
c & d & d & d & d \\
d & d & d & d & d \\
\end{array}
$$

(see [50]). The product of any four elements is d, so that 4-associativity trivially holds. But since $(a \cdot a) \cdot a = b \cdot a = d$ and $a \cdot (a \cdot a) = a \cdot b = c$, 3-associativity is violated.

For a *partial* binary operation on a set $\mathcal{M}$, i.e., a *partial groupoid* (sometimes called *halfgroupoid* [51] or *partial magma*), 2-associativity can be generalized as follows: if $(ab)c$ or $a(bc)$ exists, then both exist and we have $(ab)c = a(bc)$. A partial groupoid is *n-associative* if all monomials obtained from a word with $n+1$ elements of $\mathcal{M}$ by correct binary bracketing are equal. It is *associative* if it is *n-associative* for all $n \geq 2$.[31]

In general, an associative partial groupoid cannot be completed to a semigroup. This leads to another notion of *associativity* of a partial groupoid, which Tamari later considered as more convenient: the embeddability in a semigroup [53].

---

[29] Gian-Carlo Rota (1932–1999) was born in Italy. Motivated by the Italian fascism in those days, his family emigrated to Ecuador. Rota received a Bachelor degree from Princeton University in 1953, a Master's degree in 1954 from Yale, and two years later a Ph.D. ("Extension theory of differential operators") under the supervision of Jacob T. Schwartz. After positions at the Courant Institute and Harvard, he became professor of applied mathematics (and from 1972 also of philosophy) at the MIT in Cambridge, USA. He had an amazing depth and breadth of interests and was one of the great masters of combinatorial theory. Among his students is Richard P. Stanley, another master of combinatorics.

[30] These results have much later been rediscovered in [50].

[31] Tamari also considered weaker notions of associativity of a partial groupoid in his thesis. Also see [52].

The question whether a partial or otherwise incomplete algebraic system can be completed, i.e., embedded into a corresponding complete algebraic system, is deeply connected with the so-called *word problem*. For an algebraic system defined by generators, relations, and operations, the word problem consists in deciding whether any two given words (built using the operations) are equivalent or not. The (uniform) word problem for a class of algebraic systems asks for the existence of an algorithm that can be applied to any algebraic system of the class, with a finite number of generators, relations, and operations, to decide in finitely many steps whether two words are equivalent or not. If a class of algebraic systems has the property that any *incomplete* system of this kind can be embedded in a member of the class, then the word problem can be solved for this class (Trevor Evans [54][32]). The word problem for groups appeared in work of Max Dehn in 1911 [55]. In 1947 the word problem for semigroups was proved undecidable independently by Emil Post (1897–1954) and Andrey Andreyevich Markov Jr. (1903–1979). This would imply the undecidability of the word problem for groups if semigroups were embeddable in groups, but this is not true. In 1955 Petr Sergeevich Novikov (1901–1975) proved the existence of a finitely presented (i.e., expressed in terms of finitely many symbols) group for which the word problem is undecidable. Independently, William Boone[33] proved in a different way that Dehn's problem was unsolvable. Also see [56] for the history of the word problem.

Replacing equality relations by substitution (rewrite) rules leads to the word problem for a *semi-Thue system*.[34]

Tamari showed in the early 1960s that the associativity problem, now in the sense of embeddability in a semigroup, for the class of finite partial groupoids is equivalent to the word problem for finitely presented semigroups, and thus undecidable [58] (also see [13]). During a visit to Göttingen, Bill Boone wrote[35] in August 1967 to Gödel:

> "For some time I've wanted to write to say that I'm now completely convinced of the correctness of the argument for Tamari's Theorem. It is a really elegant result with a beautiful proof." [59]

---

[32] Evans mentioned that, by a theorem of Birkhoff any poset can be embedded in a lattice such that all joins and meets are preserved. "We can interpret this as the embedding of an incomplete lattice in a lattice and so the word problem can be solved for lattices." [54]

[33] William (Bill) Boone (1920–1983) was born in Cincinnati, Ohio. He received his doctorate in 1952 from Princeton University as a student of Alonzo Church (1903–1995). Supported by the Fulbright Commision, he was able to complete his proof in 1954–56. A Guggenheim fellowship in 1957 allowed him to visit and work with several European mathematicians. From 1958 Boone was professor at the University of Illinois, Urbana-Champaign.

[34] The Norwegian mathematician Axel Thue (1863–1922) considered such systems in 1910 [57]. Tamari did not cite Thue in his thesis and later referred to his work only in the general context of the word problem [13].

[35] He also wrote that Göttingen was beautiful and that he had been working on Hilbert's Tenth Problem. In his doctoral thesis (1970), Yuri Matiyasevich proved it to be unsolvable.

Kurt Gödel wrote in May 1973 to Abraham Ginzburg[36]:

"I may add that the reduction of the word problem to the associativity problem seems to me a step in the direction of obtaining more 'significant' proofs I have not myself checked this reduction, but specialists in this area of logic have confirmed to me its correctness." [60]

Although there is no general algorithm to decide whether a given partial groupoid is embeddable in a semi-group, methods exist that are useful in many cases. Often one can show that it is not embeddable by considering the semigroup generated by it. If there is a sequence of words, related by applications of the binary operation in this semigroup, such that it connects two different generators (single letter words), then the original partial groupoid is not embeddable in a semi-group. An example is provided by the following multiplication table, taken from [13].

$$
\begin{array}{c|ccc}
\cdot & a & b & c \\
\hline
a & a & - & a \\
b & a & a & a \\
c & a & b & a
\end{array}
$$

In the corresponding semigroup we have the "two-mountain chain"

$$b = cb = ccb = ab = bbb = ba = a$$

which contradicts $a \neq b$. Therefore this partial groupoid is not embeddable in a semigroup.

A wordchain formed by using the rules of a multiplication table consists of fusions and splits. A *fusion* means evaluation of a product, thus decreasing the length of a word by 1. A *split* corresponds to a factorization increasing the length of the word by 1. Representing a fusion by two incoming and one outgoing arrow, a split by one incoming and two outgoing arrows, and assigning the corresponding elements of the partial groupoid to the arrows, one obtains a *wordchain pattern* (see Figure 3). We thus enter a "graphic associativity theory" [13, 61]. Tamari's aim was in particular to explore the obstructions to associativity given by "special standard wordchains", which begin and end with a single letter word, as in the above example. By identifying the two ends of such a chain, one obtains a 3-regular and 3-connected graph, which he called a *prototype* (of degree n, the number of splits, which for a special standard wordchain is equal to the number of fusions). Ernst Steinitz had proved in 1922 that every 3-connected planar graph can be represented as the graph of a convex polyhedron, so contact is made with polyhedral combinatorics.

This short introduction to Tamari's world of mathematics does not at all do justice to his achievements. Although the other articles in this volume shed more light on some of his ideas and results, by far not all aspects of his work are sufficiently accounted for.

---

[36] Ginzburg had asked for Gödel's opinion in the process for Tamari's promotion to full professor at the Technion.

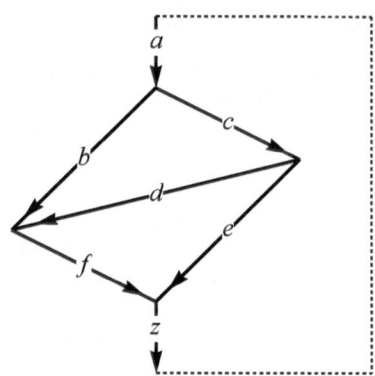

Fig. 3 A word chain pattern with two splits $a = bc$, $c = de$, and two fusions $bd = f$ and $fe = z$. The dashed line indicates the identification $a = z$. By substitution we obtain $b(de) = (bd)e$, the ordinary associativity law. The resulting graph (after the identification) forms a tetrahedron. Also see [61].

5 Israel

Around 1951 Tamari applied for a position at Hebrew University of Jerusalem, but this did not work out. In a letter to Freudenthal in June 1971, Laurent Schwartz[37] wrote:

"I have been in Israel in 1951. Tamari already had problems with the University of Jerusalem at that time, and I promised him to look for a solution. I became aware that there were three great enmities in Jerusalem against Tamari. ... However, I explained to the professors of the University that even three great personal difficulties could not justify to refuse a longtime Israeli citizen a university position in Israel, docteur es sciences and honorable mathematician." ([62], translation from the French)

Finally Schwartz addressed Professor Goldstein at the Technion (Israel Institute of Technology) in Haifa, who responded very amiably [62]. In a letter to Freudenthal in August 1971, Tamari mentioned help from Nathan Jacobson[38] in 1952 (cf. [63])

"... against political discrimination. ... At that time the Technion did not want me because of my active rightist past as a member of the 'Irgun' (Nationalist Military Underground), ..." [64]

---

[37] Laurent Schwartz (1915–2002) studied at the École Normale Supérieure in Paris, received his doctoral degree from the University of Strasbourg in 1943, held positions in Grenoble and Nancy, from 1952 at the Sorbonne in Paris, from 1959 at the Ecole Polytechnique, and from 1980 to 1983 at the Université de Paris VII. He is well known in particular for his theory of generalized functions (distributions), a work for which he received the Fields Medal in 1950. Schwartz was engaged in various political affairs.

[38] Nathan Jacobson (1910–1999) emigrated with his Jewish family from Poland to the USA in 1918. He graduated in 1930 from the University of Alabama and received his Ph.D. from Princeton University in 1934 with a thesis on "Non-commutative polynomials and cyclic algebras", supervised by Joseph Wedderburn. After a position at Bryn Mawr College (1935–36, successor of Emmy Noether), a fellowship at the University of Chicago (1936–37), he went to the University of North Carolina at Chapel Hill where he became Associate Professor in 1941. From 1943 to 1947 he held a position at Johns Hopkins University. From there he moved to Yale University in 1947, where he remained until his retirement in 1981. Jacobson was a leading algebraist of the last century.

In 1953 Tamari obtained a position at the Technion, first as a senior lecturer, then in 1955 he advanced to an associate professor. In addition to strong engagement in teaching and supervising master and doctoral students, he was elected by the faculty as a member of the Board of Governors of the Technion (1958–1960), a member of the Executive Board of the Israel Academic Union (from 1957), and president of the Technion's union of professors and lecturers (1955–1956). He was one of the leaders of a successful general strike in early 1956 against the Government of the State and the monopoly of the General Federation of Labor (Histadrut), trying to save Israel from a devastating brain drain.

Fig. 4 Dov Tamari's parents Frieda and Levi Teitler in Israel in 1955. Courtesy of Ben Tamari, son of Dov's brother Elhanan.

In 1959 Tamari's student Abraham Ginzburg submitted his doctoral thesis to the Technion.[39] Later Ginzburg and Tamari published three papers together. Since Ginzburg has a joint paper (actually two) with Paul Erdös, Tamari has 'Erdös number 2'. Ginzburg became a professor at the Technion and later served as the Vice Chancellor of the Open University in Tel Aviv. In 2011 he was 85 and lived in Tel Aviv.

Two other theses supervised by Tamari are listed in the Technion's library catalog: Michael Yoeli[40], "Mathematical theory of switching nets" (1959), Meir Steinberger[41],

---

[39] Part of Ginzburg's thesis "Multiplicative systems as homomorphic images of 'square sets' " appeared in [66].

[40] Michael Yoeli became a professor at the Department of Computer Science of the Technion.

[41] Meir (Max) Steinberger was born on January 29, 1914, in Frankfurt am Main. After the final school examination (Abitur) in 1932, he began to study chemistry, mathematics and physics in Frankfurt. But when the Nazis came to power, he decided to emigrate to Palestine. This required a practical education, so he spent a year in Paris to become a precision mechanic. In Palestine he first made his living as a building locksmith. In 1942 he received an M.Sc. degree from Hebrew

"On families of complexes in certain multiplicative systems (Monoids)" (1964), also see [70].

Another student of Tamari was Haya Friedman.[42] She started working on a doctoral thesis, but left for England with her husband Arye, who had already spent a Sabbatical at Cambridge University in 1952. He wanted to work in computer science, which was hardly possible in Israel in those days. Many years later, Tamari contacted Haya and in 1967 they published a proof of the lattice property of the associativity posets that originated in his thesis [45]. On page 220, the essential part of the proof is attributed to (unpublished) work carried out by Haya Freedman in 1958.

Fig. 5 Dov Tamari's children Yuval, Tal and Doram, in Haifa in 1959. Courtesy of Tal Tamari.

6 Via Princeton to Brazil

In the academic year 1959–60 (which started in fall 1959) Tamari was on sabbatical leave from the Technion and held a Visiting Professorship at the University of Rochester in the state of New York. This time he had obtained a visa, though almost at the last minute. In the beginning of December 1959 he applied for a stay at the Institute for Advanced Study (IAS) in Princeton and visited the IAS in the same month for a few days, where he had conversations in particular with Kurt Gödel, Deane Montgomery and André Weil (whom he had met years before in Paris) [71, 72]. From September 1960 to April 1961, Tamari indeed became a member of the IAS, on leave of absence without pay from the Technion and with a grant-in-aid of $6,000

University in Jerusalem and became a high school teacher in Afula, and from 1959 to 1979 in Haifa. Between 1969 and 1976 he also taught mathematics at the Technion. Steinberger died on March 5, 2008, in Haifa. [67, 68, 69]

[42] Haya Freedman (Friedman) was born in 1923 in Lvov (in those days in Poland, now Lviv in the Ukraine) and emigrated to Palestine in 1933. She wrote a Master of Science thesis at Hebrew University in Jerusalem, under the supervision of Jakob Levitzki who had also been the supervisor of Tamari's Master of Science thesis. In 1960 she received her Ph.D. under Kurt August Hirsch (Dr. phil., Berlin 1933; Ph.D., Cambridge 1937) at the Queen Mary College in London. In 1965 she obtained a position at the Birkbeck College in London, and in 1967 she became a member of the London School of Economics and Political Science (LSE). Haya Freedman died in 2005.

to cover the expenses of his sojourn in Princeton [73]. He did not spend the whole academic year at the IAS. In summer 1961 he went to Fortaleza (Ceará, Brazil) with a professorship award (apparently the very first one) of the Organization of American States (OAS), within Kennedy's Alliance for Progress. His application was in particular supported by Leopoldo Nachbin[43]. In July 1961 he participated in the third Colóquio Brasileiro de Matemática which took place at the Instituto de Mathemática in Fortaleza. At the institute Tamari met João Bosco Pitombeira de Carvalho, who was studying to become an engineer, and involved him in his research. This resulted in a joint paper published in 1962 [75].[44]

Any partial groupoid admits an extension, called (smallest) *symmetrization*, to a *symmetric* partial groupoid (also known as *partial inverse property loop*) which has an identity element and every element possesses a two-sided inverse (also see [52]). This is obtained by adjoining an identity element and inverse elements satisfying some obviously required relations, the involution property of the inverse, and moreover

$$ab^{-1} = \begin{cases} a_1 & a = a_1 b \\ b_1^{-1} & \text{if } b = b_1 a \\ \# & \text{otherwise} \end{cases}, \qquad a^{-1}b = \begin{cases} a_1^{-1} & a = ba_1 \\ b_1 & \text{if } b = ab_1 \\ \# & \text{otherwise}. \end{cases}$$

The symmetrization of an associative partial groupoid is not in general associative, however. For example, the partial groupoid with elements $\{a,b,c,d,f,g,h,k,l\}$ and multiplication table

$\cdot$	a	b	c	d
a	f	$-$	h	$-$
b	$-$	f	$-$	h
c	g	$-$	k	$-$
d	$-$	g	$-$	l

(no further products defined) is associative, but in its symmetrization we have $(ga^{-1})(a^{-1}(bd)) = k \neq l = (g((a^{-1}a^{-1})b))d$ [75].

The binary operation of the symmetrization of a semigroup is *not* completely defined in general, the symmetrization is a *partial* groupoid. De Carvalho and Tamari showed that it satisfies 2-associativity and also 3-associativity (where the pentagonal Tamari lattice plays a role). The *Malcev conditions*, which are necessary and sufficient

---

[43] Leopoldo Nachbin (1922–1993) was a leading Brazilian mathematician in those days. He was born in Recife, Brazil. His father was born in Poland, his mother in Austria. He moved to Rio de Janeiro in 1938, where he obtained a degree in civil engineering in 1943. Though he had already published in mathematics since 1941, it was an obstacle that he had no degree in mathematics. At the advice of his friend António Monteiro, he developed some of his previous results in topology in some detail to a thesis, which was published as the first volume of *Notas de Matemática* in 1947. From 1948 he spent two years at the University of Chicago, at that time one of the most active mathematical centres in the world. At the ICM in 1950 he met Laurent Schwartz, who had a great influence on him, and Nachbin spent the academic years from 1961 to 1963 as a Visiting Professor in Paris. Nachbin's comprehensive work is decribed in detail in [74].

[44] The work [75] with de Carvalho is the first of Tamari's publications in which the drawing of the three-dimensional associahedron from an unpublished part of his 1951 thesis appeared in print.

for the embedding of a semigroup in a group, are turned into *associativity conditions* [58].[45] They are equivalent to the (4- and higher) associativity of the symmetrization of the semigroup.

Tamari wrote to Nachbin in support of de Carvalho and his recommendation achieved that de Carvalho received a fellowship to study mathematics at the University of Chicago.[46]

Fig. 6 The photo (courtesy of Oswaldo Chateaubriand) was taken at a seminar in Fortaleza in early 1962. Tamari is in the middle. First row: Ayda Arruda, Oswaldo Chateaubriand and José Morgado. Third row: Mário Tourasse Teixeira and João Bosco Pitombeira de Carvalho.

There is another story well worth being told.[47] In fall 1961 Tamari invited António Monteiro[48] to lecture in a seminar he was organizing in Fortaleza. But Monteiro was on sabbatical leave in Buenos Aires [77] and persuaded Oswaldo Chateaubriand, a young student he met there, to participate in Tamari's seminar. With financial support

---

[45] Related ideas had appeared earlier in work of Joachim (Jim) Lambek [76]. Lambek was born in Leipzig, Germany, in 1922. In the late 1930s he left for England, but from there he was deported to Canada. After about two years in a camp, he was released and settled in Montreal. In 1950 he earned his Ph.D. degree with a dissertation consisting of two parts, "Biquaternion vectorfields over Minkowski's space" and "The immersibility of a semigroup into a group" at McGill University. His advisor was Hans Julius Zassenhaus. Lambek became a professor at McGill. Among several books he wrote, "Lectures on Rings and Modules" was particularly influential. His work also includes contributions to category theory, linguistics, logic and philosophy. In [58] Tamari mentioned that he had proposed a common paper with Lambek, which did not work out unfortunately because of "circonstances extérieures".

[46] De Carvalho finished his Ph.D. thesis in algebraic topology in 1967. His advisor was Arunas Leonardas Liulevicius (whose advisor was Saunders Mac Lane). When he returned to Brazil, he got involved in creating a new department of mathematics at the Pontifícia Universidade Católica do Rio de Janeiro, which hardly left him time to continue research. He is well known for his strong dedication to the advancement of mathematics education in Brazil.

[47] Oswaldo Chateaubriand, private communication.

[48] António Monteiro (1907–1980) graduated from the University of Lisbon in 1930 and received a doctorat d'État from the Sorbonne in Paris in 1936, where his advisor was Maurice Fréchet. Since Monteiro refused to accept fascist principles and to sign a certain supporting document, he was unable to find a position in Portugal and left for Brazil where he spent four years in Rio de Janeiro. When his contract was not renewed (perhaps due to intervention from Portugal), in 1949 he moved to Buenos Aires. In 1956 he became a professor at the newly created Universidad Nacional del Sur in Bahía Blanca, Argentina. But during the military dictatorship in Argentina, he suffered a lot [77]. Monteiro's mathematical work was devoted to topological spaces, lattices, and relations among them [78].

from an uncle, Chateaubriand arrived in January 1962 in Fortaleza. He attended the small seminar, presenting a series of lectures on Raymond Smullyan's book *Theory of Formal Systems*, which he was interested in at that time. The seminar lasted about six weeks. Also Tamari and José Morgado (a Portuguese mathematician who was teaching at the Universidade Federal de Pernambuco in Recife) gave lectures. Towards the end of the seminar, Tamari suggested recommending Chateaubriand to Alfred Tarski, one of the greatest logicians of the last century, for graduate studies at Berkeley. He was indeed admitted, in spite of the fact that he had not even an undergraduate degree at that time.[49]

Fig. 7 Dov Tamari with his wife and the two youngest children, accompanied by Giovanna Bonino (Oswaldo's mother), at Vista Chinesa, a tourist attraction in Rio. Courtesy of Oswaldo Chateaubriand.

But Tamari's stay in Brazil turned out to be a disaster in other respects. He had applied for a third year of leave of absence and was waiting for a corresponding answer from the Technion. On August 29, 1961, he received a telegram from the Technion asking for an immediate reply whether he intended to return for the coming academic year. But he ignored it, still waiting for an official answer concerning his application for extension of his leave of absence. In fact, letters had been sent to Tamari in June 1961 by the Head of the Technion's mathematics department and in July by the Technion's administration, informing him about the rejection of his application. But these letters were sent to Princeton and from there forwarded to Fortaleza, where they apparently arrived only in October. In October 1961, the official beginning of the new academic year (teaching started not before November), the Technion dismissed him summarily, without a previous warning, on the grounds that he did not return in time to his duties. It must have been like a bolt from the blue

---

[49] Chateaubriand received a Ph.D. in philosophy from Berkeley in 1971. After professorships at the University of Washington in Seattle (1967–72) and at Cornell (1972–77), in 1978 he became a professor of philosophy at the Pontifícia Universidade Católica (PUC) in Rio de Janeiro. His main areas of research include Logic, Philosophy of Language, and more generally Philosophy of Mathematics. At a congress in Rio in 1984, Chateaubriand met Hans Freudenthal. After a long chat, he asked him if he knew Tamari. In a letter to Tamari, Freudenthal wrote: "I never heard someone speaking so enthusiastically about one of his teachers" ([79], translation from the German).

when he realized that he was suddenly deprived of his livelihood, and with him his wife and their three children.

According to de Carvalho[50], Tamari came with high expectations to Brazil, but the working conditions were rather bad and after Fortaleza it seemed that he could not find any kind of position to continue research. In April 1962, Tamari visited the south of Brazil together with his family, and in particular the Instituto Nacional de Matemática Pura e Aplicada (IMPA) in Rio de Janeiro.

7 The Netherlands

On April 28, 1962, Tamari left Brazil. He stayed for a week in Paris, where he delivered a lecture, and travelled in May 1962 to Utrecht (Netherlands), where he spent five months at the Rijksuniversiteit (now Universiteit Utrecht), with the topologist and geometer Hans Freudenthal[51]. Tamari knew Freudenthal at least since the time when he worked on his thesis in Paris and from a visit to the Netherlands in spring 1953.[52] Freudenthal invited Tamari in August 1961 "not as a visiting professor, but as a visiting scientific collaborator" [81]. The invitation letter was sent to the IAS in Princeton and forwarded to Tamari's address in Brazil. Tamari accepted the invitation by telegram and in a letter to Freudenthal he added:

"I came to Brasil for this summer (here winter) and can stay indefinitely. But mathematical, cultural, and sanitary conditions on the whole ... are not too attractive.

My family and I would be very happy if you could arrange for the invitation, be it immediately for the whole coming academic year, be it for its second semester only." [82]

Freudenthal only responded in November 1961, pointing out that the Netherlands Organization for Pure Scientific Research (Nederlandse Organisatie voor Zuiver Wetenschappelijk Onderzoek, Z.W.O.) had no experience yet with the new kind of scholarship for foreign visitors and had postponed a decision several times. He asked Tamari for a "specified programme of collaboration" with Utrecht mathematicians and listed several names, also of mathematicians in Amsterdam, Delft and Eindhoven [83]. The same month Tamari sent a one-page research proposal listing the following problems:

---

[50] Private communication.

[51] Freudenthal (1905–1990) was born in Luckenwalde in Germany. He wrote his dissertation ("Über die Enden topologischer Räume und Gruppen", published in *Math. Zeitschrift* **33** (1931) 692–713) in Berlin under the supervision of Heinz Hopf (1894–1971). Just one year later, in 1931, he achieved his habilitation with a thesis about quality and quantity in mathematics. He then moved to Amsterdam where he became an assistant of L.E.J. Brower (1881–1966). During the German occupation of the Netherlands, he was expelled from the university and in 1943 deported to a labor camp in Havelte, but in 1944 he escaped with the help of his Dutch wife. From 1946 to 1975 he was professor at the University of Utrecht. Freudenthal is famous in particular for his "suspension theorems" for topological spaces, but also well known for his engagement in mathematics education.

[52] In a letter [80] to Freudenthal, Tamari refers to a discussion with him about polyhedra.

"1) Algebraic and topological embeddings or extensions of semi-groups, rings, algebras, quotient-constructions for universal enveloping (associative or not) algebras of Lie (and other) algebras.

2) Theory of general multiplicative systems, word problems, enumeration problems, analysis of associative laws, algorithmic methods.

3) Post languages and semi-group words, strategies for decoding, ev. application of linear programming.

4) Binary relations, families of such in abstractum and imposed on multiplicative systems, rings etc., ev. new abstract approach to algebraic geometry.

5) A particular problem: Find a triple of (not necessarily distinct) *global* analytical functions of one complex variable with well-defined mutual substitution (= composition by substitution for the independent variable) exhibiting non-associativity of composition (essentially a problem of analytic continuation).[53]

6) Finding and studying of interesting new examples of 'quasi-ordoform' structures and spaces." [85]

He distributed all the ten names Freudenthal had suggested as "expert knowledge" and "possible cooperation" over the above six projects. In the accompanying letter [85] he mentioned "I might have 'overdone' it by mentioning too many problems or names." Two weeks later he wrote to Freudenthal:

"I had forgotten to mention 'finite and other abstract geometries', which – I guess – might be strongly connected to my considerations of multiplicative systems and families of binary relations; no doubt that I hope to learn a lot about this from you. On the other hand, I don't think I can develop – at present – sufficient interest in the theory of codes, or even try to apply to it linear analysis and programming, except for expressing this idea vaguely." [86]

The Z.W.O. grant amounted to 9000 Florins for half a year. But not too much overlap with the summer vacations from July 15 to September 15 was allowed (there was actually no teaching already after June 1) [87], so Tamari was asked to "correct for the too long holidays" [88]. During his stay in Utrecht Tamari finished "The algebra of bracketings and their enumeration" [40]. The Z.W.O. required a scientific report about his stay in the Netherlands, but he did not send it in spite of several reminders from Freudenthal. In February 1964 he wrote to Freudenthal:

"I am sorry to have caused you some inconvenience and shall prepare a new report for the Z.W.O. in a few days … Be assured of my highest esteem and gratitude. I shall never forget what you have done for me.

May I add for my excuse – not for any justification – that a certain undeniable disorder in my affairs is not the consequence of lightheadedness or irresponsibility and certainly of no peculiar defect of mine, but of the nomadic life forced on me through the inadmissible – yea even unbelievable acts of the Technion (Israel Institute of Technology) Haifa." [89]

The personal relation between Freudenthal and Tamari was not quite that of close friends, which is evident from the formal style of their letters. But of all of Tamari's friends and colleagues it was Freudenthal who took the initiative and invested a lot of time to help him in his feud with the Technion, as we will describe in Section 9 below.

---

[53] Tamari discussed this problem later with James A. Jenkins, who published an example that indeed demonstrates such a non-associativity in analytic continuation [84].

After the stay in Utrecht, a position as a Visiting Professor at the Université de Caen in France followed, arranged by the French Government on recommendation of a group of French mathematicians who had learned about Tamari's dismissal from the Technion. Together with Roger Apéry, he organized the Séminaire Apéry-Tamari. One of the participants was Danièle Lambot de Fougères (see reference [1] in [90]). She later became Daniéle Huguet and wrote a paper with Tamari that was published in 1978 [91]. In a footnote it is pointed out that the essential results had already been obtained during the seminar in Caen in 1962–63.

8 USA

In 1963 Tamari accepted an offer for a professorship at the State University of New York (SUNY) at Buffalo, a position that he held until his retirement in 1981. In June 1964 he became the chairman of the originally very small mathematics department, with the task to build it up further. There was indeed a dramatic advancement of its stature. In particular, in 1966 Tamari hired John Myhill[54], whose presence led to quite an increase in the number of logicians and algebraists at Buffalo in the following years. Further hires included Rafael Artzy[55] (1912–2006), in foundations of geometry, Kuo-Tsai Chen[56] (1923–1987), well known for his work on formal differential equations and iterated integrals, and Federico Gaeta[57] (1923–2007), in algebraic geometry. Tamari made a great effort to help establish Buffalo as a serious research department, and also made innovations in the graduate program. This put a lot of pressure to perform on some of his colleagues at Buffalo who were not active in research. Unfortunately, due to interdepartmental problems some of the best mathematicians left after Tamari's third year of chairmanship [93].[58]

Among Tamari's students were Carlton Maxson (Ph.D. 1967, "On near-rings and near-ring modules"), later professor at Texas A&M University, John Luedeman (Ph.D. 1969, "On the embedding of topological rings", also see [95]), later professor

---

[54] John R. Myhill (1923–1987) received his Ph.D. from Harvard in 1949. From 1966 he was professor at SUNY, Buffalo. Several important results are associated with his name, in the theory of formal languages, computability theory, cellular automata, constructive set theory, logic and music theory.

[55] Artzy received his doctorate in 1945 from Hebrew University in Jerusalem. It is likely that Tamari knew him from those days. Artzy's advisor was Theodore Motzkin (1908–1970).

[56] Chen was at the IAS in 1960–61 and probably met Tamari there.

[57] Gaeta worked with Dubreil in Paris in 1950–51 and met Tamari then. He was a person that could not keep silent when something appeared to be wrong, and thus in character apparently close to Tamari. With this attitude, Gaeta had hardly a chance to get along in the difficult political atmosphere of the 1950s in Spain, his country of birth. He lost his position, left for Argentina in 1957, moved in 1960 for three years to São Paulo, Brazil, and then settled in Buffalo until Spain's 1978 transition to democracy. Bill Lawvere said about Gaeta "Oh, someone who is always involved in troubles?" [92]. Tal Tamari remembers Gaeta as "a very, very nice man".

[58] The chairmanship was passed to Tamari's friend Kenneth D. Magill [93]. In 1981 Magill published a paper that carries a dedication to Tamari on the occasion of his 70th birthday [94].

at Clemson University in South Carolina, and Kevin Ejere Osondu, who was the first black student to earn a Ph.D. (1974, "A unified theory of extensions of bins to semigroups and groups") from SUNY at Buffalo. Osondu became a professor of mathematics at the Federal University of Technology in Owerri in Nigeria, his country of origin. Among all of Tamari's students, it was Osondu whose further work continued Tamari's to a considerable extent. Tamari had turned the problem of embedding a semigroup in a group into the problem of embedding its (minimal) symmetrization in a group, and Osondu studied this further [96].

Tamari was chairman of the mathematics department at Buffalo until the academic year 1967–68 which he spent again at the IAS in Princeton. In his application letter he wrote:

"Still, I don't believe that I was born to die as a chairman or an administrator. I have still some ambition left to complete my research ..." [97]

He pointed out that his university would not pay him during a leave of absence. Tamari's stay at the IAS was supported by Kurt Gödel, who wrote to his colleagues:

"I am *definitely* in favor of granting Dov Tamari's application for membership in 1967–68 ... Tamari's project of using, what he calls 'monoids' (i.e., binary operations without associativity and *not necessarily defined for all pairs*), for a more systematic development of results about finitely presented groups, word problems, etc., seems very interesting and promising to me. His 'standard' representations of finitely presented groups by monoids (namely, as freely generated extensions of a certain kind of finite monoids) seems destined to throw light on this area. It is surprising that, apparently, nobody before Tamari has used these quite natural representations, although it is trivial that they always exist. Tamari's expectation that this approach may lead to simpler and more 'significant' proofs has, to some extent, already been borne by his work.

That monoids are interesting objects of mathematical investigation appears from the surprising fact (proved by Tamari) that even for *finite* monoids there exists no procedure for deciding whether they are associative. Moreover their representation problem (dealt with by Tamari in several papers) seems to lead to interesting questions about the composition of analytic functions and to an abstract analogue of the theory of analytic continuation ...

In my opinion Tamari should unquestionably be given a chance to restore his lost paper and to continue carrying out his ideas. Since what he is doing is primarily a very general type of abstract algebra, his work belongs more to mathematics than to logic. But it is of interest for logic in view of its implications and of its high degree of abstractness." [98]

The "lost paper" refers to the fact that, on September 9, 1962, Tamari lost a bag at the habor of Haifa with unfinished manuscripts and important personal documents.

Tamari had proved the associativity problem for partially defined binary operations to be undecidable (by reformulating it as a *word problem*), i.e., no corresponding general algorithm exists.

"In the transition to the case of partial operations, *the devil of unsolvability has slipped in through the holes of the multiplication table.*" [99]

But a classification could be approached for subclasses. Given n elements, there are $(n+1)^{n^2}$ multiplication tables, without taking possible equivalences into account. For $n = 3$ there are thus already 262144 tables.[59] The help of computers was needed and

[59] The case $n = 2$ had already been considered in [100].

Tamari contacted Patricia J. Eberlein (1923–1998), an applied mathematician with an interest in combinatorial computing. She passed the problem to Jan van Leeuwen, at that time a young assistant professor with a Ph.D. from Utrecht, interested in algorithm design and complexity theory. He had already spent the academic year 1973–74 in the computer science department at Buffalo and was back for the academic year 1975–76. A student, Paul Bunting, essentially did the programming as a M.Sc. project. The joint work resulted in Tamari's only publication with more than one coauthor [101].[60]

Fig. 8 Dov Tamari in 1964. Courtesy, University Archives, State University of New York at Buffalo, USA.

9 Tamari-Technion Affair

The summary dismissal from the Technion in 1961 had an extreme impact on Tamari's whole further life. Needless to say, this incident must have caused an enormous psychological strain on him. Besides loss of reputation, he was practically exiled from the country he loved and which he had fought for, since all his attempts even to get an unpaid visiting professorship at a university in Israel remained unsuccessful. But he did not resign himself to his fate and fought for his rights for almost 20 years. At the end he lost more than $100,000 for legal and related expenses and gained nothing. Also the loss of substantial social welfare benefits in Israel was quite disastrous since because of his relatively short periods of employment in the USA his pension entitlements were low [102].

In the hope of getting help, in 1963 Tamari addressed several members of the Board of Governors of the Technion, including Robert Oppenheimer (IAS) [103]. In May 1963 Tamari sent him a summary of the affair. In the accompanying letter he mentioned:

---

[60] Bunting left academia and became an experienced database analyst. Jan van Leeuwen became a full professor at the University of Utrecht.

"It goes without saying that the affair … takes a great part of my working time and still a
greater part of my mental energy." [104]

The same month, Oppenheimer wrote to General Yaakov Dori, then president of the
Technion, asking for the Technion's view [105]. Dori's answer in June 1963 contained
no attempt to explain the Technion's position, but only informed Oppenheimer about
the Board's decision to appoint Judge J. Herbstein (a former Justice of the Supreme
Court of Cape Town, South Africa) to investigate the matter.

"I felt that the only way to clear the air was to appoint an impartial individual who had had
no previous association with any aspects of the matter." [106]

But Tamari's newly awakened hope was squashed (cf. [107]) when Herbstein wrote
to him:

"I am *not* authorized to investigate any of your complaints against members of the Staff and
… confine myself to the one question, namely the termination of your employment." [108]

Indeed, such a purely legalistic view could not work out in favor of Tamari, simply
because he had had no official permission from the Technion for an extension of his
leave of absence in 1961.

Tamari also addressed the *International Association of University Professors and
Lecturers* (IAUPL), asking for help in his dispute with the Technion. In a letter from
the Technion's president, then A. Goldberg, to the Secretary General of the IAUPL,
A. Hacquaert, in January 1968, we read:

"It was agreed with Professor Tamari that his complaint against the Technion will be brought
up for arbitration, and both sides committed themselves in advance to adhere to the ruling of
the arbitrator." [109]

The arbitration had actually already been agreed upon in August 1967. Goldberg also
wrote that Tamari would cancel his request to the IAUPL to deal with this matter,
and that, in view of the agreement on the arbitration procedure, he would not reply to
the questions raised in a letter by Hacquaert of October 13, 1967. But Tamari had
not agreed to cancel his request [110].[61]

The Technion's Steering Committee had not been consulted *before* the announcement of Tamari's dismissal (though much later, in June 1962). The Arbitrator decided
that this was not in accordance with the Technion's constitution and therefore the
announcement of the dismissal was ineffective [112]. The crucial question was then
whether Tamari had left the Technion (in a legal sense), and the Arbitrator answered
this in the negative. Hence the decision of the Arbitrator turned out to be in favor of
Tamari. However, the Technion refused to accept it. Right after the positive arbitration, Tamari had taken a leave of absence from Buffalo and returned to Israel with his
family in the hope to be able to resume his position there. In a letter to Freudenthal
in April 1971, he mentioned that one of his friends said "You have won the case, but
you cannot collect" [113].

---

[61] In January 1972 Freudenthal tried to reactivate the IAUPL concerning Tamari's case [111].

9.1 Freudenthal's initiative

In May 1971 Freudenthal wrote to Tamari:

> "La meilleure action me semble une démarche collective d'un groupe de chercheurs scientifique et je serais prêt d'organiser telle chose." [114]

The same month, Freudenthal composed a letter that was sent to quite a number of well-known mathematicians to ask them

> "... to join me in an action in favour of a distinguished scholar who has harshly and injustly been treated for many years by his academic institution. ...
>
> In 1961 while on study leave in USA, Tamari, a tenured member of Technion, was dismissed with no previous notice and has since been subjected to continual harassments by Technion.
>
> In 1967, under pressure from abroad, in particular by the International Association of University Professors and Lecturers (IAUPL), Technion agreed to submit the case to binding arbitration. By a decision of the Court of Arbitration of April 15, 1970, signed by Professor R. Yaron, Dean of the Law School of Hebrew University, Jerusalem, Tamari has been rehabilitated as *a member in full, standing of the Academic Staff of Technion who never had ceased to be so*.
>
> Notwithstanding this decision, Technion still refuses to restore Tamari in his rights. Though there can be little doubt that the decision will be confirmed in the courts, nobody knows how long this will take. Tamari is now near the age of sixty. ...
>
> I have drafted a letter to be addressed to Technion and to be signed by scholars from different countries. May I ask you for your signature? ...
>
> I take the liberty to add some pieces of personal information on Tamari. He is a distinguished mathematician ... Tamari is not at all a litigator or a crackpot, nor is he a fanatic. On the contrary he is a cheerful man and good company. Even under the stress of his struggle against Technion he continued his research and used to publish 2–3 papers a year." [115]

Freudenthal attached a seven page summary [116] that Tamari had prepared about the affair, and a draft of a letter to be sent to the Technion's president:

> "We have learned and are deeply concerned about the misadventures of your staff member Dov Tamari. Even after the binding decision of the Court of Arbitration in Tamari's favour, Technion perseveres in its harsh and unjust policy against Tamari by refusing to restore him in his rights.
>
> It is our firm conviction that an academic institution of world-wide renown as is Technion cannot afford to provoke the slightest doubts with respect to their own credibility. ...
>
> As long as the case Tamari will continue to exist, it will attract the full light of the academic publicity. ...
>
> Please let us know with no delay that our fears are unfounded and that the Tamari case finally ceased to exist." [117]

Henri Cartan responded:

> "Je connais Dov TAMARI depuis de longues années, et je suis plein de sympathie pour lui à cause des difficultés dans lesquelles il se débat." [118]

But he hesitated to sign worrying that such an international pressure might not have a favorable effect. Freudenthal also received a response from a group of mathematicians in Aarhus, Denmark, including Tamari's friend Gabriel Dirac[62], who suggested to ask the Technion to disclose in detail the reasons for its actions [120]. In case of an unsatisfactory answer, the possibility of publicizing the matter should be considered, as a letter in the journal 'Nature', for example. Moreover, the Aarhus group suggested to raise funds for Tamari's legal expenses and enclosed a cheque for the amount of 500 guilders.

In a letter to Freudenthal, Kurt Gödel wrote:

> "... in view of the fact that a satisfactory answer can hardly be expected from the Technion, I would suggest, before taking the second step, to obtain some more information from other sources, in particular: 1. the full text of the decision by the court of arbitration, including the reasons adduced for it, 2. the exact charges brought against the integrity of the arbitrator."
> [121]

Freudenthal's action had been supported by a large number of well-known mathematicians and scientists, and other prominent figures. By July 1971 the Technion must have received quite a number of letters in support of Tamari. The first response from the Technion's president, A. Goldberg, was hardly satisfactory:

> "You are quite right in saying that this case was taken to arbitration, but we came to the conclusion, on taking legal advice, that the arbitrator was not justified in his findings, and we decided to have the award tested in court. ...
>
> We believe that the reinstatement was impossible in the Technion, as Dr. Tamari had accused so many of his colleagues of the most dreadful misdeeds." [122]

After his summary dismissal, Tamari tried to identify the main culprits, of course. Relying, as always, firmly on his conviction, he was probably not cautious enough concerning what he said or wrote. This made a reinstatement at the Technion indeed difficult.

Almost in parallel to Goldberg's letter arrived a letter from Nathan Jacobson[63] and Abraham Robinson[64] in support of the Technion:

---

[62] Gabriel Andrew Dirac (1925–1984) was born in Budapest, Hungary. He was a nephew of the physicist and Nobel prize laureate Eugene Wigner. His mother married the famous physicist and Nobel prize laureate Paul Dirac. Gabriel Dirac received his Ph.D. from the University of London in 1951 with the thesis "On the colouring of graphs: combinatorial topology of linear complexes", (advisor: Richard Rado). He held positions at the universities of Toronto, Vienna, Hamburg, Ilmenau, Dublin, Aarhus and Swansea. After 1970 he was professor of mathematics at Aarhus Universitet. Dirac is well known for his work in graph theory. He has Erdös number 1. [119]

[63] Nathan Jacobson was the President of the AMS at that time. He was familiar with Tamari's case and got involved again especially through Tamari's letter to the editors of the *Notices of the AMS*, see Section 9.2.

[64] Abraham Robinson (1918–1974) emigrated with his family from Germany to Palestine in 1933, studied in particular with Levitzki and Abraham Fraenkel at Hebrew University and received a fellowship in 1939 to continue his studies at the Sorbonne in Paris. During the German invasion of France, he escaped to England where he carried out research work in aerodynamics for the Royal Aircraft Establishment. After World War II he wrote a Master's thesis at Hebrew University, continued his studies at the University of London, and received a Ph.D. from Hebrew University in 1949. A position as a professor for applied mathematics in Toronto followed, and in 1957 he became a professor of mathematics at Hebrew University. In 1962 he moved to the University of California in Los Angeles. From 1967 till his death he was professor at Yale University. He is well

"Reading Tamari's account (p. 1) one gets the impression that it never crossed his mind that his institution might object if he stayed away for a third (!) year. However, it was common knowledge in Israel at that time that (as in other countries) it was considered undesirable to be on leave for more than one year and that an absence of more than two years was bound to raise disciplinary problems." [124]

But there were scientists at the Technion who in those days had been granted an extended leave of absence for even more than three years. In any case, a request for extension of leave should certainly not be answered with a summary dismissal, surely not without a previous warning. Freudenthal wrote back to Goldberg:

"... we asked four very precise questions, your letter answers none of them.

You admit that you 'decided to have the award tested in court'. We had asked whether Tamari is true to claim that Technion had accused the duly appointed Arbitrator of legal misconduct. Cannot you imagine that this was the very point that gravely shocked the academic community? ...

Meanwhile I fear from your vague remark that 'reinstatement was impossible in the Technion as Dr. Tamari had accused so many of his colleagues of the most dreadful misdeeds' people will draw the conclusion that rather than as for all other reasons, it is because of personal enmities that Tamari is refused reinstatement. ...

I pray you, dear Dr. Goldberg, to consider that it is not yet too late to close the case Tamari in an honorable way." [125]

A six-page answer from Goldberg followed in November, containing remarks that appear to shed some light on the affair, in particular:

"In the course of the arbitration proceedings a letter written by Tamari at the beginning of October 1961 to Prof. Ginsburg, the then acting President of the Technion, was produced in evidence. I shall quote just a few words from it ...

'We are glad to have the opportunity of writing to you. You are asking until when we shall stay abroad. The answer: until I shall have a proper position at the Technion. ...' ..." [126]

Tamari was indeed not happy that the Technion did not promote him in due time to full professor. But he pointed out that the "opportunity of writing" was a Jewish New Year greeting card from David Ginsburg and his wife, and that in the above quote Goldberg refers to a *private* letter to "David and Hemda" from "Dov and Sara", who did not know then that Dov's colleague David would become the Technion's president [127]. As a private letter, the arbitrator had to disregard it. Goldberg's answer was hardly satisfactory, rather disappointing for Freudenthal, who expressed this in a further letter in January 1972, also posing some more concrete questions [128]. However, Goldberg was not willing to provide more evidence:

"The letter to you dated 16th November 1971, gave all the relevant facts and I do not consider that this is the time to answer the various points numerated by you." [129]

In January 1972, Gabriel Dirac and John Myhill, who was then visiting Aarhus, reported to Tamari about a conversation the two had with a professor of the Technion's

---

known for his 'non-standard analysis'. In 1962 Robinson considered the problem of embedding rings in division rings [123], a subject closely related to Tamari's interests, and he acknowledged discussions with Tamari.

computer science department, who attended a conference in Aarhus [130]. This professor, let us call him P, said that President Goldberg had briefed him, since he was going to visit a place where supporters of Tamari were. His opinion was the following. From a legal point of view, the Technion was right to accuse the arbitrator of legal misconduct. From a moral view point, both sides were wrong; the Technion for having wrongfully dismissed Tamari, and Tamari for having insulted some senior members of the Technion's mathematics department. He mentioned that Goldberg said it was the biggest mistake of his life to take Tamari's case to arbitration. Goldberg was forced to do it because of the united opposition of many people against Tamari's reinstatement. P also said that the whole correspondence concerning Tamari's case was on display to all in the dean's office at the Technion. He further said that it was ridiculous to speak of political pressure on academic institutions in Israel, and in particular of any exertion of influence by the Technion on universities in Israel against Tamari.

"John and I are of the firm and clear opinion that logic, open mindedness, fairness, or even accurate information, were utterly remote from [P], ..." [130]

We close this section by quoting from letters of Ephraim Katchalski and Stephen Cole Kleene to Freudenthal.

"Many errors and unfortunate mistakes were made by both sides through the years, and it is very dubious whether the wheel can be turned back." [131]

"I would be very surprised, from the apparent extremity of the measure taken by so many functionaries in Israel, if there were not more to the case than meets the eye." [132]

9.2 AMS and CAFTES

On March 23, 1970, Tamari wrote a letter to the editors of the *Notices of the AMS*:

"... to inform my colleagues that, although I have been in Israel since May 1970, my plans to be at the Hebrew University and the Technion for the academic year 1970–71, as listed in these A.M.S. Notices, October 1970, have not materialized. I was not allowed to stay at either of these institutions even for one day." [133]

Attached was a two-page summary of what happened after the arbitrator's decision. Tamari wanted it to be published in the *Notices*. It was not published, however, since the trustees of the AMS intervened.

In a letter to Tamari in September 1971, Nathan Jacobson, at that time the President of the AMS, wrote:

"My primary reason for taking some action at this time is that I feel that the repercussions of this affair is doing serious harm to the good reputation and the high standing of Israeli academic institutions." [63]

and he urged Tamari to withdraw his letter to the *Notices*.[65] Of course, what is a single man's reputation compared with that of an institute like the Technion? But Tamari was not the person to retreat, irrespective of how mighty his opponents were.

In the course of the interaction between the editors of the *Notices* and the trustees of the AMS, the *Committee for Academic Freedom, Tenure, and Employment Security* (CAFTES) was created [136]. It was composed of Murray Gerstenhaber, Paul Mostert[66] as chairman, and Paul Sally. The committee was asked to look into the Tamari-Technion affair, to report on it to the AMS Council and to write a report suitable for publication in the *Notices*. In a letter to Tamari, in October 1972 Freudenthal wrote:

> "You know that I was not happy about your letter to the Editor though afterwards I supported it. However, now it has yielded the creation of the Commission on Academic Freedom, which is an excellent idea, at least if 'Academic Freedom' is interpreted as to cover your case." [137]

In November and December 1972 Mostert wrote in particular to Freudenthal, Goldberg (President of the Technion) and Tamari, posing a number of concrete questions and asking for copies of correspondence and (translated) documents [138, 139, 140]. Tamari's answer [141] contained many details, some of which had been written up apparently for the first time.

The international pressure on the Technion did not remain without effect. General Amos Horev, at that time president of the Technion, declared in a letter to Paul Mostert that the Technion intended to comply with the arbitrator's decision [102], so that the case appeared to be closed for CAFTES [142]. Already in April 1973, Tamari's promotion to (full) professor at the Technion had been considered.[67] Indeed, in January 1974 Tamari was promoted to Professor of Mathematics at the Technion [144]. Since June 1973 Tamari had been working at the Technion, though somewhat displaced from the location of the mathematics department. The Technion granted him leave of absence without pay from January 15 to June 15, 1974 [145]. He needed time to get his and his family's life adapted to the new situation. However, so far he had not received any compensation for his expenses and losses caused by his

---

[65] Jacobson had asked Tamari to keep the letter confidential, but Tamari passed it to his trusted friend Gabriel Dirac and after some time it bounced back to Jacobson via Saunders Mac Lane. In a letter to Tamari in June 1972 he complained about this and asked Tamari to withdraw all copies of the former letter [134]. Tamari's answer was not at all polite [135]. After all, Jacobson once helped him a lot. Jacobson was forced to make a *political* decision. Tamari apparently never accepted that, in the real world, decisions cannot always be made on the basis of (his understanding of) truth and honesty.

[66] Paul Stallings Mostert was born in 1927, received a B.S. degree from Southwestern at Memphis in 1950, a M.S. degree from the University of Chicago in 1951, and a Ph.D. ("Fiberable spaces") from Purdue University in 1953. He became a leading expert in the area of topological semigroups. He was a professor (and chairman) of the mathematics departments of Tulane University and then at the University of Kansas. Paul Mostert holds various patents and copyrights of software, in particular related to breeding and racing of horses. In 2011 he lived in Lexington, Kentucky.

[67] In this process A. Ginzburg, Tamari's former student and at that time Deputy Vice-President of the Technion, asked Gödel for his opinion [143]. Gödel's response [60] was based on his previous evaluation of Tamari [98], from which we quoted in Section 8.

dismissal and the subsequent litigation, and his financial claims surely met with displeasure on the side of the Technion's administration [146].

A CAFTES report about Tamari's case only appeared almost ten years later in 1983 [102]. In a letter to Freudenthal in 1983, Tamari wrote:

"It took more than two full years to water down the text before something could be published."
([147], translation from the German)

According to this report, Tamari returned to the Technion, but soon developed health problems, which he attributed to the living conditions, for which the Technion was responsible. He went back to the United States for medical treatment[68] and this trip was regarded by the Technion as a resignation, thus ending its obligations [102]. Tamari resumed his position at Buffalo and went again to court. The case extended at least till the end of 1980, finally without success for him, according to our knowledge. He had had the chance of ending the fight by accepting a financial settlement. But, apparently, it was more important for him not to back down, a question of honor and truth.

10 After retirement

10.1 Mathematical activities

After his retirement in 1981, Tamari held a Visiting Professorship at the California State University in Chico in 1984, and in 1986–87 he was Adjunct Professor at the California State University, Los Angeles, where he involved an undergraduate, Douglas Bowman[69], in his research about (graphic) associativity theory. In 1988 they submitted two abstracts to the 839th AMS meeting [148].

Tamari still travelled a lot and participated in several conferences. The following is not at all exhaustive.

In May 1982 he attended a meeting at the castle Rauischholzhausen near Giessen. A written version of his talk about *graphic associativity theory* appeared in the proceedings of the conference [61]. At the Third Centenary Celebrations of the Mathematische Gesellschaft Hamburg in March 1990, he presented a talk "Graphic associativity theory (G.A.T.) and 4CM".[70] His submission "Foundations of graphic associativity theory (G.A.T.) and 4CM" was not accepted for the Festschrift, however, and remained a preprint. The manuscript carries a dedication: "In Memoriam GABRIEL DIRAC, 1925–1984. Graph Theorist and Loyal Friend."

---

[68] A melanoma had been detected on his back and the cancer had already spread into lymph nodes, so he needed surgery. He was extremely lucky that the medical treatment was finally successful.

[69] Douglas Bowman is presently Associate Professor at Northern Illinois University.

[70] Here 4CM stands for 'four color map'. Gabriel Dirac drew Tamari's attention to the fact that the four color map problem (solved in 1976 by Kenneth Appel and Wolfgang Haken) can be reduced to a problem of three-coloring of edges [61], which fits into Tamari's wordchain framework.

In 1996 Zervos organized a Colloquium in Athens (Greece) in honor of Tamari, who had been in contact with Zervos for about twenty years and had visited him several times in Athens. The contributions for the Colloquium, and in particular Tamari's work "Elementary theory of special lists and their poset-lattices", were intended to be published in *Eleutheria*.[71] However, because of a serious illness caused by microbes that reside in books and old paper, Zervos was exiled from his house and Tamari's manuscript never went to press.

Another manuscript sent to Zervos in 1997 shared this fate. Actually, Tamari had combined three different manuscripts under the title "On Proofs – A Trilogy. An epilogue in three essays. What great masters thought, and editorial policies", and this was *"dedicated to Gian-Carlo Rota, Master of Combinatorial Theory, and Man of Noble Courage."* Two of these essays are about history of science and one reflects a personal experience. The first part, "200 years theorem of Gauß 1797 - 1997", deals with Gauss' Fundamental Theorem of Algebra. The second part, "The Euclidean way. Euclid, the man and the mathematician. Euclid and modern civilization. From Euclid to Pasch and Hilbert." leads from Euclid over Alexander the Great to Pasch and Hilbert. The third part, "The story: On proofs and editorial policies", is a reflection about problems Tamari faced when trying to publish his paper "Problems of associativity: a simple proof ... " jointly with his student Samuel Huang.

Although Tamari sent the above-mentioned manuscripts to some friends and colleagues, the fact that they were not published means that they are practically non-existent. Even if one learns about their existence, they are hardly accessible nowadays. We should keep in mind that, because of his blindness, Tamari was to a large extent cut off from the active mathematical world at that time.

10.2 Back to the roots

Tamari's first visit back to Germany was a short one in 1977, to Fulda and Giessen, where he met some friends from his former high school class. He was on the way to Greece (to meet Zervos in Athens) via Yugoslavia by train, and his wife Sara and daughter Tal accompanied him.

In 1982, his application to attend the meeting at the castle Rauischholzhausen, which belongs to Justus-Liebig Universität of his former home town Giessen, arrived a bit late and he was informed that the schedule was already full. When he wrote back that he was originally from Giessen, a former student of its university, and also a secretary of the blind professor Pasch, and that he would even be interested in a

---

[71] At the Burlington Mathfest in August 1995 Tamari had presented a related talk, "Elementary theory of simple special n-lists and their spindle lattices. Simple proof of **Lp**." Here **Lp** stands for lattice property.

passive participation, the mathematician Günther Pickert[72] arranged a place for him in the conference program ([1], pp. 11–12). Pickert wanted to know the man who knew Pasch personally and he later asked him to write up his reminiscences of Pasch.

In the late 1980s, Tamari became gradually blind and deaf.[73] Nevertheless, visits to his former home town Giessen followed, during which a project about Moritz Pasch developed. Much of the bibliographic work was done in spring 1998 in Giessen, in spite of his very bad health condition, with the help of 'secretaries' and various high-tech devices. It seemed that this work kept him alive.

Tamari's book about Pasch, which he never saw in print, emphasizes the fact that Pasch's achievements concerning the axiomatization of projective geometry have mostly been incorrectly attributed to David Hilbert.[74] He also "accuses" Hans Freudenthal in this respect, expressing the dilemma that in this case he had to criticize his best friend ([1], p. 6).

Dov Tamari died on August 11, 2006, in Jerusalem. He had just finished his book about Pasch. Dov Tamari is survived by his wife Sara Slutzkai-Tamari and their children: Doram, an attorney in New York, Tal, an anthropologist at CNRS in Paris, and Yuval, a businessman in Los Angeles. Sara had returned to Israel in 1985 and lives in Motza Ilith, near Jerusalem.

---

[72] Günter Pickert was born in 1917 in Eisenach, Germany, and studied mathematics and physics at the University of Göttingen and the Technische Hochschule in Danzig. In particular, in 1933–34 he attended David Hilbert's lectures on the foundations of geometry. In 1938 he earned his doctoral degree with work on the structure theory of commutative associative algebras, supervised by Helmut Hasse. He received his doctoral certificate only in 1939 because of the obligation to present a number of copies of the dissertation to the university, in his case offprints of *Math. Annalen* **116** (1939) 217–280. A position at the University of Tübingen followed where he became *apl.* Professor in 1953. From 1962 until his retirement in 1985, Pickert was professor at the University of Giessen. A large part of his work concerns finite geometry. Among several books he wrote, the one about projective planes (*Projektive Ebenen*, Springer, 1955, 2nd edition 1975) became particularly influential. Pickert took part in the presentation of Tamari's book about Pasch in May 2008 in Giessen [149].

[73] According to Zervos [7], in 1987 Tamari was invited, for the first time, to Moscow, by the mathematician Arkadii A. Maltsev (Mal'tsev), a son of Anatoly Maltsev. The Greek mathematician Efstratios Makras accompanied him. The unbelievable fact was that it was in general Tamari who showed Makras the right way. "In the open air, Tamari marched like a *commando*! Makras concluded: 'If there is a physical catastrophe and, from all men I know only one would survive, this would be Dov Tamari!'."

[74] Hilbert was actually honest in contrast to many of those who referred to his work. In *Grundlagen der Geometrie* (2nd edition), in a footnote on page 4, Hilbert points out "These axioms have first been extensively explored by M. Pasch in his *Vorlesungen über neuere Geometrie*, Leipzig 1882. In particular Axiom II 4 stems from M. Pasch with regard to contents." Also on page 84 we read "These axioms were first established and systematically investigated by M. Pasch", again with reference to Pasch's book. In a letter to the mathematician Friedrich Engel in Giessen, Hilbert wrote on 14 January 1894: "I learned non-Euclidean geometry [which in those days meant projective geometry] solely from this book [referring to Pasch's book]" ([1], p. xx).

11 Further recollections

At least after 1948 Tamari opposed the injustices the Israelis did to the Palestinians, as well as discrimination directed against Jewish immigrants from Middle-Eastern countries, and these views were not at all widely accepted in those days (cf. Doram Tamari in [1], p. 334). In a letter to Gerhard Dautzenberg (professor of theology, Giessen) in December 1989 Tamari wrote:

"I don't wonder at all that you perceived the Palestinian as more trustworthy. Regularly, I find ... the Palestinians on TV more wholehearted, more dignified etc. as the Jewish Israelis." ([150], translation from the German)

Spiros P. Zervos wrote to us:

"D. Tamari said to me: 'I was one of those who (... as young member of Irgun) sent Palestinians out of their homes.' ... True, the Dov Tamari that I knew was decidably *a great friend of Palestinians.*" [7]

With his attitude to openly express what he thought was the truth, Tamari did not always make friends.

"Je suis persuadé que Tamari est un homme dont le caractère est difficile; ..." (H. Cartan [151])

"Il me paraît probable qu'il a un caractère difficile et qu'il est mal supporté par ses collègues." (L. Schwartz [62])

"A characteristic of Dov Tamari [was] that, although not due to bad intention, [he] could temporally deceive some of his truest friends and admirers." (S.P. Zervos [7])

In his letter to us, Zervos further revealed Dov Tamari's character in a superb way.

"Dov Tamari ... had a permanent smile (sometimes sympathetically ironic) and the flame of genius in his eyes ..." [7]

"In questions that he regarded as *truly important*, Dov Tamari would never accept to *necessarily be in line with the fashion of the day.* His reaction, not in loud tones, expressed in a smile, apparently sympathetic, with an ironic comment, was deeply felt by those to who it was addressed." [7]

"There was, of course, no 'theatrical' element in Tamari's character and his ambitions had nothing to do with the ambitions of an 'actor'. I believe, however, that in the right time, the early fifties, Tamari could be an *ideal* candidate for the role of 'Phileas Fogg' in Jules Verne's [Around the World in Eighty Days]. I feel *unhappy* that I have not thought of it when he was alive." [7]

"He appreciated the fact that somebody acted, in life, in accordance with the principles (ideals, etc) he declared were his own." [7]

To substantiate the last statement, Zervos recalled a chat with Tamari about a professor in Giessen:

"According to Dov, Professor [G] was, as a research mathematician, in the middle class (but we must not forget that Dov's criteria were, generally, too strict). Since he was the only student that solved all the exercises, [G] received him at his home and treated him with special care. When [G], a Nazi, learned that Tamari was Jewish, he continued treating him in the same excellent way. Dov said 'Later, the Nazis elevated Prof. [G] in places superior to his value as a mathematician.' 'However', he added, with a feeling of respect (if not even of tacit admiration), 'when Germany collapsed he committed suicide.' " [7]

This refers to Harald Geppert (1902–1945) about whom Tamari wrote:

"Geppert, a student of Bieberbach, a brilliant lecturer and teacher, was also a fanatic, leading Nazi and Anti-Semite. … he treated me, his only openly Jewish student in Gießen, correctly, even with special care and respect." ([152], footnote 3)

Frequently we heard statements about Tamari like the following:

"As already in former days, he was well articulated and intellectually stimulating; … For me Dov Tamari remains a remarkable figure in my mathematical environment." (Karl H. Hofmann, private communication, translation from the German)

Fig. 9 Dov Tamari around 2001. Courtesy of Doram Tamari.

Acknowledgement F M-H is very grateful for information and help from Armin and Yaara Biess, Godelieve Bolten (Noord-Hollands Archief), Gerhard Dautzenberg, João Bosco Pitombeira Fernandes de Carvalho, Oswaldo Chateaubriand, Christine Di Bella and Erica Mosner (The Shelby White and Leon Levy Archives Center), Aristophanes Dimakis, Arye Freedman, Mirjam Freudenthal, Susan Gensemer, Murray Gerstenhaber, Frederik von Harbou, Ellen Heffelfinger (American Institute of Mathematics), Karlheinz Hintermeier, Karl H. Hofmann, Bill Lawvere, Gila Manusovich-Shamir (Mathematics and Computer Science Library, Hebrew University), Carl Maxson, Paul Mostert, William Offhaus (University Archives at Buffalo), Adam J. Ostaszewski, Günter Pickert, Jim Stasheff, Yoram Tamir (Museum of Underground Prisoners, Jerusalem), Ben (Ephraim) Tamari, Doram Tamari, Tal Tamari, Armin Thedy, Jan van Leeuwen, Richard E. Vesley, Michael Vinegrad (Hebrew University Central Archives) and Spiros P. Zervos. Jim Stasheff and Tal Tamari also helped greatly to polish the prefinal versions of this work.

Abbreviations used in the references:

CZA   Central Zionist Archive, Jerusalem, Israel
HUCA   Hebrew University Central Archives, Jerusalem, Israel
IAS   Institute for Advanced Study, Princeton, NJ, USA
JI   Jabotinsky Institute, New Zionist Organization, Israel
MUP   Museum of Underground Prisoners, Archive, Jerusalem, Israel
NHA   Noord-Hollands Archief, Haarlem NL, The Netherlands
NHA-F   NHA, Papers of Hans Freudenthal
NHA-F74   NHA-F, inv. nr. 74
PUL-G   Kurt Gödel Papers, SWLLAC,
                   on deposit at Princeton University Manuscripts Division,
                   Princeton University Library, Princeton, NJ, USA
SWLLAC   The Shelby White and Leon Levy Archives Center,
                   IAS, Princeton, NJ, USA
SWLLAC-T38 SWLLAC, Records of the School of Mathematics / Members,
                   Visitors, Assistants files (1978–1983) / Box 38 / Tamari,
                   Dov (Teitler)
SWLLAC-T136 SWLLAC, Records of the Office of the Director / Member
                   files / Box 136 / Tamari, Dov (1960–61; 1967–68)

References

1. D. Tamari, *Moritz Pasch (1843–1939)*, Shaker, Aachen, 2007.
2. "Memories of Dov Tamari, age 84, as told to his niece Ditta". Summary of a telephone interview with Ditta Tamari, written up by the latter. Circa 1995. Computer typescript, 7 pp.
3. D. Tamari, Biographical Note written in Paris around 1952–53; Doram Tamari, private archive.
4. W. Magnus and R. Moufang, "Max Dehn zum Gedächtnis", *Math. Annalen* **127** (1954) 215–227.
5. G. Dautzenberg, "Prof. Dr. Dov Tamari †", *uniforum*, Zeitung der Justus-Liebig-Universität Gießen, 19. Jahrgang, Nr. 4 (October 12, 2006), 11.
6. Letter, Tamari to Freudenthal (September 2, 1984), NHA-F74.
7. Letters, Zervos to Müller-Hoissen (May 12 and August 11, 2011).
8. M. Krasner, "Nombres semi-réels et espaces ultramétriques", *C. R. Acad. Sci.* **229** (1944) 433–435.
9. D. Tamari, Biographical Note written around 1952. Doram Tamari, private archive.
10. Letter, Frieda Teitler to the General Officer commanding the Forces in Palestine (March 23, 1941), CZA and MUP.
11. "Dov Tanari: list of papers", SWLLAC-T38.
12. D. Tamari, "Une théorie constructive de l'associativité générale", *Eleutheria* (1986A) 93–127.
13. D. Tamari, "Foundations of graphic associativity theory (G.A.T.) and 4CM", unpublished manuscript (1990).
14. CIA, Office of Strategic Services, "The objectives and activities of the Irgun Zvai Leumi", R&A No. 2612 (October 23, 1944), declassified document.
15. *Palestine Post* (July 14, 1939).
16. *Palestine Post* (July 16, 1939).
17. D. Rossoff, "Arie Levin, father of the Jewish prisoners", *The Jewish Magazine*, no. 18 (February 1999).

18. *Palestine Post* (May 1, 1940).
19. Letter, L. Altman to the High Commissioner of Palestine (February 24, 1941), JI, Correspondence 1 11 6.
20. Letter, Major Taylor to Altman (February 26, 1941), JI, Correspondence 1 11 6.
21. *Palestine Gazette* (November 26, 1942), 1299.
22. *Palestine Post* (July 25, 1939).
23. Letter, Rector of the Hebrew University to Burton (February 10, 1944), HUCA.
24. D. Tamari, "On a certain classification of rings and semigroups", *Bulletin AMS* **54** (1948) 153–158.
25. L. Corry, *Modern Algebra and the Rise of Mathematical Structures*, Birkhäuser, Basel, 2004.
26. L.A. Bokut, "Embeddings of rings", *Russ. Math. Surv.* **42** (1987) 105–138.
27. S.C. Coutinho and J.C. McConnell, "The quest for quotient rings (of noncommutative Noetherian rings)", *Amer. Math. Monthly* **110** (2003) 298–313; S.C. Coutinho, "Quotient rings of noncommutative rings in the first half of the 20th century", *Arch. Hist. Exact Sci.* **58** (2004) 255–281.
28. O. Ore, "Linear equations in non-commutative fields", *Ann. Math.* **31** (1931) 463–477.
29. P. Dubreil, *Algèbre*, Gauthier-Villars, Paris, 1946.
30. D. Tamari, "On the embedding of Birkhoff-Witt rings in quotient fields", *Proc. AMS* **4** (1953) 197–202.
31. A.W. Goldie, "Some aspects of ring theory", *Bull. London Math. Soc.* **1** (1969) 129–154.
32. D. Tamari, "A refined classification of semi-groups leading to generalized polynomial rings with a generalized degree concept", in *Proc. ICM, Amsterdam* 1954, North-Holland, Amsterdam, 1957, 439–440.
33. List of talks, *Zeitschr. für math. Logik und Grundlagen d. Math.* **8** (1962) 351–352.
34. *Bulletin de la S.M.F.* **79** (1951) xx.
35. P.S. Alexandrov et al, "Anatoly Ivanovich Malcev: Obituary", *Russ. Math. Surv.* **23** (1968) 157–168.
36. R. Dimitric, "Anatoly Ivanovich Maltsev", *The Math. Intell.* **14** (1992) 26–31.
37. G. Grätzer, *Lattice Theory, First Concepts and Distributive Lattices*, Freeman, San Francisco, 1971.
38. D. Tamari, "Elementary theory of special lists and their poset-lattices", unpublished manuscript (1996).
39. G.-C. Rota, "The many lives of lattice theory", *Not. AMS* **44** (1997) 1440–1445.
40. D. Tamari, "The algebra of bracketings and their enumeration", *Nieuw Archief voor Wiskunde (3)* **10** (1962) 131–146.
41. R. Stanley, *Enumerative Combinatorics*, vol. 2, Cambridge University Press, Cambridge, 1999; "Catalan addendum", *http://www-math.mit.edu/~rstan/ec/catadd.pdf* (2011).
42. J. Stasheff, "How I 'met' Dov Tamari", in *this volume*.
43. D. Tamari, "Associativity theory and the theory of lists. Their applications from abstract algebra to the four-colour map problem", *Congressus Numerantium* **54** (1986) 39–53.
44. D. Tamari, "Sur quelques problèmes d'associativité", *Ann. sci. Univ. Clermont-Ferrand 2, Série Math.* **24** (1964) 91–107.
45. H. Friedman and D. Tamari, "Problèmes d' associativité: une structure de treillis finis induite par une loi demi-associative", *Journal of Combinatorial Theory* **2** (1967) 215–242.
46. S. Huang and D. Tamari, "Problems of associativity: a simple proof for the lattice property of systems ordered by a semi-associative law", *J. Comb. Theory (A)* **13** (1972) 7–13.
47. D. Tamari, "The story: On proofs and editorial policies", unpublished manuscript (1997).
48. D. Huguet, "La structure du treillis des polyédres de parenthésages", *Algebra Universalis* **5** (1975) 82–87.
49. E.S. Ljapin and A.E. Evseev, *The Theory of Partial Algebraic Operations*, Kluwer, Dordrecht, 1997.
50. W.P. Wardlaw, "Computer aided intuition in abstract algebra", *Comput. & Education* **2** (1978) 247–257; "Finitely associative groupoids and algebras", *Houston J. Math.* **9** (1983) 587–598; "A generalized general associative law", *Math. Magazine* **74** (2001) 230–233.

51. R.H. Bruck, *A Survey of Binary Systems*, Springer, Berlin-Heidelberg-New York, 1971.
52. S. Gensemer, "Partial groupoid embeddings in semigroups", in *this volume*.
53. D. Tamari, "Le problèmes d'associativité des monoïdes et le problème des mots pour les demi-groupes; algèbres partielles et chaînes élémentaires", *Sém. Dubreil-Pisot (Algèbre et Théorie des nombres)* **24**(8) (1970/71) 1–15.
54. T. Evans, "The word problem for abstract algebras", *J. London Math. Soc.* **26**(1951) 64–71; "Embeddability and the word problem", *J. London Math. Soc.* **28** (1953) 76–80.
55. M. Dehn, "Über unendliche diskontinuierliche Gruppen", *Math. Annalen* **71** (1911) 116–144.
56. J. Berstel and D. Perrin, "The origins of combinatorics on words", *Europ. J. Comb.* **28** (2007) 996–1022.
57. A. Thue, "Die Lösung eines Spezialfalles eines generellen logischen Problems", *Kra. Vidensk. Selsk. Skrifter. I. Mat. Nat. Kl.*, no. 8 (1910).
58. D. Tamari, "Problèmes d'associativité des monoïdes et problèmes des mots pour les groupes", *Sém. Dubreil-Pisot (Algèbre et Théorie des nombres)* **16**(7) (1962/63) 1–29.
59. Letter, Boone to Gödel (August 14, 1967), PUL-G: Tamari, Dov, 1959–61, 1966–67, Box 3b, Folder 180, document no. 012723.
60. Letter, Gödel to Ginzburg (May 25, 1973), SWLLAC-T38.
61. D. Tamari, "A graphic theory of associativity and word-chain patterns", *Lecture Notes in Mathematics* **969** (1982) 302–320.
62. Letter, Schwartz to Freudenthal (June 29, 1971), NHA-F1857.
63. Letter, Jacobson to Tamari (September 15, 1971), NHA-F1857.
64. Letter, Tamari to Freudenthal (August 10, 1971), NHA-F1857.
65. D. Tamari, "The associativity problem for monoids and the word problem for semigroups and groups", in *Word Problems – Decision Problems and the Burnside Problem in Group Theory*, eds W.W. Boone, F.B. Cannonito and R.C. Lyndon, North-Holland, Amsterdam, 1973, 591–607.
66. A. Ginzburg, "Representation of groups by generalized normal multiplication tables", *Canadian J. Math.* **19** (1967) 774–791.
67. Verein ehemaliger Helmholtzschüler e.V., "Nachruf auf Dr. Meir Steinberger", *VEH-Info* Nr. 114 (2008).
68. H. Thiel, *Die jüdischen Lehrer und Schüler der Frankfurter Helmholtzschule, 1912–1936*, Schriften des Vereins Ehemaliger Helmholtzschüler, Heft 5, Frankfurt am Main, 1994.
69. *Notices of the AMS* **55** (2008) 722.
70. M. Steinberger, "On multiplicative properties of families of complexes of certain loops", *Canadian J. Math.* **25** (1973) 1066.
71. Letters, Tamari to IAS (December 7, 1959; January 9, 1960), SWLLAC-T38.
72. Letter, Tamari to Weil (December 10, 1959), SWLLAC-T136.
73. Letter, Oppenheimer to Tamari (February 1, 1960), SWLLAC-T136.
74. J. Horváth, "The life and works of Leopoldo Nachbin", *Asp. Math. Appl.* **34** (1986) 1–75.
75. J.B. de Carvalho and D. Tamari, "Sur l'associativité partielle des symétrisations de semi-groupes", *Portugaliae Mathematica* **21** (1962) 157–169.
76. J. Lambek, "The immersibility of a semigroup into a group", *Canadian J. Math.* **3** (1951) 34–43.
77. E.L. Ortiz, "Professor António Monteiro and contemporary mathematics in Argentina", *Portugaliae Mathematica* **39** (1980) XIX–XXXII.
78. R. Cignoli, "The mathematics of António Aniceto Monteiro", *Actas del IX Congreso Dr. Antonio A.R. Monteiro* (2007) 3–8.
79. Letter, Freudenthal to Tamari (August 2, 1984), NHA-F74.
80. Letter, Tamari to Freudenthal (April 1, 1953), NHA-F74.
81. Letter, Freudenthal to Tamari (August 1, 1961), NHA-F74.
82. Letter, Tamari to Freudenthal (August 31, 1961), NHA-F74.
83. Letter, Freudenthal to Tamari (November 3, 1961), NHA-F74.
84. J.A. Jenkins, "A transitivity question for analytic continuation", in *Algebraic and Differential Topology – Global Differential Geometry*, ed. G.M. Rassias, Teubner, Leipzig, 1984, 147–149.

85. Letter, Tamari to Freudenthal (November 28, 1961), NHA-F74.
86. Letter, Tamari to Freudenthal (December 11, 1961), NHA-F74.
87. Letter, Freudenthal to Tamari (January 2, 1962), NHA-F74.
88. Letter, Freudenthal to Tamari (January 29, 1962), NHA-F74.
89. Letter, Tamari to Freudenthal (February 26, 1964), NHA-F74.
90. D. Lambot de Fougères, "Propriétés géométriques liées aux parenthésages d'un produit de m facteurs. Application à une loi de composition partielle", *Séminaire Dubreil-Pisot (Algèbre et théorie des nombres)* **18** (1964/65) 1–21.
91. D. Huguet and D. Tamari, "La structure polyédrale des complexes de parenthésages", *J. Comb., Inf. & Syst. Sci.* **3** (1978) 69–81.
92. I. Sols, "Federico Gaeta, among the last classics", in *Liaison, Schottky Problem and Invariant Theory: Remembering Federico Gaeta*, eds. M.E. Alonso, E. Arrondo and R. Mallavibarrena and I. Sols, Progress in Mathematics **280**, Birkhäuser, Basel, 2010, 3–6; C. Ciliberto, "Federico Gaeta and his Italian heritage", loc. cit., 9–33.
93. K.D. Magill, Jr., "An interview with John Isbell", *Topol. Commentary* **1**(2) (1996).
94. K.D. Magill and S. Subbiah, "Regular $\mathscr{J}$-classes of semigroups of continua", *Semigroup Forum* **22** (1981) 159–179.
95. J.K. Luedeman, "On the embedding of compact domains", *Math. Z.* **115** (1970) 113–116; "On the embedding of topological domains into quotient fields", *manuscripta math.* **3** (1970) 213–226.
96. K.E. Osondu, "Symmetrizations of semigroups", *Semigroup Forum* **24** (1982) 67–75; "Malcev sequences and associative symmetrizations", *Semigroup Forum* **29** (1984) 61–73.
97. Letter, Tamari to Montgomery (January 12, 1967), SWLLAC-T38.
98. Letter, Gödel to Faculty of Math. (February 9, 1967), PUL-G: Tamari, Dov, 1959–61, 1966–67, Box 3b, Folder 180, document no. 012707; also SWLLAC-T38.
99. D. Tamari, "The equivalence of associativity and word-problems", *Queen's Papers in Pure and Appl. Math.* **25** (1970) 171–189.
100. R. Hrmová, "Partial groupoids with some associativity conditions", *Mathematica Slovaca* **21** (1971) 285–311.
101. P.W. Bunting, J. van Leeuwen and D. Tamari, "Deciding associativity for partial multiplication tables of order 3", *Math. Comp.* **32** (1978) 593–605.
102. Report from CAFTES, *Notices of the AMS* **30** (1983) 232–233.
103. Letter, Carl Alpert (Executive Vice Chairman, Board of Governors, Technion) to Oppenheimer (June 18, 1963), SWLLAC-T136.
104. Letter, Tamari to Oppenheimer (May 14, 1963), SWLLAC-T136.
105. Letter, Oppenheimer to Dori (May 29, 1963), SWLLAC-T136.
106. Letter, Dori to Oppenheimer (June 19, 1963), SWLLAC-T136.
107. Letter, Tamari to Oppenheimer (February 26, 1964), SWLLAC-T136.
108. Letter, Herbstein to Tamari (July 15, 1963), SWLLAC-T136.
109. Letter, Goldberg to Hacquaert (January 16, 1968), NHA-F1857.
110. Letter, Tamari to Hacquaert (February 15, 1968), NHA-F1857.
111. Letter, Freudenthal to Merikoski (President of IAUPL) (Jan 7, 1972), NHA-F1857.
112. R. Yaron, "Die Entscheidung im Schiedsgericht zwischen Professor Dov Tamari und dem Technion Israel Institute of Technology", German translation of the decision of the court of arbitration (original in Hebrew), April 15 (1970), NHA-F1857.
113. Letter, Tamari to Freudenthal (April 9, 1971), NHA-F74.
114. Letter, Freudenthal to Tamari (May 5, 1971), NHA-F74.
115. Letter, Freudenthal to Gödel (May 28, 1971), PUL-G: Tamari, Dov, re: Tamari vs. Technion Israel 1968, 1971, Box 3b, Folder 181, attachment no. 012729; also NHA-F1857.
116. D. Tamari, "The Affair Tamari-Technion, or About Israeli Universities and their Professors' Associations", (Febr./March 1971), PUL-G: Tamari, Dov, re: Tamari vs. Technion Israel 1968, 1971, Box 3b, Folder 181, document no. 012731; also NHA-F1857.
117. Draft of letter to the President of Technion (1971), PUL-G: Tamari, Dov, re: Tamari vs. Technion Israel 1968, 1971, Box 3b, Folder 181, document no. 012730; also NHA-F1857.

118. Letter, Cartan to Freudenthal (June 5, 1971), NHA-F1857.
119. C. Thomassen, "Gabriel Andrew Dirac", *J. Graph Theory* **9** (1985) 303–318.
120. Letter, Pedersen et al to Freudenthal (June 7, 1971), NHA-F1857.
121. Letter, Gödel to Freudenthal (July 27, 1971), PUL-G: Tamari, Dov, re: Tamari vs. Technion Israel 1968, 1971, Box 3b, Folder 181, document no. 012736; also NHA-F1857.
122. Letter, President of Technion to Freudenthal (July 11, 1971), NHA-F1857.
123. A. Robinson, "A note on embedding problems", *Fund. Math.* **50** (1962) 455–461.
124. Letter, Jacobson and Robinson to Freudenthal (July 12, 1971), NHA-F1857.
125. Letter, Freudenthal to President of Technion (August 19, 1971), NHA-F1857.
126. Letter, President of Technion to Freudenthal (November 16, 1971), NHA-F1857.
127. D. Tamari, Selected comments to Goldberg's letter to Freudenthal of Nov. 16, 1971 (December 14, 1971), NHA-F1857.
128. Letter, Freudenthal to President of Technion (January 10, 1972), NHA-F1857.
129. Letter, President of Technion to Freudenthal (January 26, 1972), NHA-F1857.
130. Letter, Dirac and Myhill to Tamari (January 11, 1972), NHA-F1857.
131. Letter, Katchalski to Freudenthal (July 19, 1971), NHA-F1857.
132. Letter, Kleene to Freudenthal (June 15, 1971), NHA-F1857.
133. Letter, Tamari to Editor of the Notices of the AMS (March 23, 1971), NHA-F74.
134. Letter, Jacobson to Tamari (June 19, 1972), NHA-F1857.
135. Letter, Tamari to Jacobson (July 3, 1972), NHA-F1857.
136. E. Pitcher, *A History of the Second Fifty Years American Mathematical Society* 1939–1988, AMS, Providence, 1988, 290–291.
137. Letter, Freudenthal to Tamari (October 2, 1972), NHA-F1857.
138. Letter, Mostert to Freudenthal (November 2, 1972), NHA-F1857.
139. Letter, Mostert to Goldberg (December 4, 1972), NHA-F1857.
140. Letter, Mostert to Tamari (December 4, 1972), NHA-F1857.
141. Letter, Tamari to Mostert (March 12, 1973), NHA-F1857.
142. Letter, Mostert to Tamari (December 11, 1973), NHA-F1857.
143. Letter, Ginzburg to Gödel (April 24, 1973), PUL-G: Tamari, Dov, 1973, undated, Box 3b, Folder 182, document no. 012747.
144. Letter, Tamari to Freudenthal (February 11, 1974), NHA-F1857.
145. Letter, Tamari to Mostert (January 6, 1974), NHA-F1857.
146. Letter, Tamari to Mostert (January 16, 1974), NHA-F1857.
147. Letter, Tamari to Freudenthal (May 15, 1983), NHA-F74.
148. D. Bowman and D. Tamari, in *Abstracts of papers presented to the AMS* **9** (1988) 415, 502.
149. D. Klein, "Lebenswege der Mathematikprofessoren kreuzten sich", *uniforum*, Zeitung der Justus-Liebig-Universität Gießen, 21. Jahrgang, Nr. 2 (May 15, 2008), 14.
150. Letter, Tamari to Dautzenberg (December 8, 1989).
151. Letter, Cartan to Freudenthal (February 27, 1972), NHA-F1857.
152. D. Tamari, "200 years theorem of Gauß 1797–1997", unpublished manuscript (1997).

On Being a Student of Dov Tamari

Carl Maxson

Abstract I reminisce about being a student of Dov Tamari at the State University of New York at Buffalo in the late 1960's.

In 1963 I was one of five members of the recently formed Mathematics Department of the New York State University College at Fredonia (now SUNY at Fredonia) and a part-time graduate student working toward a Ph.D. at SUNY at Buffalo, which had just joined the New York State University system. For fall 1964 I had registered for a real variables course to be taught by a newly hired professor by the name of Dov Tamari. Tamari was about a month late arriving for the semester. I had been reading the assigned text, *The Theory of Functions of Real Variables* by L.M. Graves but when Tamari arrived we started with logic and set theory, then into analysis, not following the text at all, much different from what I was used to. As well, often the exercises were open-ended, again a complete change from my background. However, as the semester progressed, I adjusted and very much enjoyed the second semester in which we mainly did bounded variation, continuity, differentiation and Lebesgue-Stieltjes integration from the book, *Real Analysis* by McShane and Botts. In the meanwhile, with Tamari's encouragement, I applied for and received an NSF Fellowship for two years of graduate study at SUNY at Buffalo.

During this first year of full-time graduate study I took the first year graduate courses in algebra, complex analysis and topology. (As an aside, I have always been pleased that we used the English translation of Van der Waerden's two volumes, *Modern Algebra*. Over 50 years later, I still have my notes and examinations from this algebra course.) I thought these three courses would be enough but Tamari, who was now Chairman of the Mathematics Department, strongly advised me to sign up for his seminar as well. This was to be my first introduction to research. Tamari had invited several mathematicians to speak to this seminar and I was exposed to a wide range of topics. This was just part of Tamari's goal of improving the mathematics

C.J. Maxson
Professor Emeritus, Mathematics Department, Texas A&M University, College Station, TX 77843-3368 USA, e-mail: *cjmaxson@math.tamu.edu*

at Buffalo. (Buffalo had just recently joined the New York State system with the mandate of becoming one of the main research institutions in the state.) The task for Tamari was indeed a difficult one with recruitment of a research staff to have top priority. Related to this were the many colloquium lectures during the year, job candidates as well as several established mathematicians. One incident that I recall vividly is seeing Tamari jump up from his front row seat to clean the boards for the distinguished topologist from Poland, Kazimierz Kuratowski. I have no doubts this was done out of respect for Kuratowski.

I don't recall exactly when Tamari became my supervisor and I his student. It may have been during the real variables course. But, as soon as I became a full time student in fall 1964 he advised me on courses to take, always being encouraging and pushing as much breadth as possible. He arranged for me to earn some additional money, first by grading papers for the abstract algebra course he was teaching and then having me teach two weeks in the summer for a person who was late.

For the 1965 summer session, Tamari suggested that I read the recent book *Rings and Homology* by Jans, and as usual sign up for his algebra seminar which, this summer, was based on P.M. Cohn's book *Universal Algebra*. With Tamari's guidance this was one of my most rewarding times. I continue to follow some of these lines today. Somewhere around this time he had me read his first publication "On a certain classification of rings and semigroups" and to report on this in his seminar. Thus he was introducing me to research in mathematics. A slight generalization of this first paper of Tamari appears as a chapter in my dissertation and later served as an undergraduate research project for a Texas A&M student. Having to give a talk on a paper and then having to write an exposition on this report was one of the requirements of Tamari's seminars. Again, as I look back on this, I see how beneficial it was, not only introducing one to research but also mathematical exposition.

Near the end of the 1965 summer session, Tamari asked me to accompany him to the airport to meet a new professor, Yuzo Utumi, a Japanese ring theorist. I was indeed honored to do this since I was already aware of Utumi's work. I mention this just to show how thoughtful Tamari was toward his students. He realized this would be an excellent opportunity for me to meet Utumi. Sadly, Utumi became ill the following year and spent only one year at SUNY at Buffalo.

My thoughts for the fall 1965 were that I would spend all of my time on research for my dissertation. However, as I should have known, Tamari had other ideas. In addition to the necessary research he advised that I sit in on Utumi's algebra course, take the algebraic geometry course being offered by Federico Gaeta, who had just arrived from Brazil, the graduate number theory course offered by Frank Olson, and, of course, the usual Tamari seminar. This illustrates further Tamari's philosophy on breadth in mathematics. I have carried this through with my own graduate students (not always to their liking) and still feel that it is a necessary component of a graduate student's education.

Some time early in the 1965 fall semester I met with Tamari to discuss my research program. We discussed four areas:

1) Ring like domains; generalized rings;
2) Ordered semigroups and rings;

3) Semigroups and rings – extension of classification;
4) Problems of associativity.

At that time he was working on problems of associativity and gave me a preprint of "Sur quelques problèmes d'associativité: une structure de treillis finis induite par une loi demi-associative" which he was writing with Haya Freedman. He wanted to find an axiom system for this type of algebra, which has now become known as a *Tamari lattice*. I read this preprint in detail and thought about it for a time but was unable to come up with anything. It was at this time I discovered the papers by Albrecht Fröhlich on near-rings and the recent near-ring papers of James Beidleman and thus chose the direction of ring like domains and generalized rings.

During this year (1965–66) Tamari was very busy with his position as chairman, as I mentioned above, focusing on building a research department. He was involved in many meetings and was gone on several trips away from Buffalo. It was not easy for me to meet with him, in fact, I often had to make an appointment a week or so in advance, to set up a meeting. By the time of the meeting I had often solved the problem I wished to discuss with him. I found this process somewhat frustrating at the time but over the years I have realized that I really benefited from having to search and think by myself. However, I do not advocate this procedure and I don't think Tamari would in general. It was just the events of the time! Also during the Spring 1966 semester I had to prepare and submit a degree plan. I prepared, what I thought, was a rather good draft and gave it to Tamari. Within a short time it came back to me covered in red ink suggesting several modifications. After getting over the initial shock, I realized that Tamari indeed had several excellent suggestions and I now appreciate his thoroughness, it greatly improved my exposition.

I was to return to my position at Fredonia in the fall of 1966 and while there, complete my research and finish my dissertation. However before that, in the summer session, Tamari continued on my breadth requirement by having me take the logic course offered by John Myhill, and once more the Tamari seminar.

That next year, when I was in the final stage of writing up my dissertation, I would drive the 50 miles to Buffalo and meet with Tamari in the evening at his home. These meetings occurred about every other week and were very cordial and always professional. After an hour or so we would take a break for tea and cookies, served by Mrs. Tamari. I have fond memories of these meetings. As a result of these evening sessions and with Tamari's generous help I was able to submit my dissertation on time and complete all requirements for my Ph.D.

After moving to Texas in 1969 I had very little physical contact with Tamari although we did have intermittent contact through letters. Through various colleagues, in particular, Ken Magill, I was kept informed of Tamari's activities and his many health problems. Also, at various times he would send me a packet of preprints/reprints often accompanied by a very nice letter.

What was it like to be a student of Dov Tamari? There is no doubt the answer to this question has varied over time. I still recall the frustration of being unable to meet with him when needing some (what I thought at that time) immediate advice. I have very fond memories of the pleasant evenings discussing "my" mathematics at his home. As the years passed, I have often wished that we might have had some conversations about mathematics and mathematicians in general. I would have been interested in his mathematical career, his contact with the Bourbaki group as well as other mathematicians.

I always knew he was there in support of me and my work, helping me with my writing and certainly writing recommendation letters for me. I will be forever grateful for the roads he opened for me and his supervision in starting my research career. I am very pleased that I had the opportunity to study with this well-known mathematician and I am indeed very proud to say that I was a student of Dov Tamari.

How I 'met' Dov Tamari

Jim Stasheff

Abstract Although I never met Dov Tamari, neither in person nor electronically, our work had one important intersection – the *associahedra*. This Festschrift has given me the opportunity to set the record straight: the *so-called* Stasheff polytope was in fact constructed by Tamari in 1951, a full decade before my version. Here I will indulge in recollections of some of the history of the associahedra, its generalizations and applications. Others in this Festschrift will reveal still other aspects of Tamari's vision, especially in more direct relation to the lattice/poset that bears his name.

1 Introduction/History

Thanks to the invitation to contribute to this volume, I learned Dov Tamari in his 1951 thesis [52] described the associahedra (without the name) as the realization of his poset lattice of bracketings (parenthesizations) of a 'word' with n factors. This was 10 years before my description! Tamari pictured the 1-, 2- and 3-dimensional cases (Figure 1). He drew the 3-dimensional case in what is now a common graphical representation, though I learned it from John Harer only some decades later.

Tamari partially ordered the vertices and ordered the edges, starting with the choice $a(bc) \longrightarrow (ab)c$, which, for me, corresponds to a homotopy (coincidentally, this is the direction I chose in my thesis).

Remark 1.1. Once upon a time, at a conference in the then Soviet Union, I wanted to call the operation *perestroika* and said I didn't know whether that should refer to moving (parentheses) to the left or to the right. Someone in the audience responded that they didn't know either.

Jim Stasheff
UNC-CH and U Penn, e-mail: *jds@math.upenn.edu*

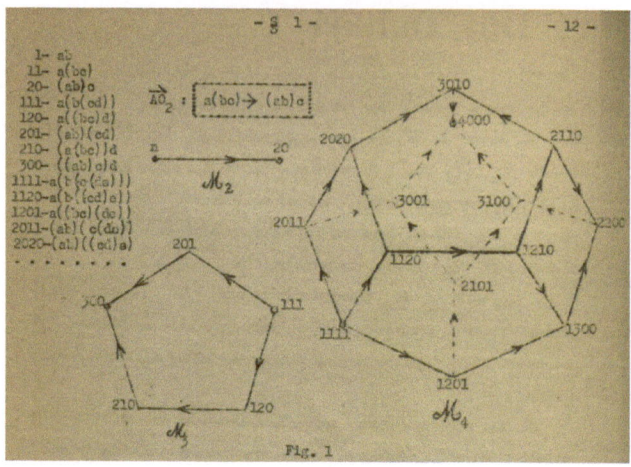

Fig. 1   Tamari's associahedron

1.1 The so-called Stasheff polytope

I use 'so-called' to emphasize that it more accurately should be called the Tamari or Tamari-Stasheff polytope. [1]

Ten years after Tamari's thesis, I drew a far from standard curvilinear representation, Figure 2, being led by my need for coordinates to generalize results of Sugawara [50]. (Although one of my undergraduate mentors at Michigan, Yuri Rainich, had written in his book on Relativity (before my birth!) that the last thing you want to do is use coordinates, I found them necessary as I'll indicate below.)

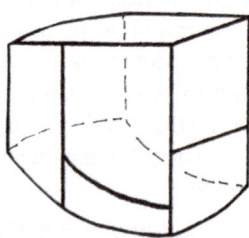

Fig. 2 My first associahedron

Recall that an *H-space* is a topological space X with a continuous multiplication $m : X \times X \to X$ with a unit $e \in X$. Sugawara defines a space F to be *group-like* if it is a homotopy associative H-space with inversion satisfying some complicated

---

[1] By a linguistic oddity, there are spherical polytopes known in Japanese handicraft as Temari balls.

assumptions involving maps $F^n \to F$, which are satisfied by loop spaces. For 'nice' spaces (technically countable CW-complexes [55, 25]), he proves:

Theorem 1.2 (Sugawara [50]). *A 'nice' space F is a group-like space if and only if there exists a 'nice' space E, containing F, and a topological space B and a continuous map p of E into B satisfying the following properties:*

- *E is contractible in itself to a vertex $\varepsilon \in F$, leaving ε fixed,*
- $p(F) = b \in B$ *and the map* $p : (E,F) \to (B,b)$ *is a weak homotopy equivalence.*

The associahedra (I was unaware of the name) were crucial ingredients in my homotopy theoretic characterization of based loops spaces. I constructed the associahedra as convex curvilinear cell complexes K_n with facets of the form $K_r \times K_s$, where $r + s = n + 1$.

Definition 1.3. An A_n-*space* (including $n = \infty$) $(X, \{m_i\})$ consists of a *coherent* family of maps

$$m_i : K_i \times X^i \to X \text{ for } i \leq n$$

where coherent means m_i restricted to any of the facets $K_r \times K_s$ of K_i $(r + s = i + 1)$ is given by an appropriate composition

$$m_r(\cdots, m_s(x_j, \ldots, x_{j+s-1}), \cdots).$$

The full definition requires that the base point $e \in X$ serves as a unit for every m_i. Fortunately for 'nice' spaces, such as CW-complexes, assuming such strict units rather than homotopy units was no problem. Bringing homotopy coherent units into the picture turned out to be a much more complicated task. The homotopy coherent notion of unital differential graded algebra was recently introduced by Fukaya, Oh, Ohta and Ono in their work on symplectic geometry in a purely algebraic way [20, 21]. The corresponding topological, in fact cellular but not polytopical, operad has quite recently been constructed by Muro and Tonks [40].

Theorem 1.4 ([47]). *A connected CW-complex X admits the structure of an A_∞-space iff it has the homotopy type of the space of based loops on a space BX.*

This was the beginning of an approach to the study of iterated loop spaces. For an overview, see [1]. As the notation BX indicates, this space is an analog of the classifying space BG of a topological group G. The latter is built from pieces $\Delta^n \times G^n$ (where Δ^n is the n-simplex) whereas BX is built from pieces $K_{n+2} \times X^n$.

That the classical $[0,1]$-parameterized based loop space admits such an A_∞-structure is a straightforward generalization of the standard proof of homotopy associativity.

In his contribution to this volume, Dehornoy establishes a relation between Tamari lattices and Thompson's group F, cf. [7]. The latter can be realized as the group of all *dyadic* order-preserving self-homeomorphisms of $[0,1]$.

Definition 1.5. A self-homeomorphism of $[0,1]$ is *dyadic* if it is piecewise linear with only finitely many breakpoints, each of which has dyadic rational coefficients (i.e., rational numbers whose denominator is a power of 2) and every slope is an integral power of 2.

He displays two typical elements

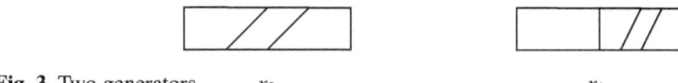

Fig. 3 Two generators x_0 x_1

which are very familiar, one from the classical proof of associativity of the fundamental group, the other from the higher homotopy associativity of the loop space. If x_0 is interpreted as an associating homotopy $h_t : (xy)z \mapsto x(yz)$ (from bottom to top), then x_1 corresponds to the edge $w\,h_t(x,y,z)$ of the higher homotopy associativity pentagon. The specific pentagonal higher homotopy illustrated in [49] involves precisely such dyadic homeomorphisms.

1.2 Giving due credit

Before we get started, I comment on some confusion in the literature as to the terms *polytope* and *polyhedron*. In his celebrated thesis of 1951, Serre refers to 'polyèdres', which I take in a topological sense, i.e., a finite cell complex. The geometric combinatorial community often speaks in terms of a convex subset of $\mathbb{R}^n$, sometimes with metric structure. In particular, Ziegler [57] uses two equivalent definitions: the convex hull of a set of points or the bounded component (if any) of the complement of a set of half-spaces. The defining points are called *vertices* though not all vertices need be extrema; see further comments below. I haven't minded over the years to have the associahedra referred to as *Stasheff polytopes*, but, thanks to the preparation of this book, I know it's time to set the record straight. I think it was Haiman [23] who first named the *associahedra*, but Tamari had constructed them as convex polytopes (though not all vertices were extrema) in his thesis of 1951 [52]. As a grad student at Princeton and then Oxford, I viewed mathematics as compartmentalized, so combinatorics did not appear in my readings. Thus it was that, in my Princeton thesis [47] under J.C. Moore, I constructed a convex curvilinear cell complex I denoted K_n with vertices corresponding to the ways of associating n variables, but with no emphasis on the purely combinatorial structure. Frank Adams called my attention to the possibility of using planar trees as labels, but, in those days, graphic trees would have had to be inserted by hand. Note n is *not* the dimension of the complex which is $n-2$. (Tamari in his thesis called $\mathscr{M}_{n-1}$ what I called K_n and he and Danièle Huguet called it P_{n-2} [28]. In [53], he used K_d to denote the number of bracketings of $d+1$ factors; there are many conventions in the literature.)

The reason for the curvilinear coordinates was the following: The generalized Hopf fibration for an H-space X which I used had total space

$$E_2 := X \times CX \cup_m X,$$

indicating that $X \times CX$ has been attached to X using the map m on the base $X \times X \subset X \times CX$. In terms of an associating homotopy $h : I \times X \times X \times X \to X$, an action of X on E_2 could be given by

$$\bar{m} : (x, (y, (t, z))) \mapsto \begin{cases} (xy, (2t, z)) \\ h(2 - st, x, y, z) \end{cases} \text{for} \quad \begin{array}{c} t \leq 1/2 \\ 1/2 \leq t \leq 1 \end{array}.$$

The next total space was $E_3 := X \times E_2 \cup_{\bar{m}} X$. Iterating this procedure to get an action X on E_3 led naturally to the curvilinear coordinates, e.g., the use of st in the above formula. Details are in [47].

In 1984, according to Carl Lee, Micha Perles asked whether the collection of all mutually non-crossing diagonals of a convex n-gon in the plane is isomorphic to the boundary complex of some $(n-3)$-dimensional simplicial polytope. Later that year, in an unpublished manuscript, Mark Haiman [23] referred to the *associahedron* as 'a mythical polytope whose face structure represents the lattice of partial parenthisizations of a sequence of variables'. He then constructs a simplicial polytope, but remarks that simplices can be aggregated into sub-polytopes.

In 1989, Lee [36] adopted Haiman's associahedron terminology and denoted the $(n-3)$-dimensional boundary polytope as Q_n. Lee gave a simpler abstract construction and geometric realization of the associahedron which exhibited its dihedral D_n symmetry. The coordinates of the vertices were given in terms of the coordinates of a standard n-gon in the complex plane.

1.3 Outline of this paper

As far as I can tell, judging by several of his papers, Tamari was interested primarily in the combinatorics of words and lists, bracketings and (partial) associativity, but, as we've seen, he did begin the display of the combinatorial *geometry* of the associahedra. That geometry for other polytopes has developed into a very rich field, cf. Ziegler's book [57] and more recent developments. In Section 2, I'll survey various realizations of the associahedra, especially in 3D, in terms of integer coordinates, graphics (including animations) and physical objects. In Section 3, I'll follow the homotopy origins of my interest in the associahedra to look at other polytopes such as the multiplihedra and composihedra relevant to morphisms of A_∞-spaces as well as others relevant to iterated loop spaces, etc. In so far as the associahedra arise in terms of words which are 'linear', the graph-associahedra are more manifestly a generalization of Tamari's work. In Section 3.3, we see many of these polytopes as compactifications of configuration spaces. In Section 4, the differential graded algebra coming from the associahedra is brought out, leading to relations with physics, both in terms of classical angular momentum and in terms of string field theory. Finally, I provide some comments on opportunities – missed or not.

2 Realization

There are now many ways of realizing the associahedra, but again Tamari led the
way. In his thesis, not only are there the drawings, but he also gave an encoding of
his poset. Loday, in his contribution to this volume, points out that the encoding
defines a set of vertices with integer coordinates such that the convex hull realizes
the associahedron. (All the edges have the same length, although some sets of 3 or
more vertices are collinear.)

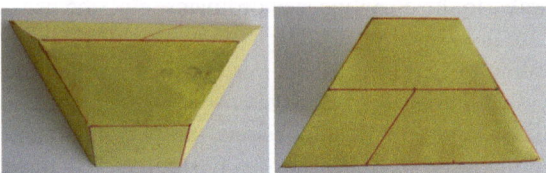

Fig. 4   Loday's model of Tamari's description

Loday's observation prompted me to realize that the coordinates of the vertices of
my representation could, after multiplication by a suitable power of 2, do the same.
For example, K_4 can be realized as the convex hull of

$$(2,4),(4,4),(4,2),(4,1),(2,2),$$

but for K_5 the bottom originally curved pentagon is realized as decomposed into
three triangles giving two spurious edges or 'creases'. (Compare Loday's remarks
about Tamari's construction realized with regular pentagons.)

However, in 1965 Danièle Lambot de Fougères had already presented in the
Séminaire Dubreil-Pisot a geometric interpretation of parenthesizations as vertices of
a polyhedron, somewhat different from Tamari's original. She calls the polyhedron H_n
for $n+2$ variables, so combinatorially equivalent to my K_{n+2}, and includes pictures
for $n = 1, 2, 3$. These polyhedra are determined as the bounded complement of a
set of hyperplanes; again not all the vertices are extrema. In particular, the bottom
facet is a pentagon but with 8 vertices on the boundary and one in the interior. She
exhibits its subdivison into two pentagons and a quadrilateral. Similarly the leftmost
facet is a pentagon with 7 vertices on the boundary subdivided into a pentagon and
a quadrilateral. (*Exercise for the reader*: Find another decomposition of H_3 as the
associahedron using the same vertices.)

Moreover, she shows that the facets of H_n are all of the form $H_{j-1} \times H_{n-j}$ and
illustrates this in her figure 5 of the 3-dimensional polyhedron. She proves again
that the number of vertices is the Catalan number (without using the name) and
also calculates the number of facets of a given form $H_{j-1} \times H_{n-j}$ and hence the
total number of facets. By induction, she gets the corresponding numbers for all the
lower-dimensional faces.

It is common to refer to a collection as consisting of *Catalan objects* if they are counted by the Catalan numbers; I suggest referring to *Tamari objects* if, as a collection, they can be put in correspondence as posets with one of the most basic examples: planar rooted trees, bracketings or non-crossing diagonals together with the basic operations: tree grafting, shifting parentheses or polygonal decompositions.

Loday [37] went further and gave integer coordinates for vertices which are all extrema. If the associahedron is to be realized by a convex polytope without further specification, then not only does this permit the realizations of Tamari and of Huguet [27] with colinear vertices, but also Shen's 'tropical' realizations [44] which include additional extrema not corresponding to parenthesizations.

On the other hand, realizations can be compared as geometric objects in the affine sense. Ceballos and Ziegler [10] compare three 'conceptual' affine realizations of the associahedra (their Ass_n is n-dimensional): 1) as a secondary polytope in the sense of Gel'fand, Zelevinsky and Kapranov, 2) as a cluster complex due to Chapoton, Fomin and Zelevinsky, 3) as a Minkowki sum of simplices after Postnikov. The secondary polytope realization is essentially that of Tamari, the cluster complex is a truncated cube and the Minkowki sum generalizes Loday's and also Buchstaber's toric version [5]. The latter two realizations were generalized by Hohlweg-Lange [26] and the cluster complex version by Ceballos, Santos and Ziegler [9]. They employ yet another *geometric* comparison in terms of linear isomorphism of the normal fans, with emphasis on parallel facets and realizations inside the standard n-dimensional cubes. From this 'normal' point of view, there are exponentially many (in terms of n) distinct realizations.

In addition to various 2-dimensional representations of the 3-dimensional associahedron, there are now several 'concrete' models, not only in cardboard but also in wood, stone and metal.

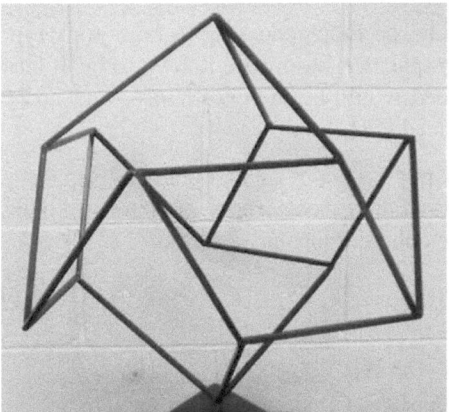

Fig. 5   Steel sculpture of a (non-convex) associahedron
Bob Szczarba, Yale

This allows the associahedron to be appreciated from literally different points of view as can also be done virtually with manipulable graphics, e.g.,

http://math.univ-lyon1.fr/~chapoton/stasheff.html and

http://www.ac-noumea.nc/maths/polyhedr/3D-img.htm

though googling 'associahedron animation' produced over 1700 other hits.

The description of a convex polytope as the bounded complement of a set of hyperplanes [57] can be used to good effect for realizations by truncating simplices or cubes, i.e., adding one or more additional hyperplanes. There are several different ways of truncating (or shaving) an n-dimensional cube to get an associahedron: one due to Devadoss [12] which generalizes the 'shave a vertex first' method, another due to Buchstaber and Volodin [6] which shaves codimension 2 faces in an appropriate order, and several presented by Ceballos, Santos and Ziegler in [9]. Instead of presenting the shaving in an appropriate order, an alternative is to specify the depth of the shaving of the various faces by carefully chosen amounts, as in [45]. Realization from the hyperplane perspective of the objects of interest here is given (in full generality to multiplihedra) by Devadoss and Forcey [11] in Section 5 of their "graph multiplihedra" paper; hyperplane equations are given in 5.2 in particular. Adding one or more additional hyperplanes can be thought of as moving them in from ∞. I hope someone will animate that perspective.

In the polytope world, I haven't seen people use blowups/blowdowns, but they do use truncation (shaving) quite commonly as well as the inverse: projections or contractions. When we talk about compactifications (namely, manifolds with corners), blowups/downs appear.

3 Other polytopes

Having characterized based loop spaces up to homotopy type, I tried to move on to iterated based loop spaces. These were known to be at least homotopy commutative/abelian, as indeed would be the case if BX were an H-space. Here again I followed Sugawara's lead [51].

Definition 3.1. A map $f : X \to Y$ of associative H-spaces X and Y is *strongly homotopy multiplicative* if there exists a coherent family of maps $\{f_i : I^{i-1} \times X^i \to Y\}$ such that $f_1 = f$. Here coherent means

$$f_i(t_1, \ldots, t_{i-1}, x_1, \ldots, x_i) = f_{i-1}(\ldots, \hat{t}_j, \ldots, x_j x_{j+1}, \ldots) \quad \text{if } t_j = 0$$

$$= f_j(t_1, \ldots, t_{j-1}, x_1, \ldots, x_j) f_{i-j}(t_{j+1}, \ldots, t_{i-1}, x_j, \ldots, x_i) \quad \text{if } t_j = 1.$$

In particular, $f_2 : I \times X \times X \to Y$ says that f is an H-map, that f respects the two multiplications up to homotopy.

Theorem 3.2 (Sugawara [51]). *If (X, m) is an associative H-space, then BX admits a multiplication if and only if the multiplication m is strongly homotopy multiplicative.*

If we peel this apart, it implies that M is, in particular, homotopy commutative, since $xy = (ex)(ye) \simeq (ey)(xe) = yx$. At the next level, we have two triangles, one the mirror image of the one displayed and both of which can be absorbed into a hexagon, see Figure 6.

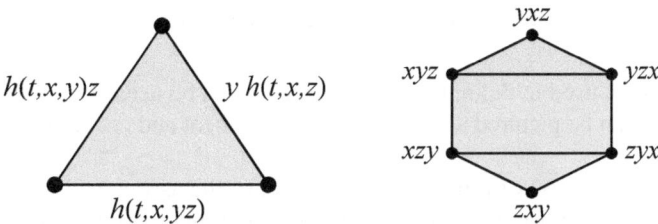

Fig. 6   Homotopy commutativity triangle and hexagon

Indeed, this hexagon is the 2-dimensional version of the *permutahedron* (also written as permutohedron), perhaps the oldest example of a polytope with algebraic origin and applications. The permutahedra have a long and illustrious history. There are currently well over 19,000 hits when googling permutahedron.

Of course, if $(X, \{m_i\})$ were A_∞ instead of associative, this hexagon would blow up to a dodecagon, having edges given alternately by the commuting homotopy and the associating homotopy. The higher-dimensional versions were first described by Kapranov [32] and are known as the *permuto-associahedra*.

3.1 Multiplihedra

The associahedra appear in another context: morphisms of an A_∞-space X to an associative space Y. Now Definition 3.1 is altered by replacing the cube I^{i-1} by the associahedron K_{i+1} with corresponding formulas for coherence.

The case of an associative space X and an A_∞-space Y is more subtle than implied in my early work. It was Forcey [18] who realized that new polytopes were needed, which he constructed and named *composihedra*.

Of course, all of these definitions are special cases of the general morphisms where both X and Y are A_∞-spaces. For this, we need further new polytopes, dubbed the *multiplihedra*. These are hinted at in my Oxford thesis and [49] with a picture of the 3-dimensional example. The 2-dimensional example is again a hexagon, the extremes of the square corresponding to $f(xyz)$ and $f(x)f(y)f(z)$ having been blown up to new edges to accommodate the associating homotopies. My hint was taken seriously by Iwase and Mimura [29], while the realization of the multiplihedra as convex polytopes was achieved by Forcey; see [17] for a definitive treatment.

For the study of the ∞-version of Hopf algebras, Saneblidze and Umble [43] have
introduced cubes with subdivided boundaries they call *biassociahedra*. At this time
they have not been realized as polytopes with all vertices as extrema.

3.2 Cyclohedra and generalized associahedra

The parentheses used in defining the associahedra can be extended to *tubes* and thus
the variables can be pictured as points on a line segment and enclosed accordingly in
tubes.

Even before the use of tubes, the analogs W_n of the associahedra corresponding
to points on a circle occurred in work of Bott and Taubes [4] in knot theory. These
I decided should be called *cyclohedra*. Once again the 2-dimensional example is a
hexagon. In general, a cyclohedron can be obtained by truncating the corresponding
associahedron; for example, a tube could enclose $(x_n x_1)$.

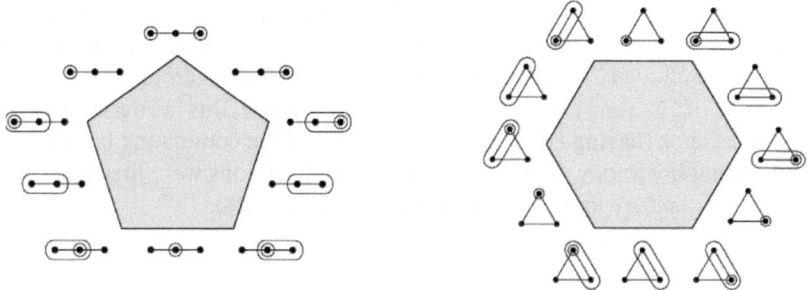

Fig. 7   Tubings – associahedron and cyclohedron

It is the cyclohedron that has perhaps the most unexpected application, what is
known as the *cyclohedron test* [41] in biology.

To circle back to Tamari who was led to the associahedra from his lattice/poset,
one can reverse the procedure and produce a poset from the face structure of any of
the graph-associahedra.

With such a plethora of named polytopes and a great variety of their representation
in 2D, it is sometimes difficult to identify different graphic versions. A great help
here can be Forcey's compilation listing vertices, triangles, squares, pentagons, etc.
See *http://www.math.uakron.edu/~sf34/hedra.htm*.

3.3 Compactified configuration spaces

Another way to approach many of these polyhedra is as compactified configuration spaces. Consider a configuration of distinct points geometrically; i.e., points are not allowed to coincide.

Definition 3.3. The *configuration space* $Conf_n(X)$ of n ordered distinct points is $X^n \setminus \Delta$ where Δ is the fat diagonal, i.e., $\{(x_1,\dots,x_n) | x_i \neq x_j$ if $i \neq j\}$.

For example, if the n letters of the word are pictured as distinct points $0 < x_1 < x_2 < \cdots < x_n < 1$ on the open unit interval I°, we see the *configuration space* $Conf_n(I^\circ) \subset (I^\circ)^n - \Delta$ which is the interior of the n-simplex. For compact X, there is the naive compactification, X^n itself, but, as Kontsevich points out [33], if points come close enough together, we can't see clearly if they coincide, so we need a magnifying glass at that resolution, which allows us to see how points approach collision. Then we may see the phenomenon repeated and need further magnification. Kontsevich describes the compactification in terms of a tree of magnifying glasses.

Another way to picture this, more relevant to physics, when X is a k-manifold is in terms of little k-spheres bubbling off X and then a whole tree of little k-spheres bubbling off little k-spheres bubbling off, etc. Figure 8 shows a picture for S^1.

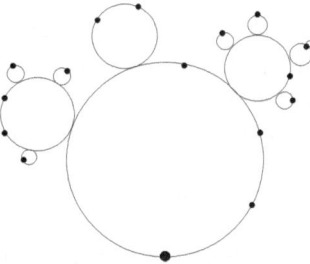

Fig. 8   Bubble tree on S^1

This compactification is precisely the real non-projective version of the Fulton-MacPherson compactification from algebraic geometry [22], cf. the Axelrod-Singer compactification [2]. For ordered points on the real line or open interval, the compactifications are precisely the associahedra. The faces of the associahedra can be regarded as blow-ups of faces of the simplex.

4 Applications

Another application outside of physics is to computer science. In [46], Sleator, Tarjan and Thurston attempt to solve the dynamic optimality conjecture concerning the performance of *splaying*. Splaying is a heuristic for modifying the structure of a binary search tree in such a way that repeatedly accessing and updating the information in the tree is efficient.

It concerns the maximum rotation distance (that is, the maximum path without repeats of the change in a subtree corresponding to an associating homotopy) between two binary trees with n-leaves.

However, most of the applications of the associahedra are through the associated algebras known as A_∞-algebras. The original definition reflected the chains on an A_∞-space:

Definition 4.1 (A_∞-algebra, strongly homotopy associative algebra [49])**.** Let A be an $\mathbb{N}$-graded vector space $A = \oplus_{r \in \mathbb{N}} A^r$ with a collection of multi-linear maps

$$\mathbf{m} := \{m_k : A^{\otimes k} \to A\}_{k \geq 1}$$

of degree $k - 2$. The pair $(A, \mathbf{m})$ is called an A_∞-*algebra* when the multi-linear maps m_k satisfy the following relations

$$\sum_{k+l=n+1} \sum_{i=1}^{k} (-1)^{a_1 + \cdots + a_{i-1}} m_k(a_1, \cdots, a_{i-1}, m_l(a_i, \cdots, a_{i+l-1}), a_{i+l}, \cdots, a_n) = 0$$

for $n \geq 1$, where a_j on (-1) denotes the degree of a_j.

If $m_1 = 0$, the interpretation in terms of homotopies is lacking, but structures with such multi-variable operations still abound. For example, the homology of an A_∞-space X inherits corresponding higher-order operations. In fact, as shown originally by Kadeishvili [30], the homology of a strictly associative H-space admits the structure of an A_∞-algebra, reflecting the homotopy type of the space X as A_∞-space, rather than just as a space. Of course, if X is finite dimensional, the operations m_i are zero for i sufficiently large. Fortunately, this had not led to a plethora of terms indicating the bound, other than A_n-algebra as for A_n-space. Contrast the analogous Lie situation where we have k-Lie algebras as distinct from Lie k-algebras.

4.1 Coderivation interpretation and curvature

Just as the structure of an A_∞-space $(X, \{m_i\})$ can be assembled into a single object BX, an A_∞-algebra structure $(A, \{m_i\})$ can be assembled into a single coderivation differential on the tensor coalgebra $T^c sA$ where sA denotes the augmentation ideal of A but with the grading shifted/suspended by 1 [48]. From that point of view, an additional generalization known as a *curved A_∞-algebra* occurs naturally.

That is, in addition to the operations $\{m_i\}, i \geq 1$, there is an operation $m_0 : k \to A$ (equivalently an element $m_0 \in A$) such that the defining relations continue to hold including m_0. Notice that m_1 is no longer a differential, since it no longer squares to 0 but rather

$$m_1 m_1 = m_2(m_0 \otimes 1 + 1 \otimes m_0).$$

The nomenclature calling m_0 the *curvature* of this curved algebra is derived from the case of the covariant derivative. Remarkably, such curved algebras occur in what physicists refer to as a 'background' for string field theory.

4.2 Pentagonal algebras

There are naturally occurring examples of what I call *pentagonal algebras* which correspond to homotopy associative algebras, but for which homotopies there is a strict identity rather than a higher homotopy. These appeared in or were motivated by physics, namely, in classical angular momentum combinatorics, quasi-Hopf algebras and open string field theories.

The (chronologically) first example I know is the Biedenharn-Elliott identities [3, 15] for what are known as Wigner 6-j *coefficients* [54]. The stage is set with the irreps (irreducible representations) of SU(2) labeled by 'spins' $j = 1/2, 1, 3/2, \ldots$, rather than positive integers since they are really irreps of Spin(3). A tensor product of two such irreps V_{j_1} and V_{j_2} decomposes as a sum of irreps:

$$V_{j_1} \otimes V_{j_2} = \bigoplus c^j_{j_1 j_2} V_j.$$

For 3 irreps V_a, V_b, V_c, a given V_j can appear in $(V_a \otimes V_b) \otimes V_c$ via some $V_k \otimes V_c$ as well as in $V_a \otimes (V_b \otimes V_c)$ via some $V_a \otimes V'_k$. Since V_j is an irrep, the two inclusions are related by a scalar, a Wigner 6-j coefficient

$$\begin{Bmatrix} a & b & k \\ c & j & k' \end{Bmatrix}.$$

The particular arrangement of the six entries is a convention (for which I've been unable to find an explanation).

The Biedenharn-Elliott identities for these 6-j coefficients express precisely the 'Mac Lane' coherence of the associativity of $\otimes$. When I learned of these in discussion with Biedenharn when he was at Duke and I was at UNC, what struck me was their form:

$$WW = WWW$$

where W denotes an appropriate 6-j coefficient; indeed further discussion revealed the relation to coherence. Seeing only the formulas and not the related tree diagrams, it took me a while to realize the depth of the analogy (see, e.g., [8]).

The Biedenharn-Elliott identities also correspond to what are known as the (2,3)-Pachner moves [42] which replace a union of 2 tetrahedra with a common face by a union of 3 tetrahedra with a common edge. (Think of the 5 tetrahedra which are the faces of the 4-simplex.) Without the name, this is pictured in a book by Biedenharn and Louck as if it were the suspension of a triangle with an additional edge between the two extreme vertices.

In such a purely algebraic/coherence setting, the analog of an associating homotopy is called an *associator*. Mac Lane treats the associativity of $\otimes$ categorically in [38]. The coherence of interest to him involves only the 2-skeleton of the associahedra so that any two ways of using an associator to go from iterating $\otimes$ always on the right to iterating $\otimes$ always on the left are the same isomorphism. That is, given a particular associator, $(V_1 \otimes (V_2 \otimes (V_3 \otimes \cdots) \cdots))$ is canonically isomorphic to $((\cdots (V_1 \otimes V_2) \otimes V_3) \otimes \cdots)$.

Associators play a crucial role in Drinfel'd's invention of *quantum groups*. In [13], to show the existence of a quantization as a coboundary quasi-Hopf quantum universal enveloping algebra, he proceeds step-by-step as is common in deformation/perturbation theory. A key point in the inductive step is to show that the obstruction to proceeding is a cycle (and hence a coboundary). The coboundary of the cocycle is realized as the boundary of the 3-dimensional associahedron K_5 and hence 'visibly' a cycle.

4.3 The pentagonal algebra of open string field theory

One version of an open string in physics is a geometric object – something like an oriented arc in a Riemannian manifold (M, g). One of the subtleties of string theory is the use of parametrized strings (paths) to obtain results that are independent of the choice of parameterization. String interactions are handled by "joining two strings to form a third" or "splitting a string in two". These are often pictured in one of three ways:

E: endpoint interaction

Here we are dealing with a space of maps $Map(I, M)$ where $I = [0, 1]$. (Physicists prefer $[0, \pi]$.) The endpoint joining E requires a reparameterization; define $X + Y : [0, 1] \to M$ by

$$\begin{cases} X(2t) & \text{for } 0 \leq t \leq 1/2 \\ Y(2t - 1) & \text{for } 1/2 \leq t \leq 1. \end{cases}$$

This operation fails to have units, inverses or associativity, though all are present "up to homotopy".

So as to restore associativity, we follow J.C. Moore [39] and consider the space PM of parametrized paths $X : [0, r] \to M$. (Fixing r can be regarded as a (partial) choice of 'gauge'.) Now we define the endpoint joining operation $X \vee Y$ for paths X, Y such that $X(r) = Y(0)$, so that $X \vee Y : [0, r+s] \to M$ by

$$\begin{cases} X(t) & \text{for } 0 \leq t \leq r \\ Y(t-r) & \text{for } r \leq t \leq r+s. \end{cases}$$

This operation is associative where defined and has units $x : [0,0] \to M$, but has inverses only up to homotopy.

Conversely, splitting of the picture of a string at an arbitrary point led naturally to the use of parametrized paths $X : [0,r] \to M$ and to what are known as 4-point functions in physics. This splitting rather than combining paths seems to have occurred first in physics in the work of Kaku and Kikkawa [31]. They remark:

> the four-string interaction corresponds to the continuous deformation of the t-channel topology into the u-channel topology... ([31], p. 1124),

in other words, an associating homotopy. In considering multiple string interactions, they remark that the number of relevant light-cone Feynman diagrams is equal to the number of non-crossing triangulations of a convex N-gon, hence the Catalan number. Further, they remark that for an N-point functions, there is a relevant $N-3$-dimensional 'surface' (what is now known as the associahedron). They sketch a reason why for their string field theory there is no need for N-point functions for $N > 4$. Translation: the pentagonal condition is satisfied on the nose, no need for higher homotopies.

M: midpoint or half overlap interaction

Here when the latter half of one path is the same as the first half of the second path but with reversed orientation, the interaction produces a third path by 'cancelling' the overlapping halves. The picture is symmetric with respect to cyclic permutations and is so interpreted.

The endpoint case E is familiar in mathematics as far back as the study of the fundamental group; the midpoint case M was considered by Witten in 1986 [56] for string field theory, although in fact it had been considered independently in mathematics by Lashof [35] in 1956.

V: variable overlap interaction

But if overlapping halves can be cancelled, why not vary the portion that is cancelled? This idea seems to have occurred first in physics in the HIKKO [2] interaction [24] of open string field theory. This (independently) uses Moore's associative endpoint interaction, but then creates inverses by cancellation as in the midpoint case. This variable overlap joining V of paths is given by the operation $X \star Y$ defined as follows: Let $u = \max\{r \mid Y(t) = X(r-t) \text{ for } 0 \leq t \leq r\}$, then $X \star Y : [0, r+s-2u] \to M$ by

---

[2] Hata, Itoh, Kugo, Kunitomo, Ogawa

$$\begin{cases} X(t) & \text{for } 0 \le t \le r - u \\ Y(t - r + 2u) & \text{for } r - u \le t \le r + s - 2u. \end{cases}$$

Again $m : [0,0] \to M$ is a unit, and the reversal $\bar{X}(t) = X(r-t)$ provides a strict inverse, but now $\star$ occasionally fails to be associative; there is a hidden destruction of associativity in what HIKKO refer to as 'horn' diagrams.

On the other hand, $(X \star Y) \star Z$ and $X \star (Y \star Z)$ are clearly homotopic – just gradually shrink the back-and-forth part of $(X \star Y) \star Z$ – and hence $\star$ is homotopy associative. Denote such a homotopy $X \star Y \star Z$. From our point of view, it is this homotopy which gives rise to the "4-string vertex" in physics. Corresponding to the ways of combining m_2 and m_3 in defining an A_∞-algebra, there are five ways of combining the operations $\star \star$ with $\star$ on W, X, Y, Z, e.g., $W \star (X \star Y) \star Z$, which remarkably add to give 0 in this string interpretation.

4.4 A_n-structures for $4 < n \le \infty$

In the topological setting, I was able to construct examples of spaces for each prime p that were A_{p-1}-spaces but not A_p-spaces, though they weren't exactly 'naturally occurring in nature'. On the other hand, the standard homotopy equivalence between the based loop space $\Omega CP(n)$ and $S^1 \times \Omega S^{2n+1}$ could be shown to be equivalent as A_n-spaces but not as A_{n+1}-spaces.

With the focus on associativity, it is reasonable to consider A_∞-structures on categories $\mathscr{C}$. If the sets (of 'objects', 'morphisms', etc) have underlying topological or differential graded structure, the definitions need to be modified only by allowing the structure maps to be partially defined, e.g., on appropriate subsets of $\mathscr{C}^n$. This is the approach of Fukaya [19]. Indeed, associativity of partially defined operations was very much part of Tamari's interests.

By contrast, quite recently Fiedorowicz, Gubkin and Vogt [16] turned the focus to A_n-structures in terms of weakening Mac Lane's monoidal category structure, as in Laplaza [34], by allowing, for example, an associator which is not an isomorphism. Key ingredients for them are both the associahedra and the Tamari posets/lattices.

Thus we have come back to Tamari's dual vision of 'my' associahedra and his lattice.

5 Opportunities – missed and not

Dyson in 1972 gave an AMS Gibbs Lecture titled 'Missed Opportunities' [14]. He talked about

> occasions on which mathematicians and physicists lost chances of making discoveries by neglecting to talk to each other . . . not to blame the mathematicians or to excuse the physicists
> . . .

He had hoped to influence the future to avoid such missed opportunities. In the physical applications I've mentioned (as well as others such as the BV and BFV formalisms, higher spin particles and quantum field theory less related to the associahedra or other polytopes), especially in direct conversation with some of the relevant physicists, we see a change in our cultures; the 'divorce' of the past has been healed and we have 'renewed our vows'.

But as Dyson himself acknowledges, the chasm can exist within one community or even one individual (himself). When I was a grad student, algebraic topology was, as Dyson said, "rushing ahead in a golden age of luxuriant growth", but combinatorics was less prominent and unknown to me. So I missed Tamari's work and his version of the associahedra. Perhaps this Festschrift can help the next generation to be more aware of mathematics (and even mathematical physics) not as disjoint compartments or even as a tree, but rather as a highly connected and coherent graph, a 'spider web' *http://www.shutterstock.com/pic.mhtml?id=19709590.*

Acknowledgement I am grateful to Devadoss, Forcey and Loday for discussions and especially for several of the graphics I have included, but, above all, to the editor-in-chief Folkert Müller-Hoissen for conceiving of this project and guiding it through.

References

1. J.F. Adams, *Infinite Loop Spaces*, Princeton University Press, Princeton, 1978.
2. S. Axelrod and I.M. Singer, "Chern-Simons perturbation theory. II", *J. Differential Geom.* **39** (1994) 173–213.
3. L.C. Biedenharn, "An identity satisfied by the Racah coefficients", *J. Math. Physics* **31** (1953) 287–293.
4. R. Bott and C. Taubes, "On the self-linking of knots", *J. Math. Phys.* **35** (1994) 5247–5287.
5. V.M. Buchstaber, "Lectures on Toric Topology", in *Toric Topology Workshop, KAIST 2008,* Trends inMathematics, Information Center of Mathematical Sciences, vol. 1, 2008, 1–55.
6. V.M. Buchstaber and V. Volodin, "Combinatorial 2-truncated cubes and applications", in *this volume.*
7. J.W. Cannon and W.J. Floyd, "What is Thompson's group", *Notices AMS* (2011) 1112–1113.
8. J.S. Carter, D.E. Flath, and M. Saito, *The classical and quantum 6j-symbols*, Mathematical Notes, vol. 43, Princeton University Press, Princeton, NJ, 1995.
9. C. Ceballos, F. Santos, and G.M. Ziegler, "Many non-equivalent realizations of the associahedron", *arxiv.org/abs/1109.5544.*
10. C. Ceballos and G.M. Ziegler, "Three non-equivalent realizations of the associahedron", *arxiv.org/abs/1006.3487.*
11. S. Devadoss and S. Forcey, "Marked tubes and the graph multiplihedron", *Algebr. Geom. Topol.* **8** (2008) 2081–2108.
12. S. Devadoss, T. Heath, and C. Vipismakul, "Deformations of bordered surfaces and convex polytopes", *Notices of the American Mathematical Society* **58** (2011) 530–541.
13. V. Drinfel'd, "Quasi-Hopf algebras", *Leningrad Math J.* **1** (1990) 1419–1457.
14. F.J. Dyson, "Missed opportunities", *Bull. Amer. Math. Soc.* **78** (1972) 635–652.
15. J. Elliott, "Theoretical studies in nuclear spectroscopy V: The matrix elements of non-central forces with an application to the 2p-shell", *Proc. Roy. Soc. A* **218** (1953) 345–370.
16. Z. Fiedorowicz, S. Gubkin, and R. Vogt, "Associahedra and weak monoidal structures on categories", *arxiv.org/abs/1005.3979.*

17. S. Forcey, "Convex Hull Realizations of the Multiplihedra", *Topology and its Applications* **156** (2008) 326–347, *arxiv.org/abs/0706.3226*.

18. S. Forcey, "Quotients of the multiplihedron as categorified associahedra", *Homology, Homotopy Appl.* **10** (2008) 227–256.

19. K. Fukaya, "Floer homology, A^∞-categories and topological field theory", in *Geometry and Physics*, J. Andersen, J. Dupont, H. Pertersen, and A. Swan, eds., Lecture Notes in Pure and Applied Mathematics, vol. 184, Marcel-Dekker, 1995, Notes by P. Seidel, 9–32.

20. K. Fukaya, Y.-G. Oh, H. Ohta, and K. Ono, *Lagrangian intersection Floer theory: anomaly and obstruction. Part I*, AMS/IP Studies in Advanced Mathematics, vol. 46, American Mathematical Society, Providence, RI, 2009.

21. _____, *Lagrangian intersection Floer theory: anomaly and obstruction. Part II*, AMS/IP Studies in Advanced Mathematics, vol. 46, American Mathematical Society, Providence, RI, 2009.

22. W. Fulton and R. MacPherson, "A compactification of configuration spaces", *Ann. Math.* **139** (1994) 183–225.

23. M. Haiman, "Constructing the associahedron", available for download at *http://math.berkeley.edu/~mhaiman/ftp/assoc/manuscript.pdf*, 1984.

24. H. Hata, K. Itoh, T. Kugo, H. Kunitomo, and K. Ogawa, "Covariant string field theory", *Phys. Rev. D* **34** (1986) 2360–2429.

25. A. Hatcher, *Algebraic Topology*, Cambridge Univeristy Press, 2002, available for download at *http://www.math.cornell.edu/~hatcher/AT/ATpage.html*.

26. C. Hohlweg and C.E.M.C. Lange, "Realizations of the associahedron and cyclohedron", *Discrete Comput. Geom.* **37** (2007) 517–543.

27. D. Huguet, "La structure du treillis des polyèdres de parenthésages", *Algebra Universalis* **5** (1975) 82–87.

28. D. Huguet and D. Tamari, "La structure polyédrale des complexes de parenthésages", *Journal of Combinatorics, Information & System Sciences* **3** (1978) 69–81.

29. N. Iwase and M. Mimura, "Higher homotopy associativity", in *Algebraic topology (Arcata, CA, 1986)*, Lecture Notes in Math., vol. 1370, Springer, Berlin, 1989, 193–220.

30. T. Kadeishvili, "On the homology theory of fibre spaces", *Russian Math. Surv.* **35:3** (1980) 231–238, math.AT/0504437.

31. M. Kaku and K. Kikkawa, "Field theorey of relativistic strings. I Trees", *Phys. Rev. D* **10** (1974) 1110–1133.

32. M.M. Kapranov, "The permutoassociahedron, Mac Lane's coherence theorem and asymptotic zones for the KZ equation", *J. Pure and Appl. Alg.* **85** (1993) 119–142.

33. M. Kontsevich, "Deformation quantization of Poisson manifolds", *Lett. Math. Phys.* **66** (2003) 157–216.

34. M.L. Laplaza, "Coherence for associativity not an isomorphism", *J. Pure Appl. Algebra* **2** (1972) 107–120.

35. R. Lashof, "Classification of fibre bundles by the loop space of the base", *Ann. of Math. (2)* **64** (1956) 436–446.

36. C. Lee, "The associahedron and triangulations of the *n*-gon", *Europ. J. Combinatorics* **10** (1989) 551–560.

37. J.-L. Loday, "Realization of the Stasheff polytope", *Arch. Math.* **83** (2004) 267–278.

38. S. Mac Lane, "Natural associativity and commutativity", *Rice Univ. Studies* **49** (1963) 28–46.

39. J.C. Moore, "Le théorème de Freudenthal, la suite exacte de James et l'invariant de Hopf généralisé", in *Séminaire Henri Cartan* **7** (1954–1955) 22-01 – 22-15.

40. F. Muro and A. Tonks, "Unital associahedra", *arxiv.org/abs/1110.1959*.

41. J. Morton, A. Shiu, L. Pachter, and B. Sturmfels, "The cyclohedron test for finding periodic genes in time course expression studies", *Stat. Appl. Genet. Mol. Biol.* **6** (2007) Art. 21, 25 pp. (electronic).

42. U. Pachner, "P.L. homeomorphic manifolds are equivalent by elementary shellings", *European J. Combin.* **12** (1991) 129–145.

43. S. Saneblidze and R. Umble, "Matrads, Biassociahedra, and A_∞-bialgebras", *Homology, Homotopy and Applications* **13** (2011) 1–57.

44. L. Shen, "Stasheff polytopes and the coordinate ring of the cluster algebra χ-variety A_n", *arxiv.org/abs/1104.3528*.
45. S. Shnider and J. Stasheff, "An operad-chik looks at configuration spaces, moduli spaces and mathematical physics, appendix B", in *Operads: Proceedings of Renaissance Conferences*, J.-L. Loday, J. Stasheff, and A.A. Voronov, eds., Contemporary Mathematics, vol. 202, Amer. Math. Soc., 1996, 75–78.
46. D.D. Sleator, R.E. Tarjan, and W.P. Thurston, "Rotation distance, triangulations, and hyperbolic geometry", *J. Amer. Math. Soc.* **1** (1988) 647–681.
47. J. Stasheff, "Homotopy associativity of H-spaces, I", *Trans. Amer. Math. Soc.* **108** (1963) 293–312.
48. J.D. Stasheff, "The intrinsic bracket on the deformation complex of an associative algebra", *JPAA* **89** (1993) 231–235, Festschrift in Honor of Alex Heller.
49. J. Stasheff, *H-spaces from a homotopy point of view*, Lecture Notes in Mathematics, Vol. 161, Springer-Verlag, Berlin, 1970.
50. M. Sugawara, "On a condition that a space is group-like", *Math. J. Okayama Univ.* **7** (1957) 123–149.
51. ———, "On the homotopy-commutativity of groups and loop spaces", *Mem. Coll. Sci. Univ. Kyoto, Ser. A Math.* **33** (1960/61) 257–269.
52. D. Tamari, "Monoides préordonnés et chaînes de Malcev", Doctorat ès-Sciences Mathématiques Thèse de Mathématique, Paris, 1951.
53. ———, "The algebra of bracketings and their enumeration", *Nieuw Archief voor Wiskunde* **10** (1962) 131–146.
54. E.P. Wigner, "On the matrices which reduce the Kronecker products of representations of S. R. groups", in *Quantum Theory of Angular Momentum*, Academic Press, New York, 1965, 87–133.
55. Wikipedia, "CW complex – Wikipedia, the free encyclopedia", 2011, [Online; accessed 9-September-2011].
56. E. Witten, "Noncommutative geometry and string field theory", *Nuclear Phys. B* **268** (1986) 253–294.
57. G.M. Ziegler, *Lectures on Polytopes*, Graduate Texts in Mathematics, vol. 152, Springer-Verlag, New York, 1995.

Dichotomy of the Addition of Natural Numbers

Jean-Louis Loday

Abstract This is an elementary presentation of the arithmetic of trees. We show how it is related to the Tamari poset. In the last part we investigate various ways of realizing this poset as a polytope (associahedron), including one inferred from Tamari's thesis.

Introduction

In this paper the addition of integers is split into two operations which satisfy some relations. These relations are taken so that they split the associativity relation of addition into three. Under these new operations, the unit 1 generates elements which are in bijection with the planar binary rooted trees. More precisely, any integer n splits as the disjoint union of the trees with n internal vertices. The Tamari poset is a partial-order structure on this set of trees. We show how the addition on trees is related to the Tamari poset structure. This first part is an elementary presentation of results contained in "Arithmetree" [19] by the author and in [24] written jointly with M. Ronco.

Prompted by an unpublished page of Tamari's thesis, we investigate various ways of realizing the Tamari poset as a polytope. In particular we show that Tamari's way of indexing the planar binary rooted trees gives rise to a hypercube-like polytope on which the associahedron is drawn.

Jean-Louis Loday
Institut de Recherche Mathématique Avancée, CNRS et Université de Strasbourg, 7 rue René-Descartes, 67084 Strasbourg Cedex, France, e-mail: *loday@math.unistra.fr*

1 About the formula $1+1=2$

The equality $3+5=8$ can be seen either as 3 acting on the left on 5, or as 5 acting on the right on 3. Since adding 3 and 5 is both, one can imagine to "split" this sum into two pieces reflecting this dichotomy. Physically, splitting the addition symbol $+$ into two pieces gives:

$$+$$
$$\dashv\vdash$$
$$\dashv\ \ \vdash$$

that is, the symbols $\dashv$ and $\vdash$. Hence, since $1+1=2$, one defines two new elements $1\dashv 1$ and $1\vdash 1$ so that

$$1\dashv 1\ \cup\ 1\vdash 1=1+1=2.$$

2 Splitting the integers into pieces

How to go on ? A priori one can form eight elements out of three copies of 1 and of the operations *left* $\dashv$ and *right* $\vdash$, that is

$$(1\dashv 1)\dashv 1\ \ ,\ \ (1\dashv 1)\vdash 1\ \ ,\ \ (1\vdash 1)\dashv 1\ \ ,\ \ (1\vdash 1)\vdash 1\ \ ,$$
$$1\dashv(1\dashv 1)\ \ ,\ \ 1\dashv(1\vdash 1)\ \ ,\ \ 1\vdash(1\dashv 1)\ \ ,\ \ 1\vdash(1\vdash 1)\ \ .$$

But we would like to keep associativity of the operation $+$, so we want that the union of the elements on the first row is equal to the union of the elements of the second row. More generally for any component r and s we split the sum as

$$r+s=r\dashv s\ \cup\ r\vdash s\ .$$

Taking again our metaphor of left action and right action, it is natural to choose the relations

$$(*)\qquad \begin{cases}(r\dashv s)\dashv t=r\dashv(s+t),\\ (r\vdash s)\dashv t=r\vdash(s\dashv t),\\ (r+s)\vdash t=r\vdash(s\vdash t),\end{cases}$$

since, by taking the union, we get readily $(r+s)+t=r+(s+t)$. The first relation says that "acting on the right by s and then by t" is the same as "acting by $s+t$". (The knowledgeable reader will remark the analogy with bimodules). Since we have three relations, our eight elements in the case $r=s=t=1$ go down to five, which are the following:

$$(1\vdash 1)\vdash 1\ ,\ (1\dashv 1)\vdash 1\ ,\ 1\vdash 1\dashv 1\ ,\ 1\dashv(1\vdash 1)\ ,\ (1\dashv 1)\dashv 1\ .$$

Indeed, since one has $(1 \vdash 1) \dashv 1 = 1 \vdash (1 \dashv 1)$, the parentheses can be discarded. On the other hand the two elements $1 \vdash (1 \vdash 1)$ and $(1 \dashv 1) \dashv 1$ can be written respectively:

$$1 \vdash (1 \vdash 1) = (1 \vdash 1) \vdash 1 \cup (1 \dashv 1) \vdash 1\,,$$
$$(1 \dashv 1) \dashv 1 = 1 \dashv (1 \vdash 1) \cup 1 \dashv (1 \dashv 1).$$

In conclusion, we have decomposed the integer 2 into two components $1 \dashv 1$ and $1 \vdash 1$ and the integer 3 into five components (see above), the integer 1 has only one component, namely itself. How about the integer n? In fact, not only would we like to decompose n into the union of some components, but we would also like to know how to add these components. The test will consist in checking that adding the components of n with all the components of m, we get back the union of all the components of $m + n$.

3 Trees and addition on trees

In order to understand the solution we introduce the notion of *planar binary rooted tree*, that we simply call tree in the sequel. Here are the first of them:

$$PBT_1 = \{\,|\,\}\,, \quad PBT_2 = \{\ \curlyvee\ \}\,, \quad PBT_3 = \left\{\ \curlyvee\ ,\ \curlyvee\ \right\}\,,$$

$$PBT_4 = \left\{\ \curlyvee\ ,\ \curlyvee\ ,\ \curlyvee\ ,\ \curlyvee\ ,\ \curlyvee\ \right\}.$$

Such a tree t is completely determined by its left part t^l and its right part t^r, which are themselves trees. The tree t is obtained by joining the roots of t^l and of t^r to a new vertex and adding a root:

This construction is called the *grafting* of t^l and t^r. One writes $t = t^l \vee t^r$.

Hence any nontrivial tree (that is different from $|$) is obtained from the trivial tree $|$ by iterated grafting. The elements of the set PBT_n are the trees with n leaves that is with $n - 1$ internal vertices. The number of elements in PBT_n is the Catalan number c_{n-1}. It is known that $c_n = \frac{(2n)!}{n!\,(n+1)!} = \frac{1}{n+1}\binom{2n}{n}$.

The solution to the splitting of natural numbers is going to be a consequence of the properties of the operations $\dashv$ and $\vdash$ on trees, which are defined as follows. For any nontrivial trees s and t one defines recursively the two operations $\dashv$ and $\vdash$ by the formulas

$$(\ddagger) \qquad s \dashv t := s^l \vee (s^r + t) \quad , \quad s \vdash t := (s + t^l) \vee t^r,$$

and the sum by

$$s + t := s \dashv t \ \cup \ s \vdash t.$$

The trivial tree is supposed to be a neutral element for the sum: $| = 0$, so $s \dashv 0 = s$ and $0 \vdash t = t$. The unique tree with one internal vertex (**Y** shape tree) represents 1. Then one gets

Notice the matching between the orientation of the leaves and the involved operations: the middle leaf of the tree , resp. , is oriented to the left, resp. right, and this tree represents the element $1 \dashv 1$, resp. $1 \vdash 1$.

The principal properties of these two operations are given by the following statement.

Proposition 3.1 ([24]). *The operations $\dashv$ and $\vdash$ satisfy the relations $(*)$ of Section 2. Any tree can be obtained from the initial tree* *by iterated application of the operations left and right.*

The solution is then the following. The integer n is the disjoint union of the elements of PBT_{n+1}, that is the trees with n internal vertices. For instance:

$$0 = |$$

The sum $+$ of integers can be extended to the components of these integers, that is to trees. Even better, the operations left and right can be extended to the trees. The above formulas $(\ddagger)$ give the algorithm to perform the computation.

4 Where we show that $1+1=2$ and $2+1=3$

Here are some computation examples:

$$1 \dashv 1 = \quad\text{}\quad , \quad 1 \vdash 1 = \quad\text{} ,$$

$$2 \dashv 1 = \quad\text{}$$

$$= \quad\text{} ,$$

$$2 \vdash 1 = \quad\text{}$$

$$= \quad\text{} .$$

Notice that $1+1 = 1 \dashv 1 \cup 1 \vdash 1 = 2$ and that $2+1 = 2 \dashv 1 \cup 2 \vdash 1 = 3$, since 3 is the union of the five trees of PBT_4. They represent the five elements which can be written with three copies of 1 (see above). Similarly one can check that

$$m+n = m \dashv n \cup m \vdash n$$
$$= \left(\bigcup_{s \in PBT_{n+1}} s \dashv \bigcup_{t \in PBT_{m+1}} t \right) \cup \left(\bigcup_{s \in PBT_{n+1}} s \vdash \bigcup_{t \in PBT_{m+1}} t \right)$$
$$= \left(\bigcup_{s \in PBT_{n+1}, t \in PBT_{m+1}} s \dashv t \right) \cup \left(\bigcup_{s \in PBT_{n+1}, t \in PBT_{m+1}} s \vdash t \right)$$
$$= \bigcup_{r \in PBT_{m+n+1}} r = m+n.$$

Finally there are two ways to look at the c_{n+1} components of the integer n: either through trees, or through n copies of 1 and the operations $\dashv$ and $\vdash$ duly parenthesized. Recall that this second presentation is not unique.

There are many families of sets whose number is the Catalan number (they are called Catalan sets). For each of them one can translate the algebraic structure unraveled above. For some of them the sum of trees has a nice description. This is the case for the "alternative Catalan tableaux" worked out in [1] by Aval and Viennot. It is interesting to notice that, in their description of the sum of two tableaux (see loc. cit. p. 6), there are two different kinds of tableaux. In fact the tableaux of one kind give the left operation and the tableaux of the other kind give the right operation.

5 The integers as molecules

Let us think of the integers as molecules and of its components as atoms. Then one would like to know of the ways the atoms are bonded in order to form the molecule. Since the molecule 2 has only two atoms, we pretend that there is a bond between the two atoms:

For our mathematical purpose it is important to see this bond as an oriented relation (it is called a covering relation):

For the molecule n one puts a bond between two atoms (i.e., trees) whenever one can obtain one of them from the other one by a local change as in the molecule 2 case. Here is what we get for $n = 3$:

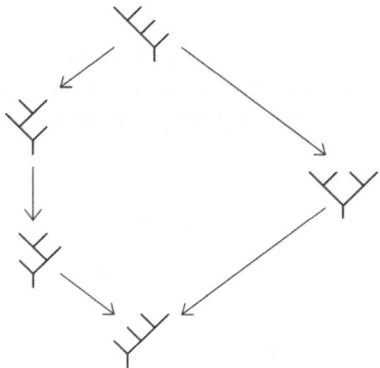

and for $n = 4$ (without mentioning the trees):

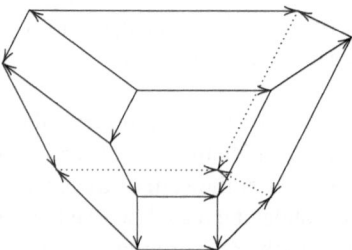

These drawings already appeared, under slightly different shape, in Dov Tamari's original thesis, defended in 1951 (see the discussion below in Section 8). In fact the covering relations on the set PBT_n make it into a "partially ordered set", usually abbreviated into "poset". This is the *Tamari poset* on trees [34]. Following Stasheff's suggestion [32] a Catalan set equipped with a partial-order isomorphic to the Tamari lattice is called a *Tamari set*. The reason for introducing this poset at this point is its strong relationship with the algebraic structure that we just described on trees. It is given by the following statement proved in a joint work with María Ronco.

Theorem 5.1 ([24]). *The sum of the trees t and s is the union of all the trees which fit in between t/s and $t\backslash s$:*

$$t+s = \bigcup_{t/s \leq r \leq t\backslash s} r$$

where t/s is obtained by grafting the root of t on the leftmost leaf of s and $t\backslash s$ is obtained by grafting the root of s on the rightmost leaf of t.

This formula makes sense because one can prove (cf. loc. cit.) that, for any trees t and s, we have $t/s \leq t\backslash s$.

6 Multiplication of trees

Multiplication of natural numbers is obtained from addition since:

$$n \times m := \underbrace{m + \cdots + m}_{n \text{ copies}}.$$

In other words, one writes n in terms of the generator 1:

$$n = \underbrace{1 + \cdots + 1}_{n \text{ copies}},$$

and then one replaces 1 by m everywhere to obtain $n \times m$.

The very same process enables us to define the product $t \times s$ of the trees t and s. First we write t in terms of 1 with the help of the left and right operations, and then we replace each occurence of 1 by s. Here are some examples:

The proof of the first case is as follows. Since we have $= 1 \vdash 1$, we can write:

It is immediate to check that if s has n internal vertices and t has m internal vertices, then $s \times t$ has nm internal vertices. Another relationship with the product of natural numbers is the following. Replacing n by the union of trees of PBT_{n+1}, and m by the union of trees of PBT_{m+1}, then $n \times m$ is actually the union of all the trees with nm internal vertices.

Some of the properties of the multiplication are preserved, but not all. The associativity holds and the distributivity with respect to the left factor also holds. But right distributivity does not (and of course commutativity does not hold). This is the price to pay for such a generalization. More properties and computation can be found in [19]. The interesting paper [5] deals with the study of prime numbers (we should say prime trees) in this framework.

Let us summarize the properties of the sum and the product of trees versus the sum and the product of integers. We let $\mathscr{P}(PBT)$ be the set of non-empty subsets of PBT_n for all n.

Proposition 6.1. *There are maps*

$$\mathbb{N} \rightarrowtail \mathscr{P}(PBT) \twoheadrightarrow \mathbb{N}$$

which are compatible with the sum and the product. The composite is the identity.

Indeed, the first map sends n to the union of all the trees in PBT_n. The second map sends a subset to the arity of its components.

7 Trees and polynomials

The algebra of polynomials (let us say with real coefficients) in one variable x admits the monomials x^n for basis. Since we know how to decompose an integer into the union of trees, we dare to write

$$x^n = \sum_{t \in PBT_{n+1}} x^t.$$

More specifically, we consider the vector space spanned by the elements x^t for any tree t. As usual, the sum of exponents gives rise to a product of factors:

$$x^{n+m} = x^n x^m \quad , \quad x^{s+t} = x^s x^t,$$

where s and t are trees. We use the notation $x^| = x^0 = 1$ and $x^Y = x^1 = x$.

In fact there is no reason to consider only polynomials and one can as well consider series since the sum and the product are well defined.

In this framework the operations

union			addition	
addition	on trees	become	multiplication	on polynomials.
multiplication			composition	

What about the operations $\dashv$ and $\vdash$? They give rise to two operations denoted $\prec$ and $\succ$ respectively, on polynomials. These two operations are bilinear and satisfy the relations:

$$(r \prec s) \prec t = r \prec (s \prec t + s \succ t),$$
$$(r \succ s) \prec t = r \succ (s \prec t),$$
$$(r \prec s + r \succ s) \succ t = r \succ (s \succ t).$$

A vector space A endowed with two bilinear operations $\prec$ and $\succ : A \otimes A \to A$, satisfying the relations just mentioned, is called a *dendriform algebra*, cf. [16, 18].

The dendriform algebras show up in many topics in mathematics: higher algebra [6, 9, 28], homological algebra [22], combinatorial algebra [1, 2, 8, 25, 27, 26], algebraic topology [7, 35, 36], renormalization theory [3, 4], quantum theory [14], to name a few. It is closely related to the notion of shuffles. In fact it could be called the theory of "non-commutative shuffles".

8 Realizing the associahedron

In Dov Tamari's seminal work "Monoides préordonnés et chaînes de Malcev" [33], which is his French doctoral thesis defended in 1951, the picture displayed in Fig. 1 appears on page 12. Unfortunately this part has not been reproduced in the published text [34] and therefore has been forgotten for all these years. It is very interesting on three grounds. First, it is the first appearance of the *Tamari poset*. Second, the Tamari poset is portrayed in dimension 2 and 3 as a polygon and a polyhedron respectively. Third, the parenthesizings has been replaced by a code that one can consider as coordinates in the euclidean space. We now analyze these three points.

8.1 Tamari poset

The Tamari poset, appearing often as the *Tamari lattice* in the literature (since it is a lattice), proved to be helpful in many places in mathematics. I mentioned earlier in this text its relevance with dendriform structures. It is playing a key role in the problem of endowing the tensor product of A_∞-algebras with an A_∞-algebra structure, cf. [22] for two reasons. First, the Tamari poset gives rise to a cell complex called

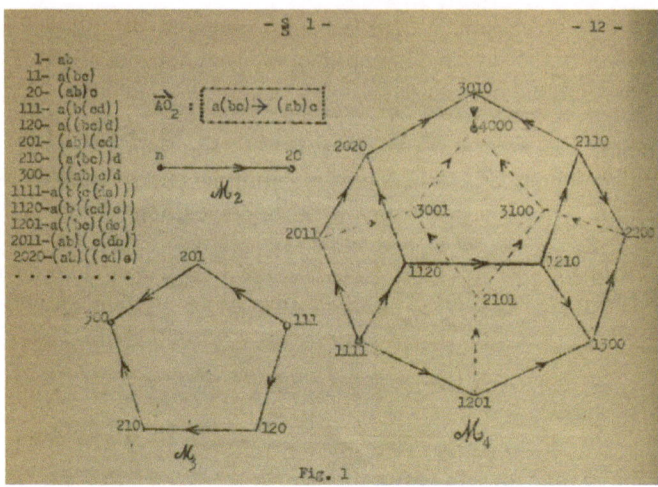

Fig. 1 Excerpt from Tamari's thesis.

the associahedron or the Stasheff polytope (see below). In 1963 Jim Stasheff showed that it encodes the notion of "associative algebra up to homotopy", now called A_∞-algebras. Let us recall that such an algebra A is equipped with a k-ary operation $m_k : A^{\otimes k} \to A, k \geq 2$, which satisfy some universal relations describing the topological structure of the associahedron. The second reason comes as follows. For a fixed integer n, the associahedron $\mathcal{K}^n$ is a cell complex of dimension n. We can prove that its cochain complex $C^\bullet(\mathcal{K}^n)$ can be endowed with a structure of A_∞-algebra. The operations m_k are trivial for $k > \frac{n(n+1)}{2}$ for $n > 1$ and can be made explicit in terms of the Tamari poset relation for $k \leq n$, cf. [22].

8.2 Associahedron and regular pentagons

The sentence following Fig. 1 in Tamari's thesis is the following
    "Généralement, on aura des hyperpolyèdres."
    But no further information is given. In fact, as we know now, we can realize the Tamari poset as a convex polytope so that each element of the poset is a vertex and each covering relation is an edge (see below). There is no harm in taking the regular pentagon in dimension 2. However, in contrast to what Fig. 1 suggests, one cannot realize the associahedron in dimension 3 with regular pentagons. What happens is the following: the four vertices corresponding to the parenthesizings $2020, 2011, 1120$ and 1111 do not lie in a common plane. It is a good trigonometric exercise for first year undergraduate students. If we take the convex hull of $\mathcal{M}_4$ of Fig. 1 (that is keeping regular pentagons), then the faces are made up of 6 pentagons, and 6 triangles instead of the 3 quadrangles. There are 3 edges which show up and which

do not correspond to any covering relation:

$$2011 - 1120, \quad 3012 - 3100, \quad 2200 - 1210.$$

8.3 Realizations of the associahedron

Though Tamari does not mention it, we can think of his clever way of indexing the parenthesizings as coordinates of points in the euclidean space $\mathbb{R}^{n+1}$. Let us recall briefly his method: given a parenthesizing (which is equivalent to a planar binary tree t) of the word $x_0 x_1 \ldots x_{n+1}$ we count the number of opening parentheses in front of x_0, then x_1, etc., up to x_n. For instance the word $((x_0 x_1)x_2)$ gives 2 0 and the word $(x_0(x_1 x_2))$ gives 1 1. Let us denote this sequence of numbers by

$$M(t) = (\alpha_0, \ldots, \alpha_n) \in \mathbb{R}^{n+1}.$$

Since the number of parentheses depends only on the length of the word, we have $\sum \alpha_i = n + 1$ and the points $M(t)$ lie in a common hyperplane. What does the convex hull look like? In dimension 2 we get the following pentagon:

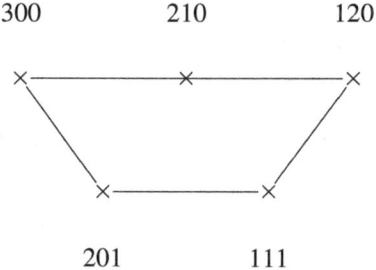

$$\begin{array}{ccc} 300 & 210 & 120 \end{array}$$

$$\begin{array}{cc} 201 & 111 \end{array}$$

As we see it is a quadrangle (that is a deformed square) with one point added on an edge.

In dimension 3 we get:

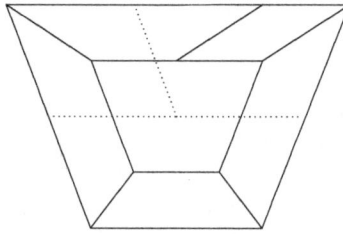

that is a deformed cube on which the associahedron has been drawn. In order to analyze the n-dimensional case, let us introduce the following notation. The convex hull of the points $M(t)$ is called the *Tamari polytope*. The *canopy* of the tree t is an element of the set $\{\pm\}^n$ corresponding to the orientation of the interior leaves. If the leaf points to the left (resp. right), then we take $-$, resp. $+$. Of course we discard the two extremal leaves, whose orientation is fixed. We denote by

$$\psi : PBT_{n+2} \to \{\pm\}^n$$

this map. Among the trees with a given canopy, we single out the tree which is constructed as follows. We first draw the outer part of the tree. Then for each occurence of $-$ we draw an edge which goes all the way to the right side of the tree (it is a left leaf). Then we complete the tree by drawing the right leaves. For instance:

This construction gives a section to ψ that we denote by

$$\sigma : \{\pm\}^n \to PBT_{n+2}.$$

Proposition 8.1. *The Tamari polytope KT^n is a hypercube shaped polytope, with extremal points $M(\sigma(\alpha))$, for $\alpha \in \{\pm\}^n$. For any tree t the point $M(t)$ lies on a face of this hypercube containing $M(\sigma\psi(t))$ (see the pictures above).*

Proof. It is easily seen by induction that the convex hull of the points $M(\sigma(\alpha))$ forms a (combinatorial) hypercube.

Up to a change of orientation, this is the cubical version of the associahedron described in [19] Section 2.5 (see also [21] Appendix 1). It is also described in [29]. $\qquad\square$

The Tamari polytope shares the following property with the standard permutohedron: all the edges have the same length.

Though Tamari himself does not consider this construction in his thesis, a close collaborator, Danièle de Fougères, worked out some variations in [13].

In 1963 Jim Stasheff [30] discovered independently the associahedron, first as a contractible cell complex, in his work on the structure of the loop spaces. It was later recognized to be realizable as a convex polytope, see for instance [31] Appendix B. In 2004 I gave in [20] an easy construction with integral coordinates as follows. It is usually described in terms of trees, but I will translate it in terms of parenthesized words.

Given a parenthesized word of length n, for instance $((x_0 x_1)(x_2 x_3))$, we associate to it a point in the euclidean space with coordinates computed as follows. The ith coordinate (i ranging from 0 to n) is the product of two numbers a_i and b_i. We consider the smallest subword which contains both x_i and x_{i+1}. Then a_i is the number of opening parentheses standing to the left of x_i and b_i is the number of closing parentheses standing to the right of x_{i+1} in the subword. In the example at hand we get 1 4 1. It gives rise to the following polytopes in low dimension:

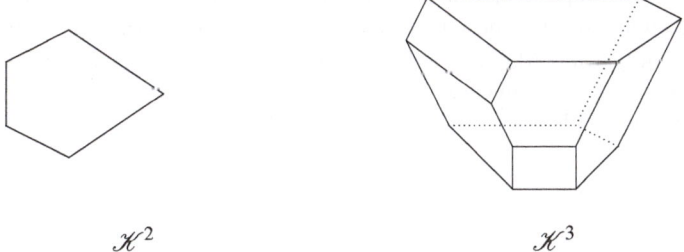

$$\mathcal{K}^2 \qquad\qquad\qquad\qquad\qquad \mathcal{K}^3$$

Since then several interesting variations for the associahedron itself and for other families of polytopes have been given along the same lines, see for instance [15, 8, 11, 12, 26, 27].

9 Associahedron and the trefoil knot

Let us end this paper with a surprizing relationship which is not so well known. If we draw a path on the 3-dimensional associahedron from the center of each quadrangle to the center of the other quadrangles via the center of the pentagons, alternating over and under as we reenter a quadrangle, then we get the trefoil knot in Fig. 2.

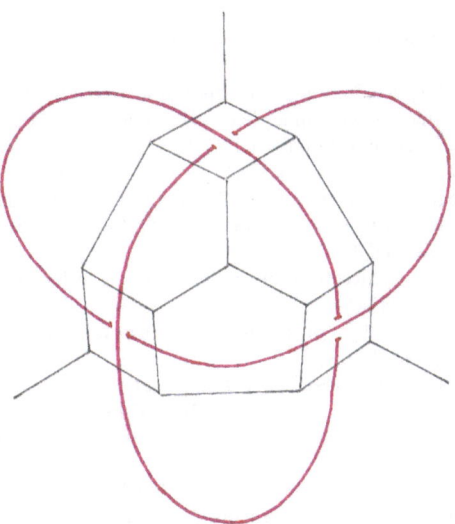

Fig. 2 Trefoil knot

The same process applied to the 3-dimensional cube gives rise to the Borromean rings. In the cube case we know how to relate the various invariants of this link: Philip Hall identity, triple Massey product and generator of the abelian free group $\mathbb{Z}^3$, cf. [17]. In the associahedron case, e.g., trefoil knot, the relevant group is the braid group with three threads.

References

1. J.-C. Aval and X. Viennot, "The product of trees in the Loday-Ronco algebra through Catalan alternative tableaux", *Sém. Lothar. Combin.* **63** (2010), Art. B63h, 8 pp.
2. M. Aguiar and F. Sottile, "Structure of the Loday-Ronco Hopf algebra of trees", *J. Algebra* **295** (2006) 473–511.
3. Ch. Brouder, "Trees, renormalization and differential equations", *BIT* **44** (2004) 425–438.
4. Ch. Brouder and A. Frabetti, "QED Hopf algebras on planar binary trees", *J. Algebra* **267** (2003) 298–322.
5. A. Bruno and D. Yazaki, "The arithmetic of trees", *arxiv.org/abs/0809.4448*.
6. E. Burgunder and M. Ronco, "Tridendriform structure on combinatorial Hopf algebras", *J. Algebra* **324** (2010) 2860–2883.
7. F. Chapoton, "Opérades différentielles graduées sur les simplexes et les permutoèdres", *Bull. Soc. Math. France* **130** (2002) 233–251.
8. S. Devadoss, "A realization of graph associahedra", *Discrete Math.* **309** (2009) 27–276.
9. V. Dotsenko, "Compatible associative products and trees", *Algebra Number Theory* **3** (2009) 567–586.
10. K. Ebrahimi-Fard, "Loday-type algebras and the Rota-Baxter relation", *Lett. Math. Phys.* **61** (2002) 139–147.
11. S. Forcey, "Quotients of the multiplihedron as categorified associahedra", *Homology, Homotopy Appl.* **10** (2008) 227–256.
12. S. Forcey, "Convex hull realizations of the multiplihedra", *Topology Appl.* **156** (2008) 326–347.
13. D. de Fougères, "Propriétés géométriques liées aux parenthésages d'un produit de m facteurs. Application à une loi de composition partielle", *Séminaire Dubreuil. Algèbre et théorie des nombres* **18**, no 2 (1964–65), exp. no 20, 1–21.
14. H. Gangl, A.B. Goncharov and A. Levin, "Multiple polylogarithms, polygons, trees and algebraic cycles", in *Algebraic geometry, Seattle* 2005, Proc. Sympos. Pure Math. **80**, Part 2, Amer. Math. Soc., Providence, RI, 2009, 547–593.
15. Ch. Hohlweg and C. Lange, "Realizations of the associahedron and cyclohedron", *Discrete Comput. Geom.* **37** (2007) 517–543.
16. J.-L. Loday, "Algèbres ayant deux opérations associatives (digèbres)", *C. R. Acad. Sci. Paris Sér. I Math.* **321** (1995) 141–146.
17. J.-L. Loday, "Homotopical syzygies", *in "Une dégustation topologique: Homotopy theory in the Swiss Alps"*, *Contemporary Mathematics* no **265** (AMS) (2000), 99–127.
18. J.-L. Loday, "Dialgebras", in *Dialgebras and related operads*, Springer Lecture Notes in Math. **1763** (2001) 7–66.
19. J.-L. Loday, "Arithmetree", *Journal of Algebra* **258** (2002) 275–309.
20. J.-L. Loday, "Realization of the Stasheff polytope", *Archiv der Mathematik* **83** (2004) 267–278.
21. J.-L. Loday, "The diagonal of the Stasheff polytope", in *Higher structures in geometry and physics*, Progr. Math. **287**, Birkhäuser, Basel, 2011, 269–292.
22. J.-L. Loday, "Geometric diagonals for the Stasheff associahedron and products of A-infinity algebras", preprint 2011, in preparation.
23. J.-L. Loday; M.O. Ronco, "Hopf algebra of the planar binary trees", *Adv. Math.* **139** (1998) 293–309.

24. J.-L. Loday and M.O. Ronco, "Order structure and the algebra of permutations and of planar binary trees", *J. Alg. Comb.* **15** (2002) 253–270.

25. J.-C. Novelli and J.-Y. Thibon, "Hopf algebras and dendriform structures arising from parking functions", *Fund. Math.* **193** (2007) 189–241.

26. A. Postnikov, "Permutohedra, associahedra, and beyond", *Int. Math. Res. Not.* **2009**, no. 6, 1026–1106.

27. V. Pilaud and F. Santos, "Multitriangulations as complexes of star polygons", *Discrete Comput. Geom.* **41** (2009) 284–317.

28. M.O. Ronco, "Primitive elements in a free dendriform algebra", in *New trends in Hopf algebra theory (La Falda, 1999)*, Contemp. Math. **267**, Amer. Math. Soc., Providence, RI, 2000, 245–263.

29. S. Saneblidze and R. Umble, "Diagonals on the permutahedra, multiplihedra and associahedra", *Homology Homotopy Appl.* **6** (2004) 363–411.

30. J. Stasheff, "Homotopy associativity of H-spaces. I, II", *Trans. Amer. Math. Soc.* **108** (1963) 275–292; ibid. **108** (1963) 293–312.

31. J. Stasheff, "From operads to "physically" inspired theories", in *Operads: Proceedings of Renaissance Conferences (Hartford, CT/Luminy, 1995)*, Contemp. Math. **202**, Amer. Math. Soc., Providence, RI, 1997, 53–81.

32. J. Stasheff, "How I 'met' Dov Tamari", in *Tamari Memorial Festschrift*.

33. D. Tamari, "Monoides préordonnés et chaînes de Malcev", Doctorat ès-Sciences Mathématiques Thèse de Mathématiques, Université de Paris (1951).

34. D. Tamari, "Monoides préordonnés et chaînes de Malcev", *Bull. Soc. Math. France* **82** (1954) 53–96.

35. B. Vallette, "Manin products, Koszul duality, Loday algebras and Deligne conjecture", *J. Reine Angew. Math.* **620** (2008) 105–164.

36. D. Yau, "Gerstenhaber structure and Deligne's conjecture for Loday algebras", *J. Pure Appl. Algebra* **209** (2007) 739–752.

Partial Groupoid Embeddings in Semigroups

Susan H. Gensemer

Abstract We examine a number of axiom systems guaranteeing the embedding of a partial groupoid into a semigroup. These include the Tamari symmetric partial groupoid and the Gensemer/Weinert equidivisible partial groupoid, provided they satisfy an additional axiom, weak associativity. Both structures share the one mountain property. More embedding results for partial groupoids into other types of algebraic structures are presented as well.

1 Introduction

Tamari [25] wrote:

> The complexity of the general concept of associativity of partial binary operations could hardly be better hidden than by its collapsing into the simple elementary formula $(xy)z = x(yz)$ for the all-important, yet still very special case of closed (i.e., complete) operations. Furthermore, the veil of deceptive simplicity is not lifted by the first encounters while reconnoitering the wilderness of partial operations.

An initial look at the underlying framework necessary to understand embeddings of partial groupoids into semigroups, or other related structures, is bewildering. The complex nature of the framework is discordant with the relatively simple presentations of complete algebraic systems such as semigroups. The framework, however, is considerably simpler when it conforms to the *one mountain condition* or its weak version (see Definition 2.6). The weak one mountain condition was used by Tamari in many of his papers.

Working with partial operations is a difficult, but important, problem. In particular, the problem of embedding partial groupoids into semigroups is recognized as a

Susan H. Gensemer
Department of Economics, Maxwell School of Citizenship and Public Affairs, Syracuse University, Syracuse, NY 13244-1n020, e-mail: *gensemer@maxwell.syr.edu*

challenging and vital undertaking ([16], [17] p. 105). This study utilizes the one mountain condition or its weak version to shed some light on navigating the wilderness encountered in embedding partial groupoids into semigroups. Then we present results on embedding partial groupoids, which satisfy the one mountain condition, into other types of algebraic structures.

2 Definitions and preliminary results

In this section, we present some definitions and basic results.

Definition 2.1. $(A, P, \circ)$ is a *partial groupoid* if:

1. A is a nonempty set,
2. $P \subseteq A \times A$ is a nonempty set, and
3. $\circ : P \to A$ is a binary operation.

A partial groupoid has also been called a halfgroupoid, a pargoid, or a monoid [1, 17, 24]. Throughout, we let $(A, P, \circ)$ denote a partial groupoid only.

Definition 2.2. $(A, P, \circ)$ is a *semigroup* if:

1. it is a partial groupoid where $P = A \times A$ and
2. $\circ$ is associative, i.e., for every $a, b, c \in A$, $(a \circ b) \circ c = a \circ (b \circ c)$.

A semigroup may be denoted simply by $(S, *)$.

Definition 2.3. A partial groupoid $(A, P, \circ)$ is *symmetric* if:

1. there exists an identity element, i.e., there exists $1 \in A$ such that for every $a \in A$, $(a, 1), (1, a) \in P$ and $a \circ 1 = 1 \circ a = a$.
2. every element a has an inverse, i.e., for all $a, b, c \in A$ such that $(a, b), (c, a) \in P$, there exists $a^{-1} \in A$ such that $(a^{-1}, a \circ b), (c \circ a, a^{-1}) \in P$, $a^{-1} \circ (a \circ b) = b$, and $(c \circ a) \circ a^{-1} = c$.

Symmetric partial groupoids were introduced and studied by Tamari. The inverse element is appropriately called that, since it acts as a usual inverse element in that $a \circ a^{-1}$ and $a^{-1} \circ a = 1$. Such a partial groupoid has also been called a partial inverse property loop [4, 20, 22].

Definition 2.4. A partial groupoid $(A, P, \circ)$ is *equidivisible* if for all $a, b, c, d \in A$ such that $(a, b) \neq (c, d)$, if $a \circ b = c \circ d$, then $a = c \circ f$ and $d = f \circ b$ for some $f \in A$, or $c = a \circ g$ and $b = g \circ d$ for some $g \in A$.

The concept of an equidivisible partial groupoid was introduced and studied by Gensemer and Weinert [12]. The axiom in its definition appears in [10, 11, 12]. The axiom is used to define an *equidivisible semigroup*, i.e., a semigroup which satisfies the latter axiom without the restriction that $(a, b) \neq (c, d)$ [14, 18].

We use letters a, b, c, d, e, f, and g to denote elements of A only and letters i, j, k, m, and n to denote positive integers only. The *free semigroup* on A, denoted $(F_A, \cdot)$, is the set $\{a_1 \ldots a_n : a_i \in A, i = 1, \ldots, n$, for some positive integer $n\}$, where $a_1 \ldots a_n$ is a *word* in F_A, along with the binary operation $\cdot$ defined as $a_1 \ldots a_n \cdot b_1 \ldots b_m = a_1 \ldots a_n b_1 \ldots b_m$. We use the letters u, v, w, x, y, and z to denote members of F_A only. We could call $w_i = uabv \to w_{i+1} = u(a \circ b)v$ a *direct move* from w_i to w_{i+1} [6]. Tamari [25] and others have called this a *contraction* from w_i to w_{i+1}, or an *expansion* from w_{i+1} to w_i, and we use that convention here. We also write $x \to y$ if $x = y$, or if there exist w_i, $i = 1, \ldots, n$, such that $x \to w_1 \to \cdots \to w_n \to y$, i.e., y is arrived at from x through a finite sequence of contractions (there is a *wordchain* from x to y). In displaying an arrow from x to y, we will use also arrows in a northeasterly or southeasterly direction, etc.

Throughout, given a partial groupoid $(A, P, \circ)$, let κ denote the congruence relation on $(F_A, \cdot)$ generated by the defining relations $(ab, a \circ b) \in \kappa$. Denote the congruence classes of $(F_A/\kappa, \cdot)$ by $[x]$, where $x \equiv a_1 \ldots a_n$ and where the binary operation $\cdot$ is defined (with some abuse of notation) on F_A/κ by $[x] \cdot [y] = [x \cdot y]$. Define $\varphi : A \to F_A/\kappa$ by

$$\varphi(a) = [a]. \tag{1}$$

In the next section, we present sets of axioms on a partial groupoid $(A, P, \circ)$ which guarantee that it is *embeddable* in a semigroup $(S, *)$; that is, we present axioms which guarantee that there exists a one-to-one function $v : A \to S$ such that $v(a \circ b) = v(a) * v(b)$. This will be accomplished by constructing an embedding from $(A, P, \circ)$ into $(F_A/\kappa, \cdot)$. This is done without loss of generality by the following result, which has been shown by Schmidt and Tamari, among others [12, 17, 21, 23].

Lemma 2.5. *Let $\varphi : A \to F_A/\kappa$ be defined as at* (1).

1. $(F_A/\kappa, \cdot)$ *is a universal semigroup of $(A, P, \circ)$ with corresponding homomorphism φ, i.e., $\varphi(a \circ b) = \varphi(a) \cdot \varphi(b)$.*
2. $(A, P, \circ)$ *is embeddable into a semigroup if and only if φ is injective, which holds if and only if each congruence class under κ contains at most one element of A.*

To say that $(F_A/\kappa, \cdot)$ is a *universal semigroup* of $(A, P, \circ)$ (with corresponding homomorphism φ) means that for any semigroup $(S, *)$ such that there exists a homomorphism $\chi : (A, P, \circ) \to (S, *)$ (i.e., the function $\chi : A \to S$ is such that $\chi(a \circ b) = \chi(a) * \chi(b)$), there exists a unique homomorphism $\psi : (F_A/\kappa, \cdot) \to (S, *)$ satisfying $\varphi\psi = \chi$, i.e., the following diagram commutes.

$$
\begin{array}{ccc}
(A, P, \circ) & \xrightarrow{\varphi} & (F_A/\kappa, \cdot) \\
& \chi \searrow \quad \swarrow \psi & \\
& (S, *) &
\end{array}
$$

In the following are definitions for the weak one mountain condition and the one mountain condition; if a partial groupoid satisfies the one mountain condition, it satisfies its weak version. The weak one mountain condition was studied by Tamari. The labels for both are motivated by the One Mountain Theorem in Tamari's work

[3, 23, 24]; a version of the One Mountain Theorem appears as Proposition 3.3 in the next section. The representations of the conditions in the following definition show the motivation behind the one mountain label.

Definition 2.6. A partial groupoid satisfies the *weak one mountain condition* if whenever $a, b \in A$ are such that $a\kappa b$, there exists $w \in F_A$ such that

$$
\begin{matrix}
 & w & \\
\swarrow & & \searrow \\
a & & b.
\end{matrix}
\tag{2}
$$

A partial groupoid satisfies the *one mountain condition* if whenever $w, v \in F_A$ are such that $w\kappa v$, there exists $u \in F_A$ such that

$$
\begin{matrix}
 & u & \\
\swarrow & & \searrow \\
w & & v.
\end{matrix}
$$

A partial groupoid does not necessarily satisfy even the weak one mountain condition, as the following example, from [3], illustrates. This example, with three elements, is smaller than the previous counterexample, which had six elements [3, 23, 24].

$$
\begin{array}{c|ccc}
 & a & b & c \\
\hline
a & a & - & a \\
b & a & a & a \\
c & a & b & a
\end{array}
$$

Notice that

It follows that $a\kappa b$. However, there does not exist $w \in F_A$ such that (2) holds. This follows since expansions only of b will always have one b on the right-hand side and nothing or c's on the left-hand side. Then contractions only of such words will not eliminate b, i.e., they will not yield a only.

3 Results

In this section, we present sets of axioms which guarantee that a partial groupoid can be embedded in a semigroup. Of course, not all partial groupoids can be so embedded, as the following example illustrates.

$$\begin{array}{c|ccc}
 & a & b & c \\
\hline
a & - & a & - \\
b & - & - & - \\
c & - & b & a
\end{array}$$

An embedding into a semigroup doesn't exist since $c \circ (c \circ b) = c \circ b = b \neq a = a \circ b = (c \circ c) \circ b$.

In this section, however, we do show that both the symmetric and the equidivisible partial groupoids are embeddable in a semigroup, provided that they satisfy an additional axiom, *weak associativity*. In the following axiom, we write $\pi_i(a_1 \ldots a_n) = b$ when $a_1 \ldots a_n \to b$, where the subscript i denotes a particular positioning of parentheses in a product formed by $a_1, \ldots, a_n$, in that order. For example, if the products $((a \circ b) \circ c) \circ d$ and $(a \circ b) \circ (c \circ d)$ are defined in $(A, P, \circ)$, then they may be denoted, respectively, by $\pi_1(abcd)$ and $\pi_2(abcd)$.

Axiom 3.1 (A). Whenever $\pi_1(a_1 \ldots a_n)$ and $\pi_2(a_1 \ldots a_n)$ are defined in $(A, P, \circ)$, where n is any positive integer, they are equal.

The latter axiom appears in [12, 15, 17, 23]. Even though there are many ways in which associativity could be defined in the context of working with partial groupoids, we refer to this as the weak associativity axiom ([17] pp. 16–23). It is necessarily satisfied by a partial groupoid which is embeddable in a semigroup.

The proof of the following result is given in the proof of Proposition 5.2 in [12]; we provide only a sketch of the proof here.

Proposition 3.2. *An equidivisible partial groupoid satisfies the one mountain condition.*

Proof. Assume that the axiom in Definition 2.4 is satisfied. Assume that $\ell(w) = \ell(v) = k+1$ and $\ell(z) = k$ for words $w, v, z \in F_A$ such that $w \to z$ and $v \to z$, where ℓ denotes the length of a word. We show that there exists $u \in F_A$, where $\ell(u) = k+2$, such that $u \to w$ and $u \to v$. To see this, write $w = a_1 \ldots a_{k+1}$, $v = b_1 \ldots b_{k+1}$, and $z = c_1 \ldots c_k$. Assume that $c_i = a_i \circ a_{i+1}$ and $c_j = b_j \circ b_{j+1}$, and without loss of generality, assume that $i \leq j$. If $i < j$, then

$$u = c_1 \ldots c_{i-1} a_i a_{i+1} \ldots b_j b_{j+1} c_{j+1} \ldots c_k,$$

the required u. On the other hand, if $i = j$, then $c_i = a_i \circ a_{i+1} = b_i \circ b_{i+1} = c_j$. If $(a_i, a_{i+1}) = (b_i, b_{i+1})$, then $w = v$, and we are done. If $(a_i, a_{i+1}) \neq (b_i, b_{i+1})$, then by the axiom in Definition 2.4, there exists $d \in A$ such that $a_i = b_i \circ d$ and $b_{i+1} = d \circ a_{i+1}$, or there exists $e \in A$ such that $b_i = a_i \circ e$ and $a_{i+1} = e \circ b_{i+1}$. It follows that

$$u = c_1 \ldots c_{i-1} b_i d a_{i+1} c_{i+1} \ldots c_k,$$

or

$$u = c_1 \ldots c_{i-1} a_i e b_{i+1} c_{i+1} \ldots c_k,$$

respectively. In either case, we have shown that there exists $u \in F_A$, where $\ell(u) = k+2$, such that $u \to w$ and $u \to v$.

Now assume that $\ell(w) = k+1$, $\ell(v) = k+t$, and $\ell(z) = k$ for words $w, v, z \in F_A$ and that $w \to z$ and $v \to z$. Then by induction on t, it can be argued that there exists $u \in F_A$ such that $u \to w$ and $u \to v$. By continuing induction arguments, the conclusion can be shown.   □

The proof that a symmetric partial groupoid satisfies the weak one mountain condition is in [3], and is cited there and elsewhere as the One Mountain Theorem for symmetric partial groupoids [24]. We sketch the proof here that the symmetric partial groupoid also satisfies the one mountain condition.

Proposition 3.3. *A symmetric partial groupoid satisfies the one mountain condition.*

Proof. The proof is similar to the last proof. Assume that $\ell(w) = \ell(v) = k+1$ and $\ell(z) = k$ for words $w, v, z \in F_A$ such that $w \to z$ and $v \to z$. We show that there exists $u \in F_A$ such that $u \to w$ and $u \to v$. To see this, write $w = a_1 \dots a_{k+1}$, $v = b_1 \dots b_{k+1}$, and $z = c_1 \dots c_k$. Assume that $c_i = a_i \circ a_{i+1}$ and $c_j = b_j \circ b_{j+1}$, and without loss of generality, assume that $i \leq j$. If $i < j$, then

$$u = c_1 \dots c_{i-1} a_i a_{i+1} \dots b_j b_{j+1} c_{j+1} \dots c_k.$$

If $i = j$, then $c_i = a_i \circ a_{i+1} = b_i \circ b_{i+1} = c_j$. If $(a_i, a_{i+1}) = (b_i, b_{i+1})$, then $w = v$, and we are done. If $(a_i, a_{i+1}) \neq (b_i, b_{i+1})$, then by the second axiom in Definition 2.3, $a_{i+1} = a_i^{-1} \circ (b_i \circ b_{i+1})$. It follows that

$$w = c_1 \dots c_{i-1} a_i a_i^{-1} \circ (b_i \circ b_{i+1}) c_{i+1} \dots c_k.$$

From this, derive

$$u' = c_1 \dots c_{i-1} a_i a_i^{-1} (b_i \circ b_{i+1}) c_{i+1} \dots c_k,$$

and

$$u = c_1 \dots c_{i-1} a_i a_i^{-1} b_i b_{i+1} c_{i+1} \dots c_k.$$

It then follows that

$$u \to u' \to w$$

and

$$u \to c_1 \dots c_{i-1} 1 b_i b_{i+1} c_{j+1} \dots c_k \to v,$$

which shows that there exists $u \in F_A$ such that $u \to w$ and $u \to v$.

In a way similar to the previous proof, a completed induction argument shows the conclusion of the proposition.   □

As noted by Tamari [23, 24], the One Mountain Theorem has intellectual roots in ideas expressed by Newman [19]. Specifically, Newman asked whether "if two objects α and β are 'equivalent,' then it follows that there exists a third object, ω, derivable from both α and β by positive moves only."[1] Newman called this result a theorem of confluence. The One Mountain Theorem for symmetric partial groupoids

---

[1] Altering Newman's notation, we denote his objects by Greek letters to avoid confusion with notation elsewhere this article.

and Propositions 3.2–3.3 fit into the form of this theorem of confluence when one views α, β, and ω as words (as in this framework), and one views α and β as equivalent when they are of the same congruence class (as in this framework). Also, one would want to view a positive move from an object as an expansion of a word (as in this framework). Furthermore, there are also links from the one mountain conditions of Definition 2.6 and ideas in Newman's paper to work by Church and Rosser [5]. In the terminology of the latter paper, the one mountain structures in Definition 2.6 would be said to form a peak. Tamari's contraction/expansion terminology also appears in the latter paper.

Theorems of confluence appear in other situations as well. For example, a theorem of confluence occurs in the context of studying the partial groupoid of Schmidt [21]. It is a partial groupoid which satisfies the additional condition: if $(a,b),(b,c) \in P$, then $(a \circ b, c), (a, b \circ c) \in P$ and $(a \circ b) \circ c = a \circ (b \circ c)$ (this is a form of intermediate associativity ([17] p. 89)). This partial groupoid is embeddable in a semigroup [12, 21]. Also, it can be shown that if $w \kappa v$, where $w, v \in F_A$, then there exists $u \in F_A$ such that

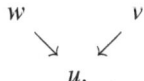

This result could be viewed as a theorem of confluence, where a positive move is a contraction. In Church and Rosser's terminology, the structure forms a valley. The proof that the Schmidt partial groupoid is embeddable in a semigroup follows essentially from this confluence and from the fact that a word can be contracted only a finite number of times. The proof given in [12] relies on Cohn's Theorem 9.3 [6].[2]

As Tamari noted, even if a partial groupoid satisfies the weak one mountain condition, this "is not of great help in 'solving' the associativity problem [i.e., deciding whether a partial groupoid can be embedded in a semigroup] since [he] showed that the problem is undecidable even for the class of symmetric [partial groupoids]" [3, 22]. However, we indicate here and in the next section that the one mountain property in some form is helpful in developing axiom systems guaranteeing an embedding of a partial groupoid into various algebraic structures. A proof of the following proposition is in [23] and is given here.

Proposition 3.4. *If a partial groupoid satisfies the weak one mountain condition and weak associativity, then it is embeddable in a semigroup.*

Proof. Assume that $a \kappa b$. Since the partial groupoid satisfies the weak one mountain condition, it follows that there exists $u = c_1 \ldots c_n \in F_A$ such that

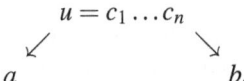

By Axiom A, $a = b$ and the result follows by Lemma 2.5.   □

---

[2] Incidentally, Cohn attributed this theorem to Newman [19] in [7]. However, there Cohn noted the qualification that Newman's lower bound condition should be replaced with Cohn's finiteness condition.

The weak one mountain structure enables an embedding into a semigroup fairly easily, with just the addition of the weak associativity axiom. In the absence of a weak one mountain structure, one could be faced with determining conditions implying that $a = b$ in a structure like the following, where $a, b \in A$ and $w_1, w_2, w_3 \in F_A$.

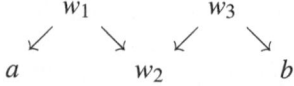

Of course, the structures are likely to be far more complicated, with far more expansions and contractions of words! They form a veritable wilderness! Determining conditions that would bear some resemblance to a condition like the strong associativity of a semigroup would be a very hard, if not intractable, problem. Tamari [3, 22] indicated that the weak one mountain condition simplifies checking for violations of associativity in partial groupoid tables in that one has to check only for violations of weak associativity in chains as in the latter proof, those with one mountain. The proof of the following is immediate given Propositions 3.2–3.4.

Corollary 3.5. *If a partial groupoid is symmetric or equidivisible, and it satisfies weak associativity, then it can be embedded into a semigroup.*

The universal semigroup $(F_A/\kappa, \cdot)$ into which Tamari's symmetric partial groupoid is embedded in the construction is a group [20]. The universal semigroup $(F_A/\kappa, \cdot)$ into which the equidivisible partial groupoid is embedded is equidivisible [12].

Finally, we indicate that while the symmetric and equidivisible partial groupoid concepts have some overlap, they also define distinct structures as well. The examples in the proof also satisfy weak associativity.

Proposition 3.6. *The classes of equidivisible and symmetric partial groupoids have a nonempty intersection, but neither class is contained in the other.*

Proof. The following example is an equidivisible partial groupoid, but not a symmetric partial groupoid.

	a	b	c	d	e	f
a	–	–	e	–	–	–
b	–	–	f	–	–	–
c	–	–	–	–	–	–
d	b	–	–	–	f	–
e	–	–	–	–	–	–
f	–	–	–	–	–	–

The example isn't a symmetric partial groupoid since there are no inverses. The example is an equidivisible partial groupoid since $b \circ c = f = d \circ e$, $(b, c) \neq (d, e)$, $b = d \circ a$, and $e = a \circ c$. To see that the example satisfies weak associativity, note that the longest products that are defined are $(d \circ a) \circ c$ and $d \circ (a \circ c)$, and that $(d \circ a) \circ c = b \circ c = f = d \circ e = d \circ (a \circ c)$.

The following example is a symmetric partial groupoid, but not an equidivisible one. The table appears as Figure 1(f) in [3], and in [8] as G(f). Both references establish the fact that the partial groupoid satisfies weak associativity. It is readily established that the structure is a symmetric partial groupoid. The partial groupoid is not equidivisible because $a \circ a = b \circ b$, but there does not exist c such that $a = b \circ c$ and $b = c \circ a$ and there does not exist c such that $b = a \circ c$ and $a = c \circ b$.

$$
\begin{array}{c|ccc}
 & 1 & a & b \\
\hline
1 & 1 & a & b \\
a & a & 1 & - \\
b & b & - & 1
\end{array}
$$

The following example is a group. Inverses and an identity element exist. Strong associativity can also be shown.

$$
\begin{array}{c|cccc}
 & 1 & a & b & c \\
\hline
1 & 1 & a & b & c \\
a & a & 1 & c & b \\
b & b & c & 1 & a \\
c & c & b & a & 1
\end{array}
$$

Since the structure is a group, it is also both a symmetric and an equidivisible partial groupoid. □

4 Further results

In this section, we illustrate how one mountain structured partial groupoids with additional axioms can be embedded into other algebraic structures.

4.1 Embedding into a partially ordered semigroup

In this subsection, we consider a more complex embedding, one of an ordered partial groupoid into a strict partially ordered semigroup. More specifically, we have the following definitions.

Definition 4.1. $\mathscr{A} = (A, \succsim, P, \circ)$ is an *ordered partial groupoid* if:

1. $(A, P, \circ)$ is a partial groupoid and
2. $\succsim$ is an antisymmetric (i.e., if $a \succsim b$ and $b \succsim a$, then $a = b$), complete (i.e., for all $a, b \in A$, $a \succsim b$ or $b \succsim a$), and transitive (i.e., if $a \succsim b$ and $b \succsim c$, then $a \succsim c$) relation on A.

If, in addition, there exists $1 \in A$ such that

3. $(a, 1), (1, a) \in P$, $a \circ 1 = 1 \circ a = a$, and $a \succsim 1$ for every $a \in A$,

then $\mathscr{A}$ is a *positively ordered partial groupoid*.

From this point on, we let $\mathscr{A}$ denote a positively ordered partial groupoid only. The fact that we consider embeddings of only positively ordered partial groupoids implies that the results of this subsection cannot apply to a symmetric partial groupoid except in the trivial case where A contains the identity element only.

Definition 4.2. $\mathscr{S} = (S, \geq^*, *)$ is a *strict partially ordered semigroup* if:

1. $\geq^*$ is an antisymmetric, reflexive (i.e., for every $a \in S$, $a \geq^* a$), and transitive relation on S,
2. $(S, *)$ is a semigroup, and
3. the strict monotony laws are satisfied, i.e., if $a >^* b$ (i.e., $a \geq^* b$ and $a \neq b$), then $a * c >^* b * c$ and $c * a >^* c * b$, where $a, b, c \in S$.

The latter definition is from [9] p. 153. If a semigroup $(S, *)$ is such that whenever $a * c = b * c$ or $c * a = c * b$, it follows that $a = b$, then $(S, *)$ is *cancellative* ([2] p. 3). Fuchs ([9] p. 154) stated a result with an immediate proof which we use later: A weak partially ordered semigroup which is cancellative is a strict partially ordered semigroup. From this point on, $\mathscr{S}$ denotes a strict partially ordered semigroup only.

In this subsection, we present a set of axioms on $\mathscr{A}$ which guarantee that it is *embeddable* in some $\mathscr{S}$; that is, we present axioms which guarantee that there exists a function $v : A \to S$ such that (i) $a \succsim b$ if and only if $v(a) \geq^* v(b)$ and (ii) $v(a \circ b) = v(a) * v(b)$.

Given $\mathscr{A} = (A, \succsim, P, \circ)$ and $a_1 \ldots a_n, b_1 \ldots b_n \in F_A$ we also use $\succsim$ to write $a_1 \ldots a_n \succsim (\succ) b_1 \ldots b_n$ if $a_i \succsim b_i$ for $i = 1, \ldots, n$ (with $a_i \succ b_i$ for at least one i); notice that $\succsim$ on F_A is antisymmetric, reflexive, and transitive. Consider the following axiom which might be satisfied by $\mathscr{A}$; it is necessarily satisfied by a positively ordered partial groupoid which can be embedded in a strict partially ordered semigroup. It can be viewed as a monotonicity axiom.

Axiom 4.3 (M). If $\pi_1(a_1 \ldots a_n)$ and $\pi_2(b_1 \ldots b_n)$ are defined in $(A, P, \circ)$ and $a_1 \ldots a_n \succ b_1 \ldots b_n$, then $\pi_1(a_1 \ldots a_n) \succ \pi_2(b_1 \ldots b_n)$.

The following axioms appear in [10, 11].

Axiom 4.4 (D).

1. If $a \succ b$, $(b', b'') \in P$, and $b = b' \circ b''$, there exists $(a', a'') \in P$ such that $(a', a'') \succ (b', b'')$ and $a = a' \circ a''$.
2. If $a \succ b$, $(a', a'') \in P$, and $a = a' \circ a''$, there exists $(b', b'') \in P$ where $(a', a'') \succ (b', b'')$ and $b = b' \circ b''$.

We present some preliminary lemmas. In this subsection, we regularly turn mountain structures onto their sides to economize on space. The proof of the following lemma is immediate.

Lemma 4.5. *Let $x \succ y$, where $x, y \in F_A$.*

1. *If $\mathscr{A}$ satisfies Axiom D1 and $y' \to y$, where $y' \in F_A$, then there exists $x' \in F_A$ such that $x' \to x$ and $x' \succ y'$.*

2. *If $\mathscr{A}$ satisfies Axiom D2 and $x' \to x$, where $x' \in F_A$, then there exists $y' \in F_A$ such that $y' \to y$ and $x' \succ y'$.*

Lemma 4.6. *If $\mathscr{A}$ satisfies Axiom M and $x \succ y$, where $x, y \in F_A$, then not*

$$x \to z, \atop \nearrow \atop y \tag{3}$$

where $z \in F_A$.

Proof. Assume that (3) holds and that $x \succ y$. Let $x \equiv a_1 \ldots a_n$, $y \equiv b_1 \ldots b_n$, and $z \equiv c_1 \ldots c_m$. We may assume that $a_i \neq 1$, $i = 1, \ldots, n$. Write $a_1 \ldots a_n$ as

$$a_1 \ldots a_{i_1} a_{i_1+1} \ldots a_{i_2} \ldots a_{i_{m-1}+1} \ldots a_{i_m}, \tag{4}$$

and $b_1 \ldots b_n$ as

$$b_1 \ldots b_{j_1} b_{j_1+1} \ldots b_{j_2} \ldots b_{j_{m-1}+1} \ldots b_{j_m}, \tag{5}$$

where $a_{i_m} \equiv a_n$ and $b_{j_m} \equiv b_n$,

$$c_1 = \pi_{i_1}(a_1 \ldots a_{i_1}) = \pi_{j_1}(b_1 \ldots b_{j_1}), \tag{6}$$

and

$$c_k = \pi_{i_k}(a_{i_{k-1}+1} \ldots a_{i_k}) = \pi_{j_k}(b_{j_{k-1}+1} \ldots b_{j_k}),$$

for $k = 2, \ldots, m$. Consider moving from left to right in the expressions at (4) and (5). As long as $a_i = b_i$, we must have $i_k = j_k$ by (M). However, when we have the first $a_i \succ b_i$, say at $i = i'$, we must have for the first i_k and j_k to the right of i', $j_{k'} > i_{k'}$, by (M). Furthermore, by (M), for $j_k > j_{k'}$ and $i_k > i_{k'}$, we must have

$$j_k > i_k. \tag{7}$$

(Otherwise, for example, if we have $j_{k'+1} \leq i_{k'+1}$, it necessarily follows by (M) that

$$\pi_{i_{k'+1}}(a_{i_{k'}+1} \ldots a_{i_{k'+1}}) \succ \pi_{j_{k'+1}}(b_{j_{k'}+1} \ldots b_{j_{k'+1}}),$$

a contradiction of (6). Therefore, $j_{k'+1} > i_{k'+1}$. We continue this type of argument to justify the statement at (7).) Given (7), we then have a contradiction since $i_m = j_m = n$. $\qquad \square$

Lemma 4.7. *Suppose that $\mathscr{A}$ satisfies Axiom M. If there exists $z \in F_A$ such that*

$$z \to xw \atop \searrow \atop yw \tag{8}$$

or

$$z \to wx$$
$$\searrow$$
$$wy,$$

where $w, x, y, z \in F_A$, then there exists $v \in F_A$ such that

$$v \to x$$
$$\searrow \qquad\qquad (9)$$
$$y.$$

Proof. Let $x \equiv a_1 \ldots a_\ell$, $y \equiv b_1 \ldots b_n$, $w \equiv c_1 \ldots c_m$, $z \equiv d_1 \ldots d_k$, and $u \equiv f_1 \ldots f_r$. We simply show the result at (8) and (9) for $m = 1$. Assume that $z \equiv d_1 \ldots d_k \in F_A$ at (8) exists, but that $u \equiv f_1 \ldots f_r$ at (9) does not exist. Without loss of generality we may assume that for some $i \le k$, $j \le k$, $i < j$,

$$d_i \cdots d_k \to c_1,$$
$$\nearrow$$
$$d_j \cdots d_k$$

where $d_s \ne 1$, $s = i, \ldots, k$. However, this contradicts (M). $\qquad\square$

Proposition 4.8. *If an equidivisible, positively ordered partial groupoid $\mathscr{A}$ satisfies Axioms A, M, D1, and D2, then it is embeddable in a strict partially ordered semigroup.*

Proof. By Lemma 2.5 and Corollary 3.5, $(A, P, \circ)$ is embeddable in $(F_A/\kappa, \cdot)$. We define an order $\ge$ on F_A/κ and show that it is reflexive, antisymmetric, and transitive. We then show the order satisfies the monotony laws and that $(F_A/\kappa, \cdot)$ is cancellative. It will follow that the order is a strict partial order. In this proof, we allow that $x \to x$.

For congruence classes $[x], [y] \in F_A/\kappa$, define $[x] \ge [y]$ if there exist $\bar{x}, \bar{y} \in F_A$ such that $\bar{x} \to x$, $\bar{y} \to y$, and $\bar{x} \succsim \bar{y}$. Note that the definition of $\ge$ is independent of the choices of the representatives of the congruence classes. That is, if $x, x' \in [x]$ and $y, y' \in [y]$, and there exist $\bar{x}, \bar{y} \in F_A$ such that $\bar{x} \to x$, $\bar{y} \to y$, and $\bar{x} \succsim \bar{y}$, then there exist $x'', y'' \in F_A$ such that $x'' \to x'$, $y'' \to y'$, and $x'' \succsim y''$. To see this, note that since $x, x' \in [x]$ and $y, y' \in [y]$, by Proposition 3.2, there exist $\hat{x}, \hat{y} \in F_A$ such that

$$\hat{x} \to x'$$
$$\searrow$$
$$\bar{x} \to x$$

and

$$\hat{y} \to y'$$
$$\searrow$$
$$\bar{y} \to y.$$

By Proposition 3.2, there exists $\tilde{y}$ as indicated in the following.

$$\tilde{y} \to \hat{y} \to y'$$
$$\searrow \qquad \searrow$$
$$\bar{y} \to y$$

But use (D1) to find that there exists $\tilde{x} \succsim \tilde{y}$ such that

$$\hat{x} \to x'$$
$$\searrow$$
$$\tilde{x} \to \bar{x} \to x$$

since $\tilde{x} \succsim \tilde{y}$ and $\bar{y} \to \tilde{y}$. Now use Proposition 3.2 to find that there exists x'' as indicated in the following.

$$x'' \to \qquad \hat{x} \to x'$$
$$\searrow \qquad \searrow$$
$$\tilde{x} \to \bar{x} \to x$$

Finally, since $\tilde{x} \succsim \tilde{y}$, use (D2) to find that there exists y'' such that $x'' \underset{\sim}{\succ} y''$ and

$$y'' \to \tilde{y} \to \hat{y} \to y'$$
$$\searrow \qquad \searrow$$
$$\bar{y} \to y.$$

We have shown there exist $x'', y'' \in F_A$ such that $x'' \to x'$, $y'' \to y'$, and $x'' \underset{\sim}{\succ} y''$, as desired. It follows that the ordering is well defined in the sense that it is independent of the choices of the representatives of the congruence classes. We use this result in the rest of the proof without comment.

We show that $\mathscr{A}$ is embeddable in $(F_A/\kappa, \geq, \cdot)$. Since $x \to x$, $\geq$ is reflexive.

We show that $\geq$ is antisymmetric. Suppose that $[x] \geq [y]$ and $[y] \geq [x]$. By the definition of $\geq$, there exist $\bar{x}, x', \bar{y}, y' \in F_A$ such that

$$\bar{x} \to x$$
$$\nearrow \qquad\qquad\qquad\qquad (10)$$
$$x'$$

and

$$\bar{y} \to y,$$
$$\nearrow \qquad\qquad\qquad\qquad (11)$$
$$y'$$

where

$$\bar{x} \succsim \bar{y} \qquad\qquad\qquad\qquad (12)$$

and

$$y' \succsim x'. \qquad\qquad\qquad\qquad (13)$$

Given (10), by Proposition 3.2, there exists $\hat{x} \in F_A$ such that

$$\hat{x} \to \bar{x} \to x.$$
$$\searrow \quad \nearrow \qquad \qquad (14)$$
$$x'$$

Given (11)–(14), by Lemma 4.5 (used as necessary), there exist $\hat{y}$ and y'' such that

$$\hat{y} \to \bar{y} \to y,$$
$$\nearrow \qquad \qquad (15)$$
$$y'' \to y'$$

where

$$y'' \succsim \hat{x} \succsim \hat{y}. \qquad (16)$$

Now if either inequality at (16) is an equality, then $x \kappa y$, which implies that $[x] = [y]$, and we are done. Otherwise, by the transitivity of $\succsim$, (16) implies that $y'' \succ \hat{y}$. But the latter inequality is impossible by Lemma 4.6, given (15). Therefore, $\geq$ is antisymmetric.

We show that $\geq$ is transitive. Assume that $[x] \geq [y]$ and $[y] \geq [z]$. Then there exist $\bar{x}, \bar{y}, y', z' \in F_A$ such that

$$\bar{x} \to x, \qquad (17)$$

$$\bar{y} \to y,$$
$$\nearrow \qquad \qquad (18)$$
$$y'$$

and

$$z' \to z, \qquad (19)$$

where

$$\bar{x} \succsim \bar{y} \qquad (20)$$

and

$$y' \succsim z'. \qquad (21)$$

Given (18), by Proposition 3.2, there exists $\hat{y} \in F_A$ such that

$$\hat{y} \to \bar{y} \to y.$$
$$\searrow \quad \nearrow \qquad \qquad (22)$$
$$y'$$

Given (17) and (19)–(22), by Lemma 4.5 (used as necessary), there exist $x', \bar{z} \in F_A$ such that

$$x' \to \bar{x} \to x \qquad (23)$$

and

$$\bar{z} \to z' \to z, \qquad (24)$$

where

$$x' \succsim \hat{y} \succsim \bar{z}. \qquad (25)$$

Given (25), by the transitivity of $\gtrsim$, $x' \gtrsim \bar{z}$. Then, given (23) and (24), by the definition of $\geq$, $[x] \geq [y]$, which establishes the transitivity of $\geq$.

Note that by definition of $\geq$, the weak monotony laws are assured. Now we show that $(F_A/\kappa, \cdot)$ is cancellative. If $[x] \cdot [z] = [y] \cdot [z]$, then $xz\kappa yz$; by Proposition 3.2, there exists w such that

$$w \to xz$$
$$\searrow$$
$$yz.$$

By Lemma 4.7 there exists v such that

$$v \to x$$
$$\searrow$$
$$y;$$

this implies that $x\kappa y$, and $[x] = [y]$. It follows that $(F_A/\kappa, \cdot)$ is right cancellative, and it can similarly be established that $(F_A/\kappa, \cdot)$ is left cancellative. Then by Fuchs' statement, the proof that $(F_A/\kappa, \geq, \cdot)$ is a strict partially ordered semigroup is complete. $\qquad\square$

The preceding results are from [11] and research cited there. Note that throughout the latter proof, the one mountain condition is heavily relied upon. Gornostaev [13] provided a general characterization of such an embedding, but did not provide axioms guaranteeing an embedding. He did provide an analogue to Lemma 2.5.

4.2 Embedding into a commutative semigroup

We conclude this section with an additional result indicating the simplification of the one mountain structure enables an articulation of conditions guaranteeing an embedding of a partial groupoid in a *commutative semigroup*, i.e., a semigroup where $a \circ b = b \circ a$. Given a word $w = a_1 \ldots a_n \in F_A$, a word consisting of a permutation σ of the letters of w will be denoted by $\sigma w = a_{i_1} \ldots a_{i_n}$. The following axiom combines notions of weak commutativity and weak associativity [10, 12]. It is necessarily satisfied by a partial groupoid which is embeddable in a commutative semigroup.

Axiom 4.9 (CA). For any positive integer n, if both products are defined, then $\pi_1(a_1 \ldots a_n) = \pi_2(a_{i_1} \ldots a_{i_n})$.

The proof of the following appears in the proofs of Theorem 6.5 and Lemma 6.6 of [12].

Proposition 4.10. *If a partial groupoid satisfies the one mountain condition and Axiom CA, then it is embeddable in a commutative semigroup.*

The proof of the following is immediate given Propositions 3.2, 3.3, and 4.10.

Corollary 4.11. *If a partial groupoid is symmetric or equidivisible, and it satisfies Axiom CA, then it can be embedded into a commutative semigroup.*

5 Conclusion

Working with partial operations is a complex, at times vexing, problem. Results here show that axioms can be added to structures satisfying the one mountain condition or its weak version to enable progress toward developing conditions which allow embeddings of partial groupoids into semigroups, as well as into other types of algebraic structures. The value of the one mountain condition in these results should not be underestimated.

Acknowledgement The comments of the editors of this volume are gratefully acknowledged. And thank you for the opportunity to contribute to this volume!

References

1. R.H. Bruck, *A Survey of Binary Systems*, Springer-Verlag, New York, 1971.
2. A.H. Clifford and G.B. Preston, *Algebraic Theory of Semigroups*, vol. 1, American Mathematical Society, Providence, Rhode Island, 1961.
3. P.W. Bunting, J. van Leeuwen, and D. Tamari, "Deciding associativity for partial multiplication tables of order 3", *Math. of Computation* **32** (1978) 593–605.
4. J.B. Carvalho and D. Tamari, "Sur l'associativité partielle des symétrisations de semi-groupes", *Portugaliae Mathematica* **21** (1962) 157–169.
5. A. Church and J.B. Rosser, "Some properties of conversion", *Trans. Amer. Math. Soc.* **39** (1936) 472–482.
6. P.M. Cohn, *Universal Algebra*, revised edition, D. Reidel Publishing Company, Dordrecht, Holland, 1981.
7. _____, "Embedding Problems for Rings and Semigroups", in *Universal Algebra and its Links with Logic, Algebra, Combinatorics and Computer Science*, Proceedings of the '25. Arbeitstagung über Allgemeine Algebra', Darmstadt, 1983, P. Burmeister et al., eds., Heldermann Verlag, Berlin, 1984.
8. D. Dekov, "Deciding embeddability of partial groupoids into semigroups", *Semigroup Forum* **58** (1999) 395–414.
9. L. Fuchs, *Partially Ordered Algebraic Systems,* Pergamon Press, Oxford, 1963.
10. S.H. Gensemer, *An Alternative Measurement Structure: The Positively Ordered Partial Semigroup with Identity*, Ph.D. Thesis, Purdue University, 1984.
11. _____, "On embedding ordered partial groupoids into partially ordered semigroups", *International Journal of Pure and Applied Mathematics* **42** (2007) 377–382.
12. S.H. Gensemer and H.J. Weinert, "On the embedding of partial groupoids into semigroups", *Bayreuther Mathematische Schriften* **28** (1989) 139–163.
13. O.M. Gornostaev, "Embeddings of ordered partial groupoids into ordered semigroups" (in Russian), in *Algebraic Actions and Orderings*, E.S. Ljapin, ed., Leningrad Gos. Ped. Inst., 1983, 21–25.
14. F.W. Levi, "On semigroups", *Bull. Calcutta Math. Soc.* **36** (1944) 141–146.
15. E.S. Ljapin, "Abstract characterization of partial groupoids of words with synonyms", in *Algebraic Theory of Semigroups, Proc. Conf. Szeged 1976, Colloq. Math. Soc. János Bolyai* **20** (1979), G. Pollak, ed., North-Holland Publishing Company, Amsterdam, 341–356.
16. _____, "Semigroup extensions of partial groupoids", *Lecture Notes in Mathematics* **1320**, Springer-Verlag, Berlin, 1988, 205–217.
17. E.S. Ljapin and A.E. Evseev, *The Theory of Partial Algebraic Operations* (revised and translated version of the 1991 Russian text, translated by M. Cole), Kluwer Academic Publishers, Dordrecht, Holland, 1997.

18. J.D. McKnight, Jr., and A.J. Storey, "Equidivisible Semigroups", *J. of Algebra* **12** (1969) 24–48.
19. M.H.A. Newman, "On theories with a combinatorial definition of equivalence", *Ann. of Math.* **43** (1942) 223–243.
20. K.E. Osondu, "Symmetrisations of semigroups", *Semigroup Forum* **24** (1982) 67–75.
21. J. Schmidt, "Universelle Halbgruppen, Kategorien, freies Produkt", *Math. Nachr.* **37** (1968) 345–358.
22. D. Tamari, "Problèmes d'associativité des monoïdes et problèmes des mots pour les groupes", *Séminaire Dubreil-Pisot* **16** (1962/63) 7.01–7.29.
23. _____, "Le problème de l'associativité des monoïdes et le problème des mots pour les demi-groupes; algèbres partielles et chaînes élémentaires", *Séminaire Dubreil-Pisot* **24** (1970/71) 8.01–8.15.
24. _____, "The associativity problem for monoids and the word problem for semigroups and groups", in *Word Problems: Decision Problems and the Burnside Problem in Group Theory*, W.W. Boone, F.B. Cannonito, and R.C. Lyndon, eds., North-Holland Publishing Company, Amsterdam, 1973, 591–607.
25. _____, "A graphic theory of associativity and wordchain patterns", in *Lecture Notes in Mathematics* **969**, Springer-Verlag, Berlin, 1982, 302–330.

Moduli Spaces of Punctured Poincaré Disks

Satyan L. Devadoss, Benjamin Fehrman, Timothy Heath, and Aditi Vashist

Abstract The Tamari lattice and the associahedron provide methods of measuring associativity on a line. The real moduli space of marked curves captures the space of such associativity. We consider a natural generalization by considering the moduli space of marked particles on the Poincaré disk, extending Tamari's notion of associativity based on nesting. A geometric and combinatorial construction of this space is provided, which appears in Kontsevich's deformation quantization, Voronov's swiss-cheese operad, and Kajiura and Stasheff's open-closed string theory.

1 Motivation from physics

Our story begins with the famous *associahedron* polytope. In his 1951 thesis, Dov Tamari described the associahedron K_n as the realization of his lattice of bracketings on n letters [23]. Independently, in his 1961 thesis, Jim Stasheff constructed a convex curvilinear version of it for use in homotopy theory in connection with associativity properties of H-spaces [22]. The vertices of K_n are enumerated by the Catalan numbers and its construction as a polytope was given independently by Haiman (unpublished) and Lee [18]. Figure 1(a) shows the example of the associahedron K_4.

Satyan L. Devadoss
Williams College, e-mail: *satyan.devadoss@williams.edu*

Benjamin Fehrman
University of Chicago, e-mail: *bfehrman@math.uchicago.edu*

Timothy Heath
Columbia University, e-mail: *timheath@math.columbia.edu*

Aditi Vashist
University of Michigan, e-mail: *avashist@umich.edu*

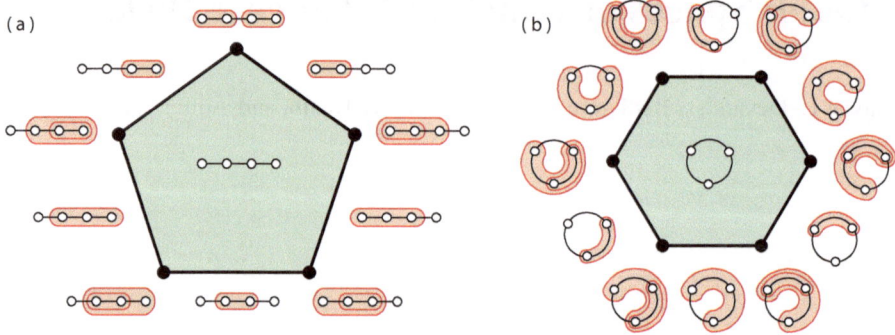

Fig. 1 (a) Associahedron K_4 and (b) cyclohedron W_3.

Definition 1.1. Let $A(n)$ be the poset of all bracketings of n letters, ordered such that $a \prec a'$ if a is obtained from a' by adding new brackets. The *associahedron* K_n is a convex polytope of dimension $n-2$ whose face poset is isomorphic to $A(n)$.

Our interests are based on associahedra as they appear in the world of algebraic geometry. The configuration space of n labeled particles on a manifold X is

$$\text{Config}^n(X) = X^n - \Delta, \quad \text{where } \Delta = \{(x_1, \dots, x_n) \in X^n \mid \exists\, i, j,\ x_i = x_j\}.$$

The Riemann moduli space $\mathcal{M}_{g,n}$ of genus g surfaces with n marked particles (sometimes called *punctures*) is an important object in mathematical physics, brought to light by Grothendieck in his famous *Esquisse d'un programme*. A larger framework, based on moduli spaces of *bordered* surfaces of arbitrary genus is considered in [9]. The special case $\mathcal{M}_{0,n}$ is defined as

$$\mathcal{M}_{0,n} = \text{Config}^n(\mathbb{CP}^1) / \mathbb{P}GL_2(\mathbb{C}),$$

the quotient of the configuration space of n labeled points on the complex projective line by $\mathbb{P}GL_2(\mathbb{C})$. There exists a Deligne-Mumford-Knudsen compactification $\overline{\mathcal{M}}_{0,n}$ of this space, which plays a crucial role in the theory of Gromov-Witten invariants, symplectic geometry, and quantum cohomology [17].

The real moduli space $\overline{\mathcal{M}}_{0,n}(\mathbb{R})$ is the set of points fixed under complex conjugation; these spaces have importance in their own right, appearing in areas such as ζ-motives [13], phylogenetic trees [10], and Lagrangian Floer theory [11]. The relationship between $\overline{\mathcal{M}}_{0,n}(\mathbb{R})$ and the associahedron is given by the subsequent important result:

Theorem 1.2 ([6, Section 3]). *The real moduli space of n-punctured Riemann spheres*

$$\mathcal{M}_{0,n}(\mathbb{R}) = \text{Config}^n(\mathbb{RP}^1) / \mathbb{P}GL_2(\mathbb{R})$$

has a Deligne-Mumford-Knudsen compactification $\overline{\mathcal{M}}_{0,n}(\mathbb{R})$, resulting in an $(n-3)$-manifold tiled by $(n-1)!/2$ copies of the K_{n-1} associahedron.

There are numerous generalizations of the associahedron currently in literature. The closest kin, however, is the *cyclohedron* polytope, originally considered by Bott and Taubes in relation to knot invariants [3]. Figure 1(b) shows the example of the 2D cyclohedron W_3.

Definition 1.3. Let $B(n)$ be the poset of all bracketings of n letters *arranged in a circle*, ordered such that $b \prec b'$ if b is obtained from b' by adding new brackets. The *cyclohedron* W_n is a convex polytope of dimension $n-1$ whose face poset is isomorphic to $B(n)$.

And just as the associahedron tiles $\overline{\mathcal{M}}_{0,n}(\mathbb{R})$, there is an analogous manifold tiled by the cyclohedron. Indeed, Armstrong et al. [1] consider a collection of such moduli spaces (based on blowups of Coxeter complexes), each tiled by different analogs of the associahedron polytope, called *graph associahedra* [4].

Theorem 1.4 ([7, Section 2]). *The moduli space $\overline{\mathcal{X}}_n$ is the (Fulton-MacPherson) compactification of* $\mathrm{Config}^n(S^1) / S^1$, *tiled by* $(n-1)!$ *copies of the cyclohedron* W_n.

Both $\overline{\mathcal{M}}_{0,n}(\mathbb{R})$ and $\overline{\mathcal{X}}_n$ consider how particles move and collide on the circle (viewed as either $\mathbb{RP}^1$ or S^1, depending on the group action). Similarly $\overline{\mathcal{M}}_{0,n}$ encapsulates particle collisions on the sphere $\mathbb{CP}^1$. In this article, we consider the Poincaré disk, a concrete playground where particles in the interior can collide similar to $\overline{\mathcal{M}}_{0,n}$ and on the boundary similar to $\overline{\mathcal{M}}_{0,n}(\mathbb{R})$ and $\overline{\mathcal{X}}_n$. Indeed, several others have considered a version of this space of punctured disks: Kontsevich in his work on deformation quantization [16], Kajiura and Stasheff from a homotopy algebra viewpoint [15], Voronov from an operadic one [24], and Hoefel from that of spectral sequences [14]. We claim that this space naturally extends the notion of Tamari's associativity, from particles on lines to particles on disks.

This is a survey article on the moduli space of punctured Poincaré disks, from a geometric and combinatorial viewpoint. Section 2 introduces the foundational setup, whereas Section 3 provides the famous Fulton-MacPherson compactification based on iterated blowups, along with discussing several examples. A local construction of this space using group actions on bubble-trees is given in Section 4, and Section 5 ends with some combinatorial results.

2 Particles on the Poincaré disk

There exists a natural action of $\mathbb{P}SL_2(\mathbb{R})$ on the upper halfplane $\mathbb{H}$ (along with the point at infinity) given by

$$\begin{pmatrix} a & b \\ c & d \end{pmatrix} \cdot x = \frac{ax+b}{cx+d}.$$

The diffeomorphism $z \to (z-i)(z+i)^{-1}$ extends this to an action of $\mathbb{P}SL_2(\mathbb{R})$ on the Poincaré disk $\mathcal{D}$, where infinity is on the boundary of $\mathcal{D}$. The classical *KAN* decomposition of $SL_2(\mathbb{R})$ is given by

$$K = SO_2(\mathbb{R}) \qquad A = \left\{ \begin{pmatrix} a & 0 \\ 0 & a^{-1} \end{pmatrix} \,\middle|\, a > 0 \right\} \qquad N = \left\{ \begin{pmatrix} 1 & x \\ 0 & 1 \end{pmatrix} \,\middle|\, x \in \mathbb{R} \right\}.$$

Analogously, the corresponding decomposition of $\mathbb{P}SL_2(\mathbb{R})$ is

$$\mathbb{P}SO_2(\mathbb{R}) \cdot A \cdot N. \tag{1}$$

The following is a classical result; see [21, Chapter 5] for details.

Proposition 2.1. *The action of $\mathbb{P}SL_2(\mathbb{R})$ preserves the boundary and the interior of $\mathcal{D}$. Moreover,*

1. *if μ is a point in the interior of $\mathcal{D}$, then for each element x in the interior of $\mathcal{D}$, there exists a unique element σ in $A \cdot N$ satisfying $\sigma \cdot x = \mu$, and*
2. *for each point y on the boundary of $\mathcal{D}$, there exists a unique element σ in $\mathbb{P}SO_2(\mathbb{R})$ satisfying $\sigma \cdot y = \infty$.*

The action of $\mathbb{P}GL_2(\mathbb{R})$ is naturally characterized by $\mathbb{P}SL_2(\mathbb{R})$: The element

$$r = \begin{pmatrix} 1 & 0 \\ 0 & -1 \end{pmatrix}$$

in $\mathbb{P}GL_2(\mathbb{R}) \setminus \mathbb{P}SL_2(\mathbb{R})$ acts on $\mathcal{D}$ as the reflection about the geodesic connecting zero to ∞. The coset partition $\{\mathbb{P}SL_2(\mathbb{R}), r \cdot \mathbb{P}SL_2(\mathbb{R})\}$ decomposes $\mathbb{P}GL_2(\mathbb{R})$ as

$$\mathbb{Z}_2 \cdot \mathbb{P}SO_2(\mathbb{R}) \cdot A \cdot N \tag{2}$$

where $\mathbb{Z}_2 = \{I, r\}$. Figure 2 shows the action of $\mathbb{P}SL_2(\mathbb{R})$ on $\mathcal{D}$ based on the decomposition given by Eq. (1), where the Poincaré disk $\mathcal{D}$ is shaded to help display the action. Part (a) shows $\mathcal{D}$ with four of its points labeled, along with the action of (b) reflection r, (c) rotation $\mathbb{P}SO_2(\mathbb{R})$, and (d) interior movement $A \cdot N$.

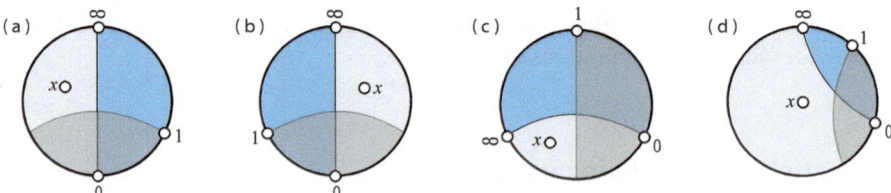

Fig. 2 Actions on the Poincaré disk.

The main interest of this paper is the configuration space

$$\mathrm{Config}^{n,m}(\mathcal{D}) = (\mathcal{D}^n \times \partial \mathcal{D}^m) - \Delta$$

such that Δ is the collection of points $(p_1, \ldots, p_n, q_1, \ldots, q_m) \in \mathcal{D}^n \times \partial \mathcal{D}^m$ where

1. *(interior collision)* there exist i, j such that $p_i = p_j$, or

2. (*boundary collision*) there exist i, j such that $q_i = q_j$, or
3. (*mixed collision*) there exists i such that $p_i \in \partial \mathscr{D}$.

Points of $\text{Config}^{n,m}(\mathscr{D})$ have n distinct labeled particles in the interior of $\mathscr{D}$ and m distinct labeled particles confined to the boundary of $\mathscr{D}$. Since each element of $\mathbb{P}SL_2(\mathbb{R})$ preserves the boundary and interior of $\mathscr{D}$, the action of $\mathbb{P}SL_2(\mathbb{R})$ on $\mathscr{D}$ naturally extends to an action on $\text{Config}^{n,m}(\mathscr{D})$. Thus, the following is well defined:

Definition 2.2. The moduli space of punctures on the Poincaré disk $\mathscr{D}$ is the quotient

$$\mathscr{K}(n,m) = \text{Config}^{n,m}(\mathscr{D})/\mathbb{P}SL_2(\mathbb{R}).$$

We will be concerned mostly with the case when $n, m \geq 1$, with at least one particle μ in the interior of $\mathscr{D}$ and one particle ∞ on the boundary of $\mathscr{D}$. Consider

$$\text{Config}^{n-1,m-1}(\mathscr{D}) - \Delta^*$$

such that Δ^* is the collection of points $(p_1, \ldots, p_{n-1}, q_1, \ldots, q_{m-1}) \in \mathscr{D}^n \times \partial \mathscr{D}^m$ where

1. (*interior collision*) there exist i, j such that $p_i = p_j$ or $p_i = \mu$, or
2. (*boundary collision*) there exist i, j such that $q_i = q_j$ or $q_i = \infty$, or
3. (*mixed collision*) there exists i such that $p_i \in \partial \mathscr{D}$.

Proposition 2.3. *The moduli space $\mathscr{K}(n,m)$ admits a natural description as the space* $\text{Config}^{n-1,m-1}(\mathscr{D}) - \Delta^*$.

Proof. It follows directly from Proposition 2.1 that each orbit under the $\mathbb{P}SL_2(\mathbb{R})$ action on $\mathscr{D}$ can be uniquely represented by a particle configuration of this form: Fixing an interior particle kills the $A \cdot N$ action, and the remaining $\mathbb{P}SO_2(\mathbb{R})$ rotation of Eq. (1) is addressed by fixing a boundary particle.   □

Remark 2.4. Although we do not discuss it here, the moduli space $\mathscr{K}(n,0)$ of n interior particles on the disk can be interpreted in terms of fixed particles, analogously to Proposition 2.3. The $\mathbb{P}SL_2(\mathbb{R})$ action fixes one of the n particles at μ using $A \cdot N$ and confines another particle to the geodesic connecting μ to ∞ using $\mathbb{P}SO_2(\mathbb{R})$. The resulting moduli space can be shown to be homeomorphic to $\mathscr{K}(n,0)$. The $\mathscr{K}(0,m)$ case is considered in Section 5.

So far, the particles in $\mathscr{K}(n,m)$ are not allowed to collide. The most natural manner of embracing collisions is by including the places of collision Δ that were removed by the configuration space. We define the *naive compactification* of $\mathscr{K}(n,m)$ as the inclusion of the diagonal Δ with $\mathscr{K}(n,m)$, denoted as $\mathscr{K}\langle n,m \rangle$. A visual notation is now introduced to label the particle collisions on $\mathscr{D}$ based on *arcs*.

Definition 2.5. An *arc* is a curve on $\mathscr{D}$ such that its endpoints are on the boundary of $\mathscr{D}$, it does not intersect itself nor the particles, and encloses at least one interior particle or two boundary particles. A *loop* is an arc with its endpoints identified, enclosing multiple interior particles.[1] Two arcs are *compatible* if there are curves in their respective isotopy classes which do not intersect.

From the definition of Δ, three types of collisions emerge. Figure 3 illustrates collisions in $\mathscr{K}\langle 7,6\rangle$, part (a) showing a diagram of the disk with the appropriate particles, with examples of an (b) interior collision, (c) boundary collision, and (d) mixed collision.

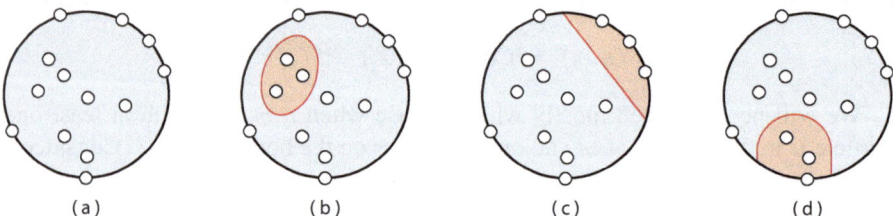

Fig. 3 Diagrams of arcs on $\mathscr{D}$ corresponding to collisions of particles.

Let us consider some low-dimensional examples, where *black* particles will represent the fixed particles μ and ∞ on the Poincaré disk.

Example 2.6. Since the action of $\mathbb{P}SL_2(\mathbb{R})$ fixes a particle on the interior and one on the boundary, the moduli space $\mathscr{K}\langle 1,1\rangle$ is a point.

Example 2.7. The left side of Figure 4 displays $\mathscr{K}\langle 1,2\rangle$, which is a circle with a vertex (a), corresponding to the free particle on the boundary colliding with the fixed particle ∞.

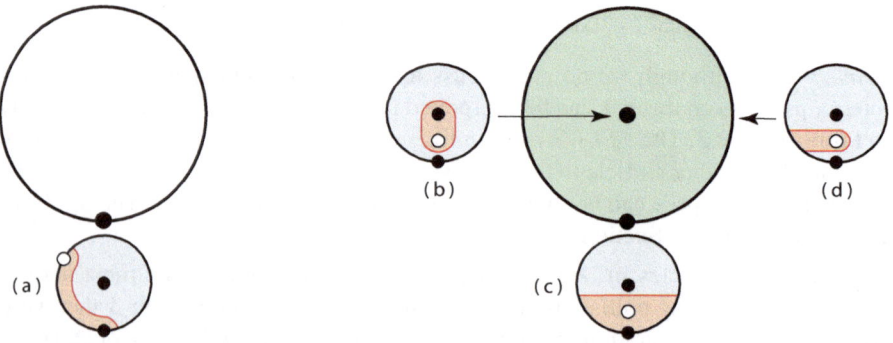

Fig. 4 $\mathscr{K}\langle 1,2\rangle$ and $\mathscr{K}\langle 2,1\rangle$

Example 2.8. The right side of Figure 4 displays $\mathscr{K}\langle 2,1\rangle$ as a disk with two vertices. Here, there are three types of collisions signified by three cells: vertex (b) is the free particle colliding with the interior fixed particle μ, vertex (c) is the free particle

---

[1] Abusing terminology, we will refer to both arcs and loops as simply "arcs".

colliding with ∞, and circle (d) is the free particle colliding with the boundary of $\mathscr{D}$. Note that this last condition is considered a *collision* since Δ^* includes $\partial\mathscr{D}$.

Example 2.9. Figure 5 shows the two-torus $\mathscr{K}\langle 1,3\rangle$. It has three circles on it, all of which are copies of $\mathscr{K}\langle 1,2\rangle$, given by collisions (a, b, c), along with a vertex (d) where all the circles meet.

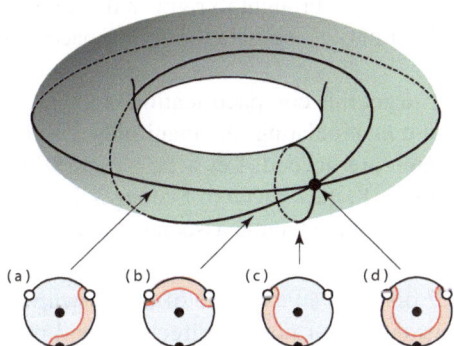

Fig. 5 The two-torus $\mathscr{K}\langle 1,3\rangle$.

Example 2.10. Figure 6 displays $\mathscr{K}\langle 2,2\rangle$ as a *solid* two-torus with diagrammatic labelings of the collisions. It has one vertex (e) where all the free particles collide with ∞, three $\mathscr{K}\langle 1,2\rangle$ circles (b, c, f), a $\mathscr{K}\langle 2,1\rangle$ disk (d), and the space $\mathscr{K}\langle 1,3\rangle$ as the two-torus boundary (a) of $\mathscr{K}\langle 2,2\rangle$.

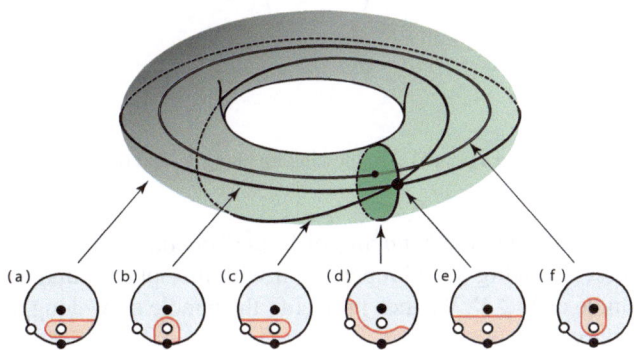

Fig. 6 The solid two-torus $\mathscr{K}\langle 2,2\rangle$.

3 The Fulton-McPherson compactification

Although the naive compactification $\mathscr{K}\langle n,m \rangle$ encapsulates particle collisions, there are other compactifications which contain more useful information. In general, compactifying $\mathscr{K}(n,m)$ enables the particles to collide and a *system* is introduced to record the *directions* particles arrive at the collision. In the work of Fulton and MacPherson [12], this method is brought to rigor in the algebro-geometric context.[2] We now care not just about particle collisions but the space of *simultaneous particle collisions*.

The language to construct this compactification is the algebro-geometric notion of a *blowup*. Since we are manipulating *real* manifolds with boundary, there are two notions of blowups which are needed: Let Y be a manifold with boundary. For a subspace X in the *interior* of Y, the *blowup* of Y along X is obtained by first removing X and replacing it with the sphere bundle associated to the normal bundle of $X \subset Y$. We then projectify the bundle.

Example 3.1. Figure 7 displays an example of the blowup of the plane Y along a point X: Part (a) shows the set of normal directions and part (b) shows the point replaced by the sphere bundle. Part (c) shows the result of the antipodal map along this bundle (in red), where X has now been replaced by $\mathbb{RP}^1$. Indeed, the blow up keeps track of the *projective direction* in which one approaches X.

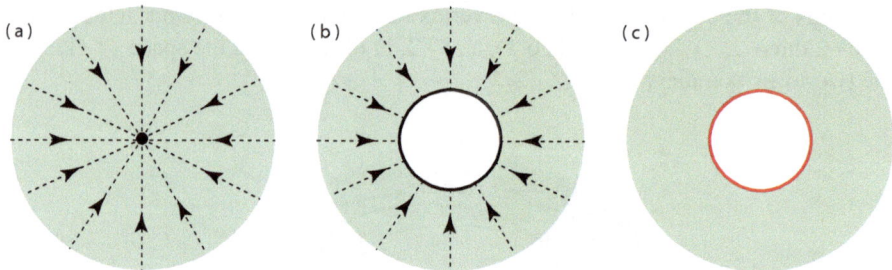

Fig. 7 (a) Set of directions to a point X, (b) blowing up along X, and (c) antipodal gluing.

For a subspace X along the *boundary* of Y, the *boundary blowup* of Y along X is obtained by first removing X and replacing it with the sphere bundle associated to the normal bundle of $X \subset Y$. We then projectify the bundle only along its intersection with the boundary.

Example 3.2. Figure 8 provides an example of the blowup of the halfplane Y along a point X on its boundary: Part (a) shows the set of normal directions and part (b) shows the point replaced by the normal bundle. Part (c) shows the result of the antipodal map along the intersection of this bundle with the boundary, in this case identifying the two highlighted points. In general, a boundary blowup keeps track of

---

[2] The real analog of the Fulton-MacPherson is given by the Axelrod-Singer compactification [2].

the directions in which one approaches X from the interior of Y, and the *projective direction* in which one approaches X along the boundary of Y.

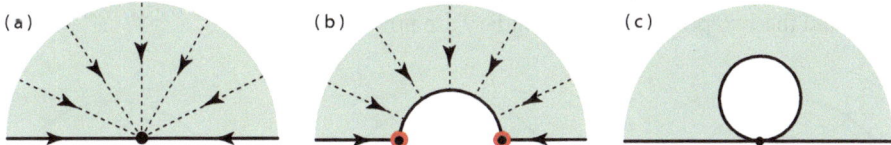

Fig. 8 (a) Set of directions, (b) blowing up along the point, and (c) gluing along the bounding hyperplane.

A general collection of blowups is usually non-commutative in nature; in other words, the order in which spaces are blown up is important. For a given hyperplane arrangement, De Concini and Procesi [5] establish the existence and uniqueness of a *minimal building set*, a collection of subspaces for which blowups commute for a given dimension. In the case of the arrangement $X^n - \mathrm{Config}^n(X)$, their procedure yields the Fulton-MacPherson compactification of $\mathrm{Config}^n(X)$.

Definition 3.3. The *minimal building set* $\mathfrak{b}(n,m)$ of $\mathcal{K}\langle n,m \rangle$ is the collection of elements in $\mathcal{K}\langle n,m \rangle$ labeled with a single arc on $\mathscr{D}$.

The elements of $\mathfrak{b}(n,m)$ are partitioned according to the magnitude of the collisions they represent. Each element represents a configuration where i interior particles and b boundary particles have collided, and the sum $2i+b$ determines the dimension of this element. One important note is that real blowups along cells which are already codimension one do not alter the topology of the manifold. Thus, we discount codimension one elements from the building set $\mathfrak{b}(n,m)$. The work of De Concini and Procesi give us the ensuing result:

Theorem 3.4. *The Fulton-McPherson compactification $\overline{\mathcal{K}}(n,m)$ is obtained from $\mathcal{K}\langle n,m \rangle$ by the iterated sequence of blowups of $\mathfrak{b}(n,m)$ in increasing order of dimension.*

In the language of algebraic geometry, the effect of a blowup replaces the cell with an *exceptional divisor* of the resulting manifold. From a topological perspective, a blowup of Y along X "promotes" X to become a codimension one cell of Y. An analog from the world of polytopes is the notion of truncation: Truncating any face of a polytope introduces a new facet, and the truncation of a current facet does not change the combinatorics of the polytope.

Let us consider some low-dimensional examples of the construction of $\overline{\mathcal{K}}(n,m)$ using iterated blowups.

Example 3.5. The left side of Figure 4 displays $\mathcal{K}\langle 1,2 \rangle$, which is a circle with a vertex (a). Since cell (a) is codimension one, no blowups are necessary, and $\overline{\mathcal{K}}(1,2)$ is the same as $\mathcal{K}\langle 1,2 \rangle$.

Example 3.6. The building set $\mathfrak{b}(2,1)$ consists of two points, given by the two vertices of Figure 9(a); compare with the labels (b) and (c) of Figure 4. The blowups along these two cells are first accomplished by replacing these points with the sphere bundle along the normal bundle. The interior vertex is replaced with a circle, and the boundary vertex with an arc, as in Figure 9(b). Performing the projective identification is shown in part (c), where the interior circle has an antipodal map (in red), and the two points on the boundary are now identified.

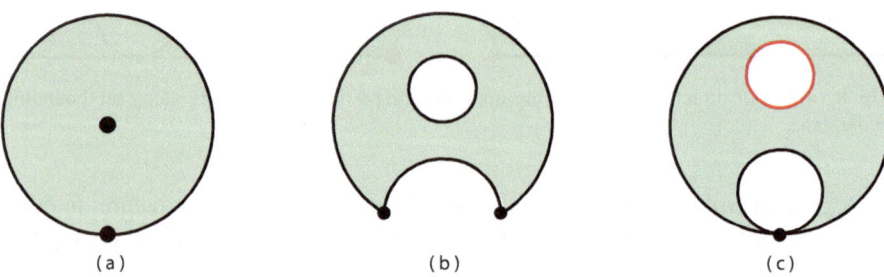

Fig. 9 (a) $\mathcal{K}\langle 2,1\rangle$, (b) the eye of Kontsevich, and (c) $\overline{\mathcal{K}}(2,1)$.

Remark 3.7. The picture in Figure 9(b) appears in the work by Kontsevich on deformation quantization [16, Figure 7], where he calls this the *"eye"*.

Example 3.8. Figure 5 shows $\mathcal{K}\langle 1,3\rangle$ as a two-torus, with the building set $\mathfrak{b}(1,3)$ containing only one vertex, labeled by (d). Blowing up this vertex results in the connected sum of the two-torus with an $\mathbb{RP}^2$, given in Figure 10(a). Part (b) of this figure displays the two tiling hexagons along with their gluing map used to construct $\overline{\mathcal{K}}(1,3)$.

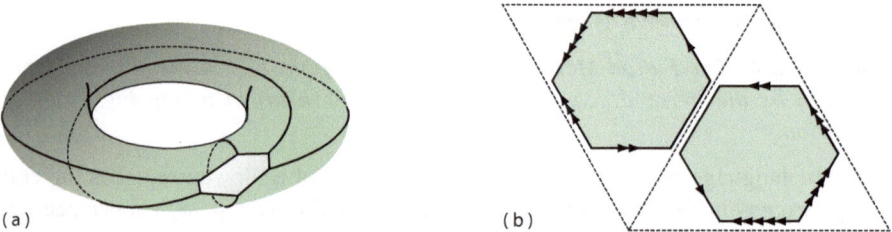

Fig. 10 (a) $\overline{\mathcal{K}}(1,3)$ obtained from (b) gluing two hexagons.

Example 3.9. The building set $\mathfrak{b}(2,2)$ consists of four elements, shown in Figure 6: three circles (b, c, f), and one vertex (e). A redrawing of this solid two-torus cut open along the disk (e) is given in Figure 11(a). The construction of $\overline{\mathcal{K}}(2,2)$ follows from iterated blowups in increasing order of dimension. Figure 11(b) shows the blowup along the vertex on the boundary, and part (c) shows the blowup along the three circles. Although not displayed in these figures, there are gluing maps from the blowups identifying faces of these cells. The combinatorics of this space is expanded upon in Figure 15.

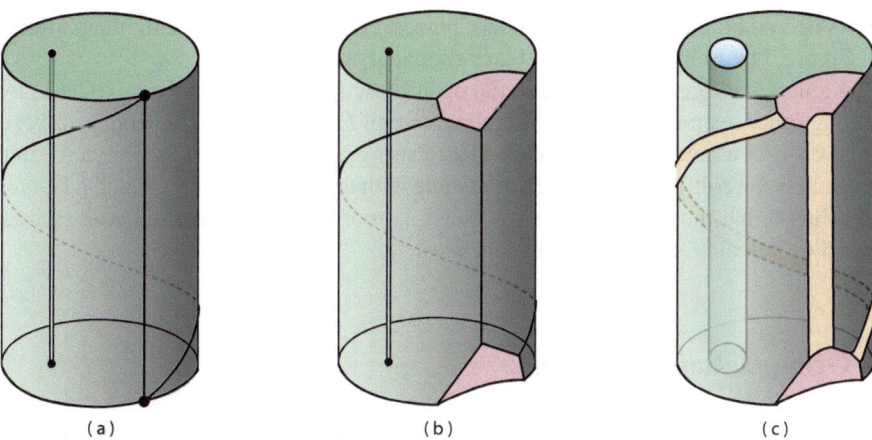

Fig. 11 Iterated truncation of $\mathcal{K}(2,2)$ resulting in $\overline{\mathcal{K}}(2,2)$.

Recall that $\mathcal{K}\langle n,m\rangle$ is stratified by compatible arcs on the disk $\mathscr{D}$ representing collisions. For the compactified moduli space $\overline{\mathcal{K}}(n,m)$, the stratification is given by compatible *nested* arcs, the notion coming from the work of Fulton-MacPherson [12]. Consider Figure 12: parts (a) and (b) show valid compatible arcs, representing cells of $\overline{\mathcal{K}}(7,5)$. Part (c) shows compatible nested arcs of the disk, representing a cell of $\overline{\mathcal{K}}(7,5)$. It is this nesting which gives rise to a generalized notion of Tamari's associativity.

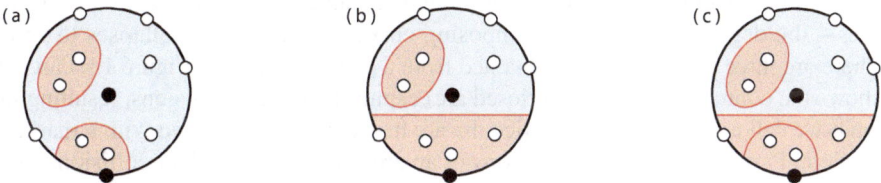

Fig. 12 Compatible arcs (a) and (b) of $\mathcal{K}\langle 7,5\rangle$, along with nested arcs (c) of $\overline{\mathcal{K}}(7,5)$.

We summarize as follows:

Theorem 3.10. *The space $\overline{\mathcal{K}}(n,m)$ of marked particles on the Poincaré disk is a compact manifold of real dimension $2n+m-3$, naturally stratified by nested compatible arcs, where k arcs on $\mathscr{D}$ correspond to a codimension k face of $\overline{\mathcal{K}}(n,m)$.*

This space can be viewed from the perspective of open-closed string field theory, where Figure 3 is reinterpreted as displaying (b) open strings, (c) closed strings, and (d) open-closed strings. There is a rich underlying operadic perspective to this space, brought to light by the works of others such as Voronov [24], Kajiura-Stasheff [15], and Hoefel [14]. The *swiss-cheese* operad structure of Voronov extends the little disks operad by incorporating boundary pieces, envisioned by replacing the particles of Figure 3 by loops (interior) and arcs (boundary).

Kajiura and Stasheff introduced the open-closed homotopy algebras (OCHA) by adding other operations to an A_∞ algebra over an L_∞ algebra. The meaning of these extra operations amounts to making a closed string become open; from our viewpoint, it signifies the collision of the interior particle with the boundary of $\mathscr{D}$. Finally, Hoefel has shown that the OCHA operad is quasi-isomorphic to the swiss-cheese operad based on calculations of spectral sequences.

4 Group actions on screens

Instead of a global perspective, based on blowups of cells in the building set, there is a local, combinatorial perspective in which to construct this moduli space. Fulton and MacPherson describe their compactification from the collision perspective as follows: As points collide along a k-dimensional manifold, they land on a k-dimensional *screen*, which is identified with the point of collision. These screens have been dubbed *bubbles*, and the compactification process as *bubbling*; see [20] for details from an analytic viewpoint. Now these particles on the screen are themselves allowed to move and collide, landing on higher level screens. Kontsevich describes this process in terms of a magnifying glass: On any given level, only a configuration of points is noticeable; but one can zoom-in on a particular point and peer into its screen, seeing the space of collided points.

As the stratification of this space is given by collections of compatible nested arcs, the compactification of this space can be obtained by the *contraction* of these arcs – the degeneration of a decomposing curve γ as its length collapses to zero. There are three possible results obtained from a contraction, as in Figure 13. Part (a) shows the contraction of a loop (closed arc) capturing interior collisions, resulting in what we call a *sphere bubble*. Notice the arc has now been identified with a marked particle. Part (b) displays contractions of an arc capturing boundary collisions; the

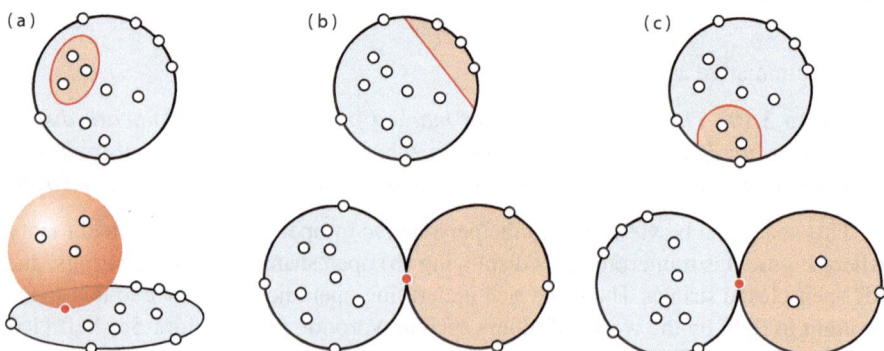

Fig. 13 (a) Sphere bubble, (b) flat bubble, and (c) punctured bubble.

resulting bubble is called a *flat bubble*. Notice that all the particles are only on the boundary of this new disk. Finally, (c) shows contractions for a mixed collision, called a *punctured bubble*.

Given a collection of compatible nested arcs, collapsing all the arcs results in several bubbles, sometimes refereed to as a *bubble tree*. There exists a natural dual perspective in which to visualize the cells of $\overline{\mathscr{K}}(n,m)$, as given by Figure 14. Based on the notation of Hoefel [14], we obtain *partially-planar* trees, where the interior particles correspond to leaves in $\mathbb{R}^3$ and boundary particles to leaves restricted to the xy-plane. The spatial edges (red zig-zags) are allowed to move freely, whereas the cyclic ordering of the planar ones (black lines) are determined by the disk. The natural operad composition maps for such trees are given in [14, Section 4]. If we choose a boundary particle to be fixed at ∞, then *rooted* trees arise.

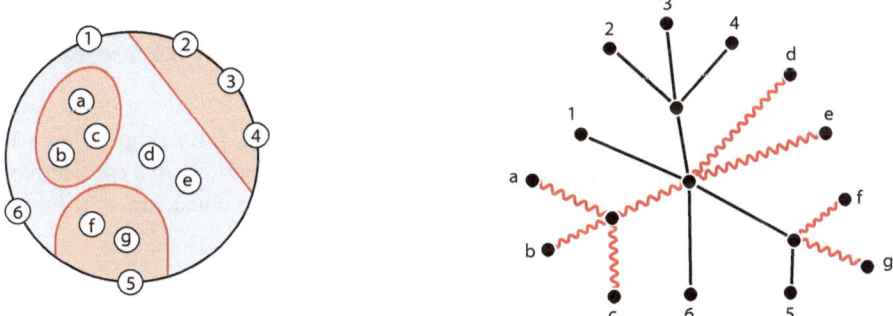

Fig. 14 Duality between arcs on the disk and partially-planar trees.

These three types of bubbles are the *screens* described by Fulton and MacPherson. According to the compactification, there is an action of an automorphism group on each screen (bubble). In order to describe the groups, a slight detour is taken. Similar to $\mathbb{PSL}_2(\mathbb{R})$, the matrix group $\mathbb{PSL}_2(\mathbb{C})$ acts on $\mathbb{CP}^1$, considered as $\mathbb{C} \cup \{\infty\}$. The decomposition of $SL_2(\mathbb{C})$ analogous to Eq. (1) is given by

$$K_{\mathbb{C}} = SO_2(\mathbb{C}) \qquad A_{\mathbb{C}} = \left\{ \begin{pmatrix} z & 0 \\ 0 & z^{-1} \end{pmatrix} \middle| z \in \mathbb{C}^{\times} \right\} \qquad N_{\mathbb{C}} = \left\{ \begin{pmatrix} 1 & z \\ 0 & 1 \end{pmatrix} \middle| z \in \mathbb{C} \right\}.$$

The corresponding decomposition of $\mathbb{PSL}_2(\mathbb{C})$ is given by

$$\mathbb{PSO}_2(\mathbb{C}) \cdot A_{\mathbb{C}} \cdot N_{\mathbb{C}}.$$

An interest of this paper is the particular group

$$G_I = \mathbb{Z}_2 \cdot \mathbb{PSO}_2(\mathbb{C}) \cdot A \cdot N_{\mathbb{C}} \tag{3}$$

acting on $\mathbb{CP}^1$. Here, $\mathbb{Z}_2 = \{I, r\}$ and $A_{\mathbb{C}} = S^1 \cdot A$, where S^1 is the group of rotations

$$\left\{ \begin{pmatrix} e^{i\theta} & 0 \\ 0 & e^{-i\theta} \end{pmatrix} \ \middle| \ \theta \in \mathbb{R} \right\}.$$

Proposition 4.1. *The action of G_I on $\mathbb{CP}^1$ is given as follows:*

1. *The element r acts on each circle of constant norm in $\mathbb{CP}^1$ as the antipodal map.*
2. *For each point x in $\mathbb{CP}^1$, there exists a unique element σ in $\mathbb{P}SO_2(\mathbb{C})$ satisfying $\sigma \cdot x = \infty$.*
3. *For any two elements x, y in $\mathbb{CP}^1 \setminus \{\infty\}$, there exists a unique element σ in $A \cdot N_{\mathbb{C}}$ satisfying $\sigma \cdot x = 0$ and $||\sigma \cdot y|| = 1$.*

Proof. Since the elements of $\mathbb{P}SO_2(\mathbb{C})$ act homeomorphically on $\mathbb{CP}^1$, the action of each element sends a unique point x in $\mathbb{CP}^1$ to ∞. Because $SO_2(\mathbb{C})$ is disconnected, this criterion characterizes each element of $SO_2(\mathbb{C})$ up to sign. For two distinct points x, y on $\mathbb{CP}^1$, each element of $A \cdot N_{\mathbb{C}}$ corresponds to a dilation and translation of $\mathbb{CP}^1 \setminus \{\infty\}$. There exists a unique positive scalar α satisfying $\alpha \cdot ||x - y|| = 1$ and a unique β in $\mathbb{C}$ satisfying $\beta + \alpha \cdot x = 0$. This specifies σ uniquely because $A \cdot N_{\mathbb{C}}$ is connected. $\square$

The action of G_I on $\mathbb{CP}^1$ extends to an action on $\mathrm{Config}^n(\mathbb{C})$. By Proposition 4.1, the action fixes two of the n particles, and restricts the third particle to a (projective) circle's worth of freedom. The following is a natural object to consider:

Definition 4.2. The moduli space $\mathscr{M}_n(\mathbb{C})$ is the quotient space

$$\mathscr{M}_n(\mathbb{C}) \ = \ \mathrm{Config}^n(\mathbb{CP}^1) \, / \, G_I.$$

Denote by $\overline{\mathscr{M}}_n(\mathbb{C})$ the natural Fulton-MacPherson compactification of this space.

Remark 4.3. There is a difference between $\overline{\mathscr{M}}_n(\mathbb{C})$ and the classical moduli space of curves $\overline{\mathscr{M}}_{0,n}$. The former is based on real blowups while the latter is based on complex blowups. Indeed, the group $\mathbb{P}GL_2(\mathbb{C})$ has six real dimensions, whereas G_I only has five. Thus, for instance, $\overline{\mathscr{M}}_3(\mathbb{C})$ is equivalent to $\mathbb{RP}^1$ whereas $\overline{\mathscr{M}}_{0,3}$ is simply a point.

Lemma 4.4. *Each type of bubble receives a different group action:*

1. *The group G_I acts on sphere bubbles.*
2. *The group $\mathbb{P}GL_2(\mathbb{R})$ acts on flat bubbles.*
3. *The group $\mathbb{P}SL_2(\mathbb{R})$ acts on punctured bubbles.*

Proof. The action of G_I on the sphere bubbles is given by Proposition 4.1 above, where it mimics the real blowup of particle collisions on the plane, as displayed in Figure 7. The action of $\mathbb{P}GL_2(\mathbb{R})$ on flat bubbles is based on the decomposition given in Eq. (2). There is a $\mathbb{Z}_2$ component which identifies each flat bubble with its mirror image about the geodesic connecting zero to ∞ on $\mathscr{D}$, replicating the boundary blowup structure shown in Figure 8. Finally, $\mathbb{P}SL_2(\mathbb{R})$ is the natural group action on the punctured bubbles (Poincaré disks) from Proposition 2.1. $\square$

Since the arcs on $\mathscr{D}$ are compatible and thus nonintersecting, each bubble-tree corresponds to a product of smaller moduli spaces. The group actions from Lemma 4.4 yield the subsequent result:

Theorem 4.5. *Each bubble gives rise to a compactified moduli space:*

1. *Sphere bubbles with n particles produce* $\overline{\mathscr{M}}_n(\mathbb{C})$.
2. *Flat bubbles with m particles produce* $\overline{\mathscr{M}}_{0,m}(\mathbb{R})$.
3. *Punctured bubbles with n interior and m boundary particles produce* $\overline{\mathscr{K}}(n,m)$.

Moreover, each diagram of $\mathscr{D}$ with compatible nested arcs corresponds to a product of moduli spaces, with each bubble contributing a factor.

Example 4.6. Consider $\overline{\mathscr{K}}(2,2)$ given in Figure 11; it is redrawn (with a twist of π of the top disk) in Figure 15 on the left. The colors and the labels of these figures are coordinated to match.

1. There is one interior collision (f) resulting in a sphere bubble adjoined to a punctured bubble, associated to the space $\overline{\mathscr{K}}(1,2) \times \overline{\mathscr{M}}_3(\mathbb{C})$. Since both $\overline{\mathscr{K}}(1,2)$ and $\overline{\mathscr{M}}_3(\mathbb{C})$ are topological circles, the resulting divisor is a two-torus.
2. There is one boundary collision (d) producing a punctured bubble adjoined to a flat bubble, giving rise to $\overline{\mathscr{K}}(2,1) \times \overline{\mathscr{M}}_{0,3}(\mathbb{R})$. Since $\overline{\mathscr{M}}_{0,3}(\mathbb{R})$ is a point, the resulting divisor is simply $\overline{\mathscr{K}}(2,1)$.
3. Two mixed collisions (b, c) produce $\overline{\mathscr{K}}(1,2) \times \overline{\mathscr{K}}(1,2)$, yielding a two-torus, the product of two circles.
4. Two mixed collisions (a, e) produce $\overline{\mathscr{K}}(1,1) \times \overline{\mathscr{K}}(1,3)$. Since $\overline{\mathscr{K}}(1,1)$ is a point, we are left with $\overline{\mathscr{K}}(1,3)$ of Figure 10(a).

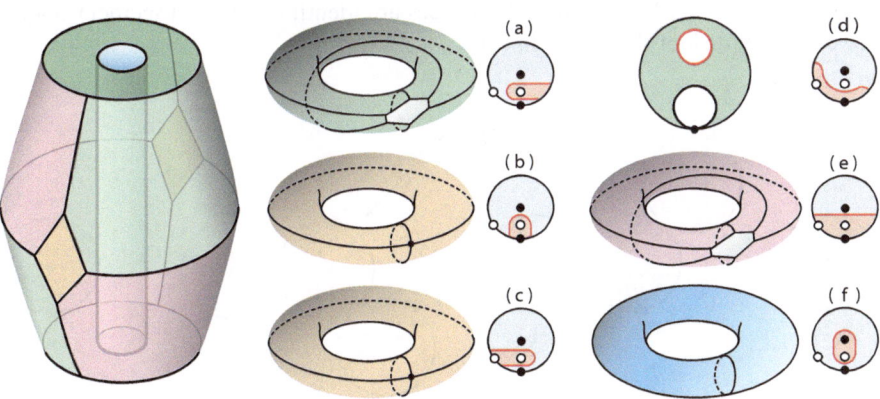

Fig. 15 $\overline{\mathscr{K}}(2,2)$ and its exceptional divisors.

5 Combinatorial results

The stratification of $\overline{\mathscr{K}}(n,m)$ based on bubble-trees leads to several inherent combinatorial structures.

Proposition 5.1. *The space $\overline{\mathscr{K}}(n,m)$ is tiled by $(m-1)!$ chambers.*

Proof. There exist $m!$ orderings of the m boundary particles on $\mathscr{D}$. By Proposition 2.1, the action of $\mathbb{P}SO_2(\mathbb{R})$ is characterized by fixing one boundary particle at ∞. The orderings of the remaining $m-1$ particles on $\partial \mathscr{D}$ represent unique equivalence classes.   □

There is a combinatorial gluing on the boundaries of these $(m-1)!$ chambers which result in $\overline{\mathscr{K}}(n,m)$, based on the following definition: A *flip* of a flat bubble in a bubble-tree is obtained by replacing it with its mirror image but preserving the remaining bubbles on the tree.

Theorem 5.2. *Two codimension k cells, each corresponding to a bubble-tree coming from a diagram of $\mathscr{D}$ with k nested arcs, are identified in $\overline{\mathscr{K}}(n,m)$ if flips along flat bubbles of one diagram result in the other.*

Proof. The group $\mathbb{P}GL_2(\mathbb{R})$ acts on flat bubbles and there is a $\mathbb{Z}_2$ component of $\mathbb{P}GL_2(\mathbb{R})$ which identifies each flat bubble with its mirror image. Since each chamber of $\overline{\mathscr{K}}(n,m)$ is identified with the cyclic ordering of the $m-1$ boundary particles, a flip identifies faces of one chamber with another.   □

Example 5.3. Figure 16(a) and (d) show the two vertices from Figure 9(b), the corners of the Kontsevich eye. Parts (b) and (c) show the bubble-trees of these cells, respectively. In $\overline{\mathscr{K}}(2,1)$, both of these trees are identified (glued together) since a *flip* of the flat disk of (b) results in (c).

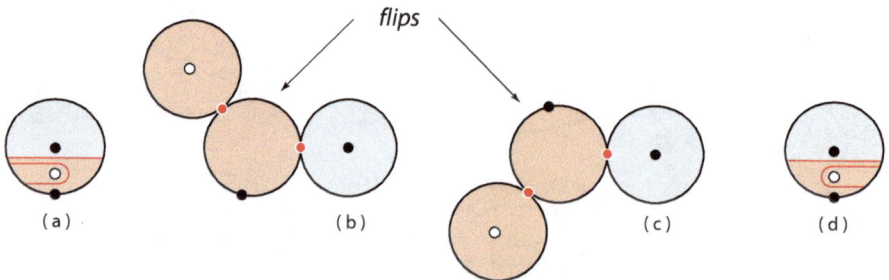

Fig. 16 Two cells of the eye identified in $\overline{\mathscr{K}}(2,1)$ by flips of flat disks.

Theorem 5.4. *The space $\overline{\mathscr{K}}(0,n)$ is isomorphic to two disjoint copies of the real moduli space $\overline{\mathscr{M}}_{0,n}(\mathbb{R})$ of curves, tiled by associahedra K_{n-1}.*

Proof. For n particles on the boundary, the action of $\mathbb{P}SL_2(\mathbb{R})$ fixes three such particles (call them $0, 1, \infty$) due to Möbius transformations. However, there are two equivalence classes of such orderings, with the three fixed particles arranged clockwise as either $0, 1, \infty$ or as $1, 0, \infty$. Each such equivalence class has particles only on the boundary, and is acted upon by flips from $\mathbb{P}GL_2(\mathbb{R})$. The result follows from Theorem 1.2. $\square$

Theorem 5.5. *The moduli space $\overline{\mathscr{K}}(1,n)$ is isomorphic to the space of cyclohedra $\overline{\mathscr{Z}}_n$.*

Proof. As $\mathbb{P}SL_2(\mathbb{R})$ fixes the interior particle and a boundary particle, $\mathscr{K}\langle 1,n \rangle$ becomes an $(m-1)$-torus, with $m-1$ labeled particles moving on the boundary of $\mathscr{D}$. Each ordering of these labels yields a simplicial chamber, all of which glue along the long diagonal of a cube, whose faces identify to form the torus. The Fulton-MacPherson compactification truncates each simplex into a cyclohedron, resulting in $\overline{\mathscr{Z}}_n$ promised by Theorem 1.4. $\square$

Remark 5.6. Figure 17(a) shows the associahedron K_4 as a tile of $\overline{\mathscr{K}}(0,5)$. The other tiles of $\overline{\mathscr{K}}(0,5)$ correspond to all ways of labeling the five boundary particles. Similarly, part (b) depicts the cyclohedron W_3 as a tile of $\overline{\mathscr{K}}(1,3)$. Compare these diagrams with Figure 1.

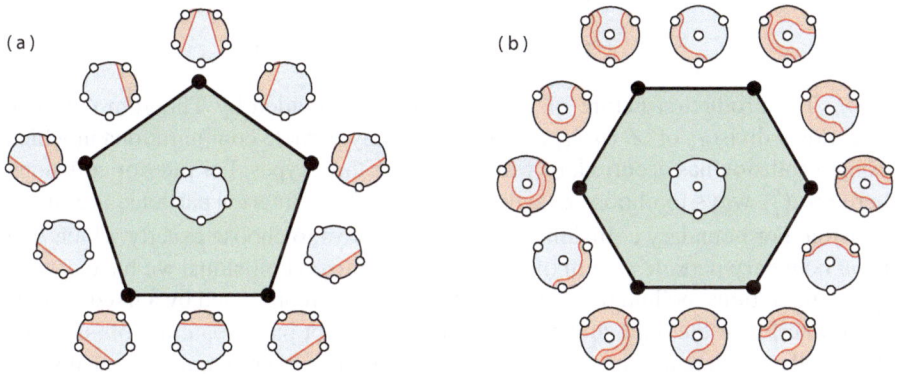

Fig. 17 (a) Associahedron K_4 of $\overline{\mathscr{K}}(0,5)$ and (b) cyclohedron W_3 of $\overline{\mathscr{K}}(1,3)$.

Example 5.7. Figure 10(b) shows two hexagonal cyclohedra W_3 tiling $\overline{\mathscr{K}}(1,3)$. Figure 18 shows the iterated truncation of the three-torus $\mathscr{K}\langle 1,4 \rangle$, yielding $\overline{\mathscr{K}}(1,4)$ tiled by six cyclohedra.

We close by enumerating the exceptional divisors (the codimension one spaces) of $\overline{\mathscr{K}}(m,n)$. With this result, one can use induction to calculate all codimension k spaces if desired.

Theorem 5.8. *Let $n, m \geq 1$. Then the exceptional divisors of $\overline{\mathscr{K}}(n,m)$ are categorized by the following classification:*

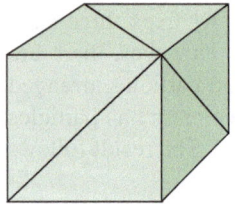 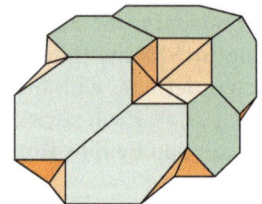 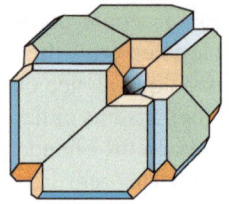

Fig. 18 Iterated blowups resulting in $\overline{\mathscr{K}}(1,4)$ tiled by six cyclohedra W_4.

1. *There are a total of $2^n - n - 1$ divisors enumerating interior collisions, with $\binom{n}{i}$ divisors where $i > 1$ interior particles collide, each topologically equivalent to*

$$\overline{\mathscr{K}}(n-i+1,m) \times \overline{\mathscr{M}}_{i+1}(\mathbb{C}).$$

2. *There are a total of $2^m - m - 1$ divisors enumerating boundary collisions, with $\binom{m}{b}$ divisors where $b \geq 2$ boundary particles collide, each topologically equivalent to*

$$\overline{\mathscr{K}}(n,m-b+1) \times \overline{\mathscr{M}}_{0,b+1}(\mathbb{R}).$$

3. *There are a total of $2^m(2^{n-1} - 1)$ divisors enumerating mixed collisions, with $\binom{n-1}{i}\binom{m}{b}$ divisors where $1 \leq i \leq n-1$ interior particles and $b \geq 0$ boundary particles collide, each topologically equivalent to*

$$\overline{\mathscr{K}}(n-i,m-b+1) \times \overline{\mathscr{K}}(i,b+1).$$

Proof. The product structure of these divisors is provided by Theorem 4.5. The exceptional divisors of $\overline{\mathscr{K}}(n,m)$ are enumerated by particle configurations in which a single collision has occurred, of which there are three types. For interior collisions, there are $\binom{n}{i}$ ways to choose exactly which i of the n interior particles are in the collision. For boundary collisions, there are $\binom{m}{b}$ ways to choose exactly which b of the m boundary particles are in the collision. For mixed collisions, we have chosen our configurations modulo the action of $\mathbb{P}SL_2(\mathbb{R})$ to be represented by a fixed particle in the interior of the disk, and so this particle may not participate in collisions on the boundary. There are then $\binom{n-1}{i}$ ways to choose i of the remaining $n-1$ interior particles and $\binom{m}{b}$ ways to choose b of the m boundary particles that are in a mixed collision. $\square$

Remark 5.9. The exceptional divisors corresponding to boundary collisions and interior collisions are in the interior of the moduli space whereas mixed collision divisors are on its boundary.

Acknowledgement We thank Eduardo Hoefel, Hiroshige Kajiura, Melissa Liu, and Cid Vipismakul for helpful conversations and clarifications, and a special thanks to Jim Stasheff for his continued encouragement. We are also grateful to Williams College and to the NSF for partially supporting this work with grant DMS-0353634.

References

1. S. Armstrong, M. Carr, S. Devadoss, E. Engler, A. Leininger, and M. Manapat, "Particle configurations and Coxeter operads", *Journal of Homotopy and Related Structures* **4** (2009) 83–109.
2. S. Axelrod and I. M. Singer, "Chern-Simons perturbation theory II", *Journal of Differential Geometry* **39** (1994) 173–213.
3. R. Bott and C. Taubes, "On the self-linking of knots", *Journal of Mathematical Physics* **35** (1994) 5247–5287.
4. M. Carr and S. Devadoss, "Coxeter complexes and graph-associahedra", *Topology and its Applications* **153** (2006) 2155–2168.
5. C. De Concini and C. Procesi, "Wonderful models of subspace arrangements", *Selecta Mathematica* **1** (1995) 459–494.
6. S. Devadoss, "Tessellations of moduli spaces and the mosaic operad", in *Homotopy Invariant Algebraic Structures*, Contemporary Mathematics **239** (1999) 91–114.
7. ——— , "A space of cyclohedra", *Discrete and Computational Geometry* **29** (2003), 61–75.
8. ——— , "Combinatorial equivalence of real moduli spaces", *Notices of the American Mathematical Society* (2004) 620–628.
9. S. Devadoss, T. Heath, and W. Vipismakul, "Deformations of bordered surfaces and convex polytopes", *Notices of the American Mathematical Society* (2011) 530–541.
10. S. Devadoss and J. Morava, "Diagonalizing the genome I", *arxiv.org/abs/1009.3224*.
11. K. Fukaya, Y-G Oh, H. Ohta, and K. Ono, "Lagrangian intersection Floer theory: anomaly and obstruction", *Kyoto Department of Mathematics* 00-17.
12. W. Fulton and R. MacPherson, "A compactification of configuration spaces", *Annals of Mathematics* **139** (1994) 183–225.
13. A. Goncharov and Y. Manin "Multiple ζ-motives and moduli spaces $\overline{\mathscr{M}}_{0,n}(\mathbb{R})$", *Compositio Mathematica* **140** (2004) 1–14.
14. E. Hoefel, "OCHA and the Swiss-cheese operad", *Journal of Homotopy and Related Structures* **4** (2009) 123–151.
15. H. Kajiura and J. Stasheff, "Open-closed homotopy algebra in mathematical physics", *Journal of Mathematical Physics* **47** (2006).
16. M. Kontsevich, "Deformation quantization of Poisson manifolds", *Letters in Mathematical Physics* **66** (2003) 157–216.
17. M. Kontsevich and Y. Manin, "Gromov-Witten classes, quantum cohomology, and enumerative geometry", *Communications in Mathematical Physics* **164** (1994) 525–562.
18. C. Lee, "The associahedron and triangulations of the n-gon", *European Journal of Combinatorics* **10** (1989) 551–560.
19. D. Mumford, J. Fogarty, and F. Kirwan, *Geometric Invariant Theory*, Springer-Verlag, New York, 1994.
20. T. Parker and J. Wolfson, "Pseudo-holomorphic maps and bubble trees", *Journal of Geometric Analysis* **3** (1993) 63–98.
21. J. Ratcliffe, *Foundations of hyperbolic manifolds*, Springer-Verlag, New York, 1994.
22. J. Stasheff, "Homotopy associativity of H-spaces", *Transactions of the American Mathematical Society* **108** (1963), 275–292.
23. D. Tamari, "Monoïdes préordonnés et chaînes de Malcev", Doctorat ès-Sciences Mathématiques Thèse de Mathématiques, Université de Paris (1951).
24. A. Voronov, "The Swiss-cheese operad", in *Homotopy Invariant Algebraic Structures*, Contemporary Mathematics **239** (1999) 365–373.

Realizing the Associahedron: Mysteries and Questions

Cesar Ceballos and Günter M. Ziegler

Abstract There are many open problems and some mysteries connected to the realizations of the associahedra as convex polytopes. In this note, we describe three – concerning special realizations with the vertices on a sphere, the space of all possible realizations, and possible realizations of the multiassociahedron.

1 Introduction

Realizing the n-dimensional associahedron as a convex polytope is a non-trivial task: You are given the *combinatorics* of a polytope, and you are supposed to produce *geometry*, namely coordinates for a correct realization – such that the vertices correspond to the triangulations of an $(n+3)$-gon, and the facets to its diagonals, and a vertex lies on a facet if and only if the triangulation uses the diagonal.

The realization problem appeared first in Tamari's thesis from 1951 [39]. It was explicitly posed by Stasheff's 1963 paper [36], and first solved somewhat "by hand": As far as we know, the n-dimensional associahedron was constructed
- 1963 by Stasheff [36] as a cellular ball,
- 1960s by Milnor for the first time as a polytope (lost),
- 1978 by Huguet & Tamari (see [18]: no proof given),
- 1984 by Haiman (unpublished, but see [15]), and finally
- 1989 by Lee (the first published realization: [23]).
Subsequently, more systematic construction methods emerged, among them
- the construction as secondary polytopes of convex $(n+3)$-gons,

Cesar Ceballos
Inst. Mathematics, FU Berlin, Arnimallee 2, 14195 Berlin, Germany,
e-mail: *ceballos@math.fu-berlin.de*

Günter M. Ziegler
Inst. Mathematics, FU Berlin, Arnimallee 2, 14195 Berlin, Germany,
e-mail: *ziegler@math.fu-berlin.de*

- the construction from cluster complexes of the root systems A_n, and
- the construction as a (weighted) Minkowski sum of faces of a simplex,

all of them described in more detail below. In recent work [7], we have discovered that the realizations produced by these three families of constructions are disjoint, and that they can be distinguished by quite remarkable, geometric properties – and moreover, that there are many more realizations that seem natural as well, including the exponentially-sized family of Hohlweg & Lange [16] [7, Sect. 4], and the even larger, Catalan-sized family of Santos [32] [7, Sect. 5].

There are many open problems and some mysteries connected to the realizations of the associahedra as convex polytopes. In this note, we describe three:

○ There are several very natural, but fundamentally different constructions of the n-dimensional associahedron, which produce disjoint parameterized families of polytopes. How do these families lie in the *realization space* (defined below) of the n-dimensional associahedron? How do they relate?

○ The associahedron constructed as the secondary polytope of $n+3$ equally-spaced points on a quadratic planar curve turns out to have all its vertices on an ellipsoid. This phenomenon extends to the permuto-associahedron and to the cyclohedron. Explain!

○ Generalization of triangulations to multitriangulations leads to multiassociahedra and, more generally, to generalized multiassociahedra. Up to now, one can show that these combinatorial objects are vertex-decomposable spheres, but (how) can they be realized as convex polytopes?

Of course this note is written with the hope to clarify the situation and to explain some observations and pieces of progress related to the problems. However, some mystery remains, and perhaps this is also natural, in view of the sentence that starts Haiman's 1984 manuscript [15]:

The associahedron is a mythical polytope

2 Realization space

As just mentioned, there are three very natural, but fundamentally different constructions of the associahedron that may be considered to be "classical" by now:

(I) as the secondary polytope of a convex $(n+3)$-gon Q by Gelfand, Kapranov & Zelevinsky [13] [14] (see also [12, Chap. 7]),

$$\mathrm{Ass}_n(Q) := \mathrm{conv}\{\sum_{i=0}^{n+2} \sum_{\sigma \in T : i \in \sigma} \mathrm{vol}(\sigma) f_i : T \text{ is a triangulation of } Q\}, \quad (1)$$

where $f_0, \ldots, f_{n+2}$ are the vertices of an $(n+2)$-simplex,

(II) via cluster complexes of the root system A_n as conjectured by Fomin & Zelevinsky [10] and constructed by Chapoton, Fomin & Zelevinsky [8],

$$\mathrm{Ass}_n(A_n) := \{x \in \mathbb{R}^{n+1} \mid x_i - x_j \leq f_{i,j} \text{ for } i - j \geq -1, \ \textstyle\sum_i x_i = 0\}$$

for suitable $f_{i,j} > 0$, and

(III) as Minkowski sums of simplices, as introduced by Postnikov in [27]

$$\mathrm{Ass}_n(\Delta_n) := \textstyle\sum_{1 \leq i < j \leq n} \alpha_{i,j} \Delta_{[i...j]},$$

for arbitrary $\alpha_{i,j} > 0$, which in various different descriptions appears in earlier work by various other authors, including Shnider & Sternberg [34], Loday [24], Rote, Santos & Streinu [31].

Some of these realizations have very striking properties, such as the vertices on a sphere (see below), or having facet normals in the root system A_n.

One would perhaps expect that "if you set the parameters right" you could get the one-and-only most beautiful realization, but a priori it is not clear, which one that would be. However, it turned out (see [7]) that these approaches yield fundamentally distinct realizations. For example, the associahedra produced as the secondary polytopes of a convex $(n+3)$-gon don't have *any* parallel facets, like the one in the figure,

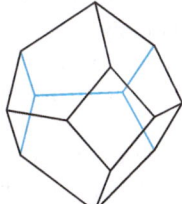

whereas the others do, typically with n pairs of parallel facets that correspond to certain pairs of intersecting diagonals:

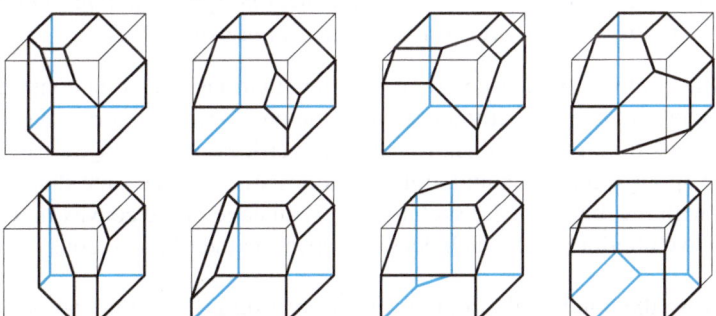

With the huge number of different realizations that are analyzed and distinguished in [7], one is led to ask a number of questions about the space of all realizations of the n-dimensional associahedron:

○ What is the structure of the space? Is it contractible (if we divide out the action of the group of affine transformations, say)? Is it even connected?
○ Do the constructions of associahedra that we know cover a large/typical part of the realization space?

○ Is there any connection between the realizations? Could we get some types as a deformation/limit of other types?

The space of all realizations of (a combinatorial type of) a convex polytope is known to be a semialgebraic set defined over $\mathbb{Z}$. There are various possible definitions, which differ somewhat; if we do not identify affinely equivalent realizations and decide to only consider realizations with the origin in the interior, then the set of all such realizations – called the *realization space* – for an n-dimensional polytope P with N facets can be identified with the semialgebraic set

$$\{C \in \mathbb{R}^{n \times N} : \text{ the inequality system } C^t x \leq 1 \text{ defines a realization of } P\}$$

We refer to Richter-Gebert [29] for an extensive treatment of realization spaces of polytopes and to [30] for an introduction. The realization space of a simple n-dimensional polytope with $N = \frac{1}{2}n(n+3)$ facets can be seen (if we again require the origin to lie in the interior, and do not divide out a group action) as an open subset of $\mathbb{R}^{n \times N}$. It is known that realization spaces of some simple polytopes are disconnected – there are sporadic examples in dimension 4 (see [2]) and systematic constructions for high-dimensional simple polytopes [19]. But for the associahedron not much is known beyond dimension 3, where Steinitz proved in 1922 ([37]; see [29]) that the realization space of any 3-dimensional polytope after dividing out the action of the affine group is a topological ball of dimension $f_1 - 6$.

Here is one observation that needs to be followed up: The secondary polytope construction produces a realization of the n-dimensional associahedron from any given convex $(n+3)$-gon. In other words, we get an associahedron from any convex configuration of $n+3$ points in the plane. The converse to this turns out to be *false*: "convex position" is sufficient, but not necessary for getting an associahedron.

Proposition 2.1. *The secondary polytope of any configuration of $n+3$ points in the plane, which consists of all the vertices of a convex polygon and at most one point in the relative interior of any edge, is an n-dimensional associahedron.*

Proof. The combinatorial structure of the triangulations of a point configuration P with these properties is exactly the same as the one for a configuration Q of points in convex position. If we cyclically label the vertices of P and Q, a triangulation of P corresponds to the triangulation of Q consisting of the same diagonals of P together with the diagonals i, j such that i, j is an edge of the convex hull of P that has a relative interior point which does no appear on the triangulation.   □

Of course the point configurations of Proposition 2.1 are limit cases of strictly-convex configurations, and thus the realizations of the associahedron obtained from them are deformations of secondary polytopes of convex $(n+3)$-gons. However, the deformations do not share all their properties: Indeed, the following configuration

of 6 points

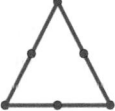

produces an associahedron that has three pairs of parallel facets – which you can never get from a hexagon [7, Thm. 3.5]. So, do we get associahedra from cluster algebras, or associahedra from weighted Minkowski sums, as limit cases of secondary polytopes? (For dimension larger than 3 we cannot expect that; cf. [7, Remark 3.6].)

And which more general, non-convex planar point configurations could still produce associahedra?

3 Vertices on a sphere

The associahedron constructed as the secondary polytope of $n+3$ equally-spaced points on a quadratic planar curve turns out to have all its vertices on an ellipsoid, or in suitable (and still natural) coordinates even on a sphere:

Theorem 3.1. *Let $p, q \in \mathbb{R}[t]$ be quadratic polynomials such that the convex curve $C = \{(p(t), q(t)) \mid t \in \mathbb{R}\}$ is not a line (that is, such that $\{p, q, 1\}$ are linearly independent), and let $v_0, \ldots, v_{n+2}$ be equally-spaced points on C, that is, $v_i := (p(ai+b), q(ai+b))$ for $a, b \in \mathbb{R}$, $a \neq 0$.*

Then the secondary polytope of $Q := conv\{v_0, \ldots, v_{n+2}\}$, constructed according to (1) with $f_i := e_1 + \cdots + e_i$, has all its vertices on a sphere around the origin.

Proof. The description/construction of the secondary polytope of a convex polygon as given here is motivated by the more general setting of *fiber polytopes* provided by Billera & Sturmfels [1] [40, Lect. 9]. Theorem 3.1 was observed for the special case $(p(t), q(t)) = (t, t^2)$ and $a = 1$ by Reiner & Ziegler [28]. The more general Theorem 3.1 follows from this by simple functoriality properties of the fiber polytopes: If a polytope projection $\Delta_{n+2} \to Q$ is composed with an affine transformation of the polygon Q, the fiber polytope $\Sigma(\Delta_{n+2}, Q)$ changes only by a multiplication by a constant factor. An affine transformation applied to the simplex Δ_{n+2} induces the same transformation on the fiber polytope. □

For the 1994 paper [28], the sphericity was discovered by chance, and established by a simple algebraic verification (with computer algebra support), establishing that the length of the *GKZ vector* (named after Gelfand, Kapranov and Zelevinsky)

$$\text{GKZ}(T) := \sum_{i=0}^{n+2} \sum_{\sigma \in T : i \in \sigma} \text{vol}(\sigma) f_i$$

does not change under flips $T \to T'$. However, a "geometric explanation" was lacking then, and is still lacking now. The true reason is still a mystery.

This is even more deplorable as the phenomenon occurs in other instances as well. First, it does quite obviously extend to the realization of the "permuto-associahedron" that had been combinatorially described by Kapranov [22]: Indeed, the length of the GKZ vector does not change under permutation of coordinates.

Moreover calculations (Ziegler 1994, unpublished) show there is a "secondary polytope like" construction of the Bott–Taubes "cyclohedron" [4] (also known as the "type B" generalized associahedron) along quite similar lines, which again shows the same phenomenon: It produces integer coordinates for the cyclohedron, with all vertices on a sphere. This can be verified in examples using, e.g., polymake by Gawrilow & Joswig [11], it can be proved algebraically, but *why* is it true?

4 Realizing the multiassociahedron

The boundary complex of the *dual associahedron* is a simplicial complex whose vertices correspond to diagonals of a convex polygon, and whose faces correspond to subsets of non-crossing diagonals. This complex can be naturally generalized to a beautiful family of simplicial complexes with remarkable combinatorial properties. Members of this family are called *simplicial multiassociahedra*.

Let $k \geq 1$ and $m \geq 2k + 1$ be two positive integers. We say that a set of $k + 1$ diagonals of a convex m-gon forms a $(k + 1)$-*crossing* if all the diagonals in this set are pairwise crossing. A diagonal is called k-*relevant* if it is contained in some $(k + 1)$-crossing, that is, if there are at least k vertices of the m-gon on each side of the diagonal. The *simplicial multiassociahedron* $\Delta_{m,k}$ is the simplicial complex of $(k + 1)$-crossing-free sets of k-relevant diagonals of a convex m-gon.

For example, the 2-relevant diagonals of a convex 6-gon (labeled as in the following figure, left picture) are $14, 25$ and 36, and the simplicial multiassociahedron $\Delta_{6,2}$ is the boundary complex of a triangle (right). The set of diagonals $\{14, 25, 36\}$ is not a face because they form a 3-crossing.

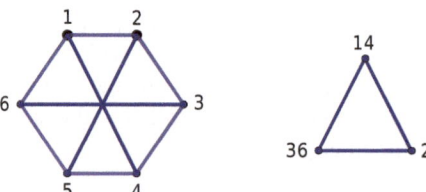

The vertices of the multiassociahedron $\Delta_{m,k}$ are given by k-relevant diagonals of the m-gon, and the facets correspond to k-*triangulations*, that is, to maximal subsets of diagonals that do not contain any $(k + 1)$-crossing. For the case of $k = 1$, the multiassociahedron is the simplicial complex of non-crossing sets of diagonals, which coincides with the boundary complex of the dual associahedron.

The combinatorial structure of the multiassociahedron has been studied by several authors. Apparently, it first appeared in work of Capoyleas & Pach [5], who showed that the maximal number of diagonals in a $(k+1)$-crossing-free set is equal to $k(2m-2k-1)$. Nakamigawa [25] introduced the flip operation on k-triangulations and proved that the flip graph is connected. Dress, Koolen & Moulton [9] obtained a reformulation of the Capoyleas–Pach result, and in particular proved that all maximal $(k+1)$-crossing-free sets of diagonals have the same number of diagonals. The results of Nakamigawa and Dress–Koolen–Moulton imply that the multiassociahedron $\Delta_{m,k}$ is a pure simplicial complex of dimension $k(m-2k-1)-1$. A more recent approach for the study of k-triangulations, using star polygons, was given by Pilaud & Santos [26]. In 2003, Jonsson [20] showed that the multiassociahedron is a piecewise linear sphere. Then, he found an explicit $k \times k$ determinantal formula of Catalan numbers counting the number of k-triangulations [21]. Additionally to the result of Jonsson about the multiassociahedron being a topological sphere, Stump [38] proved that it is a vertex-decomposable, and thus in particular shellable, simplicial sphere. See also the results by Serrano & Stump [33].

All these results suggest that the multiassociahedron $\Delta_{m,k}$ could be realized as the boundary complex of a simplicial polytope of dimension $k(m-2k-1)$. However, while for the classical associahedron we have many different construction methods (see above), all the natural approaches seem to fail for the multiassociahedron. The list of cases for which the multiassociahedron is known to be polytopal is the following. The multiassociahedron $\Delta_{m,k}$ is the boundary complex of a:

○ dual $(m-3)$-dimensional associahedron, if $k=1$;
○ point, if $m=2k+1$;
○ k-dimensional simplex, if $m=2k+2$;
○ $2k$-dimensional cyclic polytope on $2k+3$ vertices, if $m=2k+3$ [26];
○ 6-dimensional simplicial polytope, if $m=8$ and $k=2$ [3].

Currently, the smallest open case is for $m=9$ and $k=2$. Is there a simplicial polytope of dimension 8 and f-vector $(18,153,732,2115,3762,4026,2376,594)$ which realizes the multiassociahedron $\Delta_{9,2}$?

Recently, the multiassociahedron has been generalized to a family of vertex-decomposable simplicial spheres for finite Coxeter groups, by Ceballos, Labbé & Stump [6]. They suggest a family of simple polytopes called *generalized multiassociahedra*. This family includes the generalized associahedra [8] [17], and the (simple) multiassociahedra of types A and B (see [35] for the type B description). However, no polytopal realizations of generalized multiassociahedra have been found except for the Coxeter groups of type $I_2(m)$, and for some particular cases in other types. The (simple) generalized multiassociahedra of type $I_2(m)$ are given by the duals of all even-dimensional cyclic polytopes [6]. Are there polytopal realizations for generalized multiassociahedra in general?

Acknowledgement The first author was supported by DFG via the Research Training Group *Methods for Discrete Structures* and by Berlin Mathematical School. The research leading to these results has received funding from the European Research Council under the European Union's Seventh Framework Programme (FP7/2007-2013) / ERC Grant agreement no. 247029-SDModels. We are grateful to Carsten Lange and Paco Santos for many interesting discussions.

References

1. L.J. Billera and B. Sturmfels, "Fiber polytopes", *Annals of Math.* **135** (1992) 527–549.
2. J. Bokowski and A. Guedes de Oliveira, "Simplicial convex 4-polytopes do not have the isotopy property", *Portugaliae Math.* **47** (1990) 309–318.
3. J. Bokowski and V. Pilaud, "On symmetric realizations of the simplicial complex of 3-crossing-free sets of diagonals of the octagon", in *Proc. 21st Canadian Conf. Computat. Geometry (CCCG2009)*, 2009, 41–44.
4. R. Bott and C. Taubes, "On the self-linking of knots", *J. Mathematical Physics* **35** (1994) 5247–5287.
5. V. Capoyleas and J. Pach, "A Turán-type theorem on chords of a convex polygon", *J. Combinatorial Theory, Ser. B* **56** (1992) 9–15.
6. C. Ceballos, J. Labbé, and C. Stump, "Subword complexes, cluster complexes, and generalized multi-associahedra", *arxiv.org/abs/1108.1776*.
7. C. Ceballos, F. Santos, and G.M. Ziegler, "Many non-equivalent realizations of the associahedron", Preprint, September 2011, 28 pages; *arxiv.org/abs/1109.5544*.
8. F. Chapoton, S. Fomin, and A. Zelevinsky, "Polytopal realizations of generalized associahedra", *Canad. Math. Bull.* **45** (2002) 537–566.
9. A.W.M. Dress, J.H. Koolen, and V.L. Moulton, "On line arrangements in the hyperbolic plane", *European J. Combinatorics* **23** (2002) 549–557.
10. S. Fomin and A. Zelevinsky, "*Y*-systems and generalized associahedra", *Annals of Math.* **158** (2003) 977–1018.
11. E. Gawrilow and M. Joswig, "polymake: a framework for analyzing convex polytopes", in *Polytopes – Combinatorics and Computation*, G. Kalai and G.M. Ziegler, eds., Birkhäuser, 2000, 43–74.
12. I.M. Gelfand, M.M. Kapranov, and A.V. Zelevinsky, *Discriminants, Resultants, and Multidimensional Determinants*, Birkhäuser, Boston, 1994.
13. I.M. Gel'fand, A.V. Zelevinskiĭ, and M.M. Kapranov, "Newton polytopes of principal a-determinants", *Soviet Math. Doklady* **40** (1990) 278–281.
14. _____, "Discriminants of polynomials in several variables and triangulations of Newton polyhedra", *Leningrad Math. J.* **2** (1991) 449–505.
15. M. Haiman, "Constructing the associahedron", unpublished manuscript, MIT 1984, 11 pages; *math.berkeley.edu/~mhaiman/ftp/assoc/manuscript.pdf*.
16. C. Hohlweg and C.E.M.C. Lange, "Realizations of the associahedron and cyclohedron", *Discrete Comput. Geometry* **37** (2007) 517–543.
17. C. Hohlweg, C.E.M.C. Lange, and H. Thomas, "Permutahedra and generalized associahedra", *Advances in Math.* **226** (2011) 608–640.
18. D. Huguet and D. Tamari, "La structure polyédrale des complexes de parenthésages", *J. Combinatorics, Information & System Sciences* **3** (1978) 69–81.
19. B. Jaggi, P. Mani-Levitska, B. Sturmfels, and N. White, "Uniform oriented matroids without the isotopy property", *Discrete Comput. Geometry* **4** (1989) 97–100.
20. J. Jonsson, "Generalized triangulations of the *n*-gon", Report from Oberwolfach Workshop "Topological and Geometric Combinatorics", April 2003, p. 11, *www.mfo.de/occasion/0315/www_view*.
21. _____, "Generalized triangulations and diagonal-free subsets of stack polyominoes", *J. Combinatorial Theory, Ser. A* **112** (2005) 117–142.
22. M.M. Kapranov, "Permuto-associahedron, Mac Lane coherence theorem and asymptotic zones for the KZ equation", *J. Pure and Applied Algebra* **85** (1993) 119–142.
23. C.W. Lee, "The associahedron and triangulations of the *n*-gon", *European J. Combinatorics* **10** (1989) 551–560.
24. J.L. Loday, "Realization of the Stasheff polytope", *Arch. Math.* **83** (2004) 267–278.
25. T. Nakamigawa, "A generalization of diagonal flips in a convex polygon", *Theoretical Computer Science* **235** (2000) 271–282.

26. V. Pilaud and F. Santos, "Multitriangulations as complexes of star polygons", *Discrete Comput. Geometry* **41** (2009) 284–317.
27. A. Postnikov, "Permutohedra, associahedra, and beyond", *International Mathematics Research Notices* **2009** (2009) 1026–1106.
28. V. Reiner and G.M. Ziegler, "Coxeter-associahedra", *Mathematika* **41** (1994) 364–393.
29. J. Richter-Gebert, *Realization Spaces of Polytopes*, Lecture Notes in Mathematics, vol. 1643, Springer-Verlag, Berlin Heidelberg, 1996.
30. J. Richter-Gebert and G.M. Ziegler, "Realization spaces of 4-polytopes are universal", *Bulletin of the American Mathematical Society* **32** (1995) 403–412.
31. G. Rote, F. Santos, and I. Streinu, "Expansive motions and the polytope of pointed pseudo-triangulations", in *Discrete and Computational Geometry*, Algorithms and Combinatorics, vol. 25, Springer, Berlin, 2003, 699–736.
32. F. Santos, "Catalan many associahedra", lecture at the "Kolloquium über Kombinatorik (KolKom04)", Magdeburg, November 2004.
33. L. Serrano and C. Stump, "Maximal fillings of moon polyominoes, simplicial complexes, and Schubert polynomials", *arxiv.org/abs/1009.4690*.
34. S. Shnider and S. Sternberg, *Quantum groups. From coalgebras to Drinfel'd algebras: A guided tour*, Graduate Texts in Math. Physics, II, International Press, Cambridge, MA, 1993.
35. D. Soll and V. Welker, "Type-B generalized triangulations and determinantal ideals", *Discrete Math.* **309** (2009) 2782–2797.
36. J.D. Stasheff, "Homotopy associativity of *H*-spaces", *Transactions Amer. Math. Soc.* **108** (1963) 275–292.
37. E. Steinitz, "Polyeder und Raumeinteilungen", in *Encyklopädie der mathematischen Wissenschaften, Anwendungen, Dritter Band: Geometrie,* III.1.2., *Heft* 9, *Kapitel* III A B 12, W.F. Meyer and H. Mohrmann, eds., B.G. Teubner, Leipzig, 1922, 1–139.
38. C. Stump, "A new perspective on *k*-triangulations", *J. Combinatorial Theory, Ser. A* **118** (2011) 1794–1800.
39. D. Tamari, "Monoïdes préordonnés et chaînes de Malcev", Doctoral Thesis, Paris 1951, 81 pages.
40. G.M. Ziegler, *Lectures on Polytopes*, Graduate Texts in Mathematics, vol. 152, Springer-Verlag, New York, 1995, Revised edition, 1998; seventh updated printing 2007.

Permutahedra and Associahedra
Generalized associahedra from the geometry of finite reflection groups

Christophe Hohlweg

Abstract Permutahedra are a class of convex polytopes arising naturally from the study of finite reflection groups, while generalized associahedra are a class of polytopes indexed by finite reflection groups. We present the intimate links those two classes of polytopes share.

1 Introduction

The purpose of this survey is to explain the realization of generalized associahedra and Cambrian lattices (which are generalizations of the Tamari lattice) from the geometrical point of view of finite reflection groups.

The story of the associahedron starts in 1963 when J. Stasheff [30], while studying homotopy theory of loop spaces, constructed a cell complex whose vertices correspond to the possible compositions of n binary operations. This cell complex turns out to be the boundary complex of a convex polytope: *the associahedron*. Long forgotten was that D. Tamari considered in his 1951 thesis a lattice which graph is the graph of the associahedron, and for which he had a realization in dimension 3 (see in this regard J.-L. Loday's text in this volume [16, §8]).

One of the easiest ways to realize an associahedron is to cut out a standard simplex. It turns out that by cutting out this associahedron, we obtain the *classical permutahedron*. The permutahedron is a polytope that arises from the symmetric group seen as a finite reflection group. This construction owed to S. Shnider and S. Sternberg [28] (see also [27]), and later completed by J.-L. Loday [15], builds a bridge between the classical permutahedron and the associahedron. It carries many combinatorial and geometrical properties: for instance, this transformation maps the weak order on the symmetric group to the Tamari lattice.

Christophe Hohlweg
Département de mathématiques, Université du Québec, Montréal, Case postale 8888, succ. Centre-ville Montréal (Québec), H3C 3P8 Canada, e-mail: *hohlweg.christophe@uqam.ca*

A similar object, *the cyclohedron*, was later discovered by R. Bott and C. Taubes in 1994 [4] in connection with knot theory. Realizations of the cyclohedron were given by M. Markl [17], R. Simion [29] and V. Reiner [26], but none of these realizations exhibit a link with the symmetric group or other finite reflection groups. In 2003, S. Fomin and A. Zelevinsky [8] discovered, in their study of finite type cluster algebras, a family of polytopes, indexed by finite reflection groups, that contains the associahedron and the cyclohedron. These *generalized associahedra* were first realized by F. Chapoton, S. Fomin and A. Zelevinsky [7] but still, these realizations were not obtained from a permutahedron of the corresponding finite reflection group.

In 2007, C. Hohlweg and C. Lange [9], and subsequently with H. Thomas [10], described many realizations of *generalized associahedra*, that are obtained by 're-moving some facets' from the permutahedron of the corresponding finite reflection group (see Figure 1 below): we start from a finite reflection group W, a special ordering of the simple reflections and a permutahedron for W and end with a realization of a generalized associahedra. This way of realizing generalized associahedra has many benefits: it maps the weak order on W to Cambrian lattices, the vertices of these realizations are labeled by clusters, and their normal fans are Cambrian fans (from N. Reading's Cambrian lattices and fans, see [24]). The spinal cord of this construction is the Coxeter singletons that allow to pinpoint nicely the facets of the permutahedron that have to be removed.

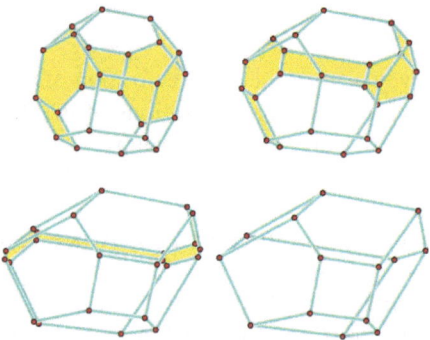

Fig. 1 The process of removing facets from a permutahedron in order to obtain a generalized associahedron, as shown in [9]. The yellow facets represent the facets in the process of being removed.

The first part of this survey is dedicated to permutahedra of finite reflection groups, and how they encode important data about the group. The second part is dedicated to realizations of generalized associahedra from a given permutahedron and how they result in a geometrical construction of Cambrian lattices and fans. Along the way, we give numerous examples and figures, and discuss open problems and further developments on the subject.

For more details on polytopes and fans, we refer to the book by G.M. Ziegler [31], from which we use the general notations. A nice presentation of finite reflection groups can be found in the book of J. Humphrey [12].

2 Permutahedra and finite reflection groups

We consider a finite-dimensional $\mathbb{R}$-Euclidean space $(V, \langle \cdot, \cdot \rangle)$.

2.1 Finite reflection groups

A *finite reflection group* is a finite subgroup of the orthogonal group $O(V)$ generated by reflections. A *reflection*[1] can be defined relative to the hyperplane it fixes pointwise, or by a normal vector to this hyperplane. Let us fix some notation: if H is an hyperplane in V and $\alpha \in V$ is a normal vector to H, the reflection s_α is the unique linear isometry which fixes $H = \alpha^\perp$ pointwise and maps α to $-\alpha$. A general formula for $s_\alpha(v)$, for a vector $v \in V$, follows:

$$s_\alpha(v) = v - 2 \frac{\langle v, \alpha \rangle}{\langle \alpha, \alpha \rangle} \alpha.$$

Example 2.1. The basic example of finite reflection groups are dihedral groups (see Figure 2). Take a regular n-gon P in the affine plane $\mathbb{R}^2$ centered in $O = (0,0)$; the symmetry group of P is the dihedral group $\mathscr{D}_n$, which is generated by the reflections it contains. Each reflection of $\mathscr{D}_n$ is determined by an axis of symmetry of P; that is, a line passing through a vertex A and the point O, or the line passing through the middle of an edge and the point O.

Example 2.2. Symmetric groups are certainly the most studied finite reflection groups, since they enjoy a particular and very detailed combinatorial representation. They appear geometrically as isometry groups of standard regular simplexes. Let us describe this representation. We consider $V = \mathbb{R}^n$, together with its canonical basis $\mathscr{B} = \{e_1, \ldots, e_n\}$ on which the symmetric group S_n acts linearly by permutation of the coordinates:

$$\sigma((x_1, \ldots, x_n)) = (x_{\sigma(1)}, \ldots, x_{\sigma(n)}).$$

The transposition $\tau_{ij} = (i \, j)$ ($1 \le i < j \le n$) acts therefore as the orthogonal reflection that fixes pointwise the hyperplane $H_{i,j} = \{(x_1, \ldots, x_n) \in \mathbb{R}^n \,|\, x_i = x_j\}$ and maps $\alpha_{i,j} := e_j - e_i$ to $-\alpha_{i,j} = e_i - e_j$. Since S_n is generated by transpositions, S_n is a finite reflection group in $O(\mathbb{R}^n)$. It is well known that this representation of S_n, as a

---

[1] We consider only orthogonal reflections in this text.

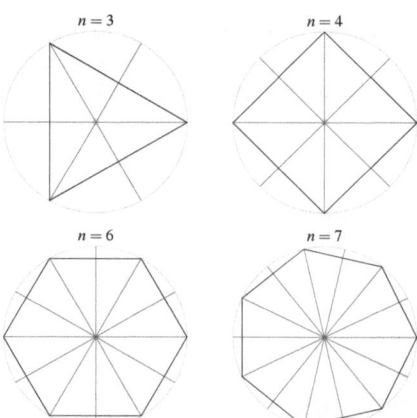

Fig. 2 The triangle, the square, the hexagon and the heptagon with their axes of symmetry. Their isometry groups are reflection groups: the groups $\mathscr{D}_3$, $\mathscr{D}_4$, $\mathscr{D}_6$ and $\mathscr{D}_7$, respectively.

subgroup of $O(\mathbb{R}^n)$, is faithful, but not irreducible, since S_n fixes the line spanned by the vector $(1,1,\ldots,1)$.

Example 2.3. Another instance of well-studied finite reflection groups are the *hyperoctahedral groups*. Hyperoctahedral groups also enjoy a particular and very detailed combinatorial representation. We consider $V = \mathbb{R}^n$, together with its canonical basis $\mathscr{B} = \{e_1,\ldots,e_n\}$. Consider the reflection $s_0 : e_1 \mapsto -e_1$ and recall the notations of Example 2.2. The group W_n generated by the reflection s_0 together with the reflections $\tau_{i,j}$ is a finite reflection group called *a hyperoctahedral group*, which contains S_n as a reflection subgroup. For more on the special combinatorics of hyperoctahedral groups, see the book by A. Björner and F. Brenti [2].

Example 2.4. A last and interesting example is the isometry group of the dodecahedron. The *regular dodecahedron* is a 3-dimensional convex polytope composed of 12 regular pentagonal faces. Its isometry group is denoted[2] by $W(H_3)$. This group is a reflection group spanned by the reflections associated to the planes of symmetry of the dodecahedron.

2.2 Permutahedra as $\mathscr{V}$-polytopes

A natural way to study a given reflection group W in $O(V)$ is to consider the W-orbit of a point $a \in V$. Here V is endowed with its natural structure of affine Euclidean space. To ensure we get all the information we want, we need to choose this point generically: $a \in V$ is *generic* if it is not fixed by any reflections in W, or equivalently,

---

[2] The notation $W(H_3)$ refers to the classification of finite reflection groups, see Remark 2.14.

if a is not a point in a hyperplane corresponding to a reflection of W. Such a choice of a point a is always possible since W is finite.

The idea behind *permutahedra* is to study W with tools from polytope theory. Instead of considering only the W-orbit of a, we consider the polytope obtained as the convex hull of this orbit.

Definition 2.5. Let a be a generic point of V. The *Permutahedron* $\mathrm{Perm}^a(W)$ is the $\mathscr{V}$-polytope[3] obtained as the convex hull of the W-orbit of a:

$$\mathrm{Perm}^a(W) = \mathrm{conv}\left\{w(a) \mid w \in W\right\}.$$

This class of polytopes is called *Coxeter permutahedra* and is sometimes referred to by W-permutahedra in the literature. As we will see, the combinatorics of $\mathrm{Perm}^a(W)$ does not depend of the choice of a, as long as this point is generic.

Since W is finite, the W-orbit of a is finite and $\mathrm{Perm}^a(W)$ is a polytope with a rich structure of faces. Before studying $\mathrm{Perm}^a(W)$ in more detail, let us cover some examples.

Example 2.6. Permutahedra of dimension 2 arise from dihedral groups[4]: a permutahedron for $\mathscr{D}_n$ is a $2n$-gon.

For instance, take $W = \mathscr{D}_4$, the symmetry group of a square (see Figure 3). The axes of symmetries, which correspond to the reflections in W, are the diagonals of the square and the lines passing through the middle of two opposite edges. We pick a point a that is not located in these lines. The red points represent the W-orbit of a, and $\mathrm{Perm}^a(W)$ is the pale red convex octagon. Observe that the number of vertices of $\mathrm{Perm}^a(W)$ is equal to $8 = |W|$, the cardinality of W.

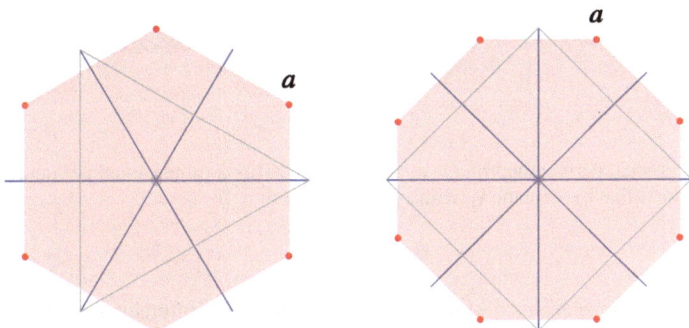

Fig. 3 The polygon on the left is a permutahedron of the dihedral group $\mathscr{D}_3$ and the polygon on the right is a permutahedron of the dihedral group $\mathscr{D}_4$

---

[3] The polytope is given as the convex hull of a set of points.

[4] The dimension of a polytope is the dimension of the affine space it spans.

Example 2.7. **The classical permutahedron**

The well-studied *classical permutahedron* is provided in the framework of symmetric groups (see for instance [31, 9] or [15, §2.2]). The classical permutahedron[5] Π_{n-1} is defined as the convex hull of all the permutations of the vector $a = (1, 2, \dots, n) \in \mathbb{R}^n$:

$$\Pi_{n-1} = \operatorname{conv} \{(\sigma(1), \dots, \sigma(n)) \in \mathbb{R}^n \mid \sigma \in S_n\}.$$

It is a $n-1$-dimensional simple[6] convex polytope which lives in the affine hyperplane:

$$V_a = \left\{ (x_1, \dots, x_n) \in \mathbb{R}^n \;\middle|\; \sum_{i=1}^n x_i = \frac{n(n+1)}{2} \right\}.$$

The vertices are naturally labeled by permutations of S_n by denoting $M(\sigma) = (\sigma(1), \dots, \sigma(n))$. The classical permutahedron Π_2 is shown in Figure 4. We observe that Π_2 is a permutahedron for the dihedral group $\mathcal{D}_3$, since S_3 acts as $\mathcal{D}_3$ on the affine hyperplane V_a.

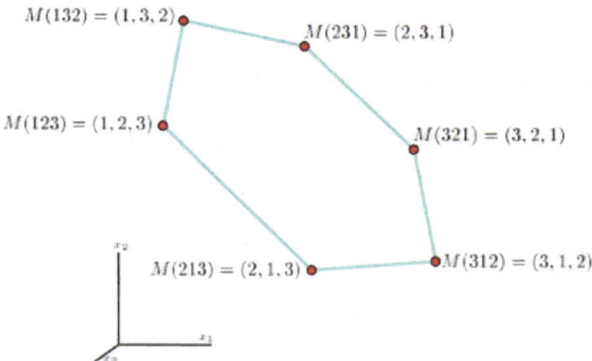

Fig. 4 The classical permutahedron Π_2 as shown in [9]. This 2-dimensional permutahedron in $\mathbb{R}^3$ is shown in the affine hyperplane V_a with $x_1 + x_2 + x_3 = 6$.

Example 2.8. There are two kinds of permutahedra of dimension 3. The first kind is obtained in $\mathbb{R}^3$ from permutahedra of dimension 2 by considering the isometry group of a regular polygonal prism: take a n-gon P in $\mathbb{R}^2 = \operatorname{span} \{e_1, e_2\}$; the convex hull of P and of $P + e_3$ is a *regular n-gonal prism*. Its isometry group W is isomorphic to $\mathcal{D}_n \times \mathcal{D}_2$ and $\operatorname{Perm}^a(W)$ is a $2n$-gonal prism.

---

[5] We use here the notation of G.M. Ziegler [31, Example 0.10].

[6] A polytope is d-dimensional simple if any vertex is contained in precisely d facets. This property is very strong: the face lattice of a simple polytope is completely determined by looking at its vertices and edges (see for instance [31, §3.4]).

The second kind of permutahedra of dimension 3 arises from the symmetric group S_4, the hyperoctahedral group W_3' and the isometry group of the dodecahedron $W(H_3)$. Examples are shown in Figure 5.

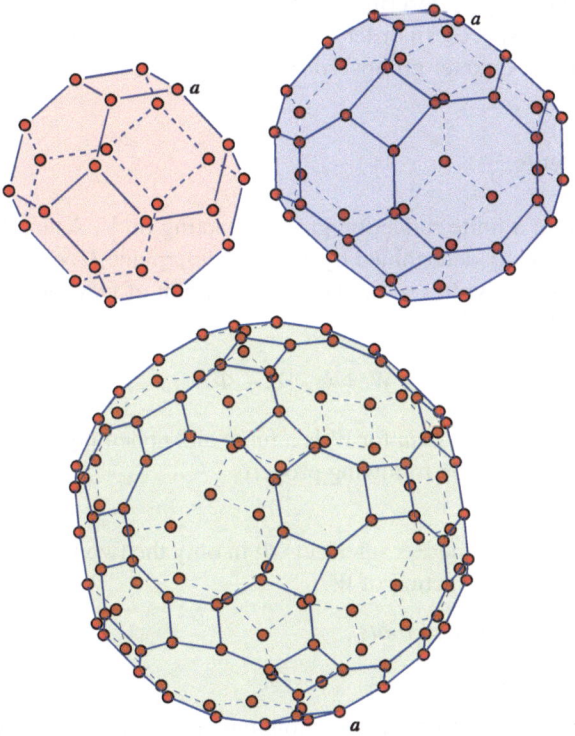

Fig. 5 Permutahedra of dimension 3 From the top left to the bottom: a permutahedron for the symmetric group S_4, a permutahedron for the hyperoctahedral group W_3 and a permutahedron for $W(H_3)$.

In both these examples, we can observe that the number of vertices of the permutahedron is the cardinality of the finite reflection group W. This observation remains true in general.

Theorem 2.9. *The W-orbit of the generic point a in V is the vertex set of* $\mathrm{Perm}^a(W)$. *In particular* $|\mathrm{vert}(\mathrm{Perm}^a(W))| = |W|$.

We observed in Figure 3 that the reflection hyperplanes of a dihedral group $W = \mathscr{D}_m$, which are lines in this case, cut the space into precisely $2m = |W|$ connected components which are polyhedral cones. In each of these cones lies a unique element of the W-orbit of a. Moreover, we observe that each edge of the permutahedron is directed by a normal vector to one of the hyperplanes of reflection. A similar phenomenon can be observed for any finite reflection group. Looking at normal vectors of the reflecting hyperplanes of W is the key to a thorough study of permutahedra, which will imply the statement in Theorem 2.9.

2.3 Root systems and permutahedra as $\mathscr{H}$-polytopes

We just described permutahedra as a $\mathscr{V}$-polytope, that is, as the convex hull of a set of points. Another way to describe a polytope is as an $\mathscr{H}$-polytope, that is, as the intersection of half-spaces. In order to do so, we introduce another very important tool related to finite Coxeter groups: *root systems*.

2.3.1 Root systems

We consider again a finite reflection group W acting on V. As explained before, a reflection s is uniquely determined by a given hyperplane H, or by a given normal vector α of H and we write $s = s_\alpha$. This second point of view brings us to consider finite subsets Φ of V such that

$$W = \langle s_\alpha \, | \, \alpha \in \Phi \rangle.$$

Definition 2.10. A *root system* for W is a finite and nonempty subset Φ of nonzero vectors of V that enjoys the following property:

1. $W = \langle s_\alpha \, | \, \alpha \in \Phi \rangle$;
2. for any $\alpha \in \Phi$, the line $\mathbb{R}\alpha$ intersects Φ in only the two vectors $-\alpha$ and α;
3. Φ is stable under the action of W.

The elements of Φ are called *roots*.

Remark 2.11.

1. Roots are normal vectors for reflection hyperplanes associated to W.
2. Root systems exist for all finite reflection groups. Indeed, the set

$$\Phi := \{\pm\alpha \, | \, s_\alpha \in W \text{ and } \langle \alpha, \alpha \rangle = 1\}$$

obviously verifies the two first properties of Definition 2.10. Moreover, it is not difficult to check that if α, β are two nonzero vectors of V then $s_\alpha s_\beta s_\alpha = s_{s_\alpha(\beta)}$. Therefore, Φ is stable under the action of W. However it is important to note that roots do not have to have the same length in general: see [12] for more details.

Example 2.12. Consider $\mathbb{R}^2$ with its canonical basis e_1, e_2. For the dihedral group $\mathscr{D}_3$, there are three axes of symmetries which are directed by vectors e_1, $e_1 + \sqrt{3}e_2$ and $-e_1 + \sqrt{3}e_2$. Let us set $\alpha_1 = \sqrt{3}e_1 + e_2$ and $\alpha_2 = -\sqrt{3}e_1 + e_2$, then a root system for $\mathscr{D}_3$ is

$$\Phi = \{\pm\alpha_1, \pm\alpha_2, \pm(\alpha_1 + \alpha_2)\}.$$

Note that the roots all have the same length (see Figure 6).

Consider now the dihedral group $\mathscr{D}_4$. Since the axes of symmetries are pairwise orthogonal, they are all directed by normal vectors corresponding to the reflections of $\mathscr{D}_4$. Therefore, a root system for $\mathscr{D}_4$ is

$$\Phi = \{\pm e_1, \pm e_2, \pm(e_1 + e_2), \pm(e_2 - e_1)\}.$$

Note that the roots do not all have the same length (illustrated by two colors in Figure 6).

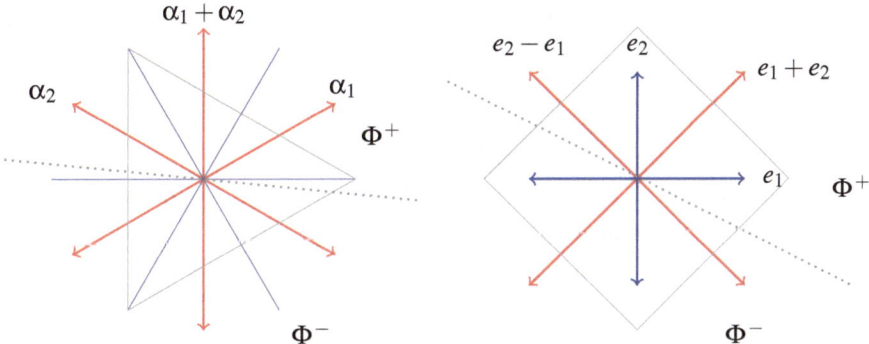

Fig. 6 The picture on the left is a root system for the dihedral group $\mathscr{D}_3$ and the picture on the right is a root system for the dihedral group $\mathscr{D}_4$.

Thanks to root systems, we are now able to find a suitable set of generators for W. The root system Φ is finite, so we can pick a hyperplane of V which does not intersect Φ (represented with a dashed line in Figure 6). This hyperplane induces a partition of Φ into two sets of the same cardinality, namely the set of *positive roots* Φ^+ and the set of *negative roots* $\Phi^- = -\Phi^+$. It is obvious that W is generated by the reflections s_α with $\alpha \in \Phi^+$. The polyhedral cone generated by Φ^+ has a unique basis $\Delta \subseteq \Phi^+$ called a *simple system* consisting of roots called *simple roots*. So any root is either a nonnegative linear combination of simple roots or a nonpositive linear combination of simple roots. In Figure 6, the simple roots for $\mathscr{D}_3$ are α_1 and α_2, while for $\mathscr{D}_4$ they are e_1 and $e_2 - e_1$. We can thus set

$$S = \{s_\alpha \mid s_\alpha \in \Delta\}.$$

The reflections in S are called *simple reflections*. The identity of W is denoted by e.

Theorem 2.13 (see [12, Chapter 1]). *Let W be a finite reflection group with root system Φ, set of positive roots Φ^+ and simple system Δ.*

1. *The simple roots are linearly independent.*
2. *Any reflection s in W is conjugate in W to a simple reflection. Moreover, for a reflection s in W, there is a unique $\beta \in \Phi^+$ such that $s = s_\beta$.*

3. (W,S) *is a* finite Coxeter system*: W is generated by S and by the relations $s^2 = e$ ($s \in S$ is a reflection) and $(st)^{o(st)} = e$ where $o(st)$ is the order of the rotation st, $s,t \in S$.*

Remark 2.14. The fact that finite reflection groups are finite Coxeter groups is crucial for their classification. Indeed, finite Coxeter groups are nicely classified through their root systems, and so are finite reflection groups, see N. Reading's text in this volume [24].

Example 2.15. For the symmetric group S_n, consider the set

$$\Phi = \left\{ e_j - e_i \mid 1 \le i \ne j \le n \right\}.$$

A positive root system of Φ is

$$\Phi^+ = \left\{ e_j - e_i \mid 1 \le i < j \le n \right\}.$$

A simple system is $\Delta = \{ e_{i+1} - e_i \mid 1 \le i < n \}$ and the corresponding simple generators are the simple transpositions $\tau_i = (i\ i+1)$, well known to generate S_n.

Example 2.16. For the hyperoctahedral group W_n, consider the set

$$\Phi = \left\{ e_j - e_i \mid 1 \le i \ne j \le n \right\} \cup \{ \pm e_i \mid 1 \le i \le n \}.$$

A positive root system of Φ is

$$\Phi^+ = \left\{ e_j - e_i \mid 1 \le i < j \le n \right\} \cup \{ e_i \mid 1 \le i \le n \}.$$

A simple system is $\Delta = \{ e_1, e_{i+1} - e_i \mid 1 \le i < n \}$ and the corresponding simple generators are the simple transpositions s_0 and $\tau_i = (i\ i+1)$.

Example 2.17. For the isometry group of the icosahedron $W(H_3)$, we have the set of simple generators $S = \{s_1, s_2, s_3\}$ with $s_i^2 = e$ and the other relations come from the order of rotations: $(s_1 s_2)^5 = (s_1 s_3)^2 = (s_2 s_3)^3 = e$. There are 15 positive roots and a simple system consists of the following vectors in $\mathbb{R}^3$:

$$(2,0,0), \qquad \left(\frac{-1-\sqrt{5}}{2}, \frac{-1+\sqrt{5}}{2}, -1 \right), \qquad (0,0,2).$$

2.3.2 Permutahedra as $\mathscr{H}$-polytope

We are now ready to complete our description of permutahedra as an $\mathscr{H}$-polytope. Fix a generic point a and consider the associated permutahedron $\text{Perm}^a(W)$. As we have seen in Example 2.7, a permutahedron can live in an affine subspace not containing the origin 0. Let us first explain this phenomenon. The subspace V_0 of V spanned by any root system Φ of W is stable under the action of W, and so is its orthogonal complement[7]. So the affine subspace $V_a = a + V_0$ of V directed by V_0 and passing through the point a is also stable under the action of W, and therefore

$$\mathrm{Perm}^a(W) \subseteq V_a.$$

Comparing Figure 3 and Figure 6, we observe that the edges of $\mathrm{Perm}^a(W)$ are directed by roots. This phenomenon holds for permutahedra of higher dimension. Therefore, the facets have to be directed by affine subspaces spanned by subsets of roots, and these subspaces have to be the boundaries of some half-spaces whose intersection is $\mathrm{Perm}^a(W)$. More precisely, choose the unique simple system[8] Δ of Φ such that $\langle a, \alpha \rangle > 0$ for all $\alpha \in \Delta$. For $\alpha \in \Delta$ a simple root, we consider the half-space in V_0

$$\mathscr{H}_0(\alpha) := \left\{ x = \sum_{\beta \in \Delta} x_\beta \beta \in V_0 \;\middle|\; x_\alpha \leq 0 \right\}$$

with its boundary $H_0(\alpha) := \mathrm{span}\,(\Delta \setminus \{\alpha\}) \subseteq V_0$. Then we consider their affine counterparts passing through the point a:

$$\mathscr{H}_a(\alpha) = a + \mathscr{H}_0(\alpha) \quad \text{and} \quad H_a(\alpha) = a + H_0(\alpha).$$

Theorem 2.18. *Let a be a generic point in V.*

1. *The permutahedron $\mathrm{Perm}^a(W)$, as an $\mathscr{H}$-polytope, is given by*

$$\mathrm{Perm}^a(W) = \bigcap_{\substack{\alpha \in \Delta \\ w \in W}} w(\mathscr{H}_a(\alpha)) \subseteq V_a.$$

2. *For $w \in W$, $w(a)$ is a vertex of $\mathrm{Perm}^a(W)$ and*

$$\{w(a)\} = \bigcap_{\alpha \in \Delta} w(H_a(\alpha)).$$

3. *The permutahedron $\mathrm{Perm}^a(W)$ is a $|\Delta|$-dimensional simple polytope. In particular, $\mathrm{Perm}^a(W)$ is full dimensional in the affine space V_a.*

Example 2.19. We illustrate this theorem in Figure 7 with the example of the permutahedron of the dihedral group $\mathscr{D}_4$.

The normal fan of a polytope P is a collection of pointed polyhedral cones indexed by the faces of P: if F is a face of P, the cone C_F in the normal fan is the polyhedral cone generated by the outer normal vectors of the facets of P containing F. Normal fans are natural objects that link polytope theory to optimization problems

---

[7] The orthogonal complement of V_0 is fixed pointwise by W since it is the intersection of all the hyperplanes associated to reflections in W.

[8] In fact, it is equivalent to choose a fundamental chamber in the Coxeter complex, which is always possible, see [12, §1.12] for more details.

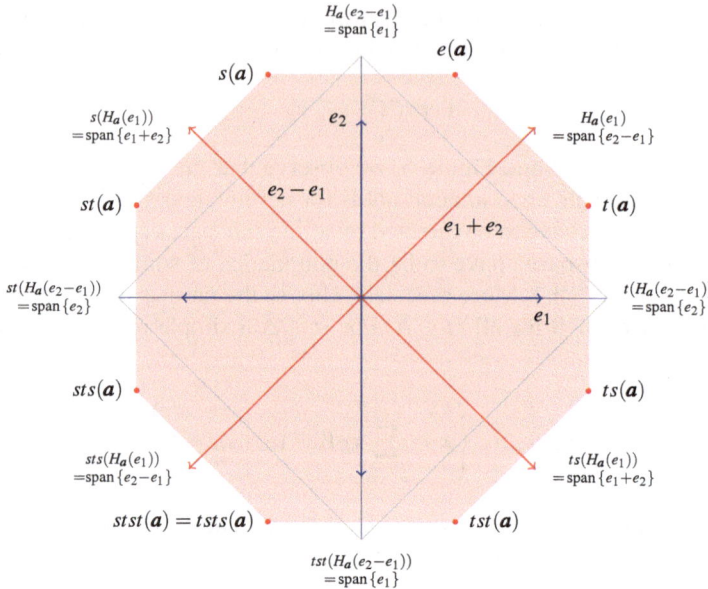

Fig. 7 A permutahedron of the dihedral group $W = \mathscr{D}_4$ given as an $\mathscr{H}$-polytope. In this picture the roots are colored in blue and red as in Figure 6; the simple reflections are $s := s_{e_1}$ and $t := s_{e_2-e_1}$, and $V_0 = V = V_a$.

or algebraic geometry. For more information on normal fans of polytopes, see [31, Chapter 7].

Definition 2.20. The normal fan $\mathscr{F}_W$ of $\mathrm{Perm}^a(W)$ is called the *Coxeter fan*.

Remark 2.21 (Note on the proof of Theorem 2.18). As far as our knowledge goes, we do not know that many references which present the construction of permutahedra of finite reflection groups. Nevertheless, it was not a new idea. The proof is based on well-known and very important properties of finite reflection groups (see [12, §1.12]): the complement of the union of all hyperplanes corresponding to reflections in W cut the space V into open convex polyhedral cones called *chambers*. The collection of cones obtained by decomposing the boundaries of the chambers is the *Coxeter fan*. Picking one of these chambers to be the *fundamental chamber*, one can show that any chamber is the image of the fundamental chamber by an element of W and that the isotropy group of a chamber is the identity. This implies that there are $|W|$ chambers and that any chamber contains exactly one point of the W-orbit of a. Then observe that the affine hyperplanes $H_a(\alpha)$ passing through a are orthogonal to the rays of the fundamental chambers and are stable under the subgroup $W_{S\setminus\{s_\alpha\}}$ generated by $\{s_\beta \mid \beta \in \Delta\setminus\{\alpha\}\}$. Finally one shows, with a bit of straightforward work, that for any $\alpha \in \Delta$ and $w \in W\setminus W_{S\setminus\{s_\alpha\}}$, the point $w(a)$ is in the interior of the half-space $\mathscr{H}_a(\alpha)$. For more details, see [3, 10].

2.4 Faces of permutahedra and the weak order

Let $\boldsymbol{a}$ be a generic point, Δ the simple system of Φ such that $\langle \boldsymbol{a}, \alpha \rangle > 0$ for all $\alpha \in \Delta$ and $S = \{s_\alpha \mid \alpha \subset \Delta\}$ the set of simple reflections generating W. Our aim in this final part of our study of permutahedra is to explain how to realize several important notions from the theory of finite reflection groups in the context of permutahedra.

2.4.1 Faces and standard parabolic subgroups

The subgroup W_I of W generated by $I \subseteq S$ is called a *standard parabolic subgroup*. Since W_I is generated by reflections, it is also a finite reflection group. Moreover, one can show that $\Delta_I := \{\alpha \in \Delta \mid s_\alpha \in I\}$ is a simple system for the root system $\Phi \cap \mathrm{span}\,(\Delta_I)$. So we can apply our construction of permutahedra to W_I: the polytope

$$F_I := \mathrm{Perm}^{\boldsymbol{a}}(W_I) = \mathrm{conv}\,\{w(\boldsymbol{a}) \mid w \in W_I\}$$

is a W_I-permutahedron. For instance, if $I = S \setminus \{s_\alpha\}$, then the hyperplane $H_{\boldsymbol{a}}(\alpha)$ is stable under the action of W_I and contains F_I, which is a facet of $\mathrm{Perm}^{\boldsymbol{a}}(W)$. More generally, F_I is a $|I|$-dimensional face of $\mathrm{Perm}^{\boldsymbol{a}}(W)$ containing $\boldsymbol{a}$ and lives in the affine space

$$\bigcap_{\alpha \in \Delta \setminus \Delta_I} H_{\boldsymbol{a}}(\alpha) = \boldsymbol{a} + \mathrm{span}\,(\Delta_I).$$

The next statement explains that the faces of $\mathrm{Perm}^{\boldsymbol{a}}(W)$ are obtained as the W-orbit of the faces containing $\boldsymbol{a}$.

Proposition 2.22.

1. *Each face of dimension k of $\mathrm{Perm}^{\boldsymbol{a}}(W)$ is a permutahedron: $w(F_I) = \mathrm{Perm}^{w(\boldsymbol{a})}(W_I)$ for $w \in W$ and $I \subseteq S$ of cardinality k.*
2. *For $w, g \in W$ and $I \subseteq S$, $w(F_I) = g(F_I)$ if and only if $wW_I = gW_I$. In other words, faces are naturally parametrized by the cosets W/W_I, $I \subseteq S$.*
3. *The face lattice of $\mathrm{Perm}^{\boldsymbol{a}}(W)$ is isomorphic to the poset of cosets of standard parabolic subgroups: $w(F_I) \subseteq g(F_J)$ if and only if $wW_I \subseteq gW_J$.*

Example 2.23. We continue the example of the permutahedron of the dihedral group $\mathscr{D}_4$ we started in §2.2. Let $s = s_{e_1}$ and $t = s_{e_2 - e_1}$; we are now able to label the vertices of $\mathrm{Perm}^{\boldsymbol{a}}(\mathscr{D}_4)$ shown in Figure 3. For $w \in \mathscr{D}_4$ we label the vertex $w(\boldsymbol{a})$ by w. In particular the vertex $\boldsymbol{a}$ is labeled by the identity e. As we can see in Figure 8, the facets, which are the edges here, are naturally labeled by the cosets of the subgroup W_s generated by s and the cosets of the subgroup W_t generated by t.

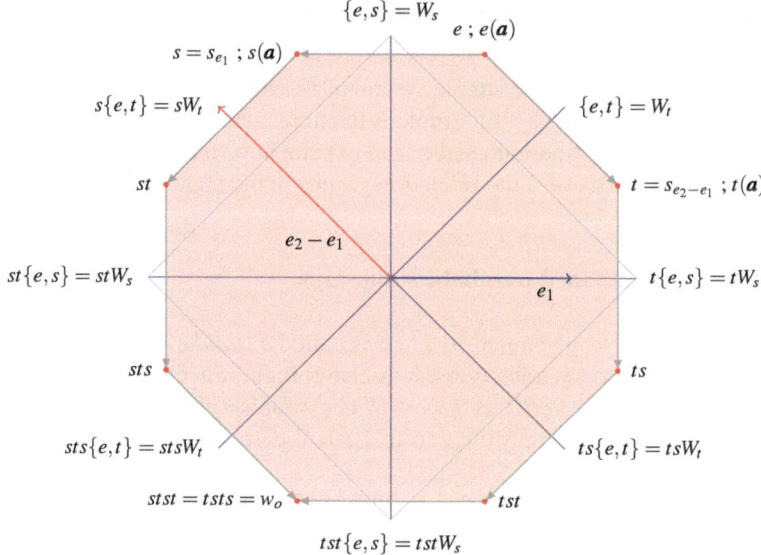

Fig. 8 The labeled permutahedron of the dihedral group $W = \mathscr{D}_4$ with nonempty faces labeled by the cosets of standard parabolic subgroups. The gray arrows, which are the oriented edges of the octagon, represent cover relations in the weak order of W.

2.4.2 Edges and weak order

We consider the length function $\ell : W \to \mathbb{N}$ mapping an element $w \in W$ to the minimal number $\ell(w)$ of letters needed to express w as a word in the alphabet S. For instance $\ell(e) = 0$ and $\ell(s) = 1$ if and only if $s \in S$. We know that an edge $w(F_s)$ of $\mathrm{Perm}^a(W)$ corresponds to a coset $wW_s = \{w, ws\}$ for $s \in S$ and $w \in W$. We orient each edge $w(F_s)$ from w to ws if $\ell(w) < \ell(ws)$, and from ws to w otherwise. See Figure 8 and Figure 9 for examples.

Proposition 2.24. *The oriented 1-skeleton[9] of* $\mathrm{Perm}^a(W)$ *is the graph of the lattice called* the right weak order *of* W. *Moreover, the minimal element is the identity* e, *the maximal element is denoted by* w_o *and each face of* $\mathrm{Perm}^a(W)$ *corresponds to an interval in the weak order.*

Remark 2.25.

1. The length $\ell(w_o)$ of w_o is $|\Phi^+|$.
2. The f-vector[10] of a permutahedron of W encloses important information about the group: f_0 is the cardinality of W, f_1 is the number of cover relations in the weak order or f_{n-1} is the number of maximal parabolic subgroups of W.

---

[9] The 1-skeleton of a polytope is the graph obtained by taking the vertices and the edges of this polytope.

[10] The k-th coordinate f_k of the f-vector of a polytope P is the number of k-dimensional faces of P, see [31].

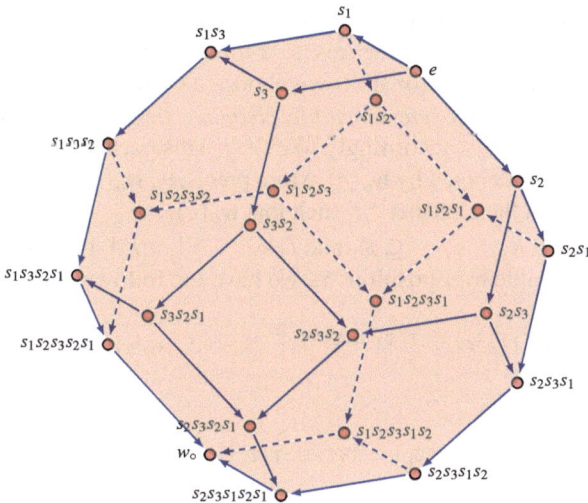

Fig. 9 A permutahedron of the group S_4 with labeled vertices and oriented edges which represent the weak order on S_4.

3 Coxeter generalized associahedra

We are now ready to build Coxeter generalized associahedra. The basic idea is to build it as an $\mathscr{H}$-polytope by identifying subsets of the half-spaces used in constructing a given permutahedron. These subsets of half-spaces will be chosen according to certain elements in W, namely *Coxeter singletons*, that are contained in the boundaries of the half-spaces.

Throughout this section, we fix a generic point $\boldsymbol{a}$, a simple system Δ such that $\langle \boldsymbol{a}, \alpha \rangle > 0$ for all $\alpha \in \Delta$ and $S = \{s_\alpha \mid \alpha \in \Delta\}$ the set of simple reflections which generates W.

3.1 Coxeter elements and Coxeter singletons

Geometrically, Coxeter singletons correspond to vertices in a permutahedron of W along certain paths between the identity e and the longest element w_o (a kind of spinal cord of the permutahedron). Let us be more precise.

Let c be a Coxeter element of W, that is, the product of the simple reflections in S taken in some order, and fix a reduced expression for c as a word in the alphabet S. For $I \subseteq S$, we denote by $c_{(I)}$ the subword of c obtained by taking only the simple reflections in I. So $c_{(I)}$ is a Coxeter element of W_I. For instance, if $W = S_4$ with $S = \{\tau_1, \tau_2, \tau_3\}$, a particular Coxeter element is $c = \tau_2 \tau_3 \tau_1$; if $I = \{\tau_1, \tau_2\} \subseteq S$, then $c_{(I)} = \tau_2 \tau_1$.

The longest element w_o of the W can be written as a reduced word on S in many ways: each word corresponds precisely to a minimal path from e to w_o on the 1-skeleton of a permutahedron for W. If we choose a Coxeter element c, N. Reading showed in his work on *Coxeter sortable elements* that we can *sort* a particular reduced expression for w_o accordingly to c [22]. This particular word is called *the c-word of w_o* and it is denoted by $\boldsymbol{w_o}(c)$. More precisely, $\boldsymbol{w_o}(c)$ is the unique reduced expression for w_o on the alphabet[11] S such that $\boldsymbol{w_o}(c) = c_{(K_1)} c_{(K_2)} \cdots c_{(K_p)}$ with non-empty $K_i \subseteq S$, $K_p \subseteq K_{p-1} \subseteq \cdots \subseteq K_1$ and $\ell(w_o) = \sum_{i=1}^{p} |K_i|$. For instance, if $W = S_4$ with S the set of simple transpositions τ_i, we have the following c-words for w_o:

$$\boldsymbol{w_o}(\tau_1 \tau_2 \tau_3) = \tau_1 \tau_2 \tau_3 . \tau_1 \tau_2 . \tau_1 = c_{(S)} c_{(\{\tau_1, \tau_2\})} c_{(\{\tau_1\})}$$

and

$$\boldsymbol{w_o}(\tau_2 \tau_3 \tau_1) = \tau_2 \tau_3 \tau_1 . \tau_2 \tau_3 \tau_1 = c_{(S)} c_{(S)}.$$

Definition 3.1. Let c be a Coxeter element and $u \in W$. We say that u is a *c-singleton* if some reduced word $\boldsymbol{u}(c)$ for u appears as a prefix of a word that can be obtained from $\boldsymbol{w_o}(c)$ by the commutation of commuting reflections of S. The word $\boldsymbol{u}(c)$ is called the *c-word* of the c-singleton u.

Example 3.2. Consider again the symmetric group $W = S_4$ together with $S = \{\tau_1, \tau_2, \tau_3\}$ the set of simple transpositions. The reader may follow this example with Figure 12 and Figure 13 in mind for an illustration on a permutahedron.

For the Coxeter element $c = \tau_1 \tau_2 \tau_3$, the c-singletons, and their c-words, are

$$e, \qquad \tau_1, \qquad \tau_1 \tau_2, \qquad \tau_1 \tau_2 \tau_3,$$
$$\tau_1 \tau_2 \tau_1, \qquad \tau_1 \tau_2 \tau_3 \tau_1, \qquad \tau_1 \tau_2 \tau_3 \tau_1 \tau_2, \qquad w_o = \tau_1 \tau_2 \tau_3 \tau_1 \tau_2 \tau_1.$$

Observe that $\tau_1 \tau_2 \tau_1$ is a not a prefix of the word $\boldsymbol{w_o}(c) = \tau_1 \tau_2 \tau_3 \tau_1 \tau_2 \tau_1$, but it is a prefix of the word $\tau_1 \tau_2 \tau_1 \tau_3 \tau_2 \tau_1$ obtained after exchanging the commuting simple transpositions τ_1, τ_3 in the word $\boldsymbol{w_o}(c)$.

For the Coxeter element $c' = \tau_2 \tau_1 \tau_3$, the c'-singletons, and their c'-words, are

$$e, \qquad \tau_2 \tau_3, \qquad \tau_2 \tau_3 \tau_1 \tau_2 \tau_3,$$
$$\tau_2, \qquad \tau_2 \tau_3 \tau_1, \qquad \tau_2 \tau_3 \tau_1 \tau_2 \tau_1, \text{ and}$$
$$\tau_2 \tau_3, \qquad \tau_2 \tau_3 \tau_1 \tau_2, \qquad w_o = \tau_2 \tau_1 \tau_3 \tau_2 \tau_1 \tau_3.$$

Observe that $\tau_2 \tau_3$ is a not a prefix of the word $\boldsymbol{w_o}(c') = \tau_2 \tau_1 \tau_3 \tau_2 \tau_1 \tau_3$, but it is a prefix of the word $\tau_2 \tau_3 \tau_1 \tau_2 \tau_3 \tau_1$ obtained after exchanging the commuting simple transpositions τ_1, τ_3 in the word $\boldsymbol{w_o}(c')$.

Problem 3.3. *Let W be a finite reflection group with set of simple reflections S. Find a formula for the number of c-singletons as a function of c. For which c's are the maximum and the minimum reached? (As we can see in the examples above, the*

---

[11] We have to make a clear distinction between words on the alphabet S and reduced expressions of elements of W. The latter are subject to relations of the Coxeter system (W, S) while the first ones are not.

number of c-singletons depends on the choice of the Coxeter element c or equivalently, on the chosen orientation of the Coxeter graph, see Remark 3.4 *below. Numerical examples are presented in* [9]).

Remark 3.4.

1. *Coxeter sortable elements.* The set of elements in W that verify the same characterization we gave above for the c-word $\boldsymbol{w_o}(c)$ of the longest element w_o are called *c-sortable elements.* They were introduced by N. Reading for studying extensions of the *Tamari lattice* called *Cambrian lattices* [22, 23]. The combinatorics of these elements is very rich (see for more details N. Reading's text in this volume [24]).
2. Coxeter elements are in bijection with *orientations of Coxeter graphs.* A Coxeter graph is a graph whose vertices are simple reflections and whose edges are labeled by the order of the product of two simple reflections (a rotation); there is no edge between two commuting simple reflections. Any Coxeter element c defines an orientation of the Coxeter graph of W: orient the edge $\{s_i, s_j\}$ from s_i to s_j if and only if s_i is to the left of s_j in any reduced word for c. The contents of the articles [9, 1, 11] makes use of this bijection, as well as N. Reading in his works on Coxeter sortable elements (see [24]).

3.2 Coxeter generalized associahedra as $\mathcal{H}$-polytopes

At this point, one should remember that the permutahedron obtained from the generic point $\boldsymbol{a}$ is described as an $\mathcal{H}$ polytope by

$$\mathrm{Perm}^{\boldsymbol{a}}(W) = \bigcap_{\substack{\alpha \in \Delta, \\ s \in W}} w\big(\mathcal{H}_{\boldsymbol{a}}(\alpha)\big) \subseteq V_{\boldsymbol{a}} \quad (\text{see §2.3.2}).$$

Definition 3.5. Let c be a Coxeter element of W. The polytope in the affine space $V_{\boldsymbol{a}}$ defined by

$$\mathrm{Asso}^{\boldsymbol{a}}_c(W) = \bigcap_{\substack{\alpha \in \Delta, \\ u \text{ is a } c\text{-singleton}}} u\big(\mathcal{H}_{\boldsymbol{a}}(\alpha)\big)$$

is called a *c-generalized associahedron.* As we will see, the combinatorics of $\mathrm{Asso}^{\boldsymbol{a}}(W)$ does not depend of the choice of $\boldsymbol{a}$, as long as this point is generic.

As a first consequence, we see that for $\alpha \in \Delta$ and $u \in W$ a c-singleton, the half-space $u(\mathcal{H}_{\boldsymbol{a}}(\alpha))$ contains the point $u(\boldsymbol{a})$ in its boundary hyperplane $u(H_{\boldsymbol{a}}(\alpha))$. We say that the half-space $w(\mathcal{H}_{\boldsymbol{a}}(\alpha))$ is *c-admissible* if its boundary $w(H_{\boldsymbol{a}}(\alpha))$ contains a c-singleton. So we can see a Coxeter generalized associahedron as obtained from a given permutahedron by keeping only the c-admissible half-spaces.

Example 3.6. Coxeter generalized associahedra of dimension 2 arise from dihedral groups: an associahedron for $\mathcal{D}_n$ is a $n+2$-gon.

For instance, take $W = \mathscr{D}_4$, the symmetry group of a square as in Example 2.6: take $s := s_{e_1}$ and $t := s_{e_2-e_1}$. Here $V_0 = V = V_a$. Consider the Coxeter element $c = ts$, then the c-singletons are $e, t, ts, tst, tsts$. Figure 10 shows the resulting associahedron $\text{Asso}^a(\mathscr{D}_4)$ containing the permutahedron $\text{Perm}^a(\mathscr{D}_4)$. Observe that the common vertices between $\text{Perm}^a(\mathscr{D}_4)$ and $\text{Asso}^a(\mathscr{D}_4)$ are precisely the c-singletons.

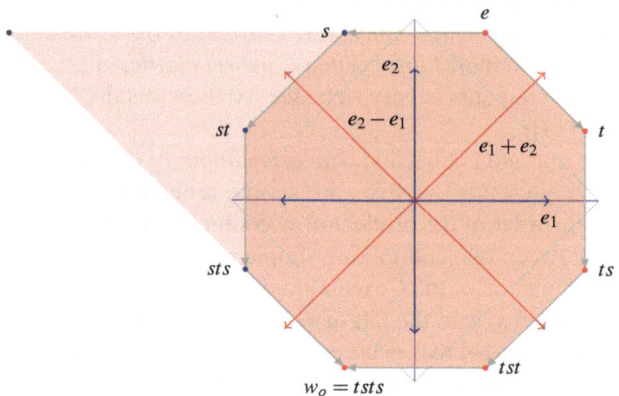

Fig. 10 A Coxeter generalized associahedron of the dihedral group $W = \mathscr{D}_4$ given as an $\mathscr{H}$-polytope and containing the permutahedron with edges oriented by the weak order. In this picture, the Coxeter singletons are in red and the boundary of the c-admissible half-spaces meet in a vertex colored black

Example 3.7. There are two kinds of Coxeter generalized associahedra of dimension 3. The first kind is obtained in $\mathbb{R}^3$ from permutahedra of dimension 2 by considering the isometry group of a regular polygonal prism as in Example 2.8. Consider a regular n-gonal prism with isometry group $W = \mathscr{D}_n \times \mathscr{D}_2$. The Coxeter singletons are the couples (u,e) and (u,s) where u is a Coxeter singleton of $\mathscr{D}_n$. Then Coxeter generalized associahedra obtained from $\text{Perm}^a(W)$ are $n+2$-gonal prisms.

The second kind of Coxeter generalized associahedra of dimension 3 arises from the symmetric group S_4 (one of those is the *classical associahedron* also called *Stasheff polytope*), the hyperoctahedral group W_3' (those are called *cyclohedron*) and the isometry group of the dodecahedron $W(H_3)$. Examples are shown in Figures 12, 13, 14, 15 and 16.

The observations we can make on these pictures are summarized in the following theorem.

Theorem 3.8 ([10]). *Let c be a Coxeter element of W.*

1. *$\text{Asso}_c^a(W)$ is a $|\Delta|$-dimensional simple convex polytope.*
2. *$\text{Perm}^a(W) \subseteq \text{Asso}_c^a(W)$.*
3. *Each facet of $\text{Asso}_c^a(W)$ is contained in the boundary of exactly one c-admissible half-space. There are $(|\Delta| + |\Phi^+|)$ facets.*

4. *The vertex sets* $\mathrm{vert}\,(\mathrm{Asso}_c^a(W))$ *and* $\mathrm{vert}\,(\mathrm{Perm}^a(W))$ *satisfy*

$$\mathrm{vert}\,(\mathrm{Asso}_c^a(W)) \cap \mathrm{vert}\,(\mathrm{Perm}^a(W)) = \{u(a)\,|\,u \text{ is a c-singleton}\};$$

and this intersection forms a distributive sublattice of the weak order.

Definition 3.9. The normal fan of $\mathrm{Asso}_c^a(W)$ is called the *c-Cambrian fan.*

Remark 3.10 (Note on the proof of Theorem 3.8). The original motivation for constructing $\mathrm{Asso}_c^a(W)$ was to show that *Cambrian fans* are normal fans of some polytopes, answering a conjecture made by N. Reading in [21]. Cambrian fans are defined as coarsening fans of the Coxeter fan, see [25, 24]. The proof the authors gave in [10] is very technical and based on some technical properties of Cambrian fans given by N. Reading and D. Speyer. The principal reason for that difficulty is that we don't have a description of $\mathrm{Asso}_c^a(W)$ as a $\mathscr{V}$-polytope, that is, as the convex hull of a given set of points. Recently, we were made aware of a very nice work in progress by V. Pilaud and C. Stump [18]: they introduce new families of polytopes called *brick polytopes* associated to finite reflection groups. A brick polytope is defined as a $\mathscr{V}$-polytope associated to a *spherical subword complex* of W, which were introduced by A. Knutson and E. Miller in [13]. They show in particular that brick polytopes contain Coxeter generalized associahedra providing a new proof of this theorem and a description of $\mathrm{Asso}_c^a(W)$ as a $\mathscr{V}$-polytope.

3.3 Faces and almost positive roots

The question now is to find a nice parameterization of the faces of $\mathrm{Asso}_c^a(W)$.

Let Φ be a root system for W, with simple system Δ. The set of almost positive roots is the set

$$\Phi_{\geq -1} := -\Delta \cup \Phi^+.$$

As stated in Theorem 3.8, there are precisely $|\Phi_{\geq -1}| = |\Delta| + |\Phi^+|$ facets. This suggests a labeling of the facets by almost positive roots. To describe this labeling of the set of facets, we define for $\alpha \in \Delta$ the *last root map* lr_α, which sends a c-singleton u to an almost positive root $\mathrm{lr}_\alpha(u) \in \Phi_{\geq -1}$, as follows:

1. if $s_\alpha \in S$ is not a letter in the c-word $u(c)$ of u, then

$$\mathrm{lr}_\alpha(u) := -\alpha \in -\Delta;$$

2. if $s_\alpha \in S$ is a letter in the word $u(c)$, we write $u(c) = u_1 s_\alpha u_2$ where u_2 is the unique largest suffix, possibly empty, of $u(c)$ that does not contain the letter s_α and

$$\mathrm{lr}_\alpha(u) = u_1(\alpha) \in \Phi^+.$$

Example 3.11. To illustrate this map, we consider again the dihedral group $W = \mathscr{D}_4$ with generators $S = \{s,t\}$ as in Example 3.6. Fix $c = ts$ as a Coxeter element. We saw in Example 3.6 that the c-singletons are: e, t, $c = ts$ and $w_o = tsts$. For the c-word $\mathbf{w_o}(c) = tsts$ and $t = s_{e_2 - e_1}$ we have $\mathbf{w_o}(c) = u_1 t u_2$ with $u_1 = ts$ and $u_2 = s$. We get $\mathrm{lr}_{e_2 - e_1}(tsts) = ts(e_2 - e_1) = e_1 + e_2$. We show in Figure 7 the last root map $\mathrm{lr}_\alpha(u)$ for all the c-singletons u and simple roots α.

Thanks to the maps lr_α, we are able, together with Theorem 3.8, to label the facets of $\mathrm{Asso}_c^a(W)$: for $\alpha \in \Phi_{\geq -1}$ denote by $F_{\mathrm{lr}_\alpha(u)}$ the unique facet of $\mathrm{Asso}_c^a(W)$ supported by the hyperplane $u(H_a(\alpha))$, where $\alpha \in \Delta$ and u is a c-singleton. Therefore, since $\mathrm{Asso}_c^a(W)$ is a $|\Delta|$-dimensional simple polytope and since any face of codimension k is the intersection of k facets, we obtain a natural labeling of the faces of $\mathrm{Asso}_c^a(W)$: any face of $\mathrm{Asso}_c^a(W)$ of codimension k is of the form F_Λ where $\Lambda = \{\alpha_1, \ldots, \alpha_k\} \subseteq \Phi_{\geq -1}$ such that

$$F_\Lambda = \bigcap_{i=1}^{k} F_{\alpha_i}.$$

Example 3.12. This labeling is shown on Figure 11 for the example of the dihedral group $W = \mathscr{D}_4$.

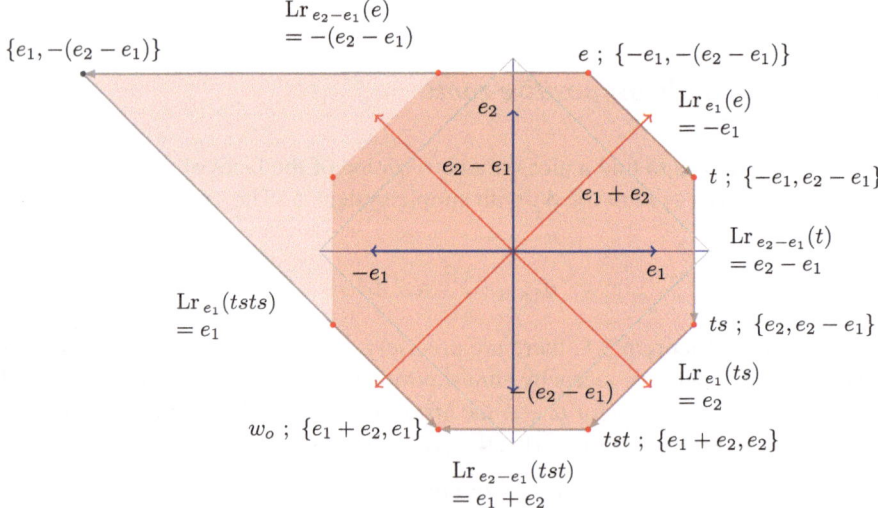

Fig. 11 A generalized associahedron of the dihedral group $W = \mathscr{D}_4$ whose faces are labeled by subsets of almost positive roots. In this picture, the Coxeter singletons are in red. The edges are oriented according to the c-Cambrian lattice, see §3.4.

Theorem 3.13.

1. *The map* $\beta \mapsto F_\beta$ *is a bijection between the set of almost positive roots* $\Phi_{\geq -1}$ *and the facets of* $\mathrm{Asso}_c^a(W)$.
2. *The vertices of* $\mathrm{Asso}_c^a(W)$ *are labeled subsets of* $\Phi_{\geq -1}$ *of cardinality* $|\Delta|$, *called* c-*clusters. The 1-skeleton of* $\mathrm{Asso}_c^a(W)$ *is called* the c-*cluster exchange graph. They are counted by the* W-*Catalan numbers.*

Catalan numbers appear in the context of symmetric groups. They count many objects such as planar binary trees, triangulations of a given polygon, noncrossing partitions, etc. They have an analog for any finite reflection group where they count, for instance, clusters of finite type, Coxeter sortable elements or generalized non-crossing partitions, see for instance [22] for more details. One way to prove that c-clusters are counted by the W-Catalan numbers is given by N. Reading in [22]: he constructs a bijection from c-clusters to noncrossing partitions via c-sortable elements. For more on the connection to cluster algebras, see N. Reading's text in this volume [24].

Problem 3.14. *Find from the vertices of a* c-*generalized associahedron a direct and uniform proof that the number of vertices is the* W-*Catalan number.*

Remark 3.15 (Notes on the proof of Theorem 3.13). The first point of this theorem was proved in [10].

In this same article, the authors showed the second point of the theorem by showing that the normal fan of this polytope is the c-*Cambrian fan* studied by N. Reading and D. Speyer in [25]. N. Reading and D. Speyer showed in particular, using results obtained by N. Reading in [22, 23], that the maximal cones of the c-Cambrian fan are the c-clusters and that the graph obtained by considering maximal cones and walls in this fan is the c-cluster exchange graph.

Recently, this connection between c-clusters and the vertices of $\mathrm{Asso}_c^a(W)$ was made entirely clear from a combinatorial point of view by C. Ceballos, J.-P. Labbé and C. Stump in [5]. The authors find the 1-skeleton of $\mathrm{Asso}_c^a(W)$ as the *facet-adjency graph of spherical subword complex* of the word $cw_0(c)$. This new object allow them, not only to recover the 1-skeleton of $\mathrm{Asso}_c^a(W)$ with the parameterizations above, but also to easily and naturally prove that this labeling of the vertices corresponds to c-clusters in the sense of N. Reading.

Remark 3.16 (The classical associahedron and the cyclohedron as Coxeter generalized associahedra). In [9], the authors present Coxeter generalized associahedra for symmetric groups and hyperoctahedral groups both as $\mathscr{V}$-polytopes and $\mathscr{H}$-polytopes. In that article, they start from the classical permutahedron Π_{n-1}. For the symmetric group S_n, the faces of generalized associahedra can be labeled by triangulations of a $n+2$-gon P. The idea is to index the vertices of P according to the chosen Coxeter element c, or equivalently to the orientation of the Coxeter graph of S_n, see Remark 3.4, and then to give a very easy combinatorial way to associate *integer coordinates* in $\mathbb{R}^n$ to any triangulation of P. These coordinates turn out to be precisely the coordinates of the vertices of the corresponding c-generalized associahedron of S_n. This particular class of generalized associahedra, obtained from the classical permutahedron, contains the 'classical' associahedron as constructed by J.-L. Loday [15] and S. Shnider-S. Sternberg [28], and the associahedron arising from cluster algebra theory as realized by F. Chapoton, S. Fomin and A. Zelevinsky [7] (see Figures 12 and 13). In regard to the last statement, we refer the reader to the very nice article by C. Ceballos, F. Santos and G.M. Ziegler [6] on the subject of comparing these realizations. Many realizations of the cyclohedron are easily obtained from this class of generalized associahedra of symmetric groups by looking at the hyperoctahedral group W_n' as a subgroup of S_{2n} (see [9] for more details).

Using this description as a $\mathscr{V}$-polytope, J. Lortie, A. Raymond and the author were able to show that the centers of the gravity of the vertices of the classical permutahedron and of all the Coxeter generalized associahedron built from it are the same [11]. This is still an open problem for an arbitrary Coxeter generalized associahedra.

Problem 3.17. *Let a be a generic point such that $\langle a, \alpha \rangle = \langle a, \beta \rangle$ for all $\alpha, \beta \in \Delta$ and W a finite reflection group. Prove that the vertices of any c-generalized associahedra $\mathrm{Asso}_c^a(W)$ and of the permutahedron $\mathrm{Perm}^a(W)$ have the same center of gravity. (There is significant computational evidence supporting this statement).*

The proof in [11] was based on the action of $\mathscr{D}_{n+2}$ on the set of vertices of the associahedron indexed by triangulations of P. It turns out that for each orbit, the center of gravity of the vertices in this orbit is also the same as that of the classical permutahedron. It leads us to ask the following question.

Problem 3.18. *For an arbitrary finite reflection group W, find an action on the vertices of a Coxeter generalized associahedron $\mathrm{Asso}_c^a(W)$ that generalizes the action of $\mathscr{D}_{n+2}$ on a regular $n+2$-gon, and such that the center of gravity of the vertices in each of the orbits is the same as that of $\mathrm{Perm}^a(W)$ if a is a generic point verifying $\langle a, \alpha \rangle = \langle a, \beta \rangle$ for all $\alpha, \beta \in \Delta$.*

3.4 Edges and Cambrian lattices

Cambrian lattices were introduced by N. Reading in [21], see also N. Reading's text in this volume [24]. The c-Cambrian lattice is a sublattice, as well as a quotient lattice, of the weak order of W. These lattices extend to the framework of finite reflection groups the celebrated *Tamari lattice* arising in the framework of symmetric groups. To be more precise: the sets of *c-sortable elements*, introduced in Remark 3.4, form a sublattice of the weak order called the *c-Cambrian lattice*. We will now explain how to recover this order on the 1-skeleton of $\mathrm{Asso}_c^a(W)$. The problem is to orient each edge $[F_\lambda, F_\mu]$ of $\mathrm{Asso}_c^a(W)$, where λ, μ are c-clusters. Unfortunately, it is not as combinatorially easy as it was for recovering the weak order from the permutahedron.

In the case where the edge $[F_\lambda, F_\mu] = [u(a), v(a)]$ with u, v are c-singletons, we just keep the orientation given by the weak order. In the case where one of the vertices of $[F_\lambda, F_\mu]$ is not a c-singleton we have to use the c-cluster map define by N. Reading in [22]. The idea is that, similar to c-singletons, any $w \in W$ has a c-word $w(c)$. We are thus able to extend the last root map lr to W and then to define the *c-cluster map* cl_c from W to the subsets of $\Phi_{\geq -1}$ as follows:

$$w \in W \mapsto \mathrm{cl}_c(w) := \{\mathrm{lr}_\alpha(w) \mid \alpha \in \Delta\} \subseteq \Phi_{\geq -1}.$$

It turns out that N. Reading shows that the restriction of this map to c-sortable elements is a bijection with the set of c-clusters. We are now able to orient the edges: orient the edge $[F_\lambda, F_\mu]$ from F_λ to F_μ if $\mathrm{cl}_c^{-1}(\lambda)$ is smaller than $\mathrm{cl}_c^{-1}(\mu)$ in the weak order. We have to note that this way of orienting the edges of $\mathrm{Asso}_c^a(W)$ is not convenient.

Problem 3.19. *Find a combinatorial way to orient the edges of* $\mathrm{Asso}_c^a(W)$ *to recover the c-Cambrian lattice without the use of the cluster map* cl_c.

A stronger statement would be to answer the following problem.

Problem 3.20. *Find a combinatorial way to label the vertices of* $\mathrm{Asso}_c^a(W)$ *by c-sortable elements without the use of the cluster map* cl_c.

Proposition 3.21 ([23, 10]). *The oriented 1-skeleton of* $\mathrm{Asso}_c^a(W)$ *is the graph of the c-Cambrian lattice. Moreover, the c-singletons form a distributive sublattice of the c-Cambrian lattice: the minimal element is the identity e, the maximal element is w_o and each face of* $\mathrm{Asso}_c^a(W)$ *corresponds to an interval in the c-Cambrian lattice.*

Examples are shown in Figures 11, 12 and 13.

Remark 3.22. As unoriented graphs, 1-skeletons of different c-generalized associahedra $\mathrm{Asso}_c^a(W)$ are isomorphic, which implies that they have the same combinatorial type. Nevertheless, in general, c-Cambrian lattices are not lattice isomorphic, and c-generalized associahedra are not isometric.

3.5 Isometry classes

A natural question we can ask is how many of the realizations are isometric? For instance, we observe that the Coxeter generalized associahedra shown in Figures 12, 13, 14, 15 and 16 are not isometric. Let us briefly explain how it works.

An automorphism of the set of simple generators S is a bijection μ on S such that the order of $\mu(s)\mu(t)$ equals the order of st for all $s, t \in S$. In particular, μ induces an automorphism on W.

Proposition 3.23 ([1]). *Let c_1, c_2 be two Coxeter elements in W. Suppose that $\langle a, \alpha \rangle = \langle a, \beta \rangle$ for all $\alpha, \beta \in \Delta$. Then the following statements are equivalent.*

1. $\mathrm{Asso}_{c_1}(W) = \varphi(\mathrm{Asso}_{c_2}(W))$ *for some linear isometry φ on V.*
2. *There is an automorphism μ of S such that $\mu(c_2) = c_1$ or $\mu(c_2) = c_1^{-1}$.*

A more general statement, that is without the conditions $\langle a, \alpha \rangle = \langle a, \beta \rangle$, can be found in [1].

3.6 Integer coordinates

An important subclass of finite reflection groups are *Weyl groups* which are linked with the theory of semi-simple Lie algebras. A finite reflection group W is a *Weyl group* if it stabilizes a lattice in V, that is, a $\mathbb{Z}$-span of a basis of V. For any Weyl group W, there are particular choices of root systems which are called *crystallographic*: a root system Φ for W is crystallographic if for any two roots $\alpha, \beta \in \Phi$ we have $s_\alpha(\beta) = \beta + \lambda\alpha$ for some $\lambda \in \mathbb{Z}$. In that case, the simple roots Δ span a $\mathbb{Z}$-lattice L. (For more details see [12, §2.8 and §2.9]). Note that not all the root systems for Weyl groups are crystallographic.

Proposition 3.24 ([10])**.** *Let Φ be a crystallographic root system for the Weyl group W and c be a Coxeter element of W. Suppose that $\mathbf{a} \in V_0 = \mathrm{span}\,(\Delta)$ has integer coordinates in Δ. Then the vertices of $\mathrm{Perm}^{\mathbf{a}}(W)$ and of $\mathrm{Asso}_c^{\mathbf{a}}(W)$ have integer coordinates.*

Problem 3.25. *If $\mathbf{a}$ is not contained in V_0 but has integer coordinates in a $\mathbb{Z}$-lattice of V, show that $\mathrm{Perm}^{\mathbf{a}}(W)$ and of $\mathrm{Asso}_c^{\mathbf{a}}(W)$ have integer coordinates.*

This problem is supported by the following example.

Example 3.26. The classical permutahedron Π_{n-1} obtained from the reflection group S_n was presented in Example 2.7: our construction applies to Π_{n-1}, see Remark 3.16. In this setting $\mathbf{a} = (1, 2, \ldots, n) \in V_{\mathbf{a}}$ has integer coordinates. The straightforward idea to apply the last theorem to this setting is to look at the orthogonal projection onto V_0, which unfortunately is $(1 - (n+1)/2, 2 - (n+1)/2, \ldots, n - (n+1)/2)$ and does not have integer coordinates. However, S_n fixes the lattice spanned by the canonical basis of $\mathbb{R}^n$, and it was shown in [9] that the coordinates of c-generalized associahedra obtained from Π_{n-1} are integers. We still cannot explain this phenomenon.

Problem 3.27. *Let W be a Weyl group and L be a lattice stable under the action of Φ. Suppose that $\mathbf{a} \in V$ has integer coordinates in L. Do $\mathrm{Perm}^{\mathbf{a}}(W)$ and $\mathrm{Asso}_c^{\mathbf{a}}(W)$ have integer coordinates?*

4 Further developments

We presented in this text many open problems. However, possible developments should be mentioned. First, Coxeter generalized associahedra arises in the theory of finite cluster algebra theory, so it is natural to ask the following question.

Problem 4.1. *Is it possible to build finite type cluster algebras from Coxeter generalized associahedra?*

Also, natural questions arising from the theory of polytopes are still open, questions such as the following.

Problem 4.2. *Compute the volume of permutahedra and of Coxeter generalized associahedra for any finite reflection group.*

Problem 4.3. *If W is a Weyl group and Φ is crystallographic, compute the number of points with integers coordinates contained in $\mathrm{Asso}_c^a(W)$ and $\mathrm{Perm}^a(W)$.*

A. Postnikov introduced in [19] the concept of *generalized permutahedra*, which provide tools to start answering Problem 4.2 and Problem 4.3. Generalized permutahedra are polytopes obtained from the classical permutahedron Π_{n-1} by 'nicely moving some facets', see [20] for details. In our setting, we are obtaining Coxeter generalized associahedra by '*re*moving some facets'. One of the interest of A. Postnikov's approach is to express the classical permutahedron, as well as the associahedron realized by J.-L. Loday and S. Shnider-S. Sternberg, as the *Minkowski sum* of faces of standard simplicies. Recently, C. Lange expressed generalized associahedra obtained from S_n as Minkowski sums *and differences* of faces of standard simplicies [14]. This result suggests extending the framework of generalized permutahedra to include subtractions of faces of standard simplicies. These questions were never explored in the context of finite reflection groups; one of the reasons is that we have, at this point, no idea what are the 'good' objects to consider in place of the standard simplex.

Problem 4.4. *For a finite reflection group W, find a suitable framework to define Coxeter generalized permutahedra and then express Coxeter generalized associahedra as Minkowski sums in this framework.*

Finally, the work of C. Ceballos and J.-P. Labbé and C. Stump [5] and V. Pilaud and C. Stump [18] on *spherical subword complexes* suggest that there is a more general family of polytopes, arising together with generalizations of the Tamari lattice, which include Coxeter generalized associahedra: they are called *generalized multi-associahedra*. This family is still awaiting polytopal realizations.

Problem 4.5 ([5]). *Give a polytopal realization of generalized multi-associahedra.*

Acknowledgement The author is more than grateful to Carsten Lange for allowing him to use some of the pictures he made for the articles [10], and also to Jean-Philippe Labbé for providing the original picture in TikZ which are the bases for the pictures of permutahedra and associahedra of the symmetric group S_4, the hyperoctahedral group W_3' and the group $W(H_3)$ presented in this article.

The author wishes also to thank Jean-Philippe Labbé, Carsten Lange, Vincent Pilaud and Christian Stump for many very interesting conversations which mostly took place at LaCIM (Laboratoire de Combinatoire et d'Informatique Mathématique) in Montréal during the summer of 2011. The author thanks Franco Saliola for his comments on a preliminary version of this text.

And finally, the author expresses his deepest gratitude to Jean-Louis Loday to have asked him the question *how to realize the cyclohedron from a permutahedron of the hyperoctahedral group* while he was working on his PhD at the *Université de Strasbourg*.

References

1. N. Bergeron, C. Hohlweg, C. Lange, and H. Thomas, "Isometry classes of generalized associahedra", *Séminaire Lotharingien de Combinatoire* **61** (2009) B61Aa.
2. A. Björner and F. Brenti, *Combinatorics of Coxeter Groups*, GTM, vol. 231, Springer, New York, 2005.

3. A.V. Borovik and A. Borovik, *Mirrors and Reflections: The Geometry of Finite Reflection Groups*, Springer, New York, 2010.
4. R. Bott and C. Taubes, "On the self-linking of knots", *J. Math. Phys.* **35** (1994) 5247–5287.
5. C. Ceballos, J.-P. Labbé, and C. Stump, "Subword complexes, cluster complexes, and generalized multi-associahedra", *preprint* (2011), *arxiv.org/abs/1108.1776*.
6. C. Ceballos, F. Santos, and G.M. Ziegler, "Many non-equivalent realizations of the associahedron", (2011), *arxiv.org/abs/1109.5544*.
7. F. Chapoton, S. Fomin, and A. Zelevinsky, "Polytopal realizations of generalized associahedra", *Canad. Math. Bull.* **45** (2003) 537–566.
8. S. Fomin and A. Zelevinsky, "Y-systems and generalized associahedra", *Annals of Math.* **158** (2003) 977–1018.
9. C. Hohlweg and C. Lange, "Realizations of the associahedron and cyclohedron", *Discrete Comput. Geom.* **37** (2007) 517–543.
10. C. Hohlweg, C. Lange, and H. Thomas, "Permutahedra and generalized associahedra", *Advances in math* **226** (2011) 608–640.
11. C. Hohlweg, J. Lortie, and A. Raymond, "The centers of gravity of the associahedron and of the permutahedron are the same", *Electronic Journal of Combinatorics* **17** (2009) R72.
12. J.E. Humphreys, *Reflection groups and Coxeter groups*, Cambridge University Press, Cambridge, 1990.
13. A. Knutson and E. Miller, "Subword complexes in Coxeter groups", *Adv. Math.* **184** (2004) 161–176.
14. C. Lange, "Minkowsky decompositions of associahedra", *preprint* (2011).
15. J.-L. Loday, "Realization of the Stasheff polytope", *Arch. Math.* **83** (2004) 267–278.
16. _____, "Dichotomy of the addition of natural numbers", in *this volume*.
17. M. Markl, "Simplex, associahedron, and cyclohedron", *Contemp. Math.* **227** (1999) 235–265.
18. V. Pilaud and C. Stump, "The brick polytope of a spherical subword complex", *in preparation* (2011).
19. A. Postnikov, "Permutohedra, associahedra and beyond", *Int. Math. Res. Not.* **6** (2009) 1026–1106.
20. A. Postnikov, V. Reiner, and L. Williams, "Faces of generalized permutohedra", *Documenta Math.* **13** (2008) 207–273.
21. N. Reading, "Cambrian lattices", *Adv. Math.* **205** (2006) 313–353.
22. _____, "Clusters, Coxeter-sortable elements and noncrossing partitions", *Trans. Amer. Math. Soc.* **359** (2007) 5931–5958.
23. _____, "Sortable elements and Cambrian lattices", *Algebra Universalis.* **56** (2007) 411–437.
24. _____, "From the Tamari lattice to Cambrian lattices and beyond", in *this volume*.
25. N. Reading and D. Speyer, "Cambrian fans", *J. Eur. Math. Soc. (JEMS)* **11** (2009) 407–447.
26. V. Reiner, "Equivariant fiber polytopes", *Documenta Math.* **7** (2002) 113–132.
27. S. Shnider and J. Stasheff, "Appendix B: Associahedra and cyclohedra as truncated simplices", in *Operads: Proceedings of Renaissance Conferences*, Contemp. Math., vol. 202, Americ. Math. Soc., March 1995 / May–June 1995 1997, 53–81.
28. S. Shnider and S. Sternberg, *Quantum Groups: From Coalgebras to Drinfeld Algebras*, Graduate texts in mathematical physics, International Press, 1994.
29. R. Simion, "A type-B associahedron", *Adv. Appl. Math.* **30** (2003) 2–25.
30. J. Stasheff, "Homotopy associativity of H-spaces I, II", *Trans. Amer. Math. Soc.* **108** (1963) 275–312.
31. G.M. Ziegler, *Lectures on Polytopes*, GTM, vol. 152, Springer, Berlin, 1995.

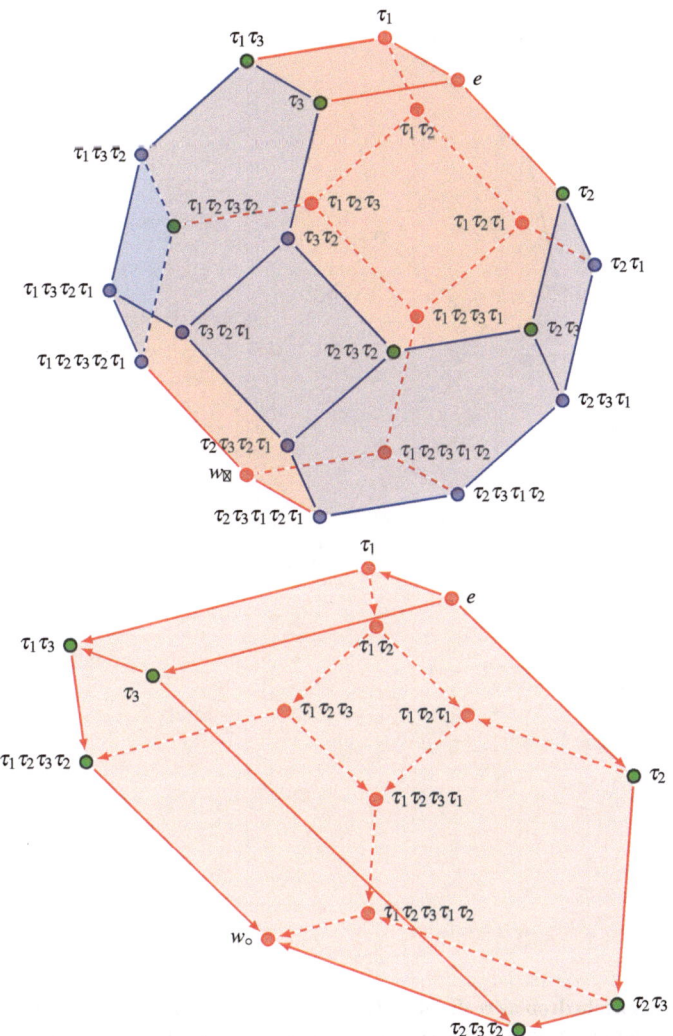

Fig. 12 The associahedron as realized by J.-L. Loday and S. Shnider-S. Sternberg. On the top is the classical permutahedron for the symmetric group S_4, see Example 2.15, and on the bottom is associahedron $\mathrm{Asso}^a_{\tau_1 \tau_2 \tau_3}(S_4)$ which corresponds to the associahedron realized by Loday and Shnider-Sternberg. The facets of $\mathrm{Asso}^a_{\tau_1 \tau_2 \tau_3}(S_4)$ containing Coxeter singletons, as well as the Coxeter singletons, are in red. In green are the vertices corresponding to the $\tau_1 \tau_2 \tau_3$-sortable elements. The orientations on the edges represent the $\tau_1 \tau_2 \tau_3$-Cambrian lattice which, is the Tamari lattice.

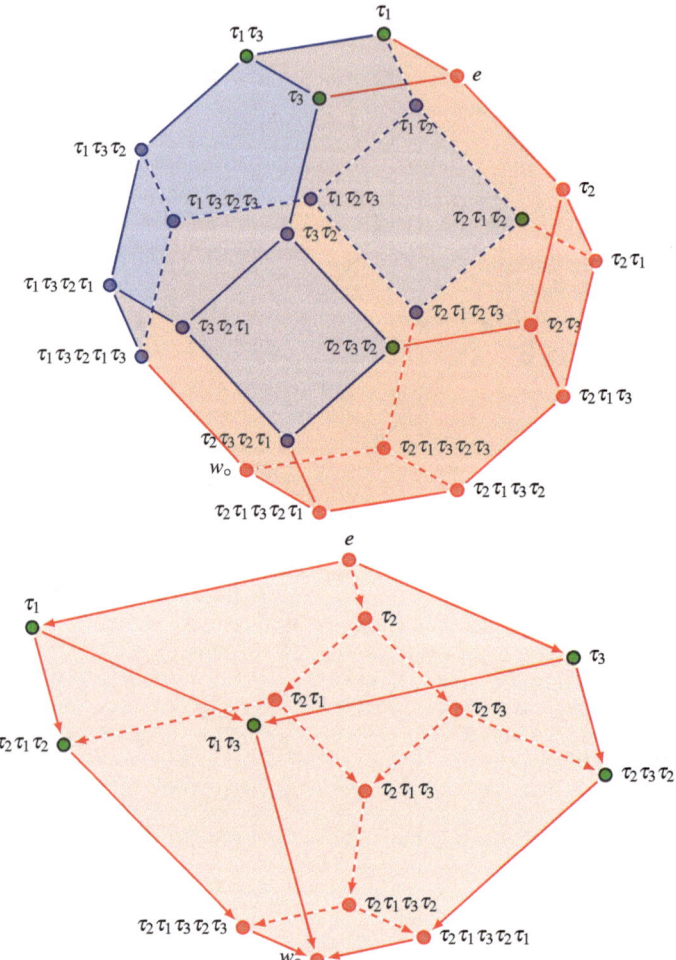

Fig. 13 The associahedron as realized by F. Chapoton, S. Fomin and A. Zelevinsky. On the top is the classical permutahedron for the symmetric group S_4, see Example 2.15, and on the bottom is associahedron $\mathrm{Asso}^{\boldsymbol{a}}_{\tau_2\tau_1\tau_3}(S_4)$ which corresponds to the associahedron realized by Chapoton, Fomin and Zelevinsky. The facets of $\mathrm{Asso}^{\boldsymbol{a}}_{\tau_2\tau_1\tau_3}(S_4)$ containing Coxeter singletons, as well as the Coxeter singletons, are in red. In green are the vertices corresponding to the $\tau_1\tau_2\tau_3$-sortable elements. The orientations on the edges represent the $\tau_2\tau_1\tau_3$-Cambrian lattice.

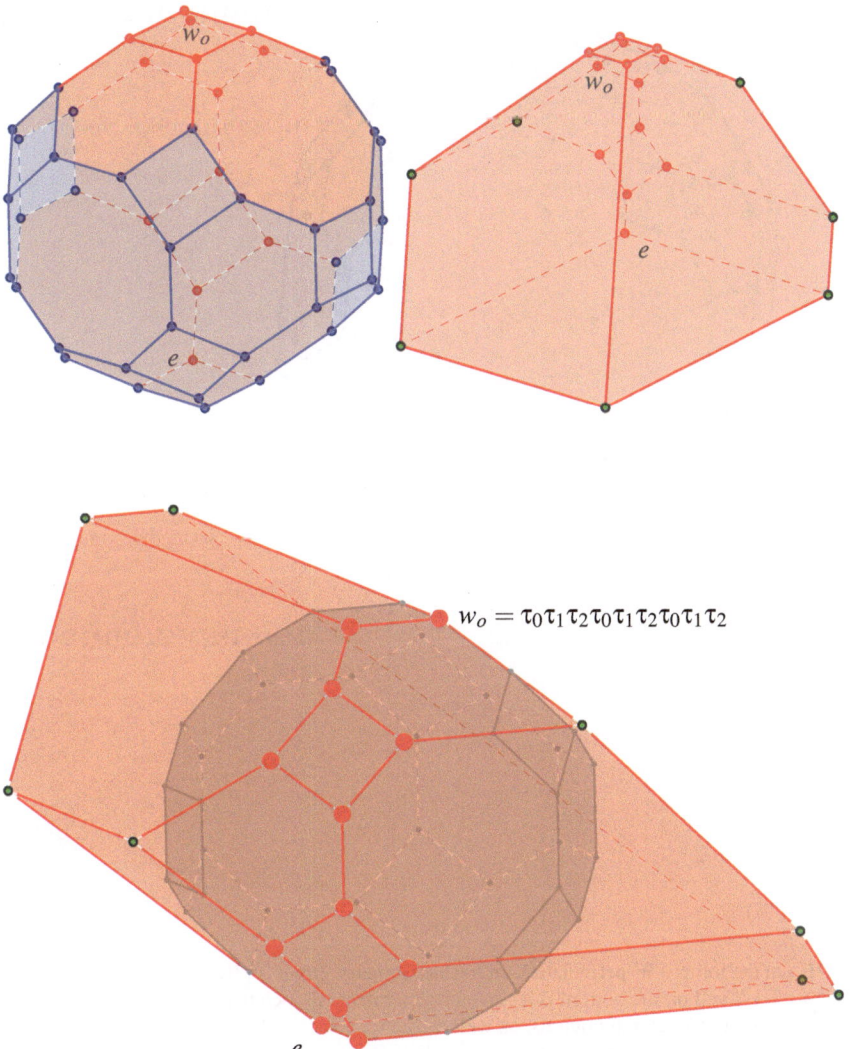

Fig. 14 The cyclohedron. Two Coxeter generalized associahedra for the hyperoctahedral W_3 (see Example 2.16). On the top left is $\mathrm{Perm}^a(W_3)$, with $w_o = \tau_1\tau_2 s_0\tau_1\tau_2 s_0\tau_1\tau_2 s_0$, and on the top right is $\mathrm{Asso}^a_{\tau_1\tau_2 s_0}(W_3)$. On the bottom we can see through a ghostly $\mathrm{Asso}^a_{s_0\tau_1\tau_2}(W_3)$ the permutahedron $\mathrm{Perm}^a(W_3)$ in gray. In both examples, the facets containing Coxeter singletons, as well as the Coxeter singletons, are in red.

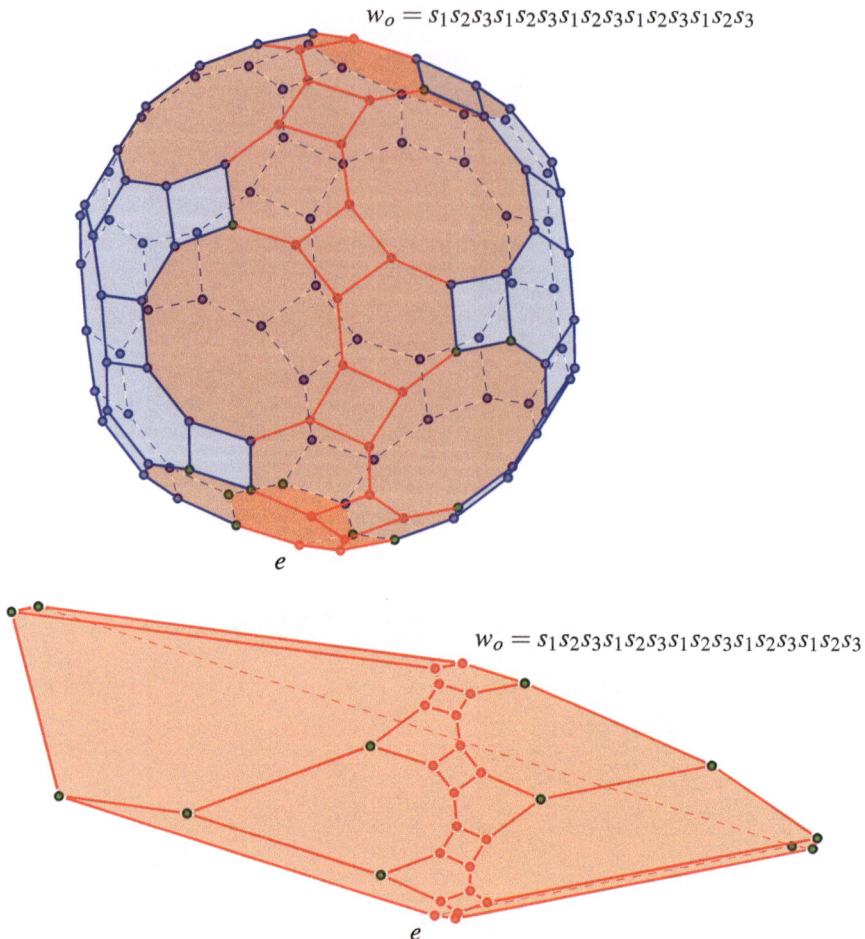

Fig. 15 On the top is a W-permutahedron for the isometry group $W(H_3)$ of the dodecahedron, see Example 2.17, and on the bottom is a Coxeter generalized associahedra $\mathrm{Asso}^a_{s_1 s_2 s_3}(W(H_3))$. The facets containing Coxeter singletons, as well as the Coxeter singletons, are in red. Note that the Coxeter generalized associahedron is reduced to half of its original size to fit in the picture; the size of the W-permutahedron is unchanged.

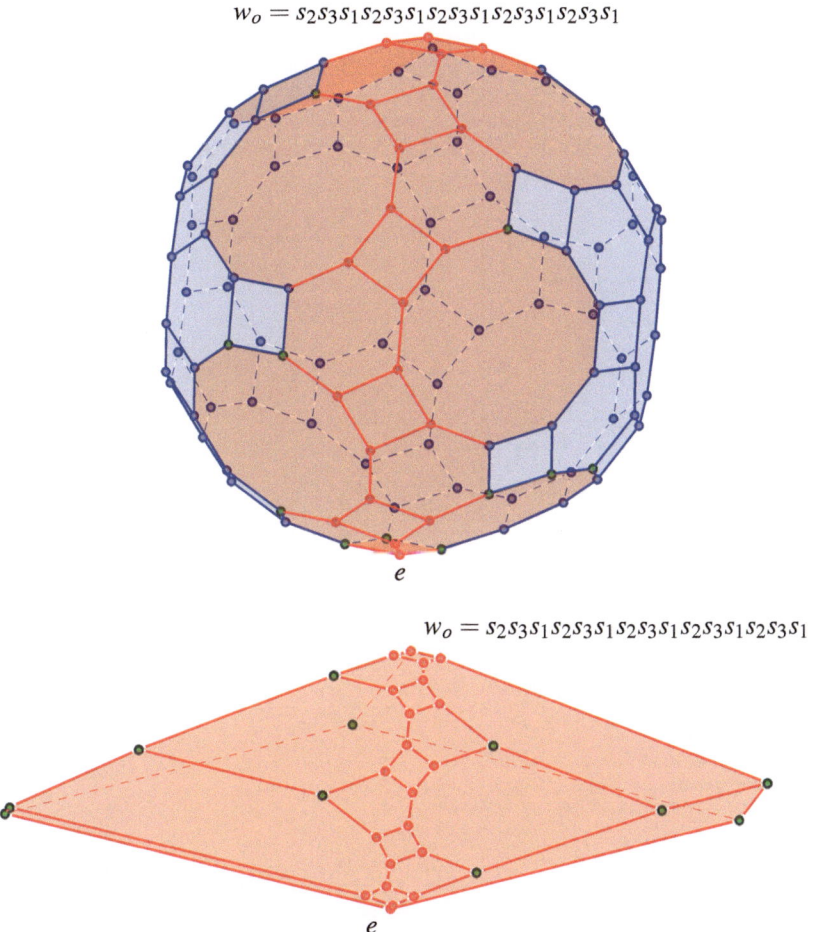

Fig. 16 On the top is a W-permutahedron for the isometry group $W(H_3)$ of the dodecahedron, see Example 2.17, and on the bottom is a Coxeter generalized associahedra $\mathrm{Asso}^a_{s_2s_3s_1}(W(H_3))$. The facets containing Coxeter singletons, as well as the Coxeter singletons, are in red. Note that the Coxeter generalized associahedron is reduced to half of its original size to fit in the picture; the size of the W-permutahedron is unchanged.

Combinatorial 2-truncated Cubes and Applications

Victor M. Buchstaber and Vadim D. Volodin

Abstract We study a class of simple polytopes, called 2-truncated cubes. These polytopes have remarkable properties and, in particular, satisfy Gal's conjecture. Well-known polytopes (flag nestohedra, graph-associahedra and graph-cubeahedra) are 2-truncated cubes.

1 Introduction

Stasheff polytopes have many geometric realizations which are not affinely equivalent. The starting point of this work is a realization of Stasheff polytopes obtained from cubes by truncations of faces of codimension 2. In the focus of our interest is the family of polytopes that are obtained from cubes by a sequence of truncations of faces of codimension 2. Such polytopes will be called 2-*truncated cubes* and a truncation of a face of codimension 2 will be called a 2-*truncation*. We will show that 2-truncated cubes have remarkable properties and this family contains classes of polytopes playing important roles in different areas of mathematics. Note that every n-dimensional 2-truncated cube is a simple flag polytope P^n and, moreover, it is an image of the moment map for some smooth toric variety M_P^{2n}. It is well known that the odd Betti numbers of M_P^{2n} are zero and the even Betti numbers are equal to the components of the h-vector of P^n. Truncation of a face of codimension 2 corresponds to a blow-up of the variety M_P^{2n} along a subvariety of complex codimension 2. For the well-known f-, h-, g- and γ-vectors, we explicitly describe their transformation under 2-truncations.

Victor M. Buchstaber
Steklov Mathematics Institute, Russian Academy of Sciences, Moscow, Russia
e-mail: *buchstab@mi.ras.ru*

Vadim D. Volodin
Steklov Mathematics Institute, Russian Academy of Sciences, Moscow, Russia
e-mail: *volodinvadim@gmail.com*

S. Gal conjectured in [14] that all components of the γ-vector of a flag general-ized homological sphere (and, therefore, a γ-vector of a flag simple polytope) are nonnegative. This was a generalization of the Charney-Davis conjecture (see [6]), which was formulated in terms of h-vectors and is equivalent to the nonnegativity of the last component of the γ-vector. In the work of R. Charney and M. Davis and later in the work of S. Gal, the operation dual to 2-truncation was used to support their conjectures. Gal's conjecture became widely known because it is connected not only with the combinatorics of sphere triangulations, but also with problems of differential geometry and the topology of manifolds. This conjecture was proved in special cases.

Using the transformation formula for the γ-vector under a 2-truncation, we obtain a proof of Gal's conjecture for our family. There is a well-known formula in toric geometry, $\sum_{i=0}^{2n}(-1)^i h_i(P^{2n}) = (-1)^n \sigma(M_P^{4n})$, where P^{2n} is the image of the moment map of a smooth toric variety M_P^{4n} and σ is the signature. The left part is equal to the last component $\gamma_n(P^{2n})$ of the γ-vector. Then our result shows that $(-1)^n \sigma(M_P^{4n})$ is nonnegative if the image of the moment map is a 2-truncated cube.

An important class of simple polytopes is that of nestohedra. These polytopes arose in the work of C. De Concini and C. Procesi (see [7]). Nestohedra were constructed as Minkowski sums of certain sets of simplices corresponding to some building set. Attention was drawn to nestohedra due to the work of A. Postnikov, V. Reiner, L. Williams (see [20]), who obtained important results about their combinatorics. In particular, Gal's conjecture was proved for chordal building sets. We show that a nestohedron is a 2-truncated cube if and only if it is a flag polytope. As a corollary, we obtain a proof of Gal's conjecture for the family of all flag nestohedra.

A wide class of flag nestohedra, the graph-associahedra, were introduced by M. Carr and S. Devadoss in [3]. Among them are the Stasheff polytopes (associahe-dra), Bott-Taubes polytopes (cyclohedra), and the permutohedra. Graph-associahedra can be described as nestohedra, where the building set is constructed in a natural way from a graph. Since every graph-associahedron is a 2-truncated cube, we obtain (see Example 8.9) realizations of Stasheff polytopes that are not equivalent to realizations described in [5], [15] and [20].

S. Fomin and A. Zelevinsky (see [13]) introduced a new class of polytopes corresponding to cluster algebras related to Dynkin diagrams. It was shown by M. Gorsky (see [16]) that the polytopes corresponding to diagrams of the D-series are not nestohedra, but each of them is a 2-truncated cube.

The face lattices of the Stasheff polytopes are the well-known Tamari lattices. Since Stasheff polytopes are 2-truncated cubes, Tamari lattices can be obtained from Boolean lattices by a special operation corresponding to the 2-truncation. As a result, we obtain a family of lattices from Boolean lattices by sequences of these operations.

Stasheff polytopes have extremal properties: their f-, h-, g- and γ- vectors are componentwise minimal in each dimension among the graph-associahedra corre-sponding to connected graphs. We describe classes of graph-associahedra among which cyclohedra, stellohedra and permutohedra having extremal properties.

We describe geometric operations that transform an n-dimensional graph-associa-hedron to an $(n + 1)$-dimensional one. It allows us to consider series of graph-associahedra and to describe their combinatorics in terms of differential and functional equations for the generating functions of face polynomials. Similar equations were obtained using the ring of simple polytopes (see [1]). For example, the famous Hopf equation describes the generating series of H-polynomials of Stasheff polytopes.

In the work of S.L. Devadoss, T. Heath and W. Vipismakul, it was shown that some moduli spaces of marked bordered surfaces have a polytopal stratification. In [9] a class of simple polytopes, called graph-cubeahedra, were introduced. They generalize polytopes associated with moduli spaces. This class contains some well-known series (for example, associahedra) and a new sequence of polytopes called halohedra.

We introduce a class of simple n-polytopes $NP(P,B)$, called nested polytopes. A member is defined by a pair (P,B), where P is a simple n-polytope with fixed order of facets and B is a building set on $[n]$. We show that the nested polytope $NP(P,B)$ is flag if both the polytope P and the nestohedron P_B are flag. The nested polytope $NP(P,B)$ is a 2-truncated cube if P is a 2-truncated cube and P_B a flag polytope.

We show that graph-cubeahedra are a special case of nested polytopes $NP(P,B)$, where P is the n-cube and B is a graphical building set. As a corollary, we obtain that any graph-cubeahedron is a 2-truncated cube.

2 Simple polytopes

A convex n-dimensional polytope P is called *simple* if each vertex belongs to exactly n facets.

A simple polytope P is called *flag* if every collection of its pairwise intersecting faces has a nonempty intersection.

2.1 Enumerative polynomials

Let f_i be the number of i-dimensional faces of an n-dimensional polytope P. The vector $(f_0,\ldots,f_n)$ is called the f-vector of P. The F-polynomial of P is defined by

$$F(P)(\alpha,t) = \alpha^n + f_{n-1}\alpha^{n-1}t + \cdots + f_1\alpha t^{n-1} + f_0 t^n,$$

The h-vector $(h_0,\ldots,h_n)$ and the H-polynomial of P are defined by

$$H(P)(\alpha,t) = F(P)(\alpha-t,t) = h_0\alpha^n + h_1\alpha^{n-1}t + \cdots + h_{n-1}\alpha t^{n-1} + h_n t^n.$$

The g-vector of a simple polytope P is the vector $(g_0,g_1,\ldots,g_{[\frac{n}{2}]})$, where $g_0 = 1$ and $g_i = h_i - h_{i-1}$ for $i > 0$.

The following formulas connect the f-, h- and g-vectors:

$$f_i(P) = \sum_{j=i}^{n} \binom{j}{i} h_{n-j}(P); \qquad h_i(P) = \sum_{j=0}^{i} g_j(P) . \tag{1}$$

According to the Dehn-Sommerville equations (see [22]), $H(P)$ is symmetric for any simple polytope (see [1]). It can therefore be expressed as a polynomial of $\alpha + t$ and αt:

$$H(P) = \sum_{i=0}^{[\frac{n}{2}]} \gamma_i (\alpha t)^i (\alpha + t)^{n-2i} . \tag{2}$$

The γ-vector of P is the vector $(\gamma_0, \gamma_1, \ldots, \gamma_{[\frac{n}{2}]})$. The γ-polynomial of P is defined by

$$\gamma(P)(\tau) = \gamma_0 + \gamma_1 \tau + \cdots + \gamma_{[\frac{n}{2}]} \tau^{[\frac{n}{2}]} .$$

The next formula (see [1]) connects the g- and the γ-vectors.

$$g_i(P) = (n - 2i + 1) \sum_{j=0}^{i} \frac{1}{n-i-j+1} \binom{n-2j}{i-j} \gamma_j(P) . \tag{3}$$

Proposition 2.1. *Let* $\gamma_i(P_1) \leq \gamma_i(P_2), i = 0, \ldots, [\frac{n}{2}]$, *where* P_1 *and* P_2 *are simple n-polytopes. Then*

1) $g_i(P_1) \leq g_i(P_2)$,
2) $h_i(P_1) \leq h_i(P_2)$,
3) $f_i(P_1) \leq f_i(P_2)$.

Proof. This follows from the nonnegativity of the coefficients in (1) and (3)   □

Gal's conjecture (see [14]) in the case of convex polytopes can be formulated as follows.

Conjecture 2.2. *Any flag simple n-polytope P satisfies* $\gamma_i(P^n) \geq 0, i = 0, \ldots, [\frac{n}{2}]$.

3 Class of 2-truncated cubes

Truncation of faces plays a key role in this work. We will use it as a combinatorial operation (dual to stellar subdivision) and as a geometric operation. We consider the combinatorial construction first, because it is stricter.

The stellar subdivision of a complex K over a simplex $\sigma \in K$ is denoted by $(v_0, \sigma)K$, where v_0 is a new vertex. The link of the simplex $\sigma \in K$ is denoted by $\mathrm{lk}(\sigma, K)$. The join of complexes K and L is denoted by $K \star L$. We will use $\approx$ to indicate combinatorial equivalence.

Recall that, for every polytope P, there exists a *dual polytope* P^*. Its partially ordered set of faces is inverse to the partially ordered set of faces of P. If P is simple, then P^* is simplicial (all its faces are simplices). If P is simplicial, then P^* is simple.

Definition 3.1. We say that a simple polytope Q is obtained from a simple polytope P by truncation of the face $G \subset P$, if the simplicial complex ∂Q^* is obtained from the simplicial complex ∂P^* by stellar subdivision over the simplex σ_G corresponding to the face G, i.e., if $\partial Q^* = (v_0, \sigma_G)\partial P^*$. The polytope Q has a facet corresponding to the vertex $v_0 \in \partial Q^*$ (added to ∂P^*).

Geometric realization of the truncation. Let $P \subset \mathbb{R}^n$ be a simple polytope containing 0 in its relative interior, and let G be a face of P. Let $l_G \in \mathbb{R}^{n*}$ be a linear function such that $l_G(P) \le 1$ and $\{x \in P : l_G x = 1\} = G$. The polytope Q, obtained from P by truncation of the face G, can be realized as $Q = \{x \in P : l_G x \le 1 - \varepsilon\}$. Here $\varepsilon > 0$ is so small that all the vertices of P, except the vertices of G, satisfy $l_G x < 1 - \varepsilon$. Informally, the polytope Q is obtained from P by shifting the support hyperplane of G inside the polytope P. The new facet F_0 of the polytope Q, corresponding to a new vertex $v_0 \in \partial Q^*$, lies in the hyperplane of the section. We will call it the *section facet* F_0.

Remark 3.2. Let P be a simple polytope and the face $G \subset P$ correspond to the simplex $\sigma_G \subset \partial P^*$. Then the complexes ∂G^* and $\mathrm{lk}(\sigma_G, \partial P^*)$ are isomorphic.

Proposition 3.3. *Let the polytope Q be obtained from the simple n-polytope P by truncation of the face G of dimension k. Then the section facet F_0 is combinatorially equivalent to $G \times \Delta^{n-k-1}$.*

Proof. Truncation of the face G corresponds to a stellar subdivision, so we have $\partial Q^* = (v_0, \sigma_G)\partial P^*$. From the properties of stellar subdivisions, we obtain

$$\mathrm{lk}(v_0, \partial Q^*) \simeq \mathrm{lk}(\sigma_G, \partial P^*) \star \partial \Delta^{n-k-1} .$$

The proof is completed by

$$\partial F_0^* \simeq \mathrm{lk}(v_0, \partial Q^*) \simeq \partial G^* \star \partial \Delta^{n-k-1} \simeq \partial (G \times \Delta^{n-k-1})^* . \qquad \square$$

Proposition 3.4. *Let the polytope Q be obtained from the simple n-polytope P by truncation of a face G of dimension k. Then*

1) $H(Q) = H(P) + \alpha t H(G) H(\Delta^{n-k-2})$,
2) $\gamma(Q) = \gamma(P) + \tau \gamma(G) \gamma(\Delta^{n-k-2})$.

Proof. The truncation removes the face G and creates the face $G \times \Delta^{n-k-1}$, so that

$$F(Q) = F(P) + t F(G) F(\Delta^{n-k-1}) - t^{n-k} F(G) .$$

Thus

$$H(Q) = H(P) + tH(G)H(\Delta^{n-k-1}) - t^{n-k}H(G)$$

$$= H(P) + tH(G)\left(\sum_{i=0}^{n-k-1} \alpha^i t^{n-i} - t^{n-k-1} \right)$$

$$= H(P) + \alpha t H(G)\left(\sum_{i=0}^{n-k-2} \alpha^i t^{n-i} \right) = H(P) + \alpha t H(G) H(\Delta^{n-k-2}),$$

and, moreoever,

$$H(Q) = \sum_{i=0}^{[\frac{n}{2}]} \gamma_i(P)(\alpha t)^i (\alpha + t)^{n-2i}$$

$$+ \alpha t \left(\sum_{i=0}^{[\frac{k}{2}]} \gamma_i(G)(\alpha t)^i (\alpha + t)^{k-2i} \right) \left(\sum_{j=0}^{[\frac{n-k-2}{2}]} \gamma_j(\Delta^{n-k-2})(\alpha t)^j (\alpha + t)^{n-k-2-2j} \right)$$

$$= \sum_{i=0}^{[\frac{n}{2}]} \gamma_i(P)(\alpha t)^i (\alpha + t)^{n-2i} + \sum_{i=0}^{[\frac{k}{2}]} \sum_{j=0}^{[\frac{n-k-2}{2}]} \gamma_i(G)\gamma_j(\Delta^{n-k-2})(\alpha t)^{i+j+1}(\alpha + t)^{n-2(i+j+1)}.$$

Then, by definition of the γ-vector, $\gamma_i(Q) = \gamma_i(P) + \sum_{p+q=i-1}\gamma_p(G)\gamma_q(\Delta^{n-k-2})$. The proof is now complete.   □

Definition 3.5. A truncation of a face of codimension 2 will be called a 2-*truncation*. A combinatorial polytope, obtained from a cube by 2-truncations, will be called a 2-*truncated cube*.

The following is the dual of a result that can be found in [14].

Corollary 3.6. *Let the polytope Q be obtained from a simple polytope P by 2-truncation of the face G. Then*

1) $H(Q) = H(P) + \alpha t H(G)$,
2) $\gamma(Q) = \gamma(P) + \tau \gamma(G)$.

Remark 3.7. Let K be a simplicial complex and $K' = (v_0, A)K$ its stellar subdivision over $A = \{v_1, \ldots, v_k\} \in K$. An arbitrary set V of vertices of K' forms a simplex iff one of the following conditions holds:

a) $v_0 \notin V$ and $\{v_1, \ldots, v_k\} \not\subseteq V$ and $V \in K$,
b) $v_0 \in V$ and $\{v_1, \ldots, v_k\} \not\subseteq V$ and $\{v_1, \ldots, v_k\} \cup (V \setminus \{v_0\}) \in K$.

Lemma 3.8. *Any 2-truncation keeps flagness.*

Proof. Let Q be obtained from P by truncation of the face $G = F_1 \cap F_2$. Then $\partial Q^* = (v_0, \sigma_G)\partial P^*$, where $\sigma_G = \{v_1, v_2\}$, and v_1, v_2 are the vertices corresponding to facets F_1, F_2. Let the vertices $V \subset \partial Q^*$ be pairwise adjacent. Note that one of the vertices v_1, v_2 is not contained in V. The vertices $V \setminus \{v_0\}$ are pairwise adjacent in the complex ∂P^*, and then $V \setminus \{v_0\} \in \partial P^*$. If $v_0 \notin V$, then $V \in \partial Q^*$ according to a) of Remark 3.7. If $v_0 \in V$, then $V \setminus \{v_0\} \in \mathrm{lk}(v_0, \partial Q^*) = \mathrm{lk}(\{v_1, v_2\}, \partial P^*)$, hence $V \in \partial Q^*$ according to b) of Remark 3.7.   □

Proposition 3.9. *Every face of a 2-truncated cube is a 2-truncated cube.*

Proof. We show that if P is a 2-truncated cube, then all the facets of P are 2-truncated cubes. The proof is by induction on the number of truncated faces. Let the polytope Q be obtained from a 2-truncated cube P by 2-truncation of a face G of codimension 2. Then the section facet has the form $F_0 \approx G \times I$, and is thus a 2-truncated cube by the induction assumption. Every other facet F' of the polytope Q is either some facet of P, or obtained from some facet F'' of P by 2-truncation of a face $G' \subseteq F''$. $\quad\square$

Proposition 3.10. *Every 2-truncated cube P satisfies $\gamma_i(P) \geq 0$, i.e., Gal's conjecture holds for 2-truncated cubes.*

Proof. According to Proposition 3.9, each face of a 2-truncated cube is a 2-truncated cube. We can therefore complete the proof by induction on the dimension of P, using the formula $\gamma(Q) = \gamma(P) + \tau\gamma(G)$. $\quad\square$

Definition 3.11. The convex polytope $P \subset \mathbb{R}^n$ is called a *Delzant polytope* if at each vertex the normal vectors of the facets through the vertex can be chosen to form a $\mathbb{Z}$-basis for the integer lattice $\mathbb{Z}^n \subset \mathbb{R}^n$.

Note that the required normal vectors can be chosen as outer integer prime normals to the corresponding facets, i.e., integer vectors directed outside of P having coprime components.

Proposition 3.12. *Let $P \subset \mathbb{R}^n$ be a Delzant polytope and $G = F_1 \cap \cdots \cap F_k$ a face. Let the polytope Q be obtained from P by truncation of G in such a way that the outer normal to the section facet F_0 is $v_0 = v_1 + \cdots + v_k$, where v_i are the outer integer prime normals to the facets F_i. Then the polytope Q is a Delzant polytope.*

Proof. It is sufficient to prove that, for each vertex $q \in F_0$ of the polytope Q, outer integer prime normals containing q form a basis of the integer lattice. By Remark 3.7, $q = F_0 \cap F_1 \cap \cdots \cap \widehat{F_i} \cap \cdots \cap F_k \cap G'$, where F_i is omitted, and the intersection $G \cap G'$ is a vertex in P (i.e., F_i is replaced by F_0 in the expression $F_1 \cap \cdots \cap F_k \cap G'$). Without loss of generality, we can assume $i = 1$.

Let $v_1, \ldots, v_k$ be outer integer prime normals to facets $F_1, \ldots, F_k$ and $v_{k+1}, \ldots, v_n$ outer integer prime normals to facets containing G'. Then $\det(v_0, v_2, \ldots, v_n) = \det(v_1 + \cdots + v_k, v_2, \ldots, v_n) = \det(v_1, v_2, \ldots, v_n) = \pm 1$. $\quad\square$

Every Delzant n-polytope P^n with m facets has a canonical characteristic function $\Lambda : \mathbb{Z}^m \to \mathbb{Z}^n$ (see [2]), which is a linear map given by the $n \times m$ matrix A which has as its j-th column the vector of components of the outer integer prime normal to the j-th facet F_j. Using Proposition 3.12, we can explicitly construct the matrix A for any 2-truncated cube P^n with m facets. The n-cube is realized by $-1 \leq x_i \leq 1, i = 1, \ldots, n$. Therefore the first n columns form the identity matrix E, the next n columns form the matrix $-E$, and every other column has the form $A_j = A_{j_1} + A_{j_2}$, where $j_1, j_2 < j$ and F_j is the section facet appearing in the $(j - 2n)$-th step after truncation of the face $F_{j_1} \cap F_{j_2}$.

4 Smooth toric varieties over 2-truncated cubes

Every 2-truncated cube is a Delzant polytope, so it is the image of the moment map for some smooth toric variety. There is a well-known formula in toric geometry.

Theorem 4.1 ([17, Theorem 3.12]). *If the n-polytope P^n is the image of the moment map for a smooth toric variety M_P^{2n}, then*

$$\sigma(M_P^{2n}) = \sum_{k=0}^{n} (-1)^k h_k(P^n) .$$

The signature $\sigma(M_P^{2n})$ is zero for odd-dimensional polytopes. Consider a polytope P^{2n} and the corresponding toric variety M_P^{4n}. Setting $\alpha = 1$ and $t = -1$ in formula (2), we obtain

$$\sigma(M_P^{4n}) = \sum_{k=0}^{2n} (-1)^k h_k(P^{2n}) = (-1)^n \gamma_n(P^{2n}) . \qquad (4)$$

Corollary 4.2. $(-1)^n \sigma(M_P^{4n}) \geq 0$ *for every 2-truncated cube P^{2n}.*

Corollary 4.3. $\sigma(M_P^{4n}) = 0$ *for every 2-truncated cube P^{2n} with less than 5n facets.*

Proof. The assumption is equivalent to the number of 2-truncations being less than n. By Corollary 3.6, we have $\gamma(Q) = \gamma(P) + \tau\gamma(G)$, where Q is obtained from P by a 2-truncation of the face G. Then, using induction on the dimension of the polytope and the number of facets, we obtain the stated result. $\qquad \square$

5 Small covers of 2-truncated cubes

In [8], the notion of a *small cover* M^n of a simple n-polytope P^n with mod 2 characteristic function on it was introduced. This is a smooth manifold with $\mathbb{Z}_2^n$ action (locally isomorphic to the standard representation of $\mathbb{Z}_2^n$ on $\mathbb{R}^n$) such that $M^n/\mathbb{Z}_2^n \simeq P^n$. The small cover M^n of any 2-truncated cube P^n is an Eilenberg-Mac Lane space $K(G, 1)$ (see [18], Proposition 4.10), since 2-truncated cubes are flag. The group G can be determined (see [8], Corollary 4.5) in terms of a right-angled Coxeter group and the characteristic function of P^n, which is explicitly constructed in §3. It was proved (see [8], Theorem 3.1) that mod 2 Betti numbers of M^n are equal to the components of the h-vector of P^n. For P^{2n-1}, the Euler characteristic of the small cover M^{2n-1} is zero. Setting $\alpha = 1$ and $t = -1$ in (2), we obtain the following formula for the Euler characteristic of the small cover M^{2n} of P^{2n}.

$$\chi(M^{2n}) = \sum_{k=0}^{2n} (-1)^k h_k(P^{2n}) = (-1)^n \gamma_n(P^{2n}) . \qquad (5)$$

We note that the Euler characteristic of the small cover of P is equal to the signature of the smooth toric variety associated with P.

Corollary 5.1. $(-1)^n \chi(M^{2n}) \geq 0$ *for every 2-truncated cube* P^{2n}.

Using Corollary 4.3, we obtain the following result.

Corollary 5.2. $\chi(M^{2n}) = 0$ *for every 2-truncated cube* P^{2n} *with less than* $5n$ *facets.*

6 Nestohedra and graph-associahedra

In this section we recall some well-known facts about nestohedra (see [12], [19], [21]).

Notation 6.1. By $[n]$ and $[i,j]$ we denote the sets $\{1,\ldots,n\}$ and $\{i,\ldots,j\}$, respectively.

Definition 6.2. A collection B of nonempty subsets of $[n+1]$ is called a *building set* on $[n+1]$ if the following conditions hold:

1) If $S_1, S_2 \in B$ and $S_1 \cap S_2 \neq \emptyset$, then $S_1 \cup S_2 \in B$;
2) $\{i\} \in B$ for every $i \in [n+1]$.

The building set B is *connected* if $[n+1] \in B$.

Remark 6.3. It is often convenient to consider an arbitrary set $A, |A| = n+1$, instead of the set $[n+1]$. Building sets are considered up to the following equivalence: a building set B_1 on A_1 is equivalent to a building set B_2 on A_2 if there exists a bijection $\sigma: A_1 \to A_2$ which induces a bijection between B_1 and B_2. We will assume that $A = [n+1]$, if not specified else.

The *restriction* of a building set B to $S \in B$ is the following building set on S:

$$B|_S = \{S' \in B: S' \subseteq S\}.$$

The *contraction* of a building set B with respect to $S \in B$ is the following building set on $[n+1] \setminus S$:

$$B/S = \{S' \subseteq [n+1] \setminus S: S' \in B \text{ or } S' \cup S \in B\} = \{S' \setminus S, S' \in B\}.$$

The *product* $B_1 \cdot B_2 = B_1 \sqcup B_2$ of building sets B_1 and B_2 on A and B, where $A \cap B = \emptyset$, is the building set on $A \sqcup B$ consisting of all elements of both building sets.

Remark 6.4. Any building set B on $[n+1]$ is the product of connected building sets: $B = B_1 \sqcup \cdots \sqcup B_k$, where B_i is a connected building set on A_i. The sets A_i are maximal in B (i.e., there is no $A_i' \in B$ containing A_i), and $B_i = B|_{A_i}$. We will denote the collection $\{A_i\}$ by $B_{\max}$. If B is connected, then $B_{\max} = \{[n+1]\}$.

Recall that a graph is called *simple* if it has no loops or multiple edges.

Definition 6.5. Let Γ be a simple graph on the node set $[n+1]$. The *graphical building set* $B(\Gamma)$ is the collection of nonempty subsets $S \subseteq [n+1]$ such that the induced subgraph $\Gamma|_S$ on the node set S is connected.

Remark 6.6. A building set $B(\Gamma)$ is connected iff Γ is connected.

Remark 6.7. Let Γ be a connected graph on $[n+1]$ and $S \in B(\Gamma)$. Then $B|_S$ and B/S are both graphical building sets corresponding to connected graphs $\Gamma|_S$ and Γ/S. The node set of Γ/S is the set $[n+1] \setminus S$. Vertices v and w are adjacent in Γ/S if they are either adjacent in Γ, or if they are both adjacent to some vertices from S in Γ.

Let M_1 and M_2 be subsets of $\mathbb{R}^n$. The *Minkowski sum* of M_1 and M_2 is the following subset of $\mathbb{R}^n$:

$$M_1 + M_2 = \{x \in \mathbb{R}^n : x = x_1 + x_2, \ x_1 \in M_1, \ x_2 \in M_2\}.$$

If M_1 and M_2 are convex polytopes, so is $M_1 + M_2$.

Definition 6.8. Let e_i be the endpoints of the basis vectors of $\mathbb{R}^{n+1}$. We define the *nestohedron* P_B corresponding to the building set B as follows,

$$P_B = \sum_{S \in B} \Delta^S, \text{ where } \Delta^S = \mathrm{conv}\{e_i, i \in S\}.$$

If $B(\Gamma)$ is a graphical building set, then $P_\Gamma := P_{B(\Gamma)}$ is called a *graph-associahedron*.

Definition 6.9. A building set B is called *flag* if the corresponding nestohedron P_B is a flag polytope.

Example 6.10. We are especially interested in the following series of graph-associahedra:

- Let L_{n+1} be the path graph on $[n+1]$. Then the polytope $P_{L_{n+1}}$ is called *associahedron* (Stasheff polytope) and denoted by As^n.
- Let C_{n+1} be the cyclic graph on $[n+1]$. Then the polytope $P_{C_{n+1}}$ is called *cyclohedron* (Bott-Taubes polytope) and denoted by Cy^n.
- Let K_{n+1} be the complete graph on $[n+1]$. Then the polytope $P_{K_{n+1}}$ is called *permutohedron* and denoted by Pe^n.
- Let $K_{1,n}$ be the complete bipartite graph on $[n+1]$. Then the polytope $P_{K_{1,n}}$ is called *stellohedron* and denoted by St^n.

Proposition 6.11 (cf. [12, §2]). *Let B be a connected building set on $[n+1]$. Then*

1) *the nestohedron P_B is a simple n-polytope given by the intersection of the hyperplane $H = \{\sum_{i=1}^{n+1} x_i = |B|\}$ with half-spaces $H_S = \{\sum_{i \in S} x_i \geq |B|_S|\}$, where $S \in B \setminus [n+1]$;*
2) *every facet of P_B has the form $F_S = P_B \cap \partial H_S$, where $S \in B \setminus [n+1]$, and it is combinatorially equivalent to the nestohedron $P_{B|_S} \times P_{B/S}$.*

Proposition 6.12 ([20, Corollary 7.2]). *Every graph-associahedron is a flag polytope, i.e., every graphical building set is flag.*

Definition 6.13. Let B be a building set on $[n+1]$. The collection $\mathscr{S} = \{S_1, \ldots, S_k\} \subseteq B$ is called a *nested set* if the following conditions hold:

1) $\forall S_i, S_j$: either $S_i \subset S_j$, or $S_i \supset S_j$, or $S_i \cap S_j = \emptyset$,
2) $\forall S_{i_1}, \ldots, S_{i_p}$ such that $S_{i_j} \cap S_{i_l} = \emptyset$: $S_{i_1} \sqcup \cdots \sqcup S_{i_p} \notin B$.

Definition 6.14. Let B be a building set on $[n+1]$. The *nested set complex* $\mathscr{N}(B)$ is the simplicial complex on the node set $B \setminus B_{\max}$ consisting of all the nested sets $\mathscr{S} \subseteq B \setminus B_{\max}$.

The next proposition allows us to describe the combinatorics of P_B in terms of elements of a building set B.

Proposition 6.15 ([19, Theorem 7.4], [12, Theorem 3.14]). *Let B be a building set on $[n]$. The nestohedron P_B is a simple polytope of dimension $n - |B_{\max}|$ and the simplicial complexes ∂P_B^* and $\mathscr{N}(B)$ are isomorphic. The facets $F_{S_1}, \ldots, F_{S_k}$ of the polytope P_B have a nonempty intersection iff $\{S_1, \ldots, S_k\} \in \mathscr{N}(B)$.*

Remark 6.16. If $B = B_1 \sqcup \cdots \sqcup B_k$, then $\mathscr{N}(B) \simeq \mathscr{N}(B_1) \star \cdots \star \mathscr{N}(B_k)$ and $P_B \approx P_{B_1} \times \cdots \times P_{B_k}$.

Notation 6.17. Occasionally, we will write "facet S" instead of "facet F_S".

7 Building set as a structure

Let us show (see Corollary 7.6) that every nestohedron is combinatorially equivalent to some nestohedron corresponding to a connected building set. Since we consider building sets up to equivalence, $[k]$ and $[l]$ will eventually denote disjoint sets consisting of k, respectively l elements.

Construction 7.1 ([10]). Let $B_1, \ldots, B_{n+1}$ be connected building sets on $[k_1]$, ..., $[k_{n+1}]$. Then, for every connected building set B on $[n+1]$, there is a connected building set $B(B_1, \ldots, B_{n+1})$ on $[k_1] \sqcup \cdots \sqcup [k_{n+1}] = [k_1 + \cdots + k_{n+1}]$, consisting of elements $S^i \in B_i$ and $\bigsqcup_{i \in S} [k_i]$, where $S \in B$.

Notation 7.2. When $B_1, \ldots, B_n$ are singletons, $\{1\}, \ldots, \{n\}$, we will simply write $B(1, 2, \ldots, n, B_{n+1})$ instead of $B(\{1\}, \{2\}, \ldots, \{n\}, B_{n+1})$.

Lemma 7.3 ([10]). *Let $B, B_1, \ldots, B_{n+1}$ be connected building sets on $[n+1]$, $[k_1]$, ..., $[k_{n+1}]$, and $B' = B(B_1, \ldots, B_{n+1})$. Then $P_{B'} \approx P_B \times P_{B_1} \times \cdots \times P_{B_{n+1}}$.*

Proof. Consider $B'' = B \sqcup B_1 \sqcup \cdots \sqcup B_{n+1}$ and the map $\varphi : B'' \to B'$ defined as follows,

$$\varphi(S) = \begin{cases} S & \text{if } S \in B_i \\ \bigsqcup_{i \in S} [k_i] & \text{if } S \in B. \end{cases}$$

Obviously, φ generates a bijection between $B'' \setminus B''_{\max}$ and $B' \setminus [n+1]$. Let $\mathscr{S} \subset B \setminus [n+1]$ and $\mathscr{S}_i \subset B_i \setminus [k_i]$. Notice that $\bigcup_{i=1}^{n+1} \varphi(\mathscr{S}_i) \cup \varphi(\mathscr{S}) \in \mathscr{N}(B')$ iff $\mathscr{S} \in \mathscr{N}(B)$

and $\mathscr{S}_i \in \mathscr{N}(B_i)$. Consequently, $\mathscr{N}(B') \simeq \mathscr{N}(B'') \simeq N(B) \star N(B_1) \star \cdots \star N(B_{n+1})$, and therefore $P_{B'} \approx P_{B''} \approx P_B \times P_{B_1} \times \cdots \times P_{B_{n+1}}$. $\qquad\qquad\square$

Example 7.4. Let B, B_1, B_2 be building sets equivalent to $\{\{1\}, \{2\}, \{1,2\}\}$ corresponding to a segment I. Let us describe the building set $B(B_1, B_2)$. In the building set $\{\{a\}, \{b\}, \{a,b\}\}$, we substitute a by $B_1 = \{\{1\}, \{2\}, \{1,2\}\}$ and b by $B_2 = \{\{3\}, \{4\}, \{3,4\}\}$. As a result, we obtain the building set B' on $[4]$, consisting of $\{i\}, \{1,2\}, \{3,4\}, [4]$.

The facet correspondence between the polytopes $P_B \times P_{B_1} \times P_{B_2}$ and $P_{B'}$ is defined as follows,

$$
\begin{aligned}
\{1\} \in B_1 &\mapsto \{1\} \in B', & \{2\} \in B_1 &\mapsto \{2\} \in B', \\
\{3\} \in B_2 &\mapsto \{3\} \in B', & \{4\} \in B_2 &\mapsto \{4\} \in B', \\
\{a\} \in B_2 &\mapsto \{1,2\} \in B', & \{b\} \in B_2 &\mapsto \{3,4\} \in B'.
\end{aligned}
$$

Example 7.5. Let $B = \{\{i\}, [n+1]\}$ be a building set corresponding to the simplex Δ^n and $B_1, \dots, B_{n+1}$ be arbitrary connected building sets on $[k_1], \dots, [k_{n+1}]$. Then $B' = B(B_1, \dots, B_{n+1}) = (B_1 \sqcup \cdots \sqcup B_{n+1}) \cup [k_1 + \cdots + k_{n+1}]$, and $P_{B'} \approx \Delta^n \times P_{B_1} \times \cdots \times P_{B_{n+1}}$.

Corollary 7.6. *Every nestohedron corresponds to some connected building set, i.e., for each nestohedron P there exists a connected building set B such that $P_B \approx P$.*

Proof. Indeed, any building set B' can be represented as $B_1 \sqcup \cdots \sqcup B_k$, where B_i are connected building sets on $[k_i + 1]$. Define a building set $B'' = B_1(1, \dots, k_1, B_2) \sqcup B_3 \sqcup \cdots \sqcup B_k$, giving the same polytope. The building set B' is a product (disjoint union) of k connected building sets and B'' is a product of $(k-1)$ connected building sets. Then we apply again a substitution to B'', and so on. At some step we obtain a connected building set B. $\qquad\qquad\square$

From now on, we can assume without loss of generality that every nestohedron corresponds to a connected building set.

Proposition 7.7. *Let B be a connected building set on $[n+1]$. Then the polytope P_B is flag iff for every element $S \in B$ with $|S| > 1$ there exist elements $S_1, S_2 \in B$ such that $S_1 \sqcup S_2 = S$.*

Proof. Suppose P_B is a flag polytope. Consider an element $S \in B$. Choose $S_1, \dots, S_k \in B \setminus \{S\}$, such that $S_1 \sqcup \cdots \sqcup S_k = S$, and k minimal among such disjoint representations of S. Note that $\forall J \subset [k], 1 < |J| < k : \bigsqcup_{j \in J} S_j \notin B$, since otherwise k can be decreased. Therefore, $k = 2$. Indeed, if $k > 2$, then the facets $F_{S_1}, \dots, F_{S_k}$ intersect pairwise, but have empty intersection.

Suppose for each element $S \in B, |S| > 1$, there exist elements $S_1, S_2 \in B$, such that $S_1 \sqcup S_2 = S$. Let $F_{S_1}, \dots, F_{S_k}$ be the minimal collection of facets that intersect pairwise but have empty intersection. By Proposition 6.15, $S_1 \sqcup \cdots \sqcup S_k = S \in B$. We will find a nontrivial subcollection of facets having empty intersection in the collection $F_{S_1}, \dots, F_{S_k}$. For that let us find a set $\widetilde{S} \in B|_S$ intersecting more than one S_i, but not intersecting every S_i. Then the collection F_{S_i}, satisfying $S_i \cap \widetilde{S} \neq \emptyset$, will be the desired subcollection, since by definition of a building set

$$\bigsqcup_{S_i:S_i \cap \widetilde{S} \neq \emptyset} S_i \in B.$$

By our assumption, $S = S' \sqcup S''$, where $S', S'' \in B$. Choose as S^1 one of the sets S' and S'', intersecting more elements S_i than the other. If S^1 intersects all the sets S_i, then $S^1 = S'^1 \sqcup S''^1$, where $S'^1, S''^1 \in B$. Choose as S^2 one of the sets S''^1 and S''^1, intersecting more elements S_i than the other. If S^2 intersects all the sets S_i, choose S^3 in the same way, and so on. The resulting sequence must be finite. Therefore, in some step we will find the desired set $\widetilde{S} \in B$ as one of the sets S'^i, S''^i.   □

The necessity of the condition in Proposition 7.7 has also been proved in [20].

Definition 7.8. A *Hasse diagram* of a partially ordered set $(X, \prec)$ is the oriented graph on the vertex set X, where vertices $x \in X$ and $y \in X$ are connected by an edge from x to y iff $x \prec y$ and no element $z \in X$ with $x \prec z \prec y$ exists.

Every building set B is partially ordered by inclusion. Its Hasse diagram will be called *B-Hasse diagram*.

Proposition 7.9. *Let B be a connected building set on $[n+1]$. The polytope P_B is combinatorially equivalent to a product of simplices iff the B-Hasse diagram is a tree. In this case $P_B \approx \Delta^{k_1-1} \times \cdots \times \Delta^{k_l-1}$, where $k_1, \ldots, k_l$ are numbers of input edges for all internal nodes of the B-Hasse diagram.*

Proof. Let the B-Hasse diagram be a tree, and $k_1, \ldots, k_l$ numbers of input edges for all its internal nodes. We prove that $P_B \approx \Delta^{k_1-1} \times \cdots \times \Delta^{k_l-1}$. For $n = 1$ there is nothing to prove. Let us assume the statement holds for $m < n$. We prove it for $m = n$. Without loss of generality, we can assume that the number of input edges for the maximal vertex $[n+1]$ is k_l and that this vertex is adjacent to the vertices $S_1, \ldots, S_{k_l}$. Therefore $S_1 \sqcup \cdots \sqcup S_{k_l} = [n+1]$. Indeed, for every vertex $\{q\}$ there exists a path to the vertex $[n+1]$, this path necessarily contains some S_i. Then $\{q\} \subset S_i$. If $\{q\} \subset S_i \cap S_j \neq \emptyset$, then there exist pathes from the vertex $\{q\}$ to the vertices S_i and S_j, but S_i and S_j are adjacent to $[n+1]$, so there is a cycle, which is a contradiction (since we consider a tree). Therefore $B = (B|_{S_1} \sqcup \cdots \sqcup B|_{S_{k_l}}) \cup [n+1]$ and the Hasse diagrams of $B|_{S_i}$ are trees, and $k_1, \ldots, k_{l-1}$ are the numbers of all their input edges. Using Example 7.5, and the induction assumption, we obtain $P_B \approx \Delta^{k_l-1} \times P_{B|_{S_1}} \times \cdots \times P_{B|_{S_{k_l}}} \approx \Delta^{k_1-1} \times \cdots \times \Delta^{k_l-1}$.

Let $P_B \approx \Delta^{k_1-1} \times \cdots \times \Delta^{k_l-1}$. We prove that each vertex S has no more than one output edge. This implies that the B-Hasse diagram is a tree. Suppose a vertex S has output edges to vertices $S', S'' \in B$. Then $S' \setminus S'' \neq \emptyset$, $S'' \setminus S' \neq \emptyset$, $S' \cap S'' \supset S$, and thus $F_{S'} \cap F_{S''} = \emptyset$. As a consequence, the facets $F_{S'}$ and $F_{S''}$ correspond to the opposite points of $\Delta^1 = I$ in the product of simplices, i.e., they are the bases of the cylinder P_B. Let us consider the collection $S_1, \ldots, S_k \in B$, such that $S_1 \sqcup \cdots \sqcup S_k = S' \setminus S''$, and k is the minimal among such disjoint representations of the set $S' \setminus S''$. Then $\bigsqcup_{j \in J} S_j \notin B$

$\forall J \subset [k], 1 < |J| < k$, since otherwise k could be decreased. Therefore the side face $\bigcap_{i=1}^{k} F_{S_i}$ does not intersect the base $F_{S''}$, which is a contradiction.   □

The last proposition implies the following result.

Corollary 7.10. *Let B be a connected building set on $[n+1]$. The polytope P_B is combinatorially equivalent to a cube iff the B-Hasse diagram is a binary tree.*

The sufficiency of the statement was proved in [20].

8 Flag nestohedra as 2-truncated cubes

We are going to construct a sequence of building sets $B_0 \subset \cdots \subset B_N = B$, where B_0 corresponds to a cube and B_i, $i > 0$, is obtained from B_{i-1} by adding one element S_i. From [11] (Theorem 4.2), it follows that, for connected building sets $B' \subset B''$, the polytope $P_{B''}$ is obtained from $P_{B'}$ by a sequence of truncations. Our goal is to show that if P_B is flag, then B_i can be chosen such that all truncated faces have codimension 2.

Proposition 8.1 ([20, Proposition 7.1]). *If B is a flag building set on $[n+1]$, then there exists a building set $B_0 \subseteq B$ such that P_{B_0} is a combinatorial cube with $\dim P_{B_0} = \dim P_B$.*

Proof. By Remarks 6.4 and 6.16, we need to consider only connected building sets. For $n = 1$, the proposition is true. Assuming that the assertion holds for $m < n$, we shall prove it for $m = n$. By Proposition 7.7, we have $[n+1] = S_1 \sqcup S_2$, where $S_1, S_2 \in B$. By the induction assumption, the building sets $B|_{S_1}$ and $B|_{S_2}$ have subsets B_1 and B_2 corresponding to cubes. The building set $B_0 = (B_1 \sqcup B_2) \cup [n+1]$ is the desired one (see Example 7.5).   □

Now we determine which faces have to be truncated in order to obtain $P_{B''}$ from $P_{B'}$, where $B' \subset B''$ are connected building sets.

Construction 8.2 (Decomposition of $S \in B_1$ by elements of B_0). Let B_0 and B_1 be building sets on $[n+1]$, $B_0 \subset B_1$, and $S \in B_1$. We define a decomposition of S by elements of B_0 as $S = S_1 \sqcup \cdots \sqcup S_k$, $S_j \in B_0$, where k is minimal among such disjoint representations of S. Denote the collection $S_1, \ldots, S_k$ by $B_0(S)$. One can see that this decomposition exists and is unique.

Lemma 8.3. *If two simplicial complexes $K \subseteq L$ are the boundaries of simplicial n-polytopes, then $K = L$.*

Proof. The lemma holds for $n = 1$. Assuming it holds for $m < n$, we shall prove it for $m = n$. Let us assume that $L \setminus K \neq \emptyset$ and choose a simplex $A \in L \setminus K$, containing a vertex $v \in K$ (which exists since L is connected). The complexes $\mathrm{lk}(v, K)$ and $\mathrm{lk}(v, L)$ are the boundaries of simplicial $(n-1)$-polytopes and $\mathrm{lk}(v, K) \subsetneq \mathrm{lk}(v, L)$, since $A \setminus \{v\} \in \mathrm{lk}(v, L) \setminus \mathrm{lk}(v, K)$. This contradicts the induction assumption.   □

The next lemma can be extracted from [11], Theorem 4.2. For convenience, we provide a simpler proof in the present more restricted context.

Lemma 8.4. *Let B_0 and B_1 be connected building sets on $[n+1]$ and $B_0 \subset B_1$. Then $\mathcal{N}(B_1)$ is obtained from $\mathcal{N}(B_0)$ by stellar subdivisions over simplices $\sigma_i = \{S_1^i, \ldots, S_{k_i}^i\}$, corresponding to decompositions $S^i = S_1^i \sqcup \cdots \sqcup S_{k_i}^i \in B_1 \setminus B_0$ of elements S^i, numbered in any order that is reverse to inclusion (i.e., $S^i \supseteq S^{i'} \Rightarrow i \leq i'$).*

Proof. We use induction on the number $N = |B_1| - |B_0|$. For $N = 1$, we have $B_1 = B_0 \cup \{S^1\}$. We show that $K \simeq \mathcal{N}(B_1)$, where $K = (S^1, B_0(S^1)) \mathcal{N}(B_0)$. By Lemma 8.3, it is sufficient to prove that $\mathcal{N}(B_1) \subseteq K$. Let $\mathscr{S} \in \mathcal{N}(B_1)$. Note that $B_0(S^1) = \{S_1^1, \ldots, S_{k_1}^1\} \not\subseteq \mathscr{S}$. If $S^1 \notin \mathscr{S}$, then $\mathscr{S} \in \mathcal{N}(B_0)$, therefore $\mathscr{S} \in K$ according to a) of Remark 3.7. If $S^1 \in \mathscr{S}$, then for each element $S \in \mathscr{S}$, intersecting with S^1, either $S^1 \subset S$, or $S^1 \supset S$ (otherwise $S \cup S^1 \in B_1$). In the last case, $\exists j : S \subset S_j^1$ (otherwise, merging S_j^1, intersecting S, we can decrease k_1, which contradicts the definition of the decomposition of S^1). Consequently, $B_0(S^1) \cup (\mathscr{S} \setminus \{S^1\}) \in \mathcal{N}(B_0)$, thus $\mathscr{S} \in K$ according to b) of Remark 3.7.

Assuming the lemma holds for $M < N$, we prove it for $M = N$. The collection of sets $B_0' = B_0 \cup \{S^1\}$ is a building set. By the induction assumption, $\mathcal{N}(B_0')$ is obtained from $\mathcal{N}(B_0)$ by stellar subdivision over the simplex corresponding to the decomposition of S^1, and $\mathcal{N}(B_1)$ is obtained from $\mathcal{N}(B_0')$ by a sequence of stellar subdivisions over simplices corresponding to decompositions of S^i, $i = 2, \ldots, N$. This completes the proof. $\square$

Corollary 8.5. *Let B_0 and B_1 be connected building sets on $[n+1]$, and $B_0 \subset B_1$. Then P_{B_1} is obtained from P_{B_0} by truncations of faces $G^i = \bigcap_{j=1}^{k_i} F_{S_j^i}$, corresponding to decompositions of $S^i = S_1^i \sqcup \cdots \sqcup S_{k_i}^i \in B_1 \setminus B_0$, numbered in any order that is reverse to inclusion (i.e., $S^i \supseteq S^{i'} \Rightarrow i \leq i'$).*

Theorem 8.6. *Let B_1 and B_2 be flag connected building sets on $[n+1]$ and $B_1 \subset B_2$. Then $\mathcal{N}(B_2)$ is obtained from $\mathcal{N}(B_1)$ by a sequence of subdivisions of edges (or, equivalently, P_{B_2} is obtained from P_{B_1} by a sequence of 2-truncations).*

Proof. We will prove the following: if B_1 and B_3 are connected flag building sets on $[n+1]$ and $B_1 \subsetneq B_3$, then there exists a building set B_2 such that $B_1 \subsetneq B_2 \subseteq B_3$ and $\mathcal{N}(B_2)$ is obtained from $\mathcal{N}(B_1)$ by a sequence of subdivisions of edges. This then implies the statement in the theorem, since stellar subdivisions over edges are dual to 2-truncations (which keep flagness).

Set $B_2 = \widehat{B_1 \cup \{S\}}$, i.e., the minimal (by inclusion) building set containing $B_1 \cup \{S\}$, where S is the minimal (by inclusion) element of $B_3 \setminus B_1$. By Proposition 7.7, there exist $I, J \in B_3$ such that $I \sqcup J = S$. From the choice of S, it follows that $I, J \in B_1$. It is easy to show that the collection of sets $B_1 \cup \{S' = S_1 \sqcup S_2 \mid S_i \in B_1, I \subseteq S_1, J \subseteq S_2\}$ is a minimal building set containing $B_1 \cup \{S\}$. Then the decomposition of any element $B_2 \setminus B_1$ consists of two elements. Therefore, $\mathcal{N}(B_2)$ is obtained from $\mathcal{N}(B_1)$ by a sequence of subdivisions of edges. $\square$

We conclude that for each flag building set B there exists a sequence of building sets $B_0 \subset B_1 \subset \cdots \subset B_N = B$, where P_{B_0} is a combinatorial cube, $B_i = B_{i-1} \cup \{S_i\}$, and the polytope P_{B_i} is obtained from the polytope $P_{B_{i-1}}$ by a 2-truncation of the face $F_{S_{j_1}} \cap F_{S_{j_2}} \subset P_{B_{i-1}}$, where $S_i = S_{j_1} \sqcup S_{j_2}$, and $S_{j_1}, S_{j_2} \in B_{i-1}$.

Using results of the last two sections, we obtain the following.

Theorem 8.7. *Every flag nestohedron is a 2-truncated cube.*

Theorem 8.8.

1) $\gamma_i(P_B) \geq 0$ *for every flag nestohedron* P_B, *i.e., Gal's conjecture holds for flag nestohedra.*
2) $\gamma_i(P_{B_1}) \leq \gamma_i(P_{B_2})$ *for connected flag building sets* B_1, B_2 *with* $B_1 \subseteq B_2$. *Moreover, equality holds for all* i *iff* $B_1 = B_2$.

Example 8.9. Let us construct a geometric realization of the 3-dimensional Stasheff polytope as a 2-truncated cube. The building set corresponding to As^3 is

$$B = \{\{1\}, \{2\}, \{3\}, \{4\}, \{1,2\}, \{2,3\}, \{3,4\}, \{1,2,3\}, \{2,3,4\}, \{1,2,3,4\}\} .$$

In order to obtain As^3 from I^3 by 2-truncations, we have to find a building set $B_0 \subset B$, such that $P_{B_0} \approx I^3$, and to order the elements of $B \setminus B_0$ such that adding a new element to the building set corresponds to a 2-truncation. On the left-hand side of Fig. 1, B_0 consists of $\{i\}, \{1,2\}, \{3,4\}, [4]$. The associahedron P_B is obtained from $P_{B_0} \approx I^3$ by stepwise truncation of the faces $F_{\{1,2\}} \cap F_{\{3\}}, F_{\{2\}} \cap F_{\{3,4\}}, F_{\{2\}} \cap F_{\{3\}}$ in this order. In the drawing, the facets $F_{\{1,2\}}$ and $F_{\{3,4\}}$ are the top and the bottom section facets, respectively, while $F_{\{2\}}$ and $F_{\{3\}}$ are the right and front facets, respectively.

On the right-hand side of Fig. 1, B_0 consists of $\{i\}, \{1,2\}, \{1,2,3\}, [4]$. First we truncate the face $F_{\{2\}} \cap F_{\{3\}}$ of P_{B_0} and derive the new facet $F_{\{2,3\}}$. Then we truncate the faces $F_{\{2,3\}} \cap F_{\{4\}}$ and $F_{\{3\}} \cap F_{\{4\}}$. In the drawing, $F_{\{2\}}, F_{\{3\}}$ and $F_{\{4\}}$ are the front, right and top facets.

In [4] it was shown that the geometric realizations of As^n that appeared in [5], [15] and [20] are not affinely equivalent. None of our realizations above is equivalent to one of those.

9 Recursion formulas

Let J be the building set $\{\{1\}, \{2\}, \{1,2\}\}$, which corresponds to the interval $P_J \approx I$. Recall that a subset V of vertices of a graph Γ is called a *clique* if the induced subgraph $\Gamma|_V$ is complete.

Construction 9.1. Let Γ_{n+1} be a connected graph on $[n+1]$ and the set $V \subseteq [n]$ of vertices adjacent to the vertex $\{n+1\}$ be a clique. Set $\Gamma_n = \Gamma_{n+1} \setminus \{n+1\}$, i.e., Γ_n is the graph obtained from Γ_{n+1} by deletion of the vertex $\{n+1\}$.

According to Lemma 7.3, the building set $B_1 = J(B(\Gamma_n), n+1) = B(\Gamma_n) \cup \{n+1\} \cup [n+1]$ corresponds to $P_{B_1} \approx P_{\Gamma_n} \times I$: the bottom and top bases of the cylinder

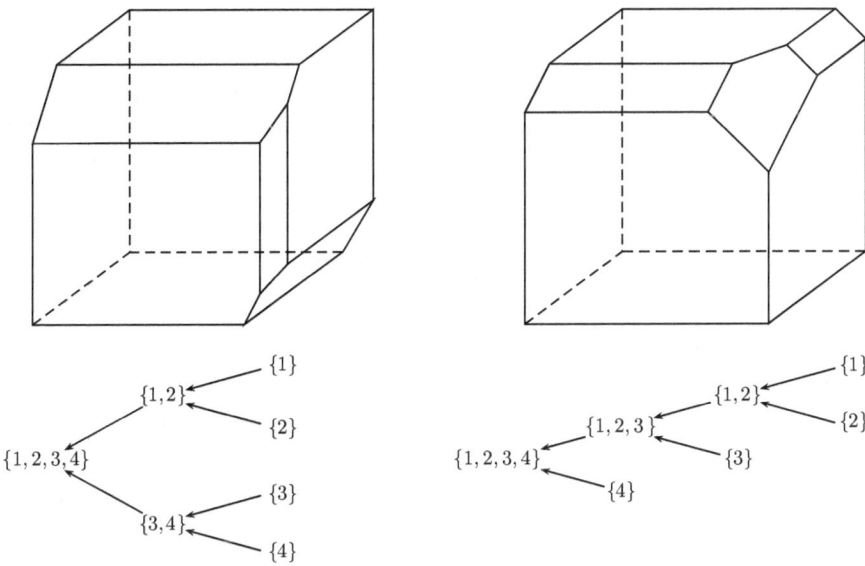

Fig. 1 Realizations of the 3-dimensional Stasheff polytope via two different choices of the building subset B_0 of B, represented by their Hasse diagrams (below the respective polytope).

$P_{\Gamma_n} \times I$ correspond to $[n]$, $\{n+1\} \in B_1$, the side facets correspond to $S \in B(\Gamma_n) \setminus [n] \subset B_1$. Thus, the side facets $F_S, S \in B_1$ of the cylinder $P_{\Gamma_n} \times I$ are naturally identified with the facets $F_S, S \in B(\Gamma_n)$ of the base P_{Γ_n}.

We have $B(\Gamma_{n+1}) \setminus B_1 = \{S \sqcup \{n+1\}, S \in \mathscr{S}\}$, where $\mathscr{S} = \{S \in B(\Gamma_n) \setminus [n]: S \cap V \neq \emptyset\}$. By Corollary 8.5, $P_{\Gamma_{n+1}}$ is obtained from $P_{\Gamma_n} \times I$ by truncations of intersections of the top base $F_{\{n+1\}}$ with side facets F_S for $S \in \mathscr{S}$. Since the top base does not change after truncation of any of its facets, the truncated faces have type $P_{\Gamma_n|_S} \times P_{\Gamma_n/S}, S \in \mathscr{S}$. By Proposition 3.6, we have

$$\gamma(P_{\Gamma_{n+1}}) = \gamma(P_{\Gamma_n}) + \tau \sum_{S \in \mathscr{S}} \gamma(P_{\Gamma_n|_S})\gamma(P_{\Gamma_n/S}), \tag{6}$$

$$H(P_{\Gamma_{n+1}}) = (\alpha + t)H(P_{\Gamma_n}) + \alpha t \sum_{S \in \mathscr{S}} H(P_{\Gamma_n|_S})H(P_{\Gamma_n/S}). \tag{7}$$

We required that V is a clique of Γ_{n+1}, because in this case every element of $B(\Gamma_{n+1}) \setminus B_1$ has a decomposition consisting of two elements ($S \sqcup \{n+1\}$, where $\{n+1\}, S \in B_1$), and we know the combinatorial type of the truncated faces of codimension 2.

9.1 Associahedra

Let us apply Construction 9.1 to the path graph L_{n+1}. After deletion of the vertex $\{n+1\}$, we obtain the graph L_n. Here $V = \{n\}$ and $\mathscr{S} = \{[i,n], i = 2,\dots n\}$. Therefore, the truncated faces have type $As^{i-1} \times As^{n-i-1}$, $i = 1,\dots,n-1$, and we obtain the recursion formulas

$$\gamma(As^n) = \gamma(As^{n-1}) + \tau \sum_{i=1}^{n-1} \gamma(As^{i-1})\gamma(As^{n-i-1}), \tag{8}$$

$$H(As^n) = (\alpha + t)H(As^{n-1}) + \alpha t \sum_{i=1}^{n-1} H(As^{i-1})H(As^{n-i-1}). \tag{9}$$

The recursion formulas for associahedra are equivalent to the equations

$$\gamma_{As}(x) = 1 + x\gamma_{As}(x) + \tau x^2 \gamma_{As}^2(x) \qquad \text{where} \qquad \gamma_{As}(x) = \sum_{n=0}^{\infty} \gamma(As^n)x^n,$$

$$H_{As}(x) = (1 + \alpha x H_{As}(x))(1 + t x H_{As}(x)) \quad \text{where} \quad H_{As}(x) = \sum_{n=0}^{\infty} H(As^n)x^n.$$

The last equation is equivalent to:

$$\frac{xH_{As}(x)}{(1 + \alpha x H_{As}(x))(1 + t x H_{As}(x))} = x.$$

Set $U(x) = xH_{As}(x)$. Then $U(0) = 0, U'(0) = 1$, and

$$\frac{U}{(1 + \alpha U)(1 + tU)} = x.$$

Applying the classical Lagrange Inversion Formula, we obtain

$$U(x) = -\frac{1}{2\pi i} \oint_{|z|=\varepsilon} \ln\left[1 - \frac{x}{z}(1 + \alpha z)(1 + tz)\right] dz$$

$$= \sum_{n=1}^{\infty} \left(\frac{1}{2\pi i} \oint_{|z|=\varepsilon} \left[\frac{(1 + \alpha z)^n(1 + tz)^n}{z^n}\right] dz\right) \frac{x^n}{n}$$

$$= \sum_{n=1}^{\infty} \left(\sum_{i+j=n-1} \binom{n}{i}\binom{n}{j}\alpha^i t^j\right) \frac{x^n}{n}.$$

Therefore,

$$H(As^n) = \frac{1}{n+1} \sum_{i+j=n} \binom{n+1}{i}\binom{n+1}{j}\alpha^i t^j = \frac{1}{n+1} \sum_{i=0}^{n} \binom{n+1}{i}\binom{n+1}{i+1}\alpha^{n-i} t^i.$$

Lemma 9.2. *For every connected graph* Γ_{n+1} *on* $[n+1]$, *we have* $\gamma_i(P_{\Gamma_{n+1}}) \geq \gamma_i(As^n)$. *Moreover, equality holds for all* i *iff* Γ_{n+1} *is a linear graph* L_{n+1}.

Proof. It is sufficient to prove the lemma for trees, since for every connected graph Γ, there exists a tree $T \subseteq \Gamma$ on the same nodes. Then $B(T) \subseteq B(\Gamma)$ and we can apply Theorem 8.8.

For $n = 1$ there is nothing to prove. Assume that the lemma holds for $m \leq n$. Let Γ_{n+1} be a tree on $[n+1]$. Without loss of generality, assume that $\{n+1\}$ is adjacent only to $\{n\}$. Then we can use Construction 9.1, setting $\Gamma_n = \Gamma_{n+1} \setminus \{n+1\}$ and $V = \{n\}$. For every $i \in [1, n-1]$, there exists a connected subgraph of Γ_n on i vertices containing $\{n\}$, i.e., there exists $S \in \mathscr{S}: |S| = i$. Therefore, comparing (6) and (8), and using the induction assumption and Remark 6.7, we obtain the proof of the stated inequalities.

Note that if Γ_{n+1} is not a linear graph, then either Γ_n is not a linear graph or, for some $i \in [1, n-1]$, there exist more than one element $S \in \mathscr{S}$ with $|S| = i$. In both cases, for some i the inequality in the lemma is strict. □

9.2 Cyclohedra

Let C_{n+1} be a cyclic graph on $[n+1]$. We apply a construction different from Construction 9.1.

Construction 9.3. According to Lemma 7.3, the building set $B_1 = B(C_n)(1, \ldots, n-1, J(n, n+1))$ corresponds to $P_{B_1} \approx Cy^{n-1} \times I$: the bottom and top bases of $Cy^{n-1} \times I$ correspond to $\{n\}, \{n+1\} \in B_1$, and the side facets correspond to elements $S \in B_1 \setminus [n+1]$ that either contain $\{n, n+1\}$, or do not intersect $\{n, n+1\}$. Thus, by contraction of the set $\{n, n+1\}$ to the point $\{n\}$, we identify side facets F_S, $S \in B_1$, of the cylinder $Cy^{n-1} \times I$ with facets $F_{S \setminus \{n+1\}}, S \setminus \{n+1\} \in B(C_n)$ of the base Cy^{n-1}.

We have

$$B(C_{n+1}) \setminus B_1 = \{S \sqcup \{n\}, S \in \mathscr{S}_n\} \cup \{S \sqcup \{n+1\}, S \in \mathscr{S}_{n+1}\},$$

where $\mathscr{S}_n = \{[i, n-1], i = 2, \ldots, n-1\}$ and $\mathscr{S}_{n+1} = \{[1, i], i = 1, \ldots, n-2\}$. By Corollary 8.5, Cy^n is obtained from $Cy^{n-1} \times I$ by truncations of intersections of the bottom base $F_{\{n\}}$ with the side facets F_S for $S \in \mathscr{S}_n$, and by truncations of intersections of the top base $F_{\{n+1\}}$ with the side facets F_S for $S \in \mathscr{S}_{n+1}$. Since the bases do not change after truncations of the facets, truncated faces of each base have the form $As^{i-1} \times Cy^{n-i-1}, i = 1, \ldots, n-1$ ($P_{C_n|S} \times P_{C_n/S}, S \in \mathscr{S}_n$ for bottom base and $P_{C_n|S} \times P_{C_n/S}, S \in \mathscr{S}_{n+1}$ for top base). By Proposition 3.6, we obtain the recursion formulas

$$\gamma(Cy^n) = \gamma(Cy^{n-1}) + 2\tau \sum_{i=1}^{n-1} \gamma(As^{i-1})\gamma(Cy^{n-i-1}), \tag{10}$$

$$H(Cy^n) = (\alpha+t)H(Cy^{n-1}) + 2\alpha t \sum_{i=1}^{n-1} H(As^{i-1})H(Cy^{n-i-1}). \tag{11}$$

The recursion formulas for cyclohedra are equivalent to the equations

$$\gamma_{Cy}(x) = 1 + x\gamma_{Cy}(x) + 2\tau x^2 \gamma_{As}(x)\gamma_{Cy}(x),$$
$$H_{Cy}(x) = 1 + (\alpha+t)xH_{Cy}(x) + 2\alpha t x^2 H_{As}(x)H_{Cy}(x),$$

where

$$\gamma_{Cy}(x) = \sum_{n=0}^{\infty} \gamma(Cy^n)x^n, \qquad H_{Cy}(x) = \sum_{n=0}^{\infty} H(Cy^n)x^n.$$

Setting $V(x) = xH_{Cy}(x)$, we obtain

$$\frac{V}{1 + (\alpha+t)V + 2\alpha t UV} = x,$$

and thus

$$\frac{U}{(1+\alpha U)(1+tU)} = \frac{V}{1 + (\alpha+t)V + 2\alpha t UV},$$

which implies

$$V = \frac{U}{1 - \alpha t U^2}.$$

Recall that a graph Γ is called *Hamiltonian* if it contains a Hamiltonian cycle, i.e., a closed loop that visits each vertex of Γ exactly once.

Lemma 9.4. $\gamma_i(P_{\Gamma_{n+1}}) \geq \gamma_i(Cy^n)$ *for any Hamiltonian graph Γ_{n+1} on $[n+1]$. Moreover, equality for all i is achieved iff Γ_{n+1} is a cyclic graph C_{n+1}.*

Proof. Since Γ_{n+1} is Hamiltonian, there exists a cyclic subgraph $C_{n+1} \subseteq \Gamma_{n+1}$. Therefore, $B(C_{n+1}) \subseteq B(\Gamma_{n+1})$ and Theorem 8.8 allows to complete the proof. $\square$

9.3 Permutohedra

Let us apply Construction 9.1 to the complete graph K_{n+1}. After deletion of the vertex $\{n+1\}$, we obtain the graph K_n. Here $V = [n]$ and $\mathscr{S} = 2^{[n]} \setminus \{\emptyset, [n]\}$. Therefore, we truncate $\binom{n}{i}$ faces of the form $Pe^{i-1} \times Pe^{n-i-1}$, $i = 1, \ldots, n-1$, and obtain the recursion formulas for permutohedra

$$\gamma(Pe^n) = \gamma(Pe^{n-1}) + \tau \sum_{i=1}^{n-1} \binom{n}{i} \gamma(Pe^{i-1})\gamma(Pe^{n-i-1}), \tag{12}$$

$$H(Pe^n) = (\alpha+t)H(Pe^{n-1}) + \alpha t \sum_{i=1}^{n-1} \binom{n}{i} H(Pe^{i-1})H(Pe^{n-i-1}). \tag{13}$$

They are equivalent to the differential equations

$$\frac{d\gamma_{Pe}(x)}{dx} = 1 + \gamma_{Pe}(x) + \tau \gamma_{Pe}^2(x) \qquad \text{where} \qquad \gamma_{Pe}(x) = \sum_{n=0}^{\infty} \gamma(Pe^n)\frac{x^{n+1}}{(n+1)!},$$

$$\frac{dH_{Pe}(x)}{dx} = (1 + \alpha H_{Pe}(x))(1 + t H_{Pe}(x)) \quad \text{where} \quad H_{Pe}(x) = \sum_{n=0}^{\infty} H(Pe^n)\frac{x^{n+1}}{(n+1)!}.$$

One can explicitly solve the last equation to obtain

$$H_{Pe}(x) = \frac{e^{\alpha x} - e^{tx}}{\alpha e^{tx} - t e^{\alpha x}}.$$

Let $A(n,k) = |\{\sigma \in \mathrm{Sym}(n) \colon \mathrm{des}(\sigma) = k\}|$. Then, by a well-known formula,

$$\frac{e^{\alpha x} - e^{tx}}{\alpha e^{tx} - t e^{\alpha x}} = \sum_{n=0}^{\infty} \left(\sum_{i=0}^{n} A(n+1,k)\alpha^k t^{n-k} \right) \frac{x^{n+1}}{(n+1)!}.$$

9.4 Stellohedra

Let us apply Construction 9.1 to the complete bipartite graph $K_{1,n}$, or n-star with apex $\{1\}$. After deletion of the vertex $\{n+1\}$, we obtain the graph $K_{1,n-1}$. Here $V = \{1\}$ and $\mathscr{S} = \{\{1\} \cup S, S \subsetneq [2,n]\}$. Therefore, we truncate $\binom{n-1}{i-1}$ faces of the form $St^{i-1} \times Pe^{n-i-1}, i = 1, \ldots, n-1$ and obtain

$$\gamma(St^n) = \gamma(St^{n-1}) + \tau \sum_{i=1}^{n-1} \binom{n-1}{i-1} \gamma(St^{i-1})\gamma(Pe^{n-i-1}), \tag{14}$$

$$H(St^n) = (\alpha+t)H(St^{n-1}) + \alpha t \sum_{i=1}^{n-1} \binom{n-1}{i-1} H(St^{i-1})H(Pe^{n-i-1}). \tag{15}$$

These recursion formulas for stellohedra are equivalent to the differential equations

$$\frac{d\gamma_{St}(x)}{dx} = \gamma_{St}(x)(1 + \tau \gamma_{Pe}(x)) \qquad \text{where} \qquad \gamma_{St}(x) = \sum_{n=0}^{\infty} \gamma(St^n)\frac{x^n}{n!},$$

$$\frac{dH_{St}(x)}{dx} = H_{St}(x)(\alpha + t + \alpha t H_{Pe}(x)) \quad \text{where} \quad H_{St}(x) = \sum_{n=0}^{\infty} H(St^n)\frac{x^n}{n!}.$$

The last equation can be solved:

$$H_{St}(x) = \frac{(\alpha - t)\, e^{(\alpha+t)x}}{\alpha e^{tx} - t e^{\alpha x}}.$$

Lemma 9.5. *For every tree Γ_{n+1} on $[n+1]$, we have $\gamma_i(P_{\Gamma_{n+1}}) \leq \gamma_i(St^n)$. Moreover, equality for all i is achieved iff Γ_{n+1} is a star graph $K_{1,n}$.*

Proof. For $n = 1$ there is nothing to prove. Assume that the Lemma holds for $m \leq n$. Let Γ_{n+1} be a tree on $[n+1]$. Without loss of generality, assume that $\{n+1\}$ is adjacent only to $\{n\}$. Then we can use Construction 9.1, setting $\Gamma_n = \Gamma_{n+1} \setminus \{n+1\}$ and $V = \{n\}$. For every $i \in [1, n-1]$, there are no more than $\binom{n-1}{i-1}$ elements $S \in \mathscr{S}$ with $|S| = i$, and for each such S we have $\gamma(P_{\Gamma_n|_S})\gamma(P_{\Gamma_n/S}) \leq \gamma(St^{i-1})\gamma(Pe^{n-i-1})$. Therefore, comparing (6) and (14), and using the induction assumption, we obtain the statement in the lemma.

We note that if Γ_{n+1} is not a star graph, then either Γ_n is not a star graph or, for some $i \in [1, n-1]$, the number of elements $S \in \mathscr{S}$ with $|S| = i$ is less than $\binom{n-1}{i-1}$. In both cases, for some i the inequality in the lemma is strict. $\qquad\square$

10 Bounds of face polynomials for flag nestohedra and graph-associahedra

Summarizing Lemmas 9.2, 9.4, 9.5 and Theorem 8.8, we obtain the following results.

Theorem 10.1. *For any flag n-dimmensional nestohedron P_B, we have*

1) $\gamma_i(I^n) \leq \gamma_i(P_B) \leq \gamma_i(Pe^n)$,
2) $g_i(I^n) \leq g_i(P_B) \leq g_i(Pe^n)$,
3) $h_i(I^n) \leq h_i(P_B) \leq h_i(Pe^n)$,
4) $f_i(I^n) \leq f_i(P_B) \leq f_i(Pe^n)$.

Moreover, the lower bound is achieved iff $P_B \approx I^n$, the upper bound is achieved iff $P_B \approx Pe^n$.

Theorem 10.2. *For any connected graph Γ_{n+1} on $[n+1]$, we have*

1) $\gamma_i(As^n) \leq \gamma_i(P_{\Gamma_{n+1}}) \leq \gamma_i(Pe^n)$,
2) $g_i(As^n) \leq g_i(P_{\Gamma_{n+1}}) \leq g_i(Pe^n)$,
3) $h_i(As^n) \leq h_i(P_{\Gamma_{n+1}}) \leq h_i(Pe^n)$,
4) $f_i(As^n) \leq f_i(P_{\Gamma_{n+1}}) \leq f_i(Pe^n)$.

Moreover, the lower bound is achieved iff Γ_{n+1} is a linear graph L_{n+1}, the upper bound is achieved iff Γ_{n+1} is a complete graph K_{n+1}.

Theorem 10.3. *For any Hamiltonian graph Γ_{n+1} on $[n+1]$, we have*

1) $\gamma_i(Cy^n) \leq \gamma_i(P_{\Gamma_{n+1}}) \leq \gamma_i(Pe^n)$,
2) $g_i(Cy^n) \leq g_i(P_{\Gamma_{n+1}}) \leq g_i(Pe^n)$,

3) $h_i(Cy^n) \leq h_i(P_{\Gamma_{n+1}}) \leq h_i(Pe^n)$,
4) $f_i(Cy^n) \leq f_i(P_{\Gamma_{n+1}}) \leq f_i(Pe^n)$.

Moreover, the lower bound is achieved iff Γ_{n+1} is a cyclyc graph C_{n+1}, the upper bound is achieved iff Γ_{n+1} is a complete graph K_{n+1}.

Theorem 10.4. *For any tree Γ_{n+1} on $[n+1]$, we have*

1) $\gamma_i(As^n) \leq \gamma_i(P_{\Gamma_{n+1}}) \leq \gamma_i(St^n)$,
2) $g_i(As^n) \leq g_i(P_{\Gamma_{n+1}}) \leq g_i(St^n)$,
3) $h_i(As^n) \leq h_i(P_{\Gamma_{n+1}}) \leq h_i(St^n)$,
4) $f_i(As^n) \leq f_i(P_{\Gamma_{n+1}}) \leq f_i(St^n)$.

Moreover, the lower bound is achieved iff Γ_{n+1} is a linear graph L_{n+1}, the upper bound is achieved iff Γ_{n+1} is a star graph $K_{1,n}$.

The bounds can be written explicitly, using results about the f-, h-, g- and γ-vectors of the respective series (cf. [20] and [1]):

$$h_i(I^n) = \binom{n}{i}, \quad h_i(As^n) = \frac{1}{n+1}\binom{n+1}{i}\binom{n+1}{i+1}, \quad h_i(Cy^n) = \binom{n}{i}^2,$$

$$h_i(Pe^n) = A(n+1,i), \quad h_i(St^n) = \sum_{k=i}^{n}\binom{n}{k}A(k,i-1), \ i > 0,$$

$$\gamma_i(I^n) = 0, \ i > 0, \quad \gamma_i(As^n) = \frac{1}{i+1}\binom{2i}{i}\binom{n}{2i}, \quad \gamma_i(Cy^n) = \binom{n}{i,i,n-2i}.$$

The derived bounds for the f-, h-, g- and γ-vectors determine bounds for the corresponding polynomials, for which generating functions were obtained in [1].

11 Nested polytopes and graph-cubeahedra as 2-truncated cubes

Let Γ be a connected graph, $B(\Gamma)$ its building set. Using the notation of [9], elements of $B(\Gamma)$ are called *tubes* and $\{S_1,\ldots,S_k\} \in \mathcal{N}(B(\Gamma))$ is called a *tubing*. Two tubes forming a tubing are called *compatible*.

In [9], a class of simple polytopes called graph-cubeahedra was introduced. To formulate a definition, we have to introduce the notion of 'design tubing'.

Definition 11.1. Let Γ be a connected graph. A *round tube* is a set of nodes S of Γ whose induced subgraph $\Gamma|_S$ is connected. A *square tube* is a single node $S = \{v\}$ of Γ. Such tubes are called *design tubes* of Γ. Two design tubes are *compatible* if one of the following conditions holds,

1) they are both round and compatible,
2) one of the tubes is square and does not intersect the other.

A *design tubing* U of Γ is a collection of design tubes of Γ such that every pair of tubes in U is compatible.

For a graph Γ with n nodes, we define $\square_\Gamma$ to be the n-cube where each pair of opposite facets corresponds to a particular node of Γ. Specifically, one facet in the pair represents that node as a round tube and the other represents it as a square tube. Each subset of nodes of Γ, chosen to be either round or square, corresponds to a unique face of $\square_\Gamma$, defined by the intersection of the faces associated with those nodes. The empty set corresponds to the face that is the entire polytope $\square_\Gamma$.

Definition 11.2. For a graph G, truncating faces of G that correspond to round tubes in increasing order of dimension, results in a convex polytope CG called *graph-cubeahedron*.

Theorem 11.3 ([9, Theorem 12]). *For a graph G with n nodes, the graph-cubeahedron CG is the simple convex polytope of dimension n whose face poset is isomorphic to the set of design tubings of G, ordered such that $U < U'$ if U is obtained from U' by adding tubes.*

Definition 11.4. Let B be a building set (not necessary connected) on $[n+1]$. Define the complex $\mathscr{K}(B)$ as the simplicial complex on the node set B consisting of all the nested sets $\mathscr{S} \subseteq B$.

Proposition 11.5. *For any building set B, we have $\mathscr{K}(B) \simeq K\mathscr{N}(B_1) \star \cdots \star K\mathscr{N}(B_k)$, where $B = B_1 \sqcup \cdots \sqcup B_k$ and B_i are connected. In particular, if B is connected, then $\mathscr{K}(B)$ is a cone over $\mathscr{N}(B)$.*

Proof. Let B be a connected building set on $[n+1]$, then the node set of the complex $\mathscr{K}(B)$ consists of the node set of the complex $\mathscr{N}(B)$ and the vertex $[n+1]$. By definition, $\{S_1, \ldots, S_k\} \in \mathscr{N}(B)$ iff $\{S_1, \ldots, S_k\} \in \mathscr{K}(B)$ iff $\{[n+1], S_1, \ldots, S_k\} \in \mathscr{K}(B)$. Therefore $\mathscr{K}(B) \simeq K\mathscr{N}(B)$. The proof is complete, since $\mathscr{K}(B_1 \sqcup B_2) \simeq \mathscr{K}(B_1) \star \mathscr{K}(B_2)$. $\qquad\square$

Lemma 11.6. *Let B_0 and B_1 be building sets on $[n+1]$, and $B_0 \subset B_1$. Then $\mathscr{K}(B_1)$ is obtained from $\mathscr{K}(B_0)$ by stellar subdivisions over $\sigma_i = \{S_1^i, \ldots, S_{k_i}^i\}$, corresponding to decompositions $S^i = S_1^i \sqcup \cdots \sqcup S_{k_i}^i \in B_1 \setminus B_0$ of elements S^i, numbered in any order that is reverse to inclusion (i.e., $S^i \supseteq S^{i'} \Rightarrow i \le i'$).*

Proof. We prove the lemma by induction on the number $N = |B_1| - |B_0|$.

Let $N = 1$, then $B_1 = B_0 \cup S^1$. We show that $\mathscr{K}(B_1) \simeq (S^1, B_0(S^1))\mathscr{K}(B_0)$. Let $B_0 = B_0^1 \sqcup \cdots \sqcup B_0^l$, where B_0^i are connected building sets on $[k_{i-1}+1, k_i]$. There are only two possibilities: $S^1 \subset [k_{i-1}+1, k_i]$ for some $i \in [l]$, or $S^1 = \bigsqcup_{i \in \sigma} [k_{i-1}+1, k_i]$ for some $\sigma \subseteq [l]$.

Consider the first case and assume $i = 1$. We have $B_1 = B_1^1 \sqcup \cdots \sqcup B_1^l$, where $B_1^1 = B_0^1 \cup \{S^1\}$ and $B_1^i = B_0^i$ for $i > 1$. By Lemma 8.4, $\mathscr{N}(B_1^1) \simeq (S^1, B_0^1(S^1))\mathscr{N}(B_0^1)$. Using Proposition 11.5, we obtain $\mathscr{K}(B_1) \simeq (S^1, B_0(S^1))\mathscr{K}(B_0)$.

Consider the second case. Without loss of generality, we can assume that $\sigma = [l]$. Using Example 7.5, $\mathscr{N}(B_1) = \partial \Delta^{l-1} \star \mathscr{N}(B_0^1) \star \cdots \star \mathscr{N}(B_0^l)$, and then $\mathscr{K}(B_1) = K(\partial \Delta^{l-1} \star \mathscr{N}(B_0^1) \star \cdots \star \mathscr{N}(B_0^l)) = K(\partial \Delta^{l-1}) \star \mathscr{N}(B_0^1) \star \cdots \star \mathscr{N}(B_0^l)$. This complex is the stellar subdivision of $\mathscr{K}(B_0) \simeq K\mathscr{N}(B_0^1) \star \cdots \star K\mathscr{N}(B_0^l)$ over the simplex $\{[1,k_1],\ldots,[k_{l-1}+1,n+1]\} \in \mathscr{K}(B_0)$, which corresponds to the join of apexes of the cones $\mathscr{K}(B_0^i)$.

Let us assume the lemma holds for $M < N$ and prove it for $M = N$. The collection of sets $B_0' = B_0 \cup \{S^1\}$ is a building set. By the induction assumption, $\mathscr{K}(B_0')$ is obtained from $\mathscr{K}(B_0)$ by stellar subdivision over the simplex corresponding to the decomposition of S^1, and $\mathscr{K}(B_1)$ is obtained from $\mathscr{K}(B_0')$ by a sequence of stellar subdivisions over simplices corresponding to decompositions $S^i, i = 2,\ldots,N$. This completes the proof. $\qquad\square$

Theorem 11.7. *Let B_1 and B_2 be flag building sets on $[n+1]$ and $B_1 \subset B_2$. Then $\mathscr{K}(B_2)$ is obtained from $\mathscr{K}(B_1)$ by a subdivision of edges.*

Proof. The proof repeats the proof of Theorem 8.6. $\qquad\square$

Let us denote by $B_{\Delta,n}$ the building set consisting of singletons $\{i\}$, $i \in [n]$.

Definition 11.8. Let P be a simple n-polytope with fixed order of facets, $F_1, F_2, \ldots, F_m$, such that $F_1 \cap \cdots \cap F_n \neq \emptyset$, and let B be a building set on n. By Lemma 11.6, $\mathscr{K}(B)$ is obtained from $\mathscr{K}(B_{\Delta,n}) \simeq \Delta^{n-1}$ by a sequence of stellar subdivisions. Pick the facet Δ_v of ∂P^* associated with the vertex $v = F_1 \cap \cdots \cap F_n$, identify vertices $\{i\} \in \mathscr{K}(B_{\Delta,n})$ with vertices Δ_{F_i} of Δ_v for $i = 1,\ldots,n$, and apply the sequence of stellar subdivisions to Δ_v inside the complex ∂P^*, to obtain the boundary of some simplicial polytope Q. By definition, the *nested polytope* $NP(P,B)$ is the dual of the polytope Q.

Proposition 11.9. *If $P = I^n$ and $B = B(\Gamma)$ is a connected graphical building set, then $NP(I^n,B)$ is combinatorially equivalent to the graph-cubeahedron associated with Γ.*

Proof. This follows from Lemma 11.6. $\qquad\square$

Theorem 11.10.

1) *If P is a flag polytope and B a flag building set, then $NP(P,B)$ is a flag polytope.*
2) *If P is a 2-truncated cube and B a flag building set, then $NP(P,B)$ is a 2-truncated cube.*

Proof. By Theorem 11.7, the complex $\mathscr{K}(B)$ can be obtained from Δ_{n-1} by stellar subdivisions over edges. Therefore $NP(P,B)$ can be obtained from P by 2-truncations. $\qquad\square$

Corollary 11.11. *Any graph-cubeahedron is a 2-truncated cube.*

Acknowledgement We are grateful to A. Gaifullin, S. Forcey, F. Müller-Hoissen, T. Panov and J. Stasheff for valuable discussions and comments. Our work has been supported by the Russian government project 11.G34.31.0053.

References

1. V. Buchstaber, "Ring of simple polytopes and differential equations", *Trudy Matematicheskogo Instituta imeni V.A. Steklova* **263** (2008) 18–43.
2. V.M. Buchstaber and T.E. Panov, *Torus Actions and Their Applications in Topology and Combinatorics*, University Lecture Series, American Mathematical Society, Providence, RI, 2002.
3. M. Carr and S. Devadoss, "Coxeter complexes and graph associahedra", *Topology and its Applications* **153** (2006) 2155–2168.
4. C. Ceballos and G. Ziegler, "Three non-equivalent realizations of the associahedron", *http://arXiv.org/abs/1006.3487*.
5. F. Chapoton, S. Fomin, and A. Zelevinsky, "Polytopal realizations of generalized associahedra", *Canad. Math. Bull.* **45** (2002) 537–566.
6. R. Charney and M. Davis, "The Euler characteristic of a nonpositively curved, piecewise Euclidean manifold", *Pacific J. Math.* **171** (1995) 117–137.
7. C.D. Concini and C. Procesi, "Wonderful models of subspace arrangements", *Selecta Mathematica (N.S.)* **1** (1995) 459–494.
8. M. Davis and T. Januszkiewicz, "Convex polytopes, Coxeter orbifolds and torus actions", *Duke Math. J.* **62** (1991) 417–451.
9. S.L. Devadoss, T. Heath, and W. Vipismakul, "Deformations of bordered surfaces and convex polytopes", *Notices of the AMS* **58** (2011) 530–541.
10. N. Erokhovets, presented in a seminar at Moscow State University.
11. E.-M. Feichtner and I. Mueller, "On the topology of nested set complexes", *Proceedings of American Mathematical Society* **133** (2005) 999–1006.
12. E.-M. Feichtner and B. Sturmfels, "Matroid polytopes, nested sets, and Bergman fans", *Portugaliae Mathematica (N.S.)* **62** (2005) 437–468.
13. S. Fomin and A. Zelevinsky, "Y-systems and generalized associahedra", *Annals of Math.* **158** (2003) 977–1018.
14. S. Gal, "Real root conjecture fails for five- and higher-dimensional spheres", *Discrete & Computational Geometry* **34** (2005) 269–284.
15. I.M. Gelfand, M.M. Kapranov, and A.V. Zelevinsky, *Discriminants, Resultants, and Multidimensional Determinants*, Birkhäuser, Boston, 1994.
16. M.A. Gorsky, "Proof of Gal's conjecture for the series of generalized associahedra", *Russian Math. Surveys* **65** (2010) 1178–1180.
17. T. Oda, *Convex Bodies and Algebraic Geometry: An Introduction to the Theory of Toric Varieties*, Springer-Verlag, Berlin, 1998.
18. T. Panov, N. Ray, and R. Vogt, "Colimits, Stanley-Reisner algebras, and loop spaces", in *In Categorical Decomposition Techniques in Algebraic Topology*, Progress in Mathematics, vol. 215, Birkhäuser, 2004, 261–291.
19. A. Postnikov, "Permutohedra, associahedra, and beyond", *International Mathematics Research Notices* (2009) 1026–1106.
20. A. Postnikov, V. Reiner, and L. Williams, "Faces of generalized permutohedra", *Documenta Mathematica* **13** (2008) 207–273.
21. A. Zelevinsky, "Nested set complexes and their polyhedral realizations", *Pure and Applied Mathematics Quarterly* **2** (2006) 655–671.
22. G. Ziegler, *Lectures on Polytopes*, Springer-Verlag, 1995.

Extending the Tamari Lattice to Some Compositions of Species

Stefan Forcey

Abstract An extension of the Tamari lattice to the multiplihedra is discussed, along with projections to the composihedra and the Boolean lattice. The multiplihedra and composihedra are sequences of polytopes that arose in algebraic topology and category theory. Here we describe them in terms of the composition of combinatorial species. We define lattice structures on their vertices, indexed by painted trees, which are extensions of the Tamari lattice and projections of the weak order on the permutations. The projections from the weak order to the Tamari lattice and the Boolean lattice are shown to be different from the classical ones. We review how lattice structures often interact with the Hopf algebra structures, following Aguiar and Sottile who discovered the applications of Möbius inversion on the Tamari lattice to the Loday-Ronco Hopf algebra.

1 Introduction

We will be looking at the following spectrum of polytopes:

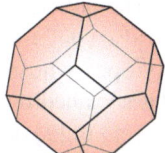 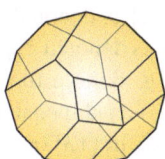 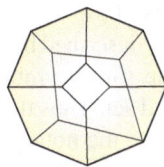 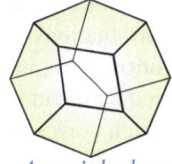 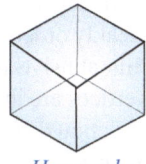

Permutohedron   *Multiplihedron*   *Composihedron*   *Associahedron*   *Hypercube*

Stefan Forcey
Department of Theoretical and Applied Mathematics
The University of Akron
Akron, OH 44325-4002,
e-mail: *sf34@uakron.edu*

Here is a *planar rooted binary tree*, often called a binary tree:

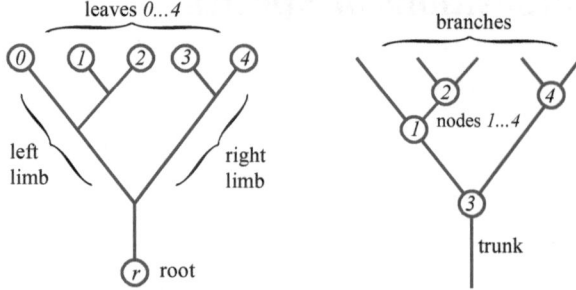

We only label the leaves and interior nodes (branch points) with their left-to-right ordering when necessary. The branches are the edges with a leaf. The nodes are also partially ordered by their proximity to the root, which is maximal; in the picture node $3 > 1 > 2$. The set of planar rooted binary trees with n nodes and $n+1$ leaves is denoted $\mathcal{Y}_n$.

1.0.1 Notation

We will choose from the current rather prolix notation used for the classical polytopes and lattices, and try to decrease the proliferation of symbols by referring to a polytope and its associated orders by a common name. The context will determine whether we are focused on the face structure, the vertices alone, or the 1-skeleton. Since the vertices are used more often, we let that be the default meaning, and add more specifics if necessary. For instance if we wish to refer to the general planar trees that index the faces of the associahedron, we'll make that clear.

The Tamari lattice is denoted either $\mathcal{Y}_n$ or $\mathbb{T}_n$. The set of painted trees with n nodes and $n+1$ leaves is denoted $\mathcal{M}_n$ and the lattice structure on that set is denoted $\mathcal{M}_n$ as well. The set of binary trees with leaves weighted by positive integers summing to $n+1$ is denoted $\mathcal{C}\mathcal{K}_n$ and the lattice structure we define on that set is denoted $\mathcal{C}\mathcal{K}_n$ as well. The Boolean lattice of subsets of $[n] = \{1,\dots,n\}$ is denoted $\mathcal{Q}_n$. The lattice of weakly ordered permutations on $[n]$, described below, is denoted $\mathfrak{S}_n$. We also continue this abusive notation by using the same symbols to denote the polytopes whose vertices are indexed by the indicated set. Thus the $(n-1)$-dimensional associahedron is denoted $\mathcal{Y}_n$, which corresponds to the notation $\mathcal{K}_n$ in [13] or $\mathcal{K}(n+1)$ in [7] or even K_{n+1} in Stasheff's original notation. The $(n-1)$-dimensional permutahedron, multiplihedron, composihedron and hypercube are denoted $\mathfrak{S}_n$, $\mathcal{M}_n$, $\mathcal{C}\mathcal{K}_n$, and $\mathcal{Q}_n$ respectively. Rather than a subscript n, we sometimes use a placeholder $\bullet$ to refer to the entire sequence at once.

The 1-skeleta of the families of polytopes $\mathfrak{S}_\bullet$, $\mathcal{M}_\bullet$, $\mathcal{Y}_\bullet$ and $\mathcal{Q}_\bullet$ are Hasse diagrams of posets. For the permutahedron $\mathfrak{S}_n$, the corresponding poset is the (left) *weak order*, which we describe in terms of permutations. A cover in the weak order has the form

$w \lessdot (k,k+1)w$, where k preceeds $k+1$ among the values of w. Figure 1 displays the weak order on $\mathfrak{S}_4$, the Tamari order on $\mathcal{Y}_4$ and the Boolean lattice $\mathcal{Q}_3$.

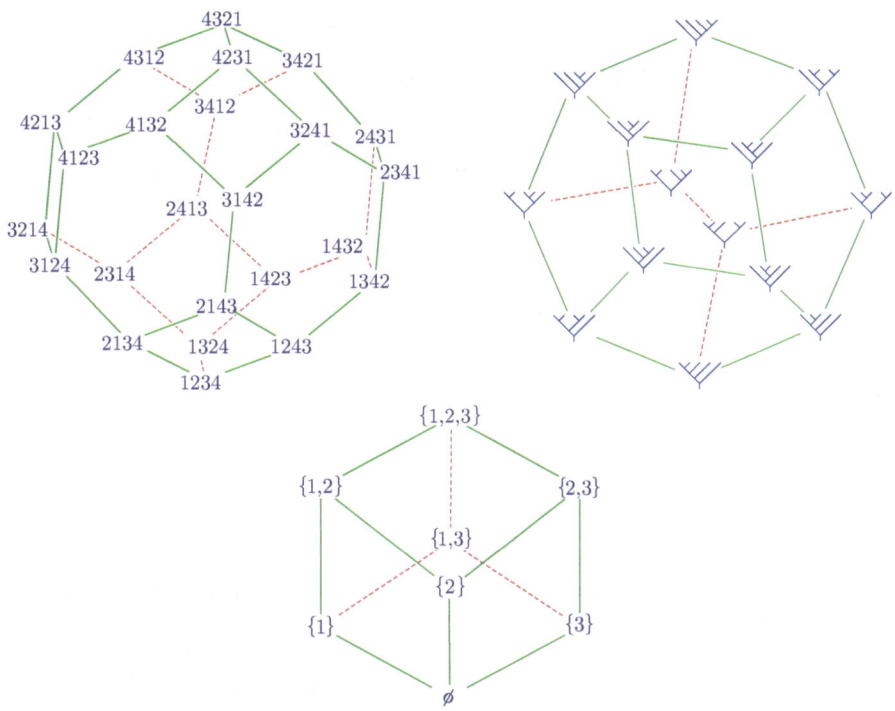

Fig. 1 Hasse diagrams of three classical lattices in this paper: Weak order on $\mathfrak{S}_4$, Tamari lattice $\mathcal{Y}_4$ and Boolean lattice $\mathcal{Q}_3$.

1.1 Species

A combinatorial *species* of sets is an endofunctor of Finite Sets with bijections.

Example 1.1. The species $\mathcal{L}$ of lists takes a set to linear orders of that set.

$$\mathcal{L}(\{a,d,h\}) = \{a<d<h,\, a<h<d,\, h<a<d,\, h<d<a,\, d<a<h,\, d<h<a\}$$

Example 1.2. The species $\mathcal{Y}$ of binary trees takes a set to trees with labeled leaves.

$$\mathcal{Y}(\{a,\, d,\, h\}) = \{\; \overset{a\;d\;h}{\curlyvee},\, \overset{a\;h\;d}{\curlyvee},\, \dots,\, \overset{a\;d\;h}{\curlyvee},\, \overset{a\;h\;d}{\curlyvee},\, \dots\}$$

We define the composition of two species, following Joyal in [12]:

$$(\mathscr{G} \circ \mathscr{H})(U) = \bigsqcup_{\pi} \mathscr{G}(\pi) \times \prod_{U_i \in \pi} \mathscr{H}(U_i)$$

where the union is over partitions of U into any number of nonempty disjoint parts.

$$\pi = \{U_1, U_2, \ldots, U_n\} \text{ such that } U_1 \sqcup \cdots \sqcup U_n = U.$$

This formula also appears to be known as the cumulant formula, the moment sequence of a random variable, and the domain for operad composition:

$$\gamma : \mathscr{F} \circ \mathscr{F} \to \mathscr{F}.$$

2 Several flavors of trees

2.1 Ordered, bi-leveled and painted trees

Many variations of the idea of the binary tree have proven useful in applications to algebra and topology. Each variation we mention can have its leaves labeled, providing an example of a set species.

An ordered tree (sometimes called leveled) has a vertical ordering of the n nodes as well as horizontal. This allows a well-known bijection between the ordered trees with n nodes and the permutations $\mathfrak{S}_n$. We will call this bijection bij_1. This bijection and all the other maps we will discuss are demonstrated in Figure 7.

As defined in 2.1 of [9], a *bi-leveled tree* $(t; \mathsf{T})$ is a planar binary tree $t \in \mathscr{Y}_n$ together with an (upper) order ideal T of its node poset, where T contains the leftmost node of t as a minimal element. (Recall that an upper order ideal is a subposet such that $x > y \in \mathsf{T}$ implies $x \in \mathsf{T}$.) We draw the underlying tree t and circle the nodes in T. By the condition on T, all nodes along the leftmost branch are circled and none are circled above the leftmost node.

Saneblidze and Umble [17] introduced bi-leveled trees in terms of equivalence classes on ordered trees. They describe a cellular projection from the permutahedra to Stasheff's multiplihedra $\mathscr{M}_\bullet$, with the bi-leveled trees on n nodes indexing the vertices $\mathscr{M}_n$.

Definition 2.1. We denote Saneblidze and Umble's map as $\beta : \mathfrak{S}_n \to \mathscr{M}_n$, as in [9], and describe it as the map which first circles all the nodes vertically ordered below and including the leftmost node, and then forgets the vertical ordering of the nodes.

Numbering the nodes in a tree $t \in \mathscr{Y}_n$ $1, \ldots, n$ from left to right, T becomes a subset of $\{1, \ldots, n\}$.

Definition 2.2. The partial order on $\mathcal{M}_n$ is defined by $(s;S) \le (t;T)$ if $s \le t$ in $\mathcal{Y}_n$ and $T \subseteq S$.

Theorem 2.3. *The poset of bi-leveled trees is a lattice.*

Proof. The unique supremum of two bi-leveled trees $(t;T)$ and $(s;S)$ is found by first taking their unique supremum $\sup\{t,s\}$ in the Tamari lattice, and then circling as many nodes of $\sup\{t,s\}$ in the intersection $T \cap S$ as are allowed by the upper-order ideal condition. That is, the circled nodes of the join comprise the largest-order ideal of nodes of $\sup\{t,s\}$ that is contained in $T \cap S$. The unique infimum is found by taking the infimum of the two trees in the Tamari lattice, the union of the two order ideals, and adding to the latter any nodes necessary to make that union an order ideal in the node poset of $\inf\{s,t\}$. That is, the meet is given by $(\inf\{t,s\}; T \cup S \cup \{x \mid x > y \in T \cup S\})$. $\qquad\qquad\Box$

The Hasse diagrams of the posets $\mathcal{M}_n$ are 1-skeleta for the multiplihedra. The Hasse diagram of $\mathcal{M}_4$ appears in Figure 4. Stasheff used a different type of tree for the vertices of $\mathcal{M}_{\bullet}$. A *painted binary tree* is a planar binary tree t, together with a (possibly empty) upper-order ideal of the node poset of t. (Recall the root node is maximal.) We indicate this ideal by painting part of a representation of t. For clarity, we stop our painting in the middle of edges (not precisely at nodes). Here are a few simple examples,

An A_n-*space* is a topological H-space with a weakly associative multiplication of points [18]. Maps between A_∞ spaces are only required to preserve the A_∞ structure up to homotopy. Stasheff [19] described these maps combinatorially using cell complexes called multiplihedra, while Boardman and Vogt [4] used spaces of painted trees. Both the spaces of trees and the cell complexes are homeomorphic to convex polytope realizations of the multiplihedra as shown in [7].

If $f: (X, \bullet) \to (Y, *)$ is an A_∞-map homotopy H-spaces, then the different ways to multiply and map n points of X are naturally represented by a painted tree, as follows. Nodes not painted correspond to multiplications in X, painted nodes correspond to multiplications in Y, and the beginning of the painting (along the edges) indicates the moment f is applied to a given point in X. (Weak associativity of X and Y justifies the use of planar binary trees, which represent the distinct associations on a set of inputs.) See Figure 2.

$$f(a) * \big(f(b \bullet c) * f(d) \big) \longleftrightarrow$$

Fig. 2 A_n-maps between H-spaces $(X, \bullet) \xrightarrow{f} (Y, *)$ are painted trees.

Figure 4 shows two versions of the three-dimensional multiplihedron as Hasse diagrams.

Bi-leveled trees having $n+1$ internal nodes are in bijection with painted trees having n internal nodes, the bijection being given by pruning: Remove the leftmost branch (and hence, node) from a bi-leveled tree to get a tree whose order ideal is the order ideal of the bi-leveled tree, minus the leftmost node. We refer to this as bij_2. This mapping and its inverse are illustrated in Figure 3. The composition of bij_2 with β is just called β. (We will often use these bijections as identities.)

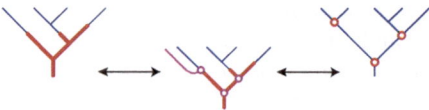

Fig. 3 Painted trees correspond to bi-leveled trees.

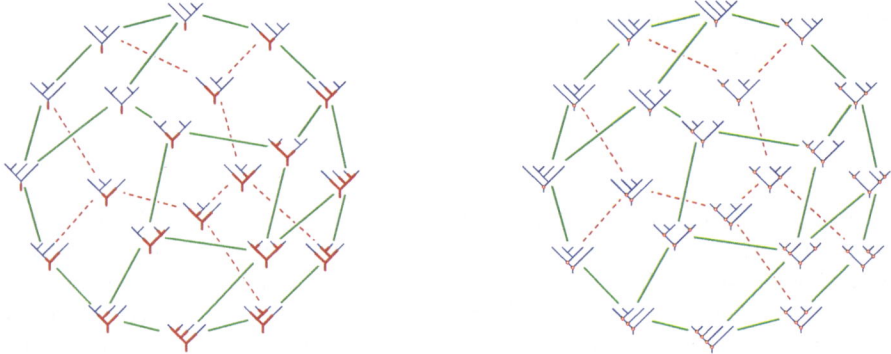

Fig. 4 Two Hasse diagrams of the multiplihedra lattice $\mathcal{M}_4$, labeled with painted and bi-leveled trees.

Remark 2.4. If the leaves of a painted binary tree are labeled by the elements of a set, it is recognizable as a structure in a certain combinatorial species: the self-composition $\mathcal{Y} \circ \mathcal{Y}$ of binary trees. The structure types of this species are the (unlabeled) binary painted trees themselves. Forgetting the painting in any painted tree is precisely the composition in the operad of binary trees.

Forgetting the levels in a bi-leveled tree (removing the circles) gives a different (from the one just remarked on) projection to binary trees. We denote this by ϕ :

$\mathscr{M}_n \to \mathscr{Y}_n$ as in [9]. Now the composition $\phi \circ \beta$ gives the Tonks projection from $\mathfrak{S}_n$ to Fy_n, denoted respectively by Θ in [20], by τ in [9] and by Ψ in [13].

In [13] Loday and Ronco define a poset map from $\mathscr{Y}_n$ to $\mathscr{Q}_n$. (They call it ϕ, here we denote it $\hat{\phi}$ to avoid duplicate naming.) This map takes a tree and gives a vertex of the hypercube $[-1, 1]^n$ by assigning either $+1$ or -1 to each of the branches not on a limb. Each branch is assigned its slope, where the tree is drawn with 45 degree angles. Further, they use a bijection (we call it bij_4) from these vertices to elements of the Boolean lattice $\mathscr{Q}_n$ defined by including the elements $i \in [n-1]$ which correspond to the coordinates $x_i = -1$. The composition of $\hat{\phi}, \phi$ and β gives the descents of the permutation.

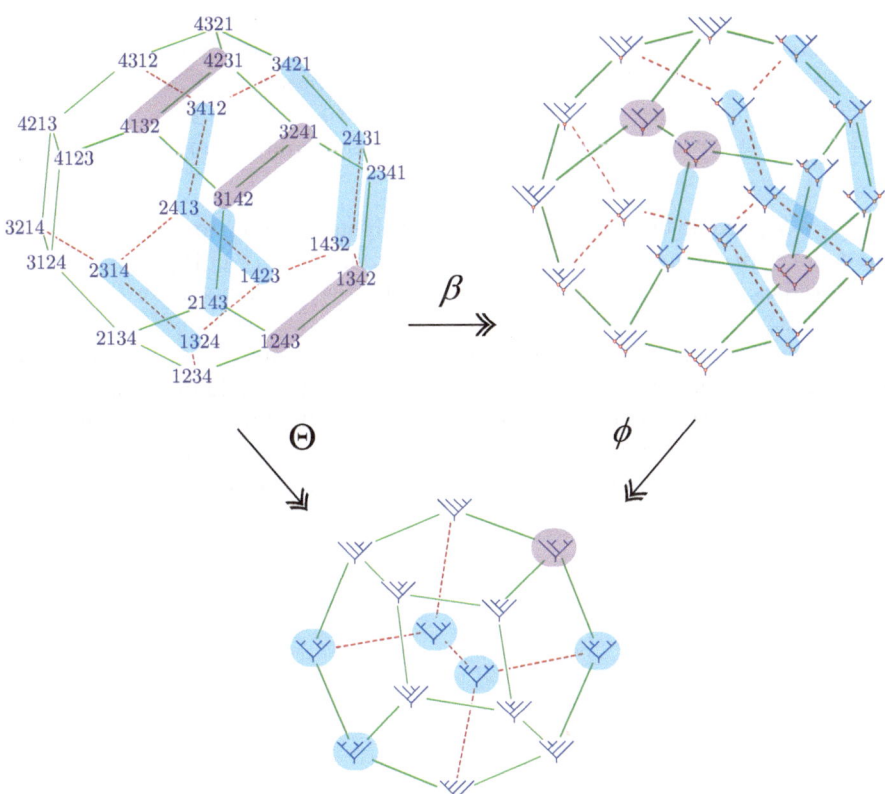

Fig. 5 The maps $\Theta = \phi \circ \beta$ shown with retracted intervals shaded.

2.2 Trees with corrollas

We will use the term corolla to describe a rooted tree with one interior node and $n+1$ leaves. In a forest of corollas attached to a binary tree, each corolla may be replaced by a positive *weight* counting the number of leaves in the corolla. (Alternately, as in [10], the corollas may be replaced by combs.) These all give *weighted trees*.

$$
\text{(tree with corollas)} = \underset{2 \quad 3 \quad 1 \quad 2}{\text{(weighted tree)}} = \text{(tree with combs)} \tag{1}
$$

Remark 2.5. By labelling leaves of a comb by the elements of a set, we define a species called $\mathfrak{C}$, which is in fact isomorphic to the species of lists. Labelled weighted trees (where we are labeling the leaves of the forest of combs grafted to a tree) are recognizable as the structures in the species composition $\mathscr{Y} \circ \mathfrak{C}$.

Let $\mathscr{C} \mathscr{K}_n$ denote the weighted trees with weights summing to $n+2$. These index the vertices of the n-dimensional *composihedron*, $\mathscr{C} \mathscr{K}(n+1)$ [8]. This sequence of polytopes parameterizes homotopy maps from strictly associative H-spaces to A_∞-spaces.

If we use right combs instead of corollas as the weights on our weighted trees, then the same relations as in $\mathscr{M}_n$ give the weighted trees a lattice structure. The joins and meets are found as for painted trees, with the final step of combing the unpainted subtrees. Figure 6 gives two pictures of the composihedron $\mathscr{C} \mathscr{K}_4$. The 2- and 3-dimensional composihedra $\mathscr{C} \mathscr{K}(n)$ also appear as the commuting diagrams in

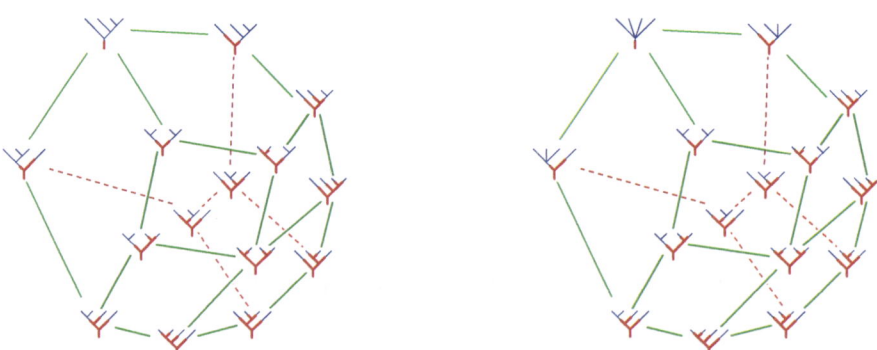

Fig. 6 The one-skeleton of the three-dimensional composihedron, as a Hasse diagram labeled by two representations of weighted trees.

enriched bicategories [8]. As a special case of enriched bicategories, these diagrams appear in the definition of pseudomonoids [1, Appendix C].

On the other hand, attaching a forest of binary trees to a single corolla is really just a way of picturing an ordered forest of binary trees, listed left to right. There is a well-known bijection from the set of ordered forests with $n+1$ total leaves to $\mathscr{Y}_n$. We call this bijection bij_3. It is described by taking the k trees of the forest in left-to-right order and attaching them to a single limb, which will be the new left limb. Thus we can recognize this set as another version of $\mathscr{Y}_n$.

Finally we consider trees with n interior nodes obtained by grafting a forest of combs to the leaves of a comb (which is painted). Analogous to (1), these are weighted combs (or corollas). As these are in bijection with number-theoretic compositions of $n+1$, we refer to them as *composition trees*.

$$\text{(tree diagram)} = \text{(tree diagram)} = \overset{3\,2\,1\,4}{\text{(tree)}} = (3,2,1,4).$$

> *Remark* 2.6. Leaf-labelled composition trees (where we are labeling the leaves of the forest of combs grafted to a comb) are recognizable as the structures in the species composition $\mathfrak{C} \circ \mathfrak{C}$.

In the next section we will describe maps from the multiplihedra to the hypercubes, but first we note that we will use a different bijection from the set of composition trees to $\mathscr{Q}_n$. A composition tree is associated by bijection bij_5 with the set of vertices that are unpainted.

3 Interval retracts

In [9] it is shown that there exists a section of the projection $\beta : \mathfrak{S}_n \to \mathscr{M}_n$ which demonstrates β to be an interval retract. We review that definition here. Recall that an interval $[a,b]$ of a poset P is a sub-poset given by $\{x \mid a \le x \le b \in \mathrm{P}.\}$

A surjective poset map $f : P \to Q$ from a finite lattice P is an *interval retract* if the fibers of f are intervals and if f admits an order-preserving section $g : Q \to P$ with $f \circ g = \mathrm{id}$. Also in [9] there is proven a useful relation between the Möbius functions of $\mathfrak{S}_{\bullet}$ and $\mathscr{M}_{\bullet}$, which is in fact established there in a general form.

Theorem 3.1 ([9]). *Let the poset map $f : P \to Q$ be an interval retract, then the Möbius functions μ_P and μ_Q of P and Q are related by the formula*

$$\mu_Q(x,y) = \sum_{\substack{f(a)=x \\ f(b)=y}} \mu_P(a,b) \qquad (\forall x,y \in Q). \tag{2}$$

The proof in [9] relies on the fact that the intersection of two intervals in a lattice is again an interval.

Here we define eight closely related maps in order to demonstrate four new interval retracts: first four projections, each associated to a corresponding section.

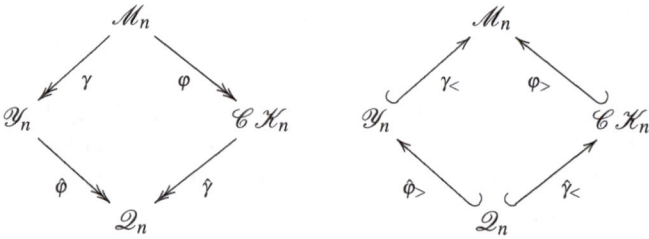

The maps γ and $\hat{\gamma}$ operate by replacing the painted portion with a corolla, while φ and $\hat{\varphi}$ replace the unpainted forest with a forest of corollas.

We define the sections $\gamma_<$ and $\hat{\gamma}_<$ by replacing painted corollas with left combs, while $\varphi_>$ and $\hat{\varphi}_>$ are defined by replacing unpainted corollas with right combs.

The main result will be that each paired projection and section between the same two lattices together define an interval retract. In figure 7 we demonstrate all the projections and bijections described above.

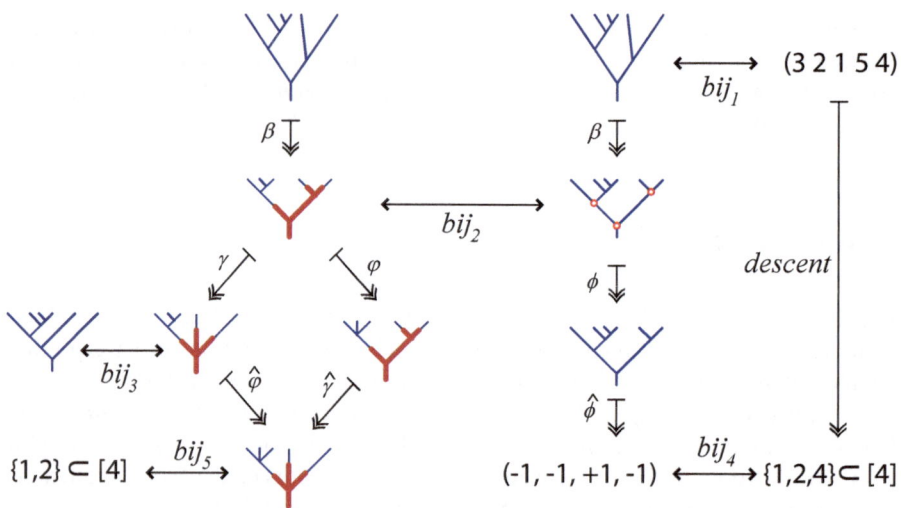

Fig. 7 Projections and bijections in this paper.

Theorem 3.2. *The map γ is an interval retract from $\mathcal{M}_n$ to $\mathcal{Y}_n$.*

Proof. We use the section $\gamma_<$. Thus we need to show four things: that $\gamma^{-1}(t)$ is an interval in $\mathcal{M}_n$ for any $t \in \mathcal{Y}_n$; that γ and $\gamma_<$ both preserve order, and that $\gamma \circ \gamma_<$ is

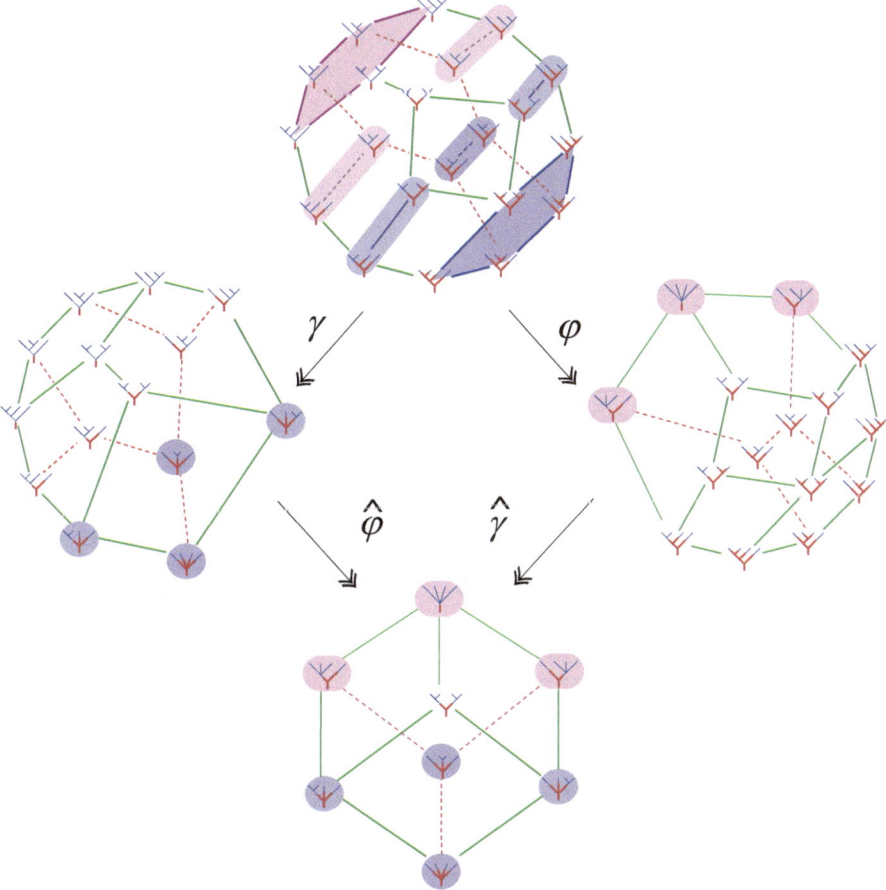

Fig. 8 The four projections in dimension 3, with shaded intervals retracted.

the identity map. This last fact is straightforward, since it constitutes first removing and then replacing an unpainted forest on its painted comb.

Consider $t \in \mathscr{Y}_n$ with the k-forest f of subtrees with initial nodes on the left limb of t. To show that inverse images of γ are intervals, we point out that $\gamma^{-1}(t)$ is the set of painted trees with unpainted forest f and any painted portion. This is an interval since its elements comprise all those between a unique min and max given by minimizing and maximizing the painted portion. In fact, the interval is isomorphic to a copy of the Tamari lattice $\mathscr{Y}_{k-1}$.

Next to show that γ preserves the order, we let $a < b \in \mathscr{M}_n$. This means that $t_a \leq t_b$ in the Tamari order, and that $P_a \supseteq P_b$. Now we may visualize the action of γ as a series of smaller steps: first we make all possible Tamari moves in the painted region of a that each yield sequentially lesser trees. Then we attach a new branch

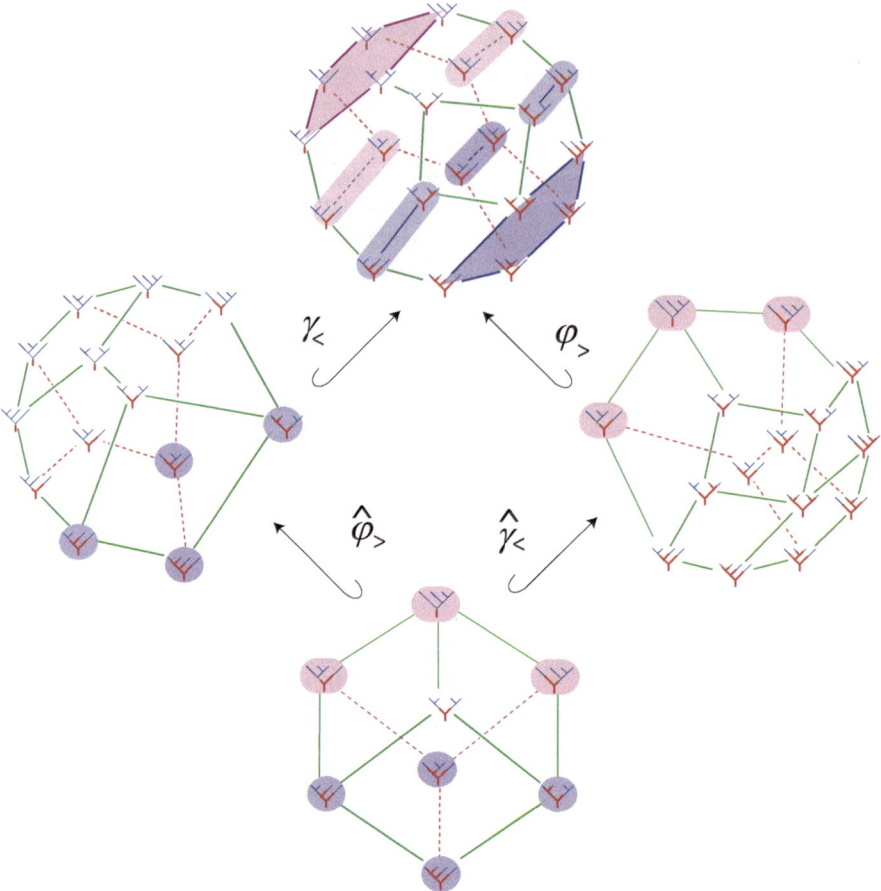

Fig. 9 The four sections in dimension 3. Here we label the elements of $\mathscr{Y}_n$, $\mathscr{C}\mathscr{K}_n$ and $\mathscr{Q}_n$ using left and right combs so that the sections can be seen as inclusions.

to the left-most painted point of a, and finally forget the painting altogether. The same basic steps are performed to find $\gamma(b)$. Since $P_b \subseteq P_a$ we can see the relation $\gamma(b) \geq \gamma(a)$ by the series of Tamari moves to get from b to a followed by more moves resulting from the possibly additional painted nodes of a.

To show that if $q < s \in \mathscr{Y}_n$ implies that $\gamma_<(q) \leq \gamma_<(s)$, we consider the string of Tamari covering moves that relate q to s. Recall that $\gamma_<$ takes the k-forest f of sub-trees attached along the left limb of q and instead attaches them to a minimal painted k-tree, that is, they are grafted to a painted left comb with k branches. Alternately this is described by simply pruning away the leftmost leaf of q and painting the nodes of q along the leftmost branch. We see that $t_{\gamma_<(q)} \leq t_{\gamma_<(s)}$ by noting that the moves between them are the same as those from q to s. Then we note that we have

$P_{\gamma_<(q)} \supseteq P_{\gamma_<(s)}$ since any move from q to s either subtracts from the set of painted nodes in the eventual image (if the move involves a node on the leftmost branch of q) or leaves that set unchanged. □

Theorem 3.3. *The map φ is an interval retract from $\mathcal{M}_n$ to $\mathscr{C}\mathscr{K}_n$.*

Proof. We use the section $\varphi_>$. Again we need to show four things: that $\varphi^{-1}(t)$ is an interval in $\mathcal{M}_n$ for any $w \in \mathscr{C}\mathscr{K}_n$; that φ and $\varphi_>$ both preserve order, and that $\varphi \circ \varphi_>$ is the identity map. This last fact is straightforward, since both maps will be the identity in this case – $\varphi_>$ will always be and φ will be the identity when applied to a painted tree with a forest of unpainted right combs.

Recall that φ involves replacing an unpainted forest f with a forest of right combs. (Sometimes alternately drawn as corollas or just a number.) Thus the fiber $\varphi^{-1}(w)$ for $w \in \mathscr{C}\mathscr{K}_n$ is a collection of painted trees in $\mathcal{M}_n$ which share the same set of painted nodes, and the same binary tree as the subtree made up of those painted nodes – but which may have any forest of unpainted trees agreeing with those facts. Thus the fiber is an interval bounded by choosing that forest to be all left or all right combs. In fact this is a cartesian product of associahedra.

Next to show that φ preserves the order, we let $a < b \in \mathcal{M}_n$. This means that $t_a \leq t_b$ in the Tamari order, and that $P_a \supseteq P_b$. First note that if $P_a = P_b$, then $\varphi(a) < \varphi(b)$ by Tamari moves in the painted nodes. If $P_a \supset P_b$, consider the forest of unpainted right combs of $\varphi(b)$. Each of these combs has a leftmost node k. If the corresponding node k of $\varphi(a)$ is painted, then we can see the relation as first performing Tamari moves on the right comb of $\varphi(b)$ until we have the binary tree supported by node k of $\varphi(a)$, and then allowing the paint level to rise to cover node k and any additional nodes to match $\varphi(a)$. These moves performed on each unpainted comb of $\varphi(b)$ give us the result.

The section $\varphi_>$ is very simple; it merely returns us to the maximum of the fiber. In fact, if we are using right combs for our weighted trees then this section is the identity map, and so order is clearly preserved. □

Theorem 3.4. *The map $\hat{\varphi}$ is an interval retract from $\mathcal{Y}_n$ to $\mathcal{Q}_n$.*

Proof. It is easiest to see this when viewing $\mathcal{Y}_n$ in its incarnation as ordered forests of binary trees, grafted onto painted left combs. Then the ordering of $\mathcal{Y}_n$ is directly inherited from $\mathcal{M}_n$, and the map $\hat{\varphi}$ is the same as φ. Thus the facts that the fibers are intervals and that $\hat{\varphi} \circ \hat{\varphi}_>$ is the identity are already proven.

Now the elements of $\mathcal{Q}_n$ are being drawn as composite trees (using combs or corollas) but we need to check that the usual ordering by inclusion of subsets (of unpainted nodes) agrees with the tree order. That is, if $p < q$ as elements of $\mathcal{Y}_n$ (each drawn as an unpainted forest grafted to a left comb) then $\hat{\varphi}(p) \leq \hat{\varphi}(q)$. By viewing two elements of $\mathcal{Q}_n$ as forests of right combs grafted to painted left combs, we see that the only relation inherited from $\mathcal{M}_n$ is that of the containment of the sets of painted nodes. Thus since in $\mathcal{M}_n$ a larger set of painted nodes is a lesser element, here a smaller set of unpainted nodes is the lesser element.

Finally the section $\hat{\phi}_>$ is given by inclusion of the composite tree as a forest of right combs grafted to a painted left comb, which ensures that the ordering is preserved. □

Theorem 3.5. *The map $\hat{\gamma}$ is an interval retract from $\mathscr{C}\mathscr{K}_n$ to $\mathscr{Q}_n$.*

Proof. We already view an element of $\mathscr{C}\mathscr{K}_n$ as a forest of right combs grafted to a painted binary tree. Viewing $\hat{\gamma}$ as replacing the painted nodes with a left comb, we see the proof proceeds just as for $\hat{\phi}$. □

4 Lattices and polytopes

Next we point out that the four interval retracts just defined extend to well-known cellular projections of the polytopes. These projections are not the same ones that appear in the work of Reading [15], Loday and Ronco [13], or Tonks [20]. Rather they are found implicitly in the work of Boardman and Vogt on maps of homotopy H-spaces [4].

Recall that the combinatorial lattices of trees we have discussed here all occur conveniently as the labels of vertices on convex polytopes. The Hasse diagrams are specific drawings of the 1-skeleton of each polytope. The polytopes are associated to another, entirely different, lattice: the poset of their faces, with the empty set adjoined as least element. It is an open question as far as I know whether there is any describable relationship between the two lattices, for instance between the Tamari lattice and the face-poset of the associahedron.

As indicated in the introduction (by our use of the same symbol for both polytope and lattice) we have the following correspondence: binary trees label vertices of the associahedra; ordered binary trees the permutahedra; painted binary trees the multiplihedra, weighted trees the composihedra; composition trees the hypercubes. The higher-dimensional faces of these polytopes are all associated to further generalizations of the trees in question, by allowing more non-binary nodes and by allowing painting to end precisely on a node.

The Hopf algebras, ordered trees and Boolean subsets have all been extended to larger Hopf algebras on the faces of the corresponding polytopes. This was achieved by Chapoton in [6]. It is the topic of future study that similar expansions exist for the multiplihedra and composihedra.

Here we show how the projections discussed in this paper appear as collapsing of the faces of the polytopes whose vertices they act upon. Figure 10 shows an alternate "above" view of the permutahedron. This is given in order to facilitate contrasting the various projection maps.

Figure 11 offers contrast and comparison of our new maps to the classic projections, showing the faces that are retracted. We use the "above" view of the posets. In the pictured 3d case it is apparent that the two compositions of maps, $\hat{\phi} \circ \phi \circ \beta$ (which is the Tonks projection) and $\hat{\phi} \circ \gamma \circ \beta$, have quite different actions on the

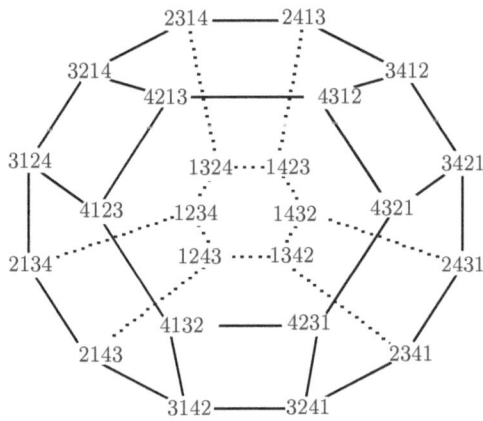

Fig. 10 The 3d permutahedron, alternate view.

permutahedron. The number of collapsed cells is the same in both composite projections, but in the first the image of 4 hexagons and 4 rectangles is a copy of S^1 where in the second the image of 2 hexagons and 6 rectangles consists of two disjoint star graphs. Next, for comparison, is the Tonks projection again, factored through the cyclohedron as in [11]. Finally for further contrast we include the projection η defined by Reading in his theory of Cambrian lattices, as seen in this volume [14].

The factorization of the Tonks projection, $\Theta = \theta_2 \circ \theta_1$, through the cyclohedron seen in Figure 11 deserves some special mention. First, this factorization is defined in greater generality [11] in terms of *tubings* of simple graphs.

Definition 4.1. Let G be a finite connected simple graph, with n numbered nodes. A *tube* is a set of nodes of G whose induced graph is a connected subgraph of G. Two tubes u and v may interact on the graph as follows:

1. Tubes are *nested* if $u \subset v$.
2. Tubes are *far apart* if $u \cup v$ is not a tube in G, that is, the induced subgraph of the union is not connected, or none of the nodes of u are adjacent to a node of v.

Tubes are *compatible* if they are either nested or far apart. We call G itself the *universal tube*. A *tubing* T of G is a set of tubes of G such that every pair of tubes in T is compatible; moreover, we force every tubing of G to contain (by default) its universal tube. By the term *k-tubing* we refer to a tubing made up of k tubes, for $k \in \{1, \ldots, n\}$.

Theorem 4.2 ([5, Section 3]). *For a graph G with n nodes, the* graph associahedron *$\mathcal{K}\,G$ is a simple, convex polytope of dimension $n-1$ whose face poset is isomorphic to the set of tubings of G, ordered such that $T \prec T'$ if T is obtained from T' by adding tubes.*

The vertices of the graph associahedron are the n-tubings of G. Faces of dimension k are indexed by $(n-k)$-tubings of G.

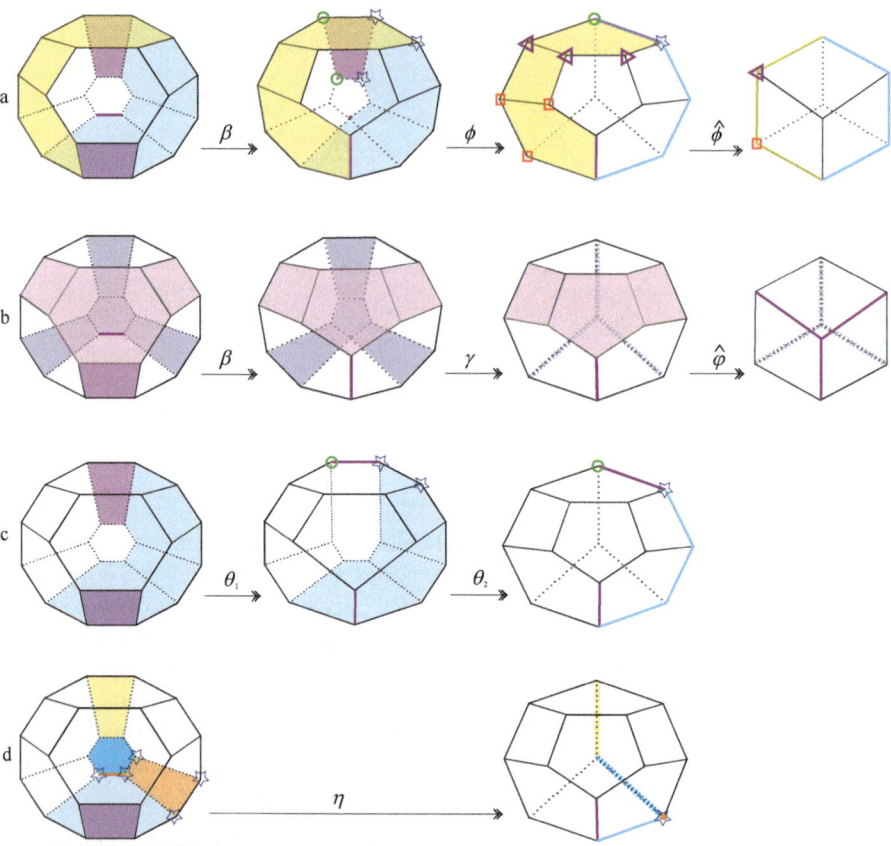

Fig. 11 In a, b, c and d the permutahedron is oriented as in Figure 10. The shaded facets and edges are collapsed in succession to similarly shaded edges and points. In a the circled and starred vertices in the domain of ϕ are mapped one and all to the circled (respectively starred) vertex in the range, and likewise for the squared and triangled vertices in the domain and range of $\hat{\phi}$. In b both γ and $\hat{\varphi}$ collapse a pentagon in their respective domains to a single vertex in their respective ranges. In c, the central polytope is the 3d cyclohedron. Finally in d η takes all the starred vertices to a single vertex.

As seen in [5], the permutahedron $\mathfrak{S}_n = \mathscr{K}G$ where G is the complete graph on n nodes; the associahedron $\mathscr{Y}_n = \mathscr{K}G$ where G is the path graph on n nodes; the cyclohedron is $\mathscr{W}_n = \mathscr{K}G$ when G is the cycle on n nodes; and the stellohedron is $\mathscr{K}G$ when G is the star graph on n nodes.

The question might be asked: how easily may the weak order on permutations and the Tamari order be generalized to n-tubings on a graph with nodes numbered $1,\ldots,n$? In order to describe the ordering we give the covering relations. We can use the same notation as when comparing tubings in the poset of faces of the graph associahedron since in that poset the n-tubings are not comparable.

Definition 4.3. Two n-tubings T, T' are in a covering relation $T' \prec T$ if they have all the same tubes except for one differing pair. We actually compare the outermost nodes, one from each of the pair of differing tubes. The outermost node of a tube is the node that is included in no other smaller sub-tube of the tubing. If the number of that node is greater for T, then T covers T'.

Note that each such covering relation corresponds to a unique $(n-1)$-tubing: the one resulting from removing the differing tubes. Thus the covering relations correspond to the edges of the graph associahedron.

For example, in Figure 12 we show a covering relation between two tubings on the complete graph on four numbered nodes. This figure also demonstrates the bijection between n-tubings and permutations of $[n]$. The nodes are the inputs for the permutation, and the output is the relative tube size. E.g., in the left-hand permutation the image of 2 is 1, and so we put the smallest tube around 2. To see the relation via tubes, we write down the sets of nodes in each tube. Only one pair of tubes differs. We compare the two numbered nodes of these which are in no smaller tubes. Here $(3124) \prec (4123)$ since $1 < 4$.

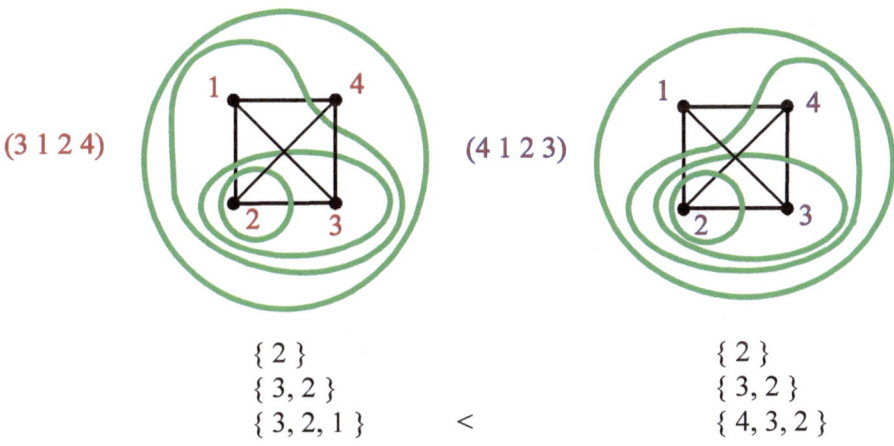

Fig. 12 A covering relation in the weak order on permutations.

It turns out that the relation generated by these covering relations of tubings has been independently demonstrated to be a poset by Ronco, in this volume [16]. In her article, the poset we have described on n-tubings of a graph is seen as the restriction of a larger poset on all the tubings of a graph.

Figure 13 shows the lattice that results from the cycle graph, rocovering the cyclohedron in dimension 3. The Hasse diagram is combinatorially equivalent to the 1-skeleton of the cyclohedron. Notice that this is quite different from the type B_3 Cambrian lattice described by Reading in this volume [14], despite the fact that the latter also is combinatorially equivalent to the 1-skeleton of the cyclohedron. Figure 14 shows the corresponding lattice on 4-tubings of the star graph on 4 nodes. This Hasse diagram is combinatorially equivalent to the 1-skeleton of the 3d stellohedron. Figure 15 shows both the cyclohedron and stellohedron lattices again, unlabeled, with a different view of each polytope for comparison.

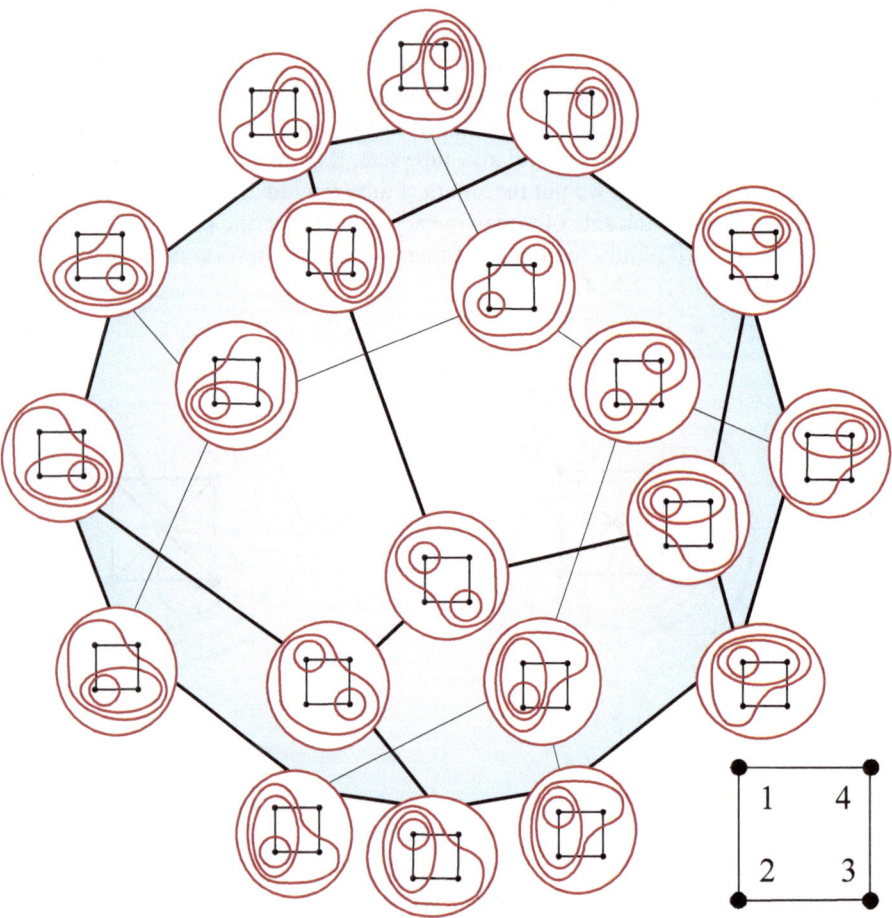

Fig. 13 This Hasse diagram is labeled by tubings of the cycle graph, with nodes numbered 1–4. The covering relations are also a picture of the edges of the cyclohedron.

We note that as seen in Ronco's article [16], the Tamari lattice is found as the lattice of n-tubings on the path graph with nodes numbered $1, \ldots, n$ in the order that they are connected by edges. Several open questions present themselves: for one, we notice that the 3-dimensional graph associahedra pictured here have associated posets which upon inspection prove to be lattices – it is not clear that they always are.

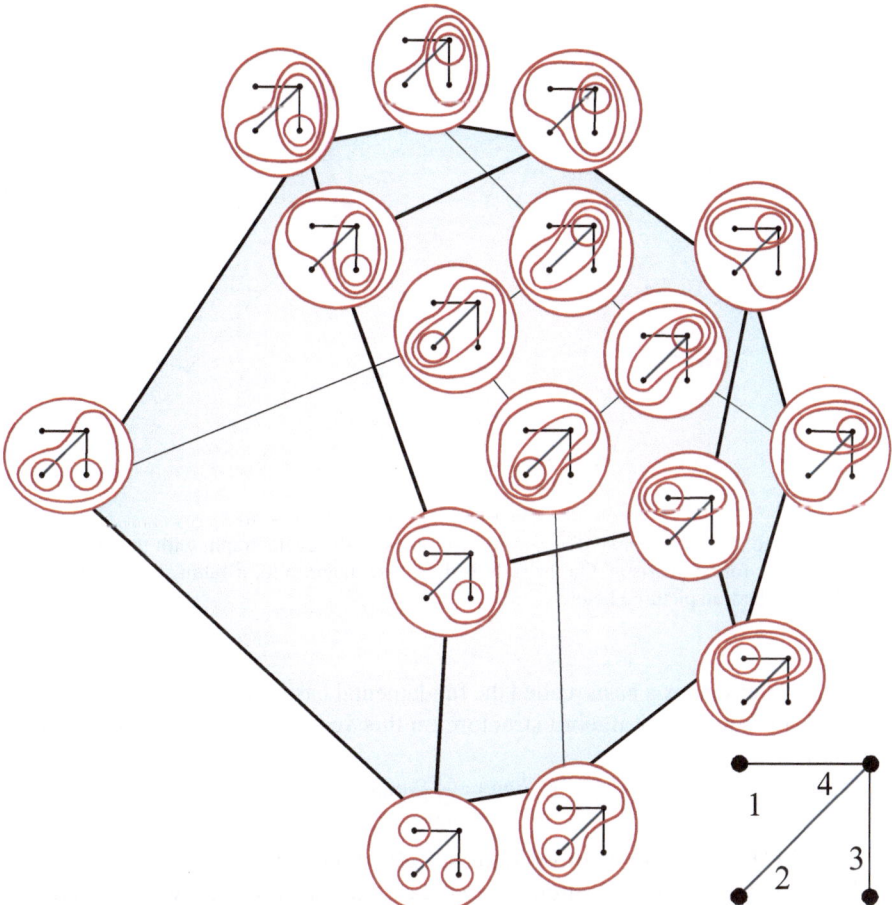

Fig. 14 This Hasse diagram is labeled by tubings of the star graph, with nodes numbered 1–4. The covering relations are also a picture of the edges of the stellohedron.

5 Algebraic implications of interval retracts

Finally we point out the importance of these lattices to the Hopf algebras defined as spans of their elements. For each of the lattices studied here, there is a graded vector space given by the direct product of the spans of the vertices of the n-dimensional polytope. For instance a vector space of binary trees is defined as:

$$\mathscr{Y}Sym = \bigoplus_{n \geq 0} \operatorname{span} \mathscr{Y}_n$$

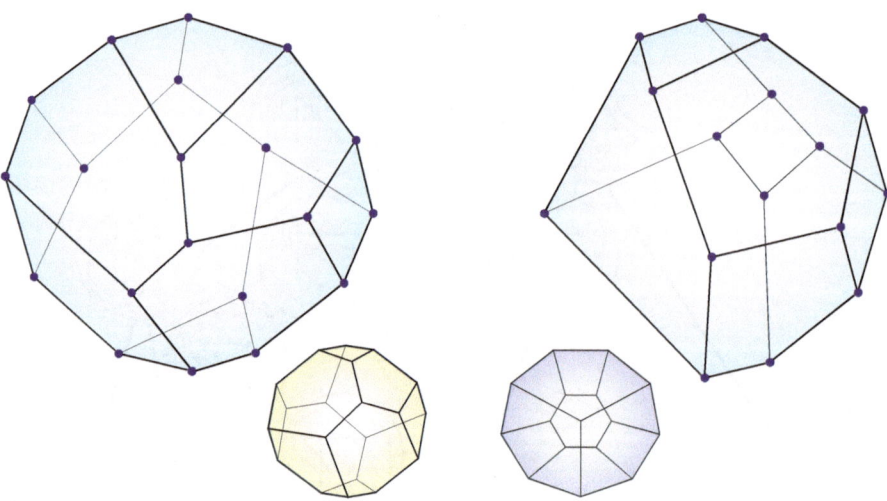

Fig. 15 On the left is the Hasse diagram for n-tubings of the cycle graph, with the cyclohedron pictured below for comparison. On the right is the Hasse diagram for n-tubings of the star graph, with the stellohedron pictured below.

The binary trees index a basis, called the fundamental basis and denoted $\{F_t \mid t \in \mathscr{Y}_n\}$. There is a graded Hopf algebra structure on this vector space, well-studied in [3]. Similarly

$$\mathfrak{S}Sym = \bigoplus_{n \geq 0} \operatorname{span} \mathfrak{S}_n$$

is a graded Hopf algebra on ordered trees, well-studied in [2].

Here we will restrict our attention to the coalgebra structures, which interact in important ways with the lattice structures. The remainder of this section is taken in part from [10]. It is included in order to demonstrate the algebraic importance of the lattice structures.

5.1 Coalgebras of trees

We define *splitting* a binary tree w along the path from a leaf to the root to yield a pair of binary trees,

$$\vcenter{\hbox{\includegraphics{trees}}} \longrightarrow \vcenter{\hbox{\includegraphics{trees2}}} \xrightarrow{\curlyvee} \left(\vcenter{\hbox{\includegraphics{treeleft}}} , \vcenter{\hbox{\includegraphics{treeright}}} \right).$$

Write $w \xrightarrow{\curlyvee} (w_0, w_1)$ when the pair of trees (w_0, w_1) is obtained by splitting w.

Definition 5.1 (Coproduct on $\mathscr{Y}Sym$). Given a binary tree t, define the coproduct in the fundamental basis by

$$\Delta(F_t) = \sum_{t \xrightarrow{Y} (t_0, t_1)} F_{t_0} \otimes F_{t_1}.$$

Here is an example:

$$\Delta(F_{\curlyvee}) = 1 \otimes F_{\curlyvee} + F_{\curlyvee} \otimes F_{\curlyvee} + F_{\curlyvee} \otimes 1.$$

5.2 Cofree composition of coalgebras

The following is excerpted with edits from [10]. Let $\mathscr{C}$ and $\mathscr{D}$ be graded coalgebras. We form a new coalgebra $\mathscr{E} = \mathscr{D} \circ \mathscr{C}$ on the vector space

$$\mathscr{D} \circ \mathscr{C} := \bigoplus_{n \geq 0} \mathscr{D}_n \otimes \mathscr{C}^{\otimes(n+1)}. \tag{3}$$

We write $\mathscr{E} = \bigoplus_{n \geq 0} \mathscr{E}_{(n)}$, where $\mathscr{E}_{(n)} = \mathscr{D}_n \otimes \mathscr{C}^{\otimes(n+1)}$. This gives a coarse coalgebra grading of $\mathscr{E}$ by $\mathscr{D}$-degree. There is a finer grading of $\mathscr{E}$ by *total degree*, in which a decomposable tensor $c_0 \otimes \cdots \otimes c_n \otimes d$ (with $d \in \mathscr{D}_n$) has total degree $|c_0| + \cdots + |c_n| + |d|$. Write $\mathscr{E}_n$ for the linear span of elements of total degree n.

Example 5.2. This composition is motivated by a grafting construction on trees. Let $d \times (c_0, \ldots, c_n) \in \mathscr{Y}_n \times \left(\mathscr{Y}_{\bullet}^{n+1}\right)$. Define $\circ$ by attaching the forest $(c_0, \ldots, c_n)$ to the leaves of d while remembering d,

This is precisely the type of tree called a *painted tree* in Section 2. Applying this construction to the indices of basis elements of $\mathscr{C}$ and $\mathscr{D}$ and extending by multilinearity gives $\mathscr{C} \circ \mathscr{D}$.

Motivated by this example, we represent an decomposable tensor in $\mathscr{D} \circ \mathscr{C}$ as

$$(c_0 \cdots \cdots c_n) \circ d \qquad \text{or} \qquad \frac{c_0 \cdots \cdots c_n}{d}$$

to compactify notation.

5.3 The coalgebra of painted trees

Let $\mathcal{M}_n$ be the poset of painted trees on n internal nodes. Then the vector space $\mathscr{P}Sym = \mathscr{Y}Sym \circ \mathscr{Y}Sym$ may be directly given by:

$$\mathscr{P}Sym = \bigoplus_{n \geq 0} \operatorname{span} \mathcal{M}_n$$

We reproduce the compositional coproduct defined in Section 2 of [10].

Definition 5.3 (Coproduct on $\mathscr{P}Sym$). Given a painted tree p, define the coproduct in the fundamental basis $\{F_p \mid p \in \mathcal{M}_\bullet\}$ by

$$\Delta(F_p) = \sum_{p \overset{Y}{\to} (p_0, p_1)} F_{p_0} \otimes F_{p_1},$$

where the painting in p is preserved in the splitting $p \overset{Y}{\to} (p_0, p_1)$.

The counit ε satisfies $\varepsilon(F_p) = \delta_{0,|p|}$, the Kronecker delta, as usual for graded coalgebras.

5.3.1 Primitives in the coalgebras of trees and painted trees

Now for the discussion of how the lattice structure found by Tamari really impacts the algebraic structure. Recall that a primitive element x of a coalgebra is such that $\Delta x = 1 \otimes x + x \otimes 1$. Theorem 2.4 of [10] describes the primitive elements of $\mathscr{P}Sym = \mathscr{Y}Sym \circ \mathscr{Y}Sym$ in terms of the primitive elements of $\mathscr{Y}Sym$. We recall the description of primitive elements of $\mathscr{Y}Sym$ as given in [3].

Let μ be the Möbius function of $\mathscr{Y}_n$ which is defined by $\mu(t,s) = 0$ unless $t \leq s$,

$$\mu(t,t) = 1, \quad \text{and} \quad \mu(t,r) = -\sum_{t \leq s < r} \mu(t,s).$$

We define a new basis for $\mathscr{Y}Sym$ using the Möbius function. For $t \in \mathscr{Y}_n$, set

$$M_t := \sum_{t \leq s} \mu(t,s) F_s.$$

Then the coproduct for $\mathscr{Y}Sym$ with respect to this M-basis is still given by splitting of trees, but only at leaves emanating directly from the right limb above the root:

$$\Delta(M\!\!\underset{}{\vee\!\!\vee}) = 1 \otimes M\underset{}{\vee\!\vee} + M\underset{}{\vee} \otimes M\underset{}{\vee} + M\underset{}{\vee\!\vee} \otimes 1.$$

A tree $t \in \mathscr{Y}_n$ is *progressive* if it has no branching along the right branch above the root node. A consequence of the description of the coproduct in this M-basis is

Corollary 5.3 of [3] that the set $\{M_t \mid t \text{ is progressive}\}$ is a linear basis for the space of primitive elements in $\mathscr{Y}Sym$.

Now according to Theorem 2.4 of [10] the cogenerating primitives in $\mathscr{P}Sym$ are of two types:

$$\frac{1 \cdot c_1 \cdot \cdots \cdot c_{n-1} \cdot 1}{M_t} \qquad \text{and} \qquad \frac{M_t}{1} \, ,$$

where t is a progressive tree. Figure 16 shows examples.

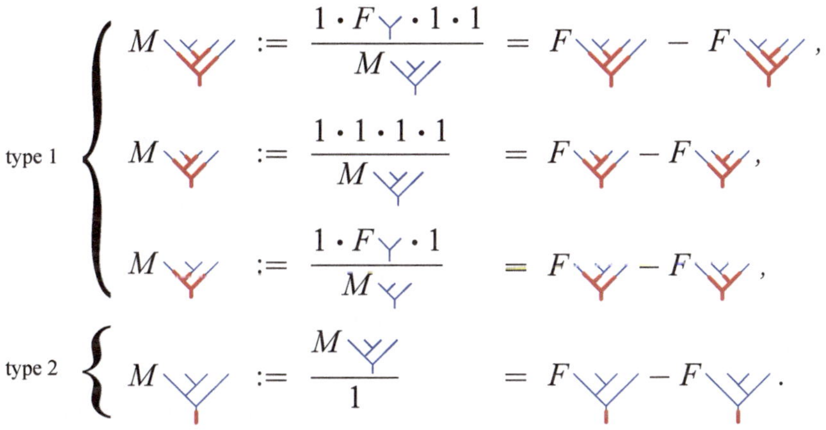

Fig. 16 Primitive elements of two types in $\mathscr{P}Sym$.

The primitives can be described in terms of Möbius inversion on certain subintervals of the multiplihedra lattice. For primitives of the first type, the subintervals are those with a fixed unpainted forest of the form $(\mid \cdot t \cdot \cdots \cdot s \cdot \mid)$. For primitives of the second type, the subinterval consists of those trees whose painted part is trivial, i.e., only the root is painted. Each subinterval of the first type is isomorphic to $\mathscr{Y}_m$ for some $m \leq n$, and the second subinterval is isomorphic to $\mathscr{Y}_n$. Figure 17 shows the multiplihedron lattice for $\mathscr{M}_4$, with these subintervals highlighted.

References

1. M. Aguiar and S. Mahajan, *Monoidal functors, species and Hopf algebras*, CRM Monograph Series, vol. 29, American Mathematical Society, Providence, RI, 2010.
2. M. Aguiar and F. Sottile, "Structure of the Malvenuto-Reutenauer Hopf algebra of permutations", *Adv. Math.* **191** (2005) 225–275.
3. ———, "Structure of the Loday-Ronco Hopf algebra of trees", *J. Algebra* **295** (2006) 473–511.
4. J.M. Boardman and R.M. Vogt, *Homotopy invariant algebraic structures on topological spaces*, Lecture Notes in Mathematics, Vol. 347, Springer-Verlag, Berlin, 1973.
5. M.P. Carr and S.L. Devadoss, "Coxeter complexes and graph-associahedra", *Topology Appl.* **153** (2006) 2155–2168.
6. F. Chapoton, "Algèbres de Hopf des permutahèdres, associahèdres et hypercubes", *Adv. Math.* **150** (2000) 264–275.
7. S. Forcey, "Convex hull realizations of the multiplihedra", *Topology Appl.* **156** (2008) 326–347.

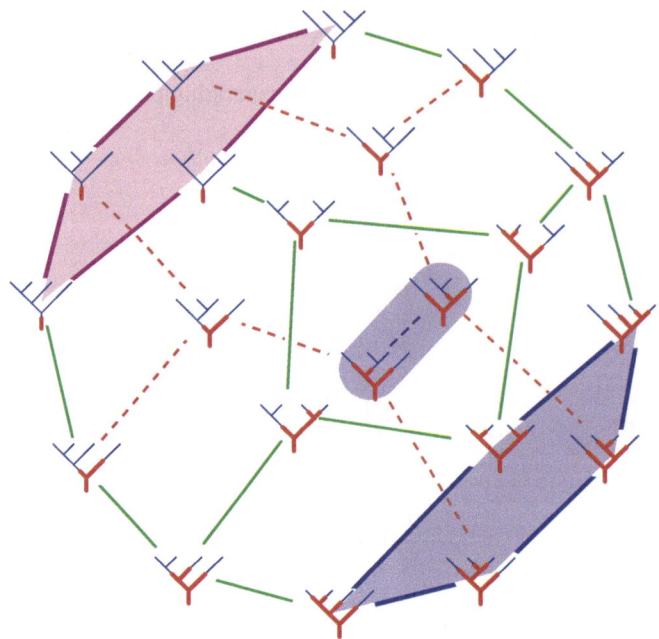

Fig. 17 The multiplihedron lattice $\mathcal{M}_4$ showing the three subintervals that yield primitives via Möbius transformation.

8. ———, "Quotients of the multiplihedron as categorified associahedra", *Homology, Homotopy Appl.* **10** (2008) 227–256.
9. S. Forcey, A. Lauve, and F. Sottile, "Hopf structures on the multiplihedra", *SIAM J. Discrete Math.* **24** (2010) 1250–1271.
10. ———, "Cofree compositions of coalgebras (extended abstract)", *DMTCS Proc. FPSAC 22* (2011) to appear.
11. S. Forcey and D. Springfield, "Geometric combinatorial algebras: cyclohedron and simplex", *J. Alg. Combin.* **32** (2010) 597–627.
12. A. Joyal, "Foncteurs analytiques et espèces de structures", in *Combinatoire enumerative*, G. Labelle and P. Leroux, eds., Lecture Notes in Math., Vol. 1234, Springer, Berlin, 1986, 126–159.
13. J.-L. Loday and M.O. Ronco, "Hopf algebra of the planar binary trees", *Adv. Math.* **139** (1998) 293–309.
14. N. Reading, "From the Tamari lattice to cambrian lattices and beyond", in *this volume*.
15. ———, "Lattice congruences, fans and Hopf algebras", *J. Combin. Theory Ser. A* **110** (2005) 237–273.
16. M. Ronco, "On some extensions of the Tamari order and their relationship with algebraic structures", in *this volume*.
17. S. Saneblidze and R. Umble, "Diagonals on the permutahedra, multiplihedra and associahedra", *Homology Homotopy Appl.* **6** (2004) 363–411 (electronic).
18. J. Stasheff, "Homotopy associativity of *H*-spaces. I, II", *Trans. Amer. Math. Soc.* **108** (1963), 275–292; *ibid.* **108** (1963) 293–312.
19. ———, *H-spaces from a homotopy point of view*, Lecture Notes in Mathematics, Vol. 161, Springer-Verlag, Berlin, 1970.
20. A. Tonks, "Relating the associahedron and the permutohedron", in *Operads: Proceedings of Renaissance Conferences (Hartford, CT/Luminy, 1995)*, Contemp. Math., vol. 202, Amer. Math. Soc., Providence, RI, 1997, 33–36.

Tamari Lattices and the Symmetric Thompson Monoid

Patrick Dehornoy

Abstract We investigate the connection between Tamari lattices and the Thompson group F, summarized in the fact that F is a group of fractions for a certain monoid F_{sym}^+ whose Cayley graph includes all Tamari lattices. Under this correspondence, the Tamari lattice meet and join are the counterparts of the least common multiple and greatest common divisor operations in F_{sym}^+. As an application, we show that, for every n, there exists a length ℓ chain in the nth Tamari lattice whose endpoints are at distance at most $12\ell/n$.

Introduction

The aim of this text is to show the interest of using monoid techniques to investigate Tamari lattices. More precisely, we shall describe the very close connection existing between Tamari lattices and a certain submonoid F_{sym}^+ of Richard Thompson's group F: equipped with the left-divisibility relation, the monoid F_{sym}^+ is a lattice that includes all Tamari lattices. Roughly speaking, the principle is to attribute to the edges of the Tamari lattices names that live in the monoid F_{sym}^+. By using the subword reversing method, a general technique from the theory of monoids, we then obtain a very simple way of reproving the existence of the lattice operations, computing them, and establishing further properties.

The existence of a connection between Tamari lattices, associativity, and the Thompson group F has been known for decades and belongs to folklore. What is specific here is the role of the monoid F_{sym}^+, which is especially suitable for formalizing the connection. Some of the results already appeared, implicitly in [7] and explicitly in [12]. Several new results are established in the current text, in particular the construction of a unique normal form in the monoid F_{sym}^+ and the group F (Subsection 3.4)

Patrick Dehornoy
Laboratoire de Mathématiques Nicolas Oresme, UMR6139, UCBN and CNRS, F-14032 Caen, France, e-mail: *patrick.dehornoy@unicaen.fr*

and the (surprising) result that the embedding of the monoid F^+_{sym} in the Thompson group F is not a quasi-isometry (Proposition 4.10). In the language of binary trees, this implies that, for every constant C, there exist chains of length ℓ whose endpoints can be connected by a path of length at most ℓ/C (Corollary 4.16).

Let us mention that a connection between the Tamari lattices and the group F is described in [28]. However both the objects and the technical methods are disjoint from those developed below. In particular, the approach of [28] does not involve the symmetric monoid F^+_{sym}, which is central here, but it uses instead the standard Thompson monoid F^+, which is not directly connected with the Tamari ordering.

The text is organized as follows. In Section 1, we recall the definition of Tamari lattices and Thompson's group F, and we establish a presentation of F in terms of some specific, non-standard generators a_α indexed by binary addresses. In Section 2, we investigate the submonoid F^+_{sym} of F generated by the elements a_α, we prove that F^+_{sym} equipped with divisibility has the structure of a lattice, and we describe the (close) connection between this lattice and Tamari lattices. Then, in Section 3, we use the Polish encoding of trees to construct an algorithm that computes common upper bounds for trees in the Tamari ordering and we deduce a unique normal form for the elements of F and F^+_{sym}. Finally, in Section 4, we gather results about the length of the elements of F with respect to the generators a_α or, equivalently, about the distance in Tamari lattices, with a specific interest on lower bounds.

1 The framework

The aim of this section is to set our notation and basic definitions. In Subsections 1.1 and 1.2, we briefly recall the definition of Tamari lattices in terms of parenthesized expressions and of binary trees, whereas Subsections 1.3 and 1.4 contain an introduction to Richard Thompson's group F and its action by rotation on trees. This leads us naturally to introducing in Subsection 1.5 a new family of generators of F indexed by binary addresses, and giving in Subsection 1.6 a presentation of F in terms of these generators.

1.1 Parenthesized expressions and associativity

Introduced by Dov Tamari in his 1951 PhD thesis, and appearing in the 1962 article [29] – also see [17] and [19] – the nth Tamari lattice, here denoted by $\mathscr{T}_n$, is, for every positive integer n, the poset (partially ordered set) obtained by considering all well-formed parenthesized expressions involving $n+1$ fixed variables and declaring that an expression E is smaller than another one E', written $E \leqslant_{\mathscr{T}} E'$, if E' may be obtained from E by applying the semi-associative law $x(yz) \to (xy)z$. As established in [29], the poset $(\mathscr{T}_n, \leqslant_{\mathscr{T}})$ is a lattice, that is, any two elements admit a least upper bound and a greatest lower bound. Moreover, $(\mathscr{T}_n, \leqslant_{\mathscr{T}})$ admits a top element, namely the expression in which all left parentheses are gathered on the left, and a bottom element, namely the expression in with all right parentheses are gathered on the right.

As associativity does not change the order of variables, we may forget about their names, and use $\bullet$ everywhere. So, for instance, there exist five parenthesized expressions involving four variables, namely $\bullet(\bullet(\bullet\bullet))$, $\bullet((\bullet\bullet)\bullet)$, $(\bullet(\bullet\bullet))\bullet$, $(\bullet\bullet)(\bullet\bullet)$, and $((\bullet\bullet)\bullet)\bullet$, and we have $\bullet((\bullet\bullet)\bullet) \leqslant_{\mathscr{T}} (\bullet(\bullet\bullet))\bullet$ in the Tamari order as one goes from the first expression to the second by applying the associativity law with $x = \bullet$, $y = \bullet\bullet$, and $z = \bullet$. The Hasse diagrams of the lattices $\mathscr{T}_3$ and $\mathscr{T}_4$ respectively are the pentagon and the 14 vertex polyhedron displayed in Figures 1 and 4 below. As is well known, the number of elements of $\mathscr{T}_n$ is the nth Catalan number $\frac{1}{n+1}\binom{2n}{n}$.

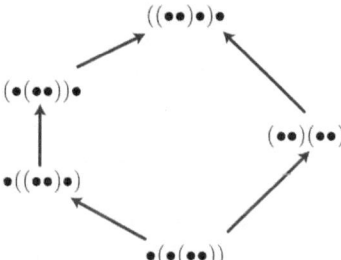

Fig. 1 The Tamari lattice $\mathscr{T}_3$ made of the five ways of bracketing a four variable parenthesized expression.

The Tamari lattice $\mathscr{T}_n$ is connected with a number of usual objects. For instance, its Hasse diagram is the 1-skeleton – that is, the graph made of the 0- and 1-cells – of the nth Mac Lane–Stasheff associahedron [21, 27]. Also $\mathscr{T}_n$ embeds in the lattice made by the symmetric group $\mathfrak{S}_n$ equipped with the weak order: $\mathscr{T}_n$ identifies with the sub-poset made by all 312-avoiding permutations (Björner & Wachs [2]).

For every n, replacing in a parenthesized expression the last (rightmost) symbol $\bullet$ with $\bullet\bullet$ defines an embedding ι_n of $\mathscr{T}_n$ into $\mathscr{T}_{n+1}$. We denote by $\mathscr{T}_\infty$ the limit of the direct system $(\mathscr{T}_n, \iota_n)$ so obtained. Note that $\mathscr{T}_\infty$ has a bottom element, namely the class of $\bullet$, which is also that of $\bullet\bullet$, $\bullet(\bullet\bullet)$, $\bullet(\bullet(\bullet\bullet))$, etc., but no top element.

1.2 Trees and rotations

There exists an obvious one-to-one correspondence between parenthesized expressions involving $n+1$ variables and size n binary rooted trees, that is, trees with n interior nodes and $n+1$ leaves, see Figure 2. In this text, we shall use both frameworks equivalently. We denote by $T_0{}^\wedge T_1$ the tree whose left-subtree is T_0 and whose right-subtree is T_1, but skip the symbol $^\wedge$ in concrete examples involving $\bullet$.

Fig. 2 Correspondence between parenthesized expressions and trees. $\bullet$ $\bullet\bullet$ $(\bullet\bullet)\bullet$ $\bullet(\bullet\bullet)$ $\bullet((\bullet\bullet)\bullet)$

When translated in terms of trees, the operation of applying associativity in the left-to-right direction corresponds to performing one left-rotation, namely replacing some subtree of the form $T_0{}^\wedge(T_1{}^\wedge T_2)$ with the corresponding tree $(T_0{}^\wedge T_1){}^\wedge T_2$, see Figure 3. So the Tamari lattice $\mathscr{T}_n$ is also the poset of size n trees ordered by the transitive closure of left-rotation. We naturally use $\leqslant_{\mathscr{T}}$ for the latter partial ordering.

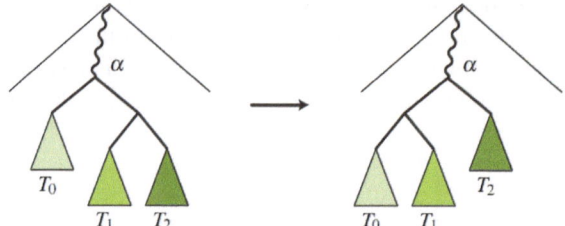

Fig. 3 Applying a left rotation in a tree: replacing some subtree of the form $T_0{}^\wedge(T_1{}^\wedge T_2)$ with the corresponding subtree $(T_0{}^\wedge T_1){}^\wedge T_2$.

In terms of trees, the bottom element of the Tamari lattice $\mathscr{T}_n$ is the size n *right-comb* (or *right-vine*) C_n recursively defined by $C_0 = \bullet$ and $C_n = \bullet{}^\wedge C_{n-1}$ for $n \geqslant 1$, whereas the top element is the size n *left-comb* (or *left-vine*) $\widetilde{C}_n$ recursively defined by $\widetilde{C}_0 = \bullet$ and $\widetilde{C}_n = \widetilde{C}_{n-1}{}^\wedge\bullet$ for $n \geqslant 1$.

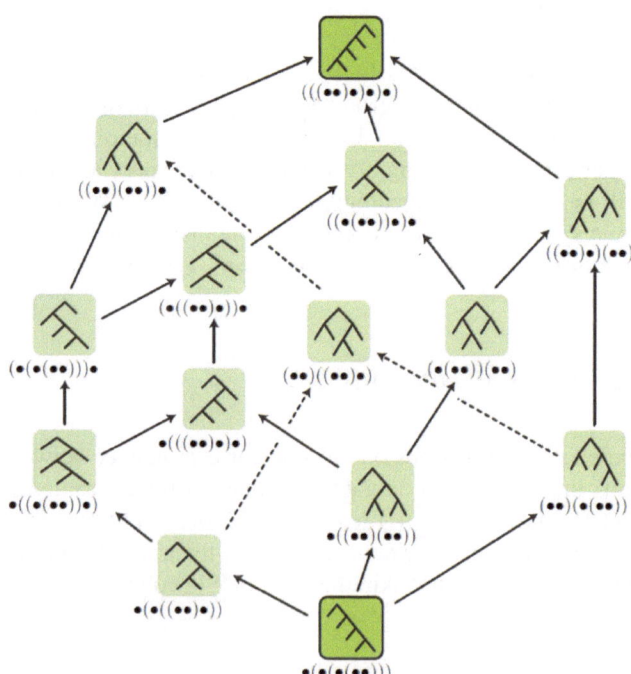

Fig. 4 The Tamari lattice $\mathscr{T}_4$, both in terms of parenthesized expressions and of binary trees.

1.3 Richard Thompson's group F

Introduced by Richard Thompson in 1965, the group F appeared in print only later, in [22] and [30]. The most common approach is to define F as a group of piecewise linear self-homeomorphisms of the unit interval $[0,1]$.

Definition 1.1. The *Thompson group F* is the group of all dyadic order-preserving self-homeomorphisms of $[0,1]$, a homeomorphism f being called *dyadic* if it is piecewise linear with only finitely many breakpoints, every breakpoint of f has dyadic rational coordinates, and every slope of f is an integral power of 2.

Typical elements of F are displayed in Figure 5. In this paper, it is convenient to equip F with reversed composition, that is, fg stands for f followed by g – using the other convention simply amounts to reversing all expressions. The notation x_0 is traditional for the element of F defined by

$$x_0(t) = \begin{cases} \frac{t}{2} & \text{for } 0 \leqslant t \leqslant \frac{1}{2}, \\ t - \frac{1}{4} & \text{for } \frac{1}{2} \leqslant t \leqslant \frac{3}{4}, \\ 2t - 1 & \text{for } \frac{3}{4} \leqslant t \leqslant 1, \end{cases}$$

and x_i is used for the element that is the identity on $[0, 1 - \frac{1}{2^i}]$ and is a rescaled copy of x_0 on $[1 - \frac{1}{2^i}, 1]$ – see Figure 5 again.

It is easy to check that F is generated by the sequence of all elements x_i, with the presentation

$$\langle x_0, x_1, \cdots \mid x_{n+1}x_i = x_i x_n \text{ for } i < n \rangle. \tag{1}$$

One deduces that F is also generated by x_0 and x_1, with the (finite) presentation

$$\langle x_0, x_1 \mid [x_0^{-1}x_1, x_0 x_1 x_0^{-1}], [x_0^{-1}x_1, x_0^2 x_1 x_0^{-2}] \rangle, \tag{2}$$

where $[x,y]$ denotes the commutator $xyx^{-1}y^{-1}$.

Fig. 5 Two representations of the elements x_0 and x_1 of the Thompson group F: above, the usual graph of a function of $[0,1]$ into itself, below, a diagram displaying the two involved dyadic decompositions of $[0,1]$, with the source above and the target below: this simplified diagram specifies the function entirely.

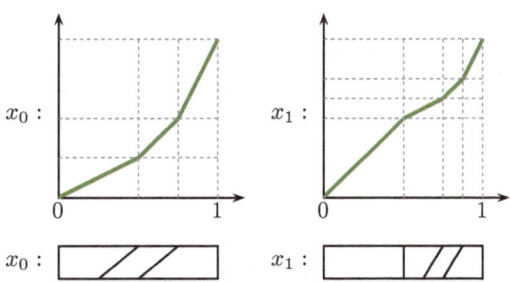

The group F has many interesting algebraic and geometric properties, see [4, 5]. Its center is trivial, the derived group $[F,F]$ is a simple group, F includes no free group of rank more than 1 (Brin–Squier [3]), its Dehn function is quadratic (Guba [18]). It is not known whether F is automatic, nor whether F is amenable. The latter question has received lot of attention as F seems to lie very close to the border between amenability and non-amenability.

Owing to the developments of Section 2 below, we mention one more (simple) algebraic result, namely that F is a group of (left)-fractions, that is, there exists a submonoid of F such that every element of F can be expressed as $f^{-1}g$ with f, g in the considered submonoid.

Proposition 1.2. *[5] Define the* Thompson monoid F^+ *to be the submonoid of F generated by the elements x_i with $i \geqslant 0$. Then, as a monoid, F^+ admits the presentation* (1), *and F is a group of left-fractions for F^+.*

Thus F^+ consists of the elements of F that admit at least one expression in terms of the elements x_i in which no factor x_i^{-1} occurs. Although easy, Proposition 1.2 is technically significant as its leads to a unique normal form for the elements of F.

1.4 The action of F on trees

An element of F is determined by a pair of dyadic decompositions of the interval $[0,1]$ specifying the intervals on which the slope has a certain value, and, from there, by a pair of trees.

To make the description precise, define a *dyadic decomposition* of $[0,1]$ to be an increasing sequence $(t_0, \ldots, t_n)$ of dyadic numbers with $t_0 = 0$ and $t_n = 1$, such that no interval $[t_i, t_{i+1}]$ may contain a dyadic number with denominator less that those of t_i and t_{i+1}: for instance, $(0, \frac{1}{2}, \frac{3}{4}, 1)$ is legal, but $(0, \frac{3}{4}, 1)$ is not. Then dyadic decompositions are in one-to-one correspondence with binary rooted trees: the decomposition associated with $\bullet$ is $(0, 1)$, whereas the one associated with $T_0{}^\wedge T_1$ is the concatenation of those associated with T_0 and T_1 rescaled to fit in $[0, \frac{1}{2}]$ and $[\frac{1}{2}, 1]$.

As the diagram representation of Figure 5 shows, every element of the group F is entirely specified by a pair of dyadic decompositions, hence by a pair of trees. Provided the resulting domains of linearity are maximal, this pair of decompositions (hence of trees) is unique. We denote by (f_-, f_+) the pair of trees associated with f. For instance, we have $1_- = 1_+ = \bullet$, and, as illustrated in Figure 6, $(x_0)_- = \bullet(\bullet\bullet)$, $(x_0)_+ = (\bullet\bullet)\bullet$, $(x_1)_- = \bullet(\bullet(\bullet\bullet))$, and $(x_1)_+ = \bullet((\bullet\bullet)\bullet)$. By construction, the trees f_- and f_+ have the same size. Moreover, we have $(f^{-1})_- = f_+$ and $(f^{-1})_+ = f_-$ as taking the inverse amounts to exchanging source and target in the diagram.

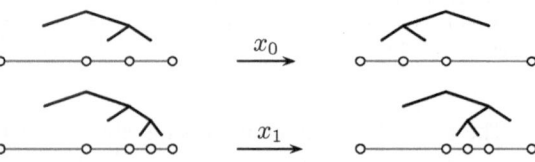

Fig. 6 Canonical pair of trees associated with an element of the group F.

We now define a partial action of the group F on finite trees. Hereafter, we denote by $\mathscr{B}$ the family of all finite, binary, rooted trees, and by $\mathscr{B}^{\#}$ the family of all (finite, binary, rooted) labeled trees whose leaves wear labels in $\mathbb{N}$. Thus $\mathscr{B}$ identifies with the family of all parenthesized expressions involving the single variable $\bullet$, and $\mathscr{B}^{\#}$ with the family of all parenthesized expressions involving variables from the list $\{\bullet_0, \bullet_1, \dots\}$. Forgetting the labels (or the indices of variables) defines a projection of $\mathscr{B}^{\#}$ onto $\mathscr{B}$; by identifying $\bullet$ with $\bullet_0$, we can see $\mathscr{B}$ as a subset of $\mathscr{B}^{\#}$. If T is a tree of $\mathscr{B}$, we denote by $T^{\#}$ the tree of $\mathscr{B}^{\#}$ obtained by attaching to the leaves of T labels $0, 1, \dots$ starting from the left.

Definition 1.3. A *substitution* is a map from $\mathbb{N}$ to $\mathscr{B}^{\#}$. If σ is a substitution and T is a tree in $\mathscr{B}^{\#}$, we define T^{σ} to be the tree obtained from T by replacing every i-labeled leaf of T by the tree $\sigma(i)$.

Formally, T^{σ} is recursively defined by the rules

$$(\bullet_i)^{\sigma} = \sigma(i), \qquad (T_0 {}^{\wedge} T_1)^{\sigma} = T_0^{\sigma} {}^{\wedge} T_1^{\sigma}.$$

For instance, if T is $\bullet_3(\bullet_0 \bullet_2)$ and we have $\sigma(0) = \bullet\bullet$ and $\sigma(2) = \sigma(3) = \bullet$, then T^{σ} is $\bullet((\bullet\bullet)\bullet)$.

Definition 1.4. If T, T' are labeled trees and f is an element of the Thompson group F, we say that $T * f = T'$ holds if we have $T = (f_-^{\#})^{\sigma}$ and $T' = (f_+^{\#})^{\sigma}$ for some substitution σ.

Example 1.5. First consider $f = 1$. Then we have $1_- = 1_+ = \bullet$, whence $1_-^{\#} = 1_+^{\#} = \bullet_0$. For every tree T, we have $T = (1_-^{\#})^{\sigma}$ for any substitution satisfying $\sigma(0) = T$, and, in this case, we have $(1_+^{\#})^{\sigma} = T$. So $T * 1$ is always defined and it is equal to T.

Consider now x_0. Then we have $(x_0)_- = \bullet(\bullet\bullet)$, whence $(x_0)_-^{\#} = \bullet_0(\bullet_1 \bullet_2)$. For a tree T, there exists a substitution satisfying $T = (\bullet_0(\bullet_1 \bullet_2))^{\sigma}$ if and only if T can be expressed as $T_0 {}^{\wedge}(T_1 {}^{\wedge} T_2)$. In this case, the tree $((\bullet_0 \bullet_1) \bullet_2)^{\sigma}$ is $(T_0 {}^{\wedge} T_1) {}^{\wedge} T_2$. So $T * x_0$ is defined if and only if T is eligible for a left-rotation and, in this case, $T * x_0$ is the tree obtained from T by that left-rotation, see Figure 3.

Consider finally x_1. Arguing as above, we see that $T * x_1$ is defined if and only if T can be expressed as $T_0 {}^{\wedge}(T_1 {}^{\wedge}(T_2 {}^{\wedge} T_3))$, in which case $T * x_1$ is the tree $T_0 {}^{\wedge}((T_1 {}^{\wedge} T_2) {}^{\wedge} T_3)$, that is, the tree obtained from T by a left-rotation at the right-child of the root.

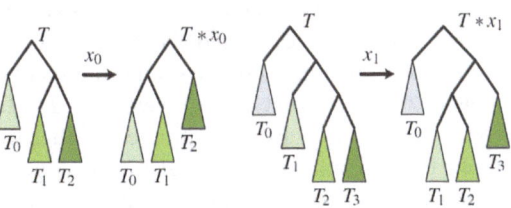

Fig. 7 Action of x_0 and x_1 on a tree: respectively applying a left-rotation at the root, and at the right-child of the root.

The above definition specifies what can naturally be called a partial action of the group F on (labeled) trees – labels are not important here as rotations do not change their order or repeat them, but they are needed for a clean definition of substitutions.

Proposition 1.6.

(i) *For every (labeled) tree T and every element f of F, there exists at most one T' satisfying $T' = T * f$.*

(ii) *For every (labeled) tree T, we have $T * 1 = T$.*

(iii) *For every (labeled) tree T and all f, g in F, we have $(T * f) * g = T * fg$, this meaning that either both terms are defined and they are equal, or neither is defined.*

Moreover, for all $f_1, \ldots, f_n$ in F, there exists T such that $T * f_i$ is defined for each i.

Sketch of the proof, see [12] for details. (i) For f in F, a given tree T can be expressed in at most one way as $(f_-^\#)^\sigma$ and, as the same variables occur on both sides of the associativity law, there is in turn at most one corresponding tree $(f_+^\#)^\sigma$.

Point (ii) has been established in Example 1.5. For (iii), the point is that there exists a simple rule for determining the pair of trees associated with fg. Indeed, we have $(fg)_- = f_-^\sigma$ and $(fg)_+ = g_+^\tau$, where σ and τ are minimal substitutions satisfying $f_+^\sigma = g_-^\tau$ – that is, (σ, τ) is a minimal identifier for f_+ and g_-.

As for the final point, it comes from the fact that, by construction, every tree $f^\#$ has pairwise distinct labels and, therefore, a tree T can be expressed as $(f_-^\#)^\sigma$ if and only if the skeleton of T (as defined in Definition 1.8 below) includes the skeleton of $f_-^\#$. Then, for $f_1, \ldots, f_n$ in F, one can always find a tree T whose skeleton includes those of $(f_1)_-^\#, \ldots, (f_n)_-^\#$. $\qquad\square$

Proposition 1.7. *For all (labeled) trees T, T', the following are equivalent:*

(i) *One can go from T to T' using a finite sequence of left and right rotations – that is, by applying associativity;*

(ii) *The trees T and T' have the same size, and the left-to-right enumerations of the labels in T and T' coincide;*

(iii) *There exists f in F satisfying $T' = T * f$.*

In this case, the element f involved in (iii) is unique.

Sketch of the proof, see [12] for details. The equivalence of (i) and (ii) follows from the syntactic properties of the terms occurring in the associativity law, namely that the same variables occur on both sides, in the same order.

Next, assume that T and T' are equal size trees. Then T and T' determine dyadic decompositions of $[0, 1]$, and there exists a dyadic homeomorphism f, hence an element of F, that maps the first onto the second. Provided the enumerations of the labels in T and T' coincide, we have $T' = T * f$. So (ii) implies (iii).

Conversely, we saw in Example 1.5 that the action of x_0 and x_1 is a rotation. On the other hand, we know that x_0 and x_1 generate F. Therefore, the action of an arbitrary element of f is a finite product of rotations. So (iii) implies (ii).

Finally, the uniqueness of the element f possibly satisfying $T' = T * f$ follows from the fact that the pair (T, T') determines a unique pair of dyadic decompositions of $[0, 1]$, so it directly determines the graph of the dyadic homeomorphism f. $\qquad\square$

Proposition 1.7 states that F is the *geometry group* of associativity in the sense of [12]. A similar approach can be developed for every algebraic law, and more generally every family of algebraic laws, leading to a similar geometry monoid (a group in good cases). In the case of associativity together with commutativity, the geometry group happens to be the Thompson group V, whereas, in the case of the left self-distributivity law $x(yz) = (xy)(xz)$, the geometry group is a certain ramified extension of Artin's braid group B_∞ [9] – also see the case of $x(yz) = (xy)(yz)$ in [10]. In the latter cases, the situation is more complicated than with associativity as, in particular, the counterparts of (i) and (ii) in Proposition 1.7 fail to be equivalent.

1.5 The generators a_α

Considering the action of the group F on trees invites us to introducing, beside the standard generators x_i, a new, more symmetric family of generators for F.

In order to define these elements, we need an index system for the subtrees of a tree. A common solution consists in describing the path connecting the root of the tree to the root of the considered subtree using (for instance) 0 for "forking to the left" and 1 for "forking to the right".

Definition 1.8. A finite sequence of 0's and 1's is called an *address*; the empty address is denoted by $\emptyset$. For T a tree and α a short enough address, the α-*subtree* of T is the part of T that lies below α. The set of all α's for which the α-subtree of T exists is called the *skeleton* of T.

Formally, the α-subtree is defined by the following recursive rules: the $\emptyset$-subtree of T is T, and, for $\alpha = 0\beta$ (*resp.* 1β), the α-subtree of T is the β-subtree of T_0 (*resp.* T_1) if T is $T_0{}^\wedge T_1$, and it is undefined otherwise.

Example 1.9. For $T = \bullet((\bullet\bullet)\bullet)$ (the rightmost example in Figure 2), the 10-subtree of T is $\bullet\bullet$, while the 01- and 111-subtrees are undefined. The skeleton of T consists of the seven addresses $\emptyset$, 0, 1, 10, 100, 101, and 11.

By definition, applying associativity in a parenthesized expression or, equivalently, applying a rotation in a tree T consists in choosing an address α in the skeleton of T and either replacing the α-subtree of T, supposed to have the form $T_0{}^\wedge(T_1{}^\wedge T_2)$, by the corresponding $(T_0{}^\wedge T_1)^\wedge T_2$, or vice versa, see Figure 3 again. By Proposition 1.7, this rotation corresponds to a unique element of F.

Definition 1.10. For every address α, we denote by a_α the element of F whose action is a left-rotation at α. We denote by $\boldsymbol{A}$ the family of all elements a_α for α an address.

According to Example 1.5 and Figure 7, the action of x_0 is a left-rotation at the root of the tree, and, therefore, we have $x_0 = a_\emptyset$. Similarly, x_1 is left-rotation at the right-child of the root, that is, at the node with address 1, and, therefore, we have $x_1 = a_1$. More generally, all elements a_α can be expressed in terms of the generators x_i, as will be done in Subsection 1.6 below. For the moment, we simply note that iterating the argument for x_1 gives for every $i \geqslant 1$ the equality $x_i = a_{1^{i-1}}$, where 1^{i-1} denotes $11 \cdots 1$, $i-1$ times 1.

The trees T such that $T * a_\alpha$ is defined are easily characterized. Indeed, a necessary and sufficient condition for $T * a_\alpha$ to exist is that the α-subtree of T is defined and a left-rotation can be applied to that subtree, that is, it can be expressed as $T_0{}^\wedge(T_1{}^\wedge T_2)$. This is true if and only if the addresses $\alpha 0$, $\alpha 10$, and $\alpha 11$ lie in the skeleton of T, hence actually if and only if $\alpha 10$ lies in the skeleton of T since $\beta 0$ may lie in the skeleton of a tree only if $\beta 1$ and β do. Symmetrically, $T * a_\alpha^{-1}$ is defined if and only if $\alpha 01$ lies in the skeleton of T. As a tree has a finite skeleton, there exist for every tree T finitely many addresses α such that $T * a_\alpha^{\pm 1}$ is defined, see Figure 8.

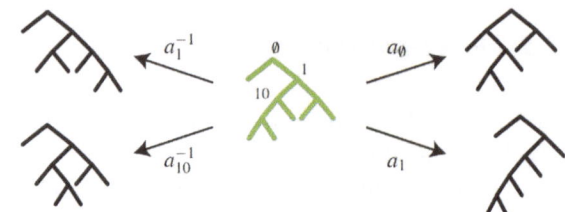

Fig. 8 The four elements $a_\alpha^{\pm 1}$ such that $T * a_\alpha^{\pm 1}$ is defined in the case $T = \bullet(((\bullet\bullet)\bullet)(\bullet\bullet))$.

Before proceeding, we note that the forking nature of the family $\mathbf{A}$ naturally gives rise to a large family of shift endomorphisms of the group F.

Lemma 1.11. *For every address α, there exists a (unique) shift endomorphism sh_α of F that maps a_β to $a_{\alpha\beta}$ for every β.*

Proof. For f in F, let $\mathrm{sh}_1(f)$ denote the homeomorphism obtained by rescaling f, applying it in the interval $[\frac{1}{2}, 1]$, and completing with the identity on $[0, \frac{1}{2}]$. Then sh_1 is an endomorphism of F, and it maps x_i to x_{i+1} for every i. Moreover, for every β, the element $\mathrm{sh}_1(a_\beta)$ is the rescaled version of a_β applied in the interval $[\frac{1}{2}, 1]$. By definition, this is $a_{1\beta}$.

Symmetrically, for f in F, let $\mathrm{sh}_0(f)$ denote the homeomorphism obtained by rescaling f, applying it in the interval $[0, \frac{1}{2}]$, and completing with the identity on $[\frac{1}{2}, 1]$. Then sh_0 is an endomorphism of F, and, for every β, the element $\mathrm{sh}_0(a_\beta)$ is the rescaled version of a_β applied in the interval $[0, \frac{1}{2}]$, hence it is $a_{0\beta}$.

Finally, we recursively define sh_α for every α by $\mathrm{sh}_\emptyset = \mathrm{id}_F$ and, for $i = 0, 1$, $\mathrm{sh}_{i\alpha}(f) = \mathrm{sh}_i(\mathrm{sh}_\alpha(f))$. By construction, $\mathrm{sh}_\alpha(a_\beta) = a_{\alpha\beta}$ holds for all α, β. $\qquad\square$

1.6 Presentation of F in terms of the elements a_α

As the family $\{x_0, x_1\}$, which is $\{a_\emptyset, a_1\}$, generates the group F, the family $\mathbf{A}$ generates F as well. By using the presentation (1) or (2), we could easily deduce a presentation of F in terms of the elements a_α. However, we can obtain a more natural and symmetric presentation by coming back to trees and associativity, and exploiting the geometric meaning of the elements of $\mathbf{A}$.

Lemma 1.12. *Say that two addresses* α, β *are* orthogonal, *written* $\alpha \perp \beta$, *if there exists* γ *such that* α *begins with* $\gamma 0$ *and* β *begins with* $\gamma 1$, *or vice versa. Then all relations of the following family* $\mathbf{R}$ *are satisfied in* F:

$$a_\alpha a_\beta = a_\beta a_\alpha \qquad \text{for } \alpha \perp \beta, \tag{3}$$

$$a_{\alpha 11\beta} a_\alpha = a_\alpha a_{\alpha 1\beta}, \quad a_{\alpha 10\beta} a_\alpha = a_\alpha a_{\alpha 01\beta}, \quad a_{\alpha 0\beta} a_\alpha = a_\alpha a_{\alpha 00\beta}, \tag{4}$$

$$a_\alpha^2 = a_{\alpha 1} a_\alpha a_{\alpha 0}. \tag{5}$$

Proof. By Proposition 1.7, in order to prove that two elements f, f' of F coincide, it is enough to exhibit a tree T such that $T * f$ and $T * f'$ are defined and equal.

The commutation relations of type (3) are trivial. If α and β are orthogonal, the α- and β-subtrees are disjoint, and the result of applying rotations (as well as any transformations) in each of these subtrees does not depend on the order. So we have

$$\mathrm{sh}_\alpha(f)\,\mathrm{sh}_\beta(g) = \mathrm{sh}_\beta(g)\,\mathrm{sh}_\alpha(f)$$

for all transformations f, g and, in particular, $a_\alpha a_\beta = a_\beta a_\alpha$.

The quasi-commutation relations of type (4) are more interesting. Assume that T, T' are trees and $a_\emptyset$ maps T to T'. Then, by definition, the 1-subtree of T' is a copy of the 11-subtree of T. Now, assume that f is a (partial) mapping of $\mathscr{B}$ to itself. Then, starting from T, first applying $a_\emptyset$ and then applying f to the 11-subtree leads to the same result as first applying f to the 1-subtree and then applying $a_\emptyset$, see Figure 9. Moreover, if f is a partial mapping, the result of one operation is defined if and only if the result of the other is. So, in all cases, we have

$$a_\emptyset\,\mathrm{sh}_{11}(f) = \mathrm{sh}_1(f)\,a_\emptyset.$$

Applying this to $f = a_\beta$ then gives $a_\emptyset a_{11\beta} = a_{1\beta} a_\emptyset$. Shifting by α this relation, we obtain $a_\alpha a_{\alpha 11\beta} = a_{\alpha 1\beta} a_\alpha$, the first relation of (4). Arguing similarly with the 0- and 10-subtrees in place of the 11-subtree, one obtains the other relations of (4).

Finally, the relations of (5) stem from the pentagon of Figure 1. As Figure 10 shows, the relation $a_\emptyset^2 = a_1 a_\emptyset a_0$ is satisfied in F and, therefore, so is its shifted version $a_\alpha^2 = a_{\alpha 1} a_\alpha a_{\alpha 0}$ for every address α. $\qquad\square$

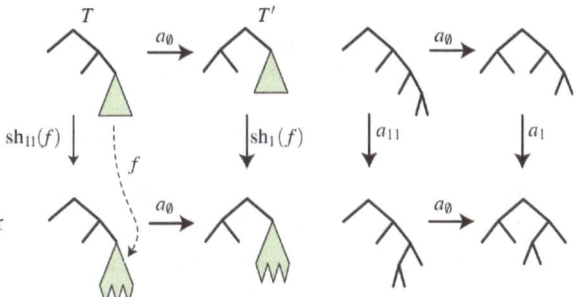

Fig. 9 Quasi-commutation relation in F: the general scheme and one example.

Fig. 10 Pentagon relation in the group F.

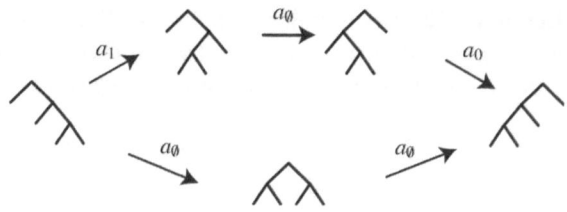

It is then easy to check that the above relations actually exhaust the relations connecting the elements a_α in the group F.

Proposition 1.13 ([7, 12])**.** *The group F admits the presentation $\langle A|R\rangle$.*

Proof. By Lemma 1.12, the relations of R are valid in F. Conversely, to prove that these relations make a presentation, it is sufficient to show that they include the relations of a previously known presentation. This is what happens as, for $1 \leqslant i < n$, the relation $a_{1^n}a_{1^{i-1}} = a_{1^{i-1}}a_{1^{n-1}}$, which is a reformulation of the relation $x_{n+1}x_i = x_i x_n$ of (1), occurs in R as the first relation of (4) with $\alpha = 1^{i-1}$ and $\beta = 1^{n-i}$. $\square$

As an application, we compute the elements a_α in terms of the generators x_i.

Proposition 1.14. *If α is an address containing at least one 0, say*

$$\alpha = 1^i 0^{1+i_0} 10^{i_1} \cdots 10^{i_m}$$

with $m \geqslant 0$ and $i, i_0, \ldots, i_m \geqslant 0$, then, putting $g = x_{i+m+1}^{i_m+1} \cdots x_{i+2}^{i_1+1} x_{i+1}^{i_0+1}$, we have

$$a_\alpha = g^{-1} x_{i+m+2}^{-1} x_{i+m+1} g. \tag{6}$$

Proof. It is sufficient to establish the formula in the case $i = 0$ as, then, applying sh_1^i gives the general case. We use induction on (m, i_0) with respect to the lexicographical (well)-order, that is, (m', i_0') is smaller than (m, i_0) if and only if we have either $m' < m$, or $m' = m$ and $i_0' < i_0$.

Assume first $(m, i_0) = (0, 0)$, that is, $\alpha = 0$. Then the pentagon relation at $\emptyset$ gives

$$a_\alpha = a_0 = a_\emptyset^{-1} a_1^{-1} a_\emptyset^2 = x_1^{-1}(x_2^{-1}x_1)x_1,$$

which is the expected instance of (6). Assume now $m \geqslant 1$ and $i_0 = 0$, that is $\alpha = 010^{i_1} \cdots 10^{i_m}$. Then the quasi-commutation relation for $\emptyset$ and 01β gives

$$a_\alpha = a_{010^{i_1} \cdots 10^{i_m}} = a_\emptyset^{-1} a_{10^{1+i_1} 10^{i_2} \cdots 10^{i_m}} a_\emptyset = x_1^{-1}(a_{10^{1+i_1} 10^{i_2} \cdots 10^{i_m}})x_1. \tag{7}$$

The number of non-initial symbols 1 in $0^{1+i_1} 10^{i_2} \cdots 10^{i_m}$ is $m - 1$. As $(m - 1, i_1)$ is smaller than $(m, 0)$, the induction hypothesis gives $a_{0^{1+i_1} 10^{i_2} \cdots 10^{i_m}} = g^{-1} x_{m+1}^{-1} x_m g$ with $g = x_m^{i_m+1} \cdots x_2^{i_2+1} x_1^{i_1+1}$. Applying sh_1, we find $a_{10^{1+i_1} 10^{i_2} \cdots 10^{i_m}} = h^{-1} x_{m+2}^{-1} x_{m+1} h$ with $h = x_{m+1}^{i_m+1} \cdots x_3^{i_2+1} x_2^{i_1+1}$. Merging with (7), we deduce the expected value for a_α.

Assume finally $i_0 \geqslant 1$. Then the quasi-commutation relation for $\emptyset$ and 00β gives

$$a_\alpha = a_{0^{1+i_0} 10^{i_1} \ldots 10^{i_m}} = a_\emptyset^{-1} a_{0^{i_0} 10^{i_1} \ldots 10^{i_m}} a_\emptyset = x_1^{-1} a_{0^{i_0} 10^{i_1} \ldots 10^{i_m}} x_1.$$

The pair $(m, i_0 - 1)$ is smaller than the pair (m, i_0), so the induction hypothesis gives $a_\alpha = x_1^{-1}(g^{-1}x_{m+2}^{-1}x_{m+1}g)x_1$ with $g = x_{m+1}^{i_m+1} \cdots x_2^{i_1+1} x_1^{i_0}$, again the expected instance of (6). So the induction is complete.   □

Example 1.15. Consider $\alpha = 01100$, which corresponds to $m = 2$, and $i = i_0 = i_1 = 0$, and $i_2 = 2$. Then we find $a_\alpha = g^{-1}x_4^{-1}x_3 g$ with $g = x_3^3 x_2 x_1$, that is, a_{01100} is equal to $x_1^{-1}x_2^{-1}x_3^{-3}x_4^{-1}x_3^4 x_2 x_1$.

2 A lattice structure on the Thompson group F

Here comes the core of our study, namely the investigation of the submonoid F_{sym}^+ of F generated by the elements a_α. The main result is that F_{sym}^+ has the structure of a lattice when equipped with its divisibility relation, and that this lattice is closely connected with the Tamari lattices, which occur as initial sublattices.

These results are not trivial, as, in particular, determining a presentation of F_{sym}^+ is not so easy. Our approach relies on using subword reversing, a general method of combinatorial group theory that turns out to be well suited for F_{sym}^+. One of the outcomes is a new proof (one more!) of the fact that Tamari posets are lattices.

The section is organized as follows. The symmetric Thompson monoid F_{sym}^+ is introduced in Subsection 2.1, and it is investigated in Subsection 2.2 using subword reversing. The lattice structure on F_{sym}^+ and its connection with the Tamari lattices are described in Subsection 2.3. Finally, a few results about the algorithmic complexity of the reversing process are gathered in Subsection 2.4.

2.1 The symmetric Thompson monoid F_{sym}^+

Once new generators a_α of the Thompson group F have been introduced, it is natural to investigate the submonoid generated by these elements.

Definition 2.1. The *symmetric Thompson monoid* F_{sym}^+ is the submonoid of F generated by the elements a_α with α a binary address.

The family **A** of all elements a_α is a sort of closure of the family of standard generators x_i under all local left-right symmetries, so the above terminology is natural. Another option could be to call F_{sym}^+ the *dual Thompson monoid* as the relation of F^+ and F_{sym}^+ is reminiscent of the relation of the standard braid monoids and the dual braid monoids generated by the Birman–Ko–Lee braids.

Although straightforward, the following connection is essential for our purpose:

Lemma 2.2. *For all trees T, T', the following are equivalent*

(i) *We have $T \leqslant_{\mathscr{T}} T'$ in the Tamari order;*
(ii) *There exists f in F_{sym}^{+} satisfying $T' = T * f$.*

Proof. By definition, $T \leqslant_{\mathscr{T}} T'$ holds if there exists a finite sequence of left-rotations transforming T into T'. Now applying the left-rotation at α is letting a_{α} act. □

In order to investigate the monoid F_{sym}^{+} and its connection with the Tamari lattices, it will be necessary to first know a presentation of F_{sym}^{+}. Owing to Propositions 1.2 and 1.13, the following result should not be a surprise.

Proposition 2.3 ([7, 12]). *The monoid F_{sym}^{+} admits the presentation $\langle \boldsymbol{A} | \boldsymbol{R} \rangle^{+}$, and F is a group of right-fractions for F_{sym}^{+} (that is, every element of F can be expressed as fg^{-1} with f, g in F_{sym}^{+}).*

However, the proof of Proposition 2.3 is more delicate than the proof of Proposition 1.2, and no very simple argument is known.

Sketch of the proof developed in [7, 12]. In order to prove that the relations of $\boldsymbol{R}$ generate all relations connecting the elements a_{α} in the monoid F^{+}, one introduces, for every size n tree T, an explicit sequence c_T of elements a_{α} satisfying $C_n * c_T = T -$ as will be made in the proof of Proposition 2.13 below. The point is then to show that, if $T' = T * w$ holds, then the relations of $\boldsymbol{R}$ are sufficient to establish the equivalence of $c_{T'}$ and $c_T w$. Then, if two $\boldsymbol{A}$-words u, v represent the same element of F_{sym}^{+}, and T is a tree such that both $T * u$ and $T * v$ are defined, the above argument shows that $c_T u$ and $c_T v$ are $\boldsymbol{R}$-equivalent, since both are $\boldsymbol{R}$-equivalent to $c_{T * u}$. Provided $\boldsymbol{R}$-equivalence is known to allow left-cancellation, one deduces that u and v are $\boldsymbol{R}$-equivalent, as expected. □

Here we shall propose a new proof, which is more lattice-theoretic in that it exclusively relies on the so-called subword reversing method, which we shall see below is directly connected with the Tamari lattice operations. Instead of working with F_{sym}^{+}, we investigate the abstract monoid $\langle \boldsymbol{A} | \boldsymbol{R} \rangle^{+}$ defined by the presentation $(\boldsymbol{A}, \boldsymbol{R})$ of Proposition 1.13. A priori, as F_{sym}^{+} is generated by $\boldsymbol{A}$ and satisfies the relations of $\boldsymbol{R}$, we only know that F_{sym}^{+} is a quotient of $\langle \boldsymbol{A} | \boldsymbol{R} \rangle^{+}$.

Definition 2.4. Assume that M is a monoid. For f, g in M, we say that f *left-divides* g, or that g is a *right-multiple* of f, written $f \preccurlyeq g$, if $fg' = g$ holds for some g' of M. We use $\mathrm{Div}(f)$ for the family of all left-divisors of f.

It is standard that the left-divisibility relation is a partial pre-ordering. Moreover, if M contains no invertible element except 1, this partial pre-ordering is a partial ordering, that is, the conjunction of $f \preccurlyeq g$ and $g \preccurlyeq f$ implies $f = g$.

Lemma 2.5. *In order to establish Proposition 2.3, it is sufficient to prove that the monoid $\langle \boldsymbol{A} | \boldsymbol{R} \rangle^{+}$ is cancellative and any two elements admit a common right-multiple.*

Proof. A classical result of Ore (see for instance [6]) says that, if a monoid M is cancellative and any two elements of M admit a common right-multiple, then M embeds in a group of right-fractions G. Moreover, if M admits the presentation $\langle A|R\rangle^+$, then G admits the presentation $\langle A|R\rangle$. So, if the hypotheses of the lemma are satisfied, then the monoid $\langle A|R\rangle^+$ embeds in a group of fractions that admits the presentation $\langle A|R\rangle$. By Proposition 1.13, the group $\langle A|R\rangle$ is the group F. Therefore, $\langle A|R\rangle^+$ is isomorphic to the submonoid of F generated by A, that is, to F_{sym}^+. Hence F_{sym}^+ admits the expected presentation, and F is a group of right-fractions for F_{sym}^+. $\square$

2.2 Subword reversing

In order to apply the strategy of Lemma 2.5, we have to prove that the presented monoid $\langle A|R\rangle^+$ is cancellative and any two elements of $\langle A|R\rangle^+$ admit a common right-multiple. The subword reversing method [8, 11] proves to be relevant. We recall below the basic notions, and refer to [14] or [15, Section II.4] for a more complete description.

Hereafter, words in an alphabet A are called *(positive) A-words*, whereas words in the alphabet $A \cup A^{-1}$, where A^{-1} consists of a copy a^{-1} for each letter a of A, are called *signed A-words*. We say that a group presentation (A, R) is *positive* if all relations in R have the form $u = v$ where u and v are nonempty positive A-words. We denote by $\langle A|R\rangle^+$ and by $\langle A|R\rangle$ the monoid and the group presented by (A, R), respectively, and we use $\equiv_R^+$ *(resp.* $\equiv_R$*)* for the congruence on positive A-words *(resp.* on signed A-words) generated by R. Finally, for w a signed A-word, we denote by $\overline{w}$ the element of $\langle A|R\rangle$ represented by w, that is, the $\equiv_R$-class of w.

Definition 2.6. Assume that (A, R) is a positive presentation. If w, w' are signed A-words, we say that w is *right-R-reversible* to w' in one step if w' is obtained from w either by deleting some length 2 subword $a^{-1}a$ or by replacing some length 2 subword $a^{-1}b$ with a word vu^{-1} such that $av = bu$ is a relation of R. We write $w \curvearrowright_R w'$ if w is right-R-reversible to w' in finitely many steps.

The principle of right-R-reversing is to use the relations of R to push the negative letters (those with exponent -1) to the right, and the positive letters (those with exponent $+1$) to the left. The process can be visualized in diagrams as in Figure 11.

Example 2.7. Consider the presentation (A, R), which is positive. Let w be the signed A-word $a_1^{-1}a_\emptyset a_{00}^{-1}a_1$. Then w contains two negative-positive length 2 subwords, namely $a_1^{-1}a_\emptyset$ and $a_{00}^{-1}a_1$. There is in R a unique relation of the form $a_1 \ldots = a_\emptyset \ldots$, namely $a_1 a_\emptyset a_0 = a_\emptyset^2$, and a unique relation $a_{00} \ldots = a_1 \ldots$, namely $a_{00}a_1 = a_1 a_{00}$. Therefore, there exists two ways to right-R-reverse w, namely replacing $a_1^{-1}a_\emptyset$ with $a_\emptyset a_{00}^{-1}a_0^{-1}$ and obtaining $w_1 = a_\emptyset a_0 a_\emptyset^{-1}a_1$, or replacing $a_{00}^{-1}a_1$ with $a_1 a_{00}^{-1}$ and obtaining $w_1' = a_1^{-1}a_\emptyset a_1 a_{00}^{-1}$. The words w_1 and w_1' each contain a unique negative-positive length 2 subword, and reversing it leads in both cases to $w_2 = a_\emptyset a_0 a_\emptyset^{-1}a_1 a_{00}^{-1}$. The word w_2 contains a unique negative-positive length two subword and reversing it leads to $w_3 = a_\emptyset a_0 a_\emptyset a_0^{-1} a_\emptyset^{-1} a_{00}^{-1}$. As the latter word contains no negative-positive subword, no further right-reversing is possible. See Figure 11.

Fig. 11 Right-$\boldsymbol{R}$-reversing of the signed $\boldsymbol{A}$-word $a_1^{-1} a_\emptyset a_{00}^{-1} a_1$: we draw the initial word as a zigzag path (here in dark green) from SW to NE by associating with every letter a a horizontal arrow labeled a and every letter a^{-1} a vertical arrow labeled a (crossed in the wrong direction); then reversing $a^{-1}b$ to vu^{-1} corresponds to closing the open pattern made by a vertical a-arrow and a horizontal b-arrow with the same source by adding horizontal arrows labeled v and vertical arrows labeled u; the final word corresponds to the rightmost path from the SW corner to the NE corner, here $a_\emptyset a_0 a_\emptyset a_0^{-1} a_\emptyset^{-1} a_{00}^{-1}$ (light green).

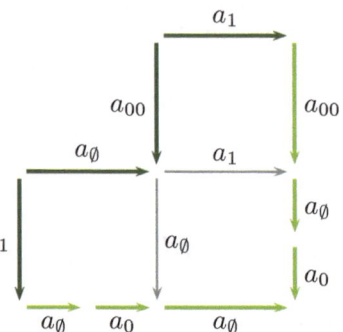

It is easy to see that, if (A,R) is a positive presentation and w, w' are signed A-words, then $w \curvearrowright_R w'$ implies $w \equiv_R w'$ and that, if u, v, u', v' are positive A-words, then $u^{-1}v \curvearrowright_R v'u'^{-1}$ implies $uv' \equiv_R^+ vu'$. In particular, using ε for the empty word,

$$u^{-1}v \curvearrowright_R \varepsilon \quad \text{implies} \quad u \equiv_R^+ v. \tag{8}$$

In general, (8) need not be an equivalence, but it turns out that this is the interesting situation, in which case the presentation (A,R) is said to be *complete with respect to right-reversing*. Roughly speaking, a presentation is complete with respect to right-reversing if right-reversing always detects equivalence. The important point here is that the presentation $(\boldsymbol{A},\boldsymbol{R})$ has this property.

Lemma 2.8 ([7]). *The presentation $(\boldsymbol{A},\boldsymbol{R})$ is complete with respect to right-reversing.*

Sketch of the proof. By [14, Proposition 2.9], a sufficient condition for a positive presentation (A,R) to be complete with respect to right-reversing is that (A,R) satisfies

(i) There exists a $\equiv_R^+$-invariant map λ from positive A-words to $\mathbb{N}$ satisfying $\lambda(uv) \geqslant \lambda(u) + \lambda(v)$ for all u, v and $\lambda(a) \geqslant 1$ for a in A, and
(ii) For all a, b, c in A and all positive A-words u, v, if $a^{-1}cc^{-1}b \curvearrowright_R vu^{-1}$ holds, then $v^{-1}a^{-1}bu \curvearrowright_R \varepsilon$ holds as well. $\tag{9}$

We claim that $(\boldsymbol{A},\boldsymbol{R})$ satisfies (9). As for (i), we cannot use for λ the length of words, as it is not $\equiv_R^+$-invariant: in the pentagon relation, the length 2 word $a_\emptyset^2$ is $\equiv_R^+$-equivalent to the length 3 word $a_1 a_\emptyset a_0$. Now, for T a tree, let $\mu(T)$ be the total number of 0's occurring in the addresses of the leaves of T: for instance, we have $\mu(\bullet(\bullet\bullet)) = 2$ and $\mu((\bullet\bullet)\bullet) = 3$, as the leaves of $\bullet(\bullet\bullet)$ have addresses 0, 10, 11, with two 0's, and those of $(\bullet\bullet)\bullet$ have addresses 00, 01, 1, with three 0's. Then put

$$\lambda(w) = \mu(\overline{w}_+) - \mu(\overline{w}_-). \tag{10}$$

For instance, if w is a_0, the trees $\overline{w}_-$ and $\overline{w}_+$ are $\bullet(\bullet\bullet)$ and $(\bullet\bullet)\bullet$, and one finds $\lambda(a_0) = 3-2 = 1$. A similar argument gives $\lambda(a_\alpha) = 1$ for every address α. More generally, one easily checks that $T \leqslant_{\mathscr{T}} T'$ implies $\mu(T') \geqslant \mu(T)$. Hence the function λ takes values in $\mathbb{N}$. Moreover, a counting argument shows that, in the previous situation, $\mu(T'^\sigma) - \mu(T^\sigma) \geqslant \mu(T') - \mu(T)$ holds for every substitution σ. If u and v are positive A-words, then, as seen in the proof of Proposition 1.6, we have $\overline{uv}_- = \overline{u}^\sigma_-$ and $\overline{uv}_+ = \overline{v}^\tau_+$ for some substitutions σ, τ satisfying $\overline{u}^\sigma_+ = \overline{v}^\tau_-$. We deduce

$$\lambda(uv) = \mu(\overline{uv}_+) - \mu(\overline{uv}_-) = \mu(\overline{v}^\tau_+) - \mu(\overline{u}^\sigma_-) = \mu(\overline{v}^\tau_+) - \mu(\overline{v}^\tau_-) + \mu(\overline{u}^\sigma_+) - \mu(\overline{u}^\sigma_-)$$
$$\geqslant \mu(\overline{v}_+) - \mu(\overline{v}_-) + \mu(\overline{u}_+) - \mu(\overline{u}_-) = \lambda(v) + \lambda(u).$$

As for (ii), the problem is to check that, whenever α, β, γ are addresses and the signed word $a_\alpha^{-1} a_\gamma a_\gamma^{-1} a_\beta$ is right-R-reversible to some positive-negative word vu^{-1}, then $v^{-1} a_\alpha^{-1} a_\beta u$ is right-R-reversible to the empty word. The systematic verification seems tedious. Actually it is not. First, what matters is the mutual position of the addresses α, β, γ with respect to the prefix ordering, and only finitely many patterns may occur. Next, for every pair of addresses α, β, there exists in R exactly one relation of the form $a_\alpha \cdots = a_\beta \ldots$, which implies that, for every signed A-word w, there exists at most one pair of positive A-words u, v such that w is right-R-reversible to vu^{-1}. Finally, all instances involving quasi-commutation relations turn out to be automatically verified. So, the only critical cases are those corresponding to the triple of addresses $\emptyset, 1, 11$ and its translated and permuted copies, and a direct verification is then easy. For instance, the reader can see on Figure 12 that we have $a_\emptyset^{-1} a_1 a_1^{-1} a_{11} \curvearrowright_R a_\emptyset^2 a_{00}^{-1} a_0^{-1} a_\emptyset^{-1} a_{01}^{-1} a_1^{-1}$ and $a_\emptyset^{-3} a_{11} a_1 a_{10} a_0 a_0 a_{00} \curvearrowright_R \varepsilon$.   □

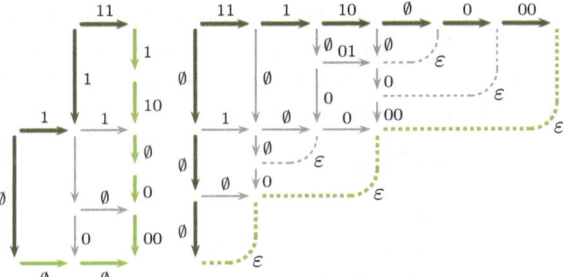

Fig. 12 Proof of Lemma 2.8: $a_\emptyset^{-1} a_1 a_1^{-1} a_{11}$ is right-R-reversible to $a_\emptyset^2 a_{00}^{-1} a_0^{-1} a_\emptyset^{-1} a_{01}^{-1} a_1^{-1}$ (left), and $a_\emptyset^{-3} a_{11} a_1 a_{10} a_0 a_0 a_{00}$ is right-R-reversible to the empty word; dotted lines represent the empty word that appears when a pattern $a_\alpha^{-1} a_\alpha$ is reversed.

Once a positive presentation is known to be complete with respect to right-reversing, it is easy to deduce properties of the associated monoid.

Proposition 2.9. *The monoid* $\langle A|R \rangle^+$ *is left-cancellative.*

Proof. By [14, Proposition 3.1], if (A, R) is a positive presentation that is complete with respect to right-reversing, a sufficient condition for the monoid $\langle A|R \rangle^+$ to be left-cancellative is that

R contains no relation of the form $au = av$ with a in A and $u \neq v$.   (11)

By definition, R satisfies (11). Hence the monoid $\langle A|R \rangle^+$ is left-cancellative.   □

As for right-cancellation, no new computation is needed as we can exploit the symmetries of R. First, we introduce a counterpart of right-reversing where the roles of positive and negative letters are exchanged.

Definition 2.10. Assume that (A, R) is a positive presentation. If w, w' are signed A-words, we say that w is *left-R-reversible* to w' in one step if w' is obtained from w either by deleting some length 2 subword aa^{-1} or by replacing some length 2 subword ab^{-1} with a word $u^{-1}v$ such that $ua = vb$ is a relation of R. We write $w \curvearrowright_R w'$ if w is left-R-reversible to w' in finitely many steps.

Of course, properties of left-reversing are symmetric to those of right-reversing.

Proposition 2.11. *The monoid* $\langle A|R \rangle^+$ *is right-cancellative.*

Proof. The argument is symmetric to the one for Proposition 2.9, and relies on first proving that (A, R) is, in an obvious sense, complete with respect to left-reversing. Due to the symmetries of R, this is easy. Indeed, for w a signed A-word, let $\widetilde{w}$ denote the word obtained by reading the letters of w from right to left, and exchanging 0 and 1 everywhere in the indices of the letters a_α. For instance, $\widetilde{a_{110}a_0}$ is $a_0 a_{001}$. A direct inspection shows that the family $\widetilde{R}$ of all relations $\widetilde{u} = \widetilde{v}$ for $u = v$ in R is R itself. It follows that, for all signed A-words w, w', the relations $w \curvearrowright_R w'$ and $\widetilde{w} \curvearrowleft_R \widetilde{w}'$ are equivalent. Then, as $w \mapsto \widetilde{w}$ is an alphabetical anti-automorphism, the completeness of (A, R) with respect to right-reversing implies the completeness of $(A, \widetilde{R})$, hence of (A, R), with respect to left-reversing. As the right counterpart of (11) is satisfied, we deduce that the monoid $\langle A|R \rangle^+$ is right-cancellative. $\square$

In order to complete the proof of Proposition 2.3 using the strategy of Lemma 2.5, we still need to know that any two elements of the monoid $\langle A|R \rangle^+$ admit a common right-multiple. Using the action on trees, it is easy to prove that result in F_{sym}^+. But this is not sufficient here as we do not know yet that F_{sym}^+ is isomorphic to $\langle A|R \rangle^+$. We appeal to right-reversing once more.

Proposition 2.12. *Any two elements of* $\langle A|R \rangle^+$ *admit a common right-multiple.*

Proof. If (A, R) is a positive presentation, say that right-R-reversing is *terminating* if, for all positive A-words u, v, there exist positive A-words u', v' satisfying $u^{-1}v \curvearrowright_R v'u'^{-1}$. We noted that the latter relation implies $uv' \equiv_R^+ vu'$, thus implying that, in the monoid $\langle A|R \rangle^+$, the elements represented by u and v admit a common right-multiple. So, in order to establish the proposition, it is sufficient to prove that right-R-reversing is terminating, a non-trivial question as, because of the pentagon relations, the length of the words may increase under right-reversing, and there might exist infinite reversing sequences – try for instance the right-reversing of $a^{-1}ba$ in the presentation $(a, b, ab = b^2a)$.

Now, by [14, Proposition 3.11], if (A, R) is a positive presentation, a sufficient condition for right-R-reversing to be terminating is that (A, R) satisfies

(i) For all a, b in A, there is exactly one relation $a \ldots = b \ldots$ in R, and
(ii) There exists a family $\widehat{A}$ of positive A-words that includes A and is
    closed under right-R-reversing, this meaning that, for all u, v in $\widehat{A}$, there
    exist u', v' in $\widehat{A} \cup \{\varepsilon\}$ satisfying $u^{-1}v \curvearrowright_R v'u'^{-1}$.

(12)

We claim that (A, R) satisfies (12). Indeed, (i) follows from an inspection of R. As
for (ii), let us put

$$\hat{a}_{\alpha, r} = a_\alpha a_{\alpha 0} \cdots a_{\alpha 0^{r-1}}$$

for α an address and $r \geqslant 1$, see Figures 14 and 19 for an illustration of the action
of $\hat{a}_{\alpha, r}$ on trees. Then the family $\widehat{A}$ of all words $\hat{a}_{\alpha, r}$ includes A as we have $a_\alpha = \hat{a}_{\alpha, 1}$
for every α, and it is closed under right-R-reversing as we find

$$\hat{a}_{\beta,s}^{-1} \hat{a}_{\alpha,r} \curvearrowright_R \begin{cases} \hat{a}_{0^{s-r}, s-r} & \text{for } \beta = \alpha \text{ with } r < s, \\ \hat{a}_{\alpha, r} \hat{a}_{\beta, s}^{-1} & \text{for } \beta \perp \alpha, \\ \hat{a}_{\alpha, r} \hat{a}_{\alpha 0^{r+1}\gamma, s}^{-1} & \text{for } \beta = \alpha 0 \gamma, \\ \hat{a}_{\alpha, r} \hat{a}_{\alpha 0^r 1 \gamma, s}^{-1} & \text{for } \beta = \alpha 10^r \gamma, \\ \hat{a}_{\alpha, r} \hat{a}_{\alpha 0^i 1 \gamma, s}^{-1} & \text{for } \beta = \alpha 10^i 1 \gamma \text{ with } i < r, \\ \hat{a}_{\alpha, r+s} \hat{a}_{\alpha 0^i, s}^{-1} & \text{for } \beta = \alpha 10^i \text{ with } i < r, \end{cases}$$

(13)

see Figure 13. Note that $\widehat{A}$ is the smallest family that includes A and is closed under
right-R-reversing as the last type of relation in (13) inductively forces every such
family to contain $\hat{a}_{\alpha, r}$ for every r.   □

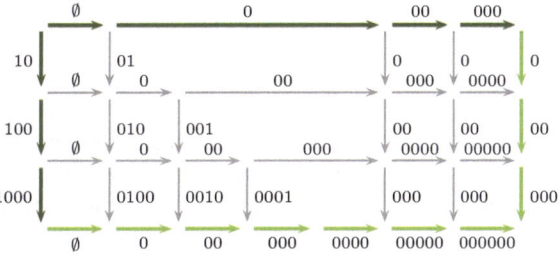

Fig. 13 Closure of the family $\widehat{A}$ under right-reversing: $\hat{a}_{10,3}^{-1} \hat{a}_{0,4}$ reverses to $\hat{a}_{0,7} \hat{a}_{0,3}^{-1}$, which corresponds to the last relation in (13) with $\alpha = \emptyset$, $r = 4$, $s = 3$, $i = 1$ (the letter "a" has been skipped everywhere).

In terms of the generators $\hat{a}_{\alpha, r}$, the pentagon relation can be expressed as $a_\emptyset^2 = a_1 \hat{a}_{\emptyset, 2}$, with both sides of length 2. The last type in (13) corresponds to an extended
pentagon relation $\hat{a}_{\alpha, r} \hat{a}_{\alpha 0^i, s} = \hat{a}_{\alpha 10^i, s} \hat{a}_{\alpha, r+s}$ for all r, s, i with $i < r$, whose counterpart
in terms of tree rotations is displayed in Figure 14.

We thus established that the monoid $\langle A \mid R \rangle^+$ satisfies the conditions of Lemma 2.5
and, therefore, the proof of Proposition 2.3 is complete.

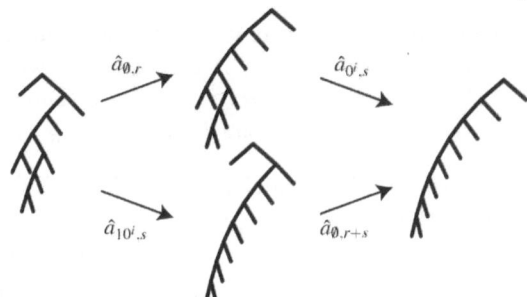

Fig. 14 Extended pentagon relation
$\hat{a}_{\alpha,r}\,\hat{a}_{\alpha0^i,s} = \hat{a}_{\alpha10^i,s}\,\hat{a}_{\alpha,r+s}$, here for
$\alpha = \emptyset, r = 4, s = 3, i = 2$.

2.3 The lattice structure of F_{sym}^+

Here comes the central point, namely the connection between the right-divisibility relation of the monoid F_{sym}^+, which we now know admits the presentation $\langle A | R \rangle^+$, and the Tamari posets. We recall that $\preccurlyeq$ denotes the left-divisibility relation, and that $\mathrm{Div}(f)$ denotes the family of all left-divisors of f.

Proposition 2.13. *For every $n \geqslant 1$, the subposet $(\mathrm{Div}(a_\emptyset^{n-1}), \preccurlyeq)$ of $(F_{sym}^+, \preccurlyeq)$ is isomorphic to the Tamari poset $(\mathcal{T}_n, \leqslant_\mathcal{T})$. The poset $(\bigcup_n \mathrm{Div}(a_\emptyset^n), \preccurlyeq)$ is isomorphic to the Tamari poset $(\mathcal{T}_\infty, \leqslant_\mathcal{T})$.*

Proof. An immediate induction gives the equality $C_n * a_\emptyset^{n-1} = \widetilde{C}_n$ for every n, that is, the element $a_\emptyset^{n-1}$ of F_{sym}^+ maps the right-comb C_n to the left-comb $\widetilde{C}_n$. Hence $C_n * f$ is defined for every element f of F_{sym}^+ that left-divides $a_\emptyset^{n-1}$. Thus, as C_n belongs to $\mathcal{T}_n$ and the action of F_{sym}^+ preserves the size of the trees, we obtain a well-defined map

$$I_n : f \mapsto C_n * f \tag{14}$$

of $\mathrm{Div}(a_\emptyset^{n-1})$ into $\mathcal{T}_n$. By Proposition 1.7, the map I_n is injective. On the other hand, we claim that I_n is surjective. To prove it, it suffices to exhibit, for every size n tree T, an element of F_{sym}^+ that maps the right-comb C_n to T. Now, for every tree T, define two elements c_T, c_T' of F_{sym}^+ by the recursive rules:

$$c_T = \begin{cases} 1 \\ c_{T_0}' \, \mathrm{sh}_1(c_{T_1}')\, a_\emptyset \end{cases} \quad c_T' = \begin{cases} 1 & \text{for } T \text{ of size } 0, \\ c_{T_0}' \, \mathrm{sh}_1(c_{T_1}')\, a_\emptyset & \text{for } T = T_0{}^\wedge T_1. \end{cases} \tag{15}$$

For every size n tree T and every $p \geq 1$, we have $C_n * c_T = T$ and $C_{n+p} * c_T' = T^\wedge C_p$, as shows an induction on T: everything is obvious for $T = \bullet$, and, for $T = T_0{}^\wedge T_1$, it suffices to follow the diagrams of Figure 15. Note that introducing both c_T and c_T' is necessary for the induction. However, the connection $c_T' = c_T\, a_{1i-1} \cdots a_1 a_\emptyset$, where i is the length of the rightmost branch in T, is easy to check.

Thus I_n is a bijection of $\mathrm{Div}(a_\emptyset^{n-1})$ onto $\mathcal{T}_n$. Moreover, I_n is compatible with the orderings. Indeed, assume $f \preccurlyeq g$, say $fg' = g$. Then, by Proposition 1.6, we have $(C_n * f) * g' = C_n * g$, whence $C_n * f \leqslant_\mathcal{T} C_n * g$ by Lemma 2.2. This completes the proof that $(\mathrm{Div}(a_\emptyset^{n-1}), \preccurlyeq)$ is isomorphic to the Tamari poset $(\mathcal{T}_n, \leqslant_\mathcal{T})$.

As for $\mathcal{T}_\infty$, we observe that, for every n, we have $C_{n+1} = C_n^{\#\sigma}$ where σ is the substitution that maps $0,\ldots,n-1$ to $\bullet$ and n to $\bullet\bullet$. On the other hand, by definition, $\mathrm{Div}(a_\emptyset^{n-1})$ is an initial segment of $\mathrm{Div}(a_\emptyset^n)$ and, for every f in $\mathrm{Div}(a_\emptyset^{n-1})$, we have

$$C_{n+1} * f = C_n^{\#\sigma} * f = (C_n * f)^{\#\sigma},$$

hence $I_{n+1}(f) = \iota_n(I_n(f))$. It follows that the family $(I_n)_{n\geqslant 1}$ induces a well-defined map I_∞ of $\bigcup_n \mathrm{Div}(a_\emptyset^n)$ into $\mathcal{T}_\infty$. The map I_∞ is injective because I_n is, it is surjective as, by definition, $\mathcal{T}_\infty$ is the limit of the directed system $(\mathcal{T}_n, \iota_n)$, and it preserves the orderings as I_n does. $\qquad\square$

Fig. 15 For T a size n tree, c_T describes how to construct T from the right-comb C_n, and c_T' describes how to construct $T^\wedge C_p$ from C_{n+p}; the figure illustrates the recursive definition of c_T' (above) and c_T (below) for $T = T_0^\wedge T_1$, with n_1 denoting the size of T_1.

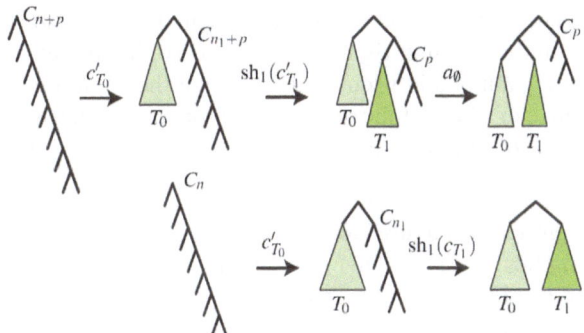

Remark 2.14. The subset $\bigcup_n \mathrm{Div}(a_\emptyset^n)$ involved in Proposition 2.13 is a proper subset of F_{sym}^+ as, for instance, it contains no a_α with 0 occurring in α: indeed, in this case, $C_n * a_\alpha$ is not defined, whereas $C_n * f$ is defined for every f left-dividing $a_\emptyset^{n-1}$.

The connection of Proposition 2.13 can be used in both directions. If we take for granted that the Tamari posets are lattices, we deduce that the subsets $(\mathrm{Div}(a_\emptyset^{n-1}), \preccurlyeq)$ of $(F_{\mathrm{sym}}^+, \preccurlyeq)$ must be lattices as well, that is, with the usual terminology of left-divisibility relation, that any two elements of $\bigcup_n \mathrm{Div}(a_\emptyset^n)$ admit a *least common right-multiple*, or *right-lcm*, and a *greatest common left-divisor*, or *left-gcd*.

On the other hand, if we have a direct proof that $(F_{\mathrm{sym}}^+, \preccurlyeq)$ is a lattice, then the isomorphism of Proposition 2.13 provides a new proof of the lattice property for the Tamari posets. This is what happens.

Proposition 2.15. *The poset* $(F_{\mathrm{sym}}^+, \preccurlyeq)$ *is a lattice.*

Corollary 2.16. *For every* n, *the Tamari poset* $(\mathcal{T}_n, \leqslant)$ *is a lattice.*

To establish Proposition 2.15, we once again appeal to subword reversing.

Proof of Proposition 2.15. By [14, Proposition 3.6], if (A, R) is a positive presentation that is complete with respect to right-reversing, a sufficient condition for any two elements of $\langle A | R\rangle^+$ that admit a common right-multiple to admit a right-lcm is that (A, R) satisfies Condition (i) of (12); moreover, in this case, the right-lcm of the elements represented by two A-words u, v is represented by uv' and vu', where u', v' are the positive A-words for which $u^{-1}v \curvearrowright_R v'u'^{-1}$ holds.

Now, as already noted, (A, R) satisfies (12). Hence any two elements of F_{sym}^+ that admit a common right-multiple admit a right-lcm. On the other hand, by Proposition 2.12, any two elements of $\langle A|R \rangle^+$, that is, of F_{sym}^+, admit a common right-multiple. Hence any two elements of F_{sym}^+ admit a right-lcm. In other words, any two elements in the poset $(F_{\text{sym}}^+, \preccurlyeq)$ admit a least upper bound.

As for left-gcd's, we can argue as follows. Let $\widetilde{\preccurlyeq}$ denote the right-divisibility relation, so that $f \widetilde{\preccurlyeq} g$ holds if and only if we have $g'f = g$ for some g' (the difference with $\preccurlyeq$ is that, here, f appears on the right and not on the left). Then we have the derived notions of a left-lcm and a right-gcd. An easy general result says that, if f, g, f', g' are elements of a monoid satisfying $fg' = gf'$, then f and g admit a left-gcd if and only if f' and g' admit a left-lcm. By Proposition 2.12, any two elements of F_{sym}^+ admit a common right-multiple and so, it suffices to show that any two elements of F_{sym}^+ admit a left-lcm to deduce that they admit a left-gcd. Now, the existence of left-lcm's in F_{sym}^+ follows from the properties of left-R-reversing, which we saw in the proof of Proposition 2.11 are similar to those of right-R-reversing. □

Remark 2.17. Another way of deducing the existence of left-gcd's from that of right-lcm's is to use Noetherianity properties. The existence of the function λ of (10) implies that a set $\text{Div}(f)$ contains no infinite increasing sequence $f_1 \prec f_2 \prec \ldots$ in F_{sym}^+. For all f, g, the family $\text{Div}(f) \cap \text{Div}(g)$ is nonempty as it contains 1, and, by Noetherianity, it contains a $\preccurlyeq$-maximal element, which must be a left-gcd of f and g.

2.4 Computing the operations

We conclude this section with results about the algorithmic complexity of subword reversing in F_{sym}^+. Here we concentrate on space complexity, namely bounds on the length of words; it would be easy to state analogous bounds on the number of reversing steps, hence for time complexity. We use $|w|$ for the length of a word w.

Proposition 2.18. *If w, w' are signed A-words, $w \curvearrowright_R w'$ implies $|w'| \leqslant |w|^2/4 + |w|$. More precisely, we have $|w'| \leqslant p + q + pq$ if w contains p positive letters and q negative letters. These bounds are sharp.*

Proof. By construction, the R-reversing steps in the right-R-reversing of w to w' can be gathered into $\widehat{R}$-reversing steps, which are at most pq in number. Consider the sum of the indices r of the involved generators $\hat{a}_{\alpha, r}$. Each R-reversing step increases this sum by 1 at most (in the case of a pentagon relation), so the total sum in the final $p + q$ generators $\hat{a}_{\alpha, r}$ is at most $p + q + pq$. So, when the generators $\hat{a}_{\alpha, r}$ are decomposed as products of a_α's, at most $p + q + pq$ of the latter occur.

The bound is sharp, as an easy induction gives

$$(a_{1p-1} \cdots a_1 a_0)^{-1} \, a_{1p}^q \; \curvearrowright_R \; a_0^q \, (\hat{a}_{1p-1, q} \cdots \hat{a}_{1, q} \hat{a}_{0, q})^{-1},$$

a word of length $p + q$ that is right-R-reversible to a word of length $p + q + pq$. □

Other upper bounds can be obtained by using the action of F_{sym}^+ on trees. To state the result, it is convenient to introduce the following natural terminology.

Definition 2.19. For every signed $\boldsymbol{A}$-word w, the *right-numerator* $N_r(w)$ and the *right-denominator* $D_r(w)$ of w are the unique $\boldsymbol{A}$-words satisfying $w \curvearrowright_{\boldsymbol{R}} N_r(w)D_r(w)^{-1}$. Symmetrically, the *left-numerator* $N_\ell(w)$ and the *left-denominator* $D_\ell(w)$ of w are the unique $\boldsymbol{A}$-words satisfying $w \curvearrowleft_{\boldsymbol{R}} D_\ell(w)^{-1}N_\ell(w)$.

As left- and right-$\boldsymbol{R}$-reversings are terminating, the positive $\boldsymbol{A}$-words $N_r(w)$, $D_r(w)$, $N_\ell(w)$, and $D_\ell(w)$ exist for every signed $\boldsymbol{A}$-word w.

Proposition 2.20. *Assume that w is a signed $\boldsymbol{A}$-word and $T * w$ is defined for some size n tree T. Then we have*

$$\max(|N_\ell(w)| + |D_r(w)|, |N_r(w)| + |D_r(w)|) \leqslant (n-1)(n-2)/2. \tag{16}$$

In order to establish Proposition 2.20, we need a preliminary result about the action of $\boldsymbol{A}$-words on trees. First, if T is a tree and w is a signed $\boldsymbol{A}$-word, we say that $T * w$ is defined if $T * \overline{u}$ is defined for every prefix u of w. Now, if two signed $\boldsymbol{A}$-words w, w' represent the same element of F, the hypothesis that $T * w$ is defined for some tree T does not guarantee that $T * w'$ is also defined: for instance, $T * \varepsilon$ is always defined, but $T * a_\alpha^{-1} a_\alpha$ is not. However, this cannot happen with reversing.

Lemma 2.21. *Assume that w, w' are signed-$\boldsymbol{A}$-words and w is right- or left-$\boldsymbol{R}$-reversible to w'. Then, for every tree T such that $T * w$ is defined, $T * w'$ is defined as well.*

Proof. The problem with arbitrary equivalences is that new pairs $a_\alpha^{-1} a_\alpha$ or $a_\alpha a_\alpha^{-1}$ may be created. This however is impossible in the case of (right- or left-) reversing, as we can only delete such pairs, but not create them. A complete formal proof requires to check all possible cases: this is easy, and we skip the details. □

Proof of Proposition 2.20. Let $T' = T * w$. By definition, w is right-$\boldsymbol{R}$-reversible to $N_r(w)D_r(w)^{-1}$, and left-$\boldsymbol{R}$-reversible to $D_\ell(w)^{-1}N_\ell(w)$. By Lemma 2.21, this implies that $T * N_r(w)D_r(w)^{-1}$ and $T * D_\ell(w)^{-1}N_\ell(w)$ are defined. Put $T_\ell = T * D_\ell(w)^{-1}$ and $T_r = T * N_r(w)$. By hypothesis, the terms T, T', T_ℓ, and T_r all have size n. Hence there exists a positive $\boldsymbol{A}$-word u (namely c_{T_ℓ}) mapping the right comb C_n to T_ℓ. By symmetry, there exists a positive $\boldsymbol{A}$-word v mapping T_r to the left comb $\widetilde{C}_n$. Then $uN_\ell(w)D_r(w)v$ and $uD_\ell(w)N_r(w)v$ are $\boldsymbol{R}$-equivalent positive $\boldsymbol{A}$-words, and both map C_n to $\widetilde{C}_n$, see Figure 16. Now $a_\emptyset^{n-2}$ also maps C_n to $\widetilde{C}_n$. Hence, by Proposition 1.7, we must have

$$a_\emptyset^{n-2} \equiv_{\boldsymbol{R}}^+ uN_\ell(w)D_r(w)v \equiv_{\boldsymbol{R}}^+ uD_\ell(w)N_r(w)v. \tag{17}$$

Then the function λ of (10) provides an upper bound for the lengths of the words $\boldsymbol{R}$-equivalent to a given word. In the current case, we have $\lambda(a_\emptyset^{n-2}) = (n-1)(n-2)/2$, and (16) follows. □

The upper bound of (16) is close to sharp: for $w = (a_{1^{p-1}} \cdots a_1 a_\emptyset)^{-1} a_{1^p}^q$, the word $D_r(w)$ is $\hat{a}_{1^{p-1},q} \cdots \hat{a}_{1,q} \hat{a}_{\emptyset,q}$, which has length pq in the alphabet $\boldsymbol{A}$, so the sum

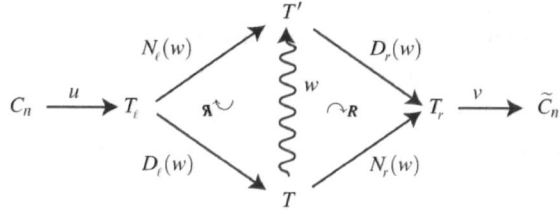

Fig. 16 Bounding the lengths of the left- and right-numerators and denominators of a signed **A**-word w in terms of the size of a term T such that $T * w$ is defined.

of the lengths of $N_\ell(w)$ and $D_r(w)$ is $p + pq$, while the minimal size of a term T such that $T * w$ is defined is $p + q + 2$.

To conclude with subword reversing, we mention one more result that involves both left- and right-reversing. The example of ε and $a_0 a_0^{-1}$ shows that **R**-equivalent words need not have **R**-equivalent numerators and denominators: the right-numerator of ε is ε, whereas the right-numerator of $a_0 a_0^{-1}$ is a_0. This cannot happen when left- and right-numerators are mixed in a double reversing.

Proposition 2.22. *For w a signed **A**-word, define $N_{\ell r}(w) = N_r(D_\ell(w)^{-1} N_\ell(w))$ and $D_{\ell r}(w) = D_r(D_\ell(w)^{-1} N_\ell(w))$. Then $w \equiv_R w'$ implies both $N_{\ell r}(w') \equiv_R^+ N_{\ell r}(w)$ and $D_{\ell r}(w') \equiv_R^+ D_{\ell r}(w)$.*

We first observe that $N_{\ell r}(w) D_{\ell r}(w)^{-1}$ is a minimal fractionary expression of $\overline{w}$.

Lemma 2.23. *If w, w' are **R**-equivalent signed **A**-words, there exist a positive **A**-word u satisfying*

$$N_r(w') \equiv_R^+ N_{\ell r}(w) u \quad and \quad D_r(w') \equiv_R^+ D_{\ell r}(w) u. \tag{18}$$

Proof. By construction, the word w is **R**-equivalent to $D_\ell(w)^{-1} N_\ell(w)$, and the latter word is right-**R**-reversible to $N_{\ell r}(w) D_{\ell r}(w)^{-1}$. Hence we have

$$D_\ell(w) N_{\ell r}(w) \equiv_R^+ N_\ell(w) D_{\ell r}(w). \tag{19}$$

Moreover, as mentioned in the proof of Proposition 2.15, the element of F_{sym}^+ represented by $D_\ell(w) N_{\ell r}(w)$ and $N_\ell(w) D_{\ell r}(w)$ is the right-lcm of $\overline{N_\ell(w)}$ and $\overline{D_\ell(w)}$.

On the other hand, $N_r(w') D_r(w')^{-1}$ is **R**-equivalent to w', hence to w, and therefore to $D_\ell(w)^{-1} N_\ell(w)^{-1}$. We deduce $D_\ell(w) N_r(w') \equiv_R N_\ell(w) D_r(w')$, whence

$$D_\ell(w) N_r(w') \equiv_R^+ N_\ell(w) D_r(w') \tag{20}$$

since F_{sym}^+ embeds in F. As $\overline{D_\ell(w) N_{\ell r}(w)}$ is the right-lcm of $\overline{N_\ell(w)}$ and $\overline{D_\ell(w)}$, comparing (19) and (20) implies the existence of u satisfying (18). $\square$

Proof of Proposition 2.22. By Lemma 2.23, there exist positive **A**-words u and u' satisfying $N_{\ell r}(w') \equiv_R^+ N_{\ell r}(w) u$ and $N_{\ell r}(w) \equiv_R^+ N_{\ell r}(w') u'$, whence $N_{\ell r}(w) \equiv_R^+ N_{\ell r}(w) u u'$. As F_{sym}^+ is left-cancellative, we deduce $\varepsilon \equiv_R^+ u u'$. The only possibility is then that u and u' are empty. $\square$

We now return to Tamari lattices, and show how to use right-reversing to compute lowest upper bounds in the Tamari poset $\mathcal{T}_n$ appealing to the words c_T of (15). Of course, left-reversing can be used symmetrically to compute greatest lower bounds.

Proposition 2.24. *Assume that T, T' are size n trees. Then the least upper bound T'' of T and T' in the Tamari lattice $\mathcal{T}_n$ is determined by*

$$T'' = T * N_r(c_T^{-1} c_{T'}) = T' * D_r(c_T^{-1} c_{T'}).$$

Proof. As mentioned in the proof of Proposition 2.15, the words $c_T N_r(c_T^{-1} c_{T'})$ and $c_{T'} D_r(c_T^{-1} c_{T'})$ both represent the right-lcm of $\overline{c_T}$ and $\overline{c_{T'}}$ in F_{sym}^+. Hence, owing to Proposition 2.13, the image of $\overline{c_T N_r(c_T^{-1} c_{T'})}$ under I_n, which, by definition, is $C_n * c_T N_r(c_T^{-1} c_{T'})$, that is, $T * N_r(c_T^{-1} c_{T'})$, is the least upper bound in $\mathcal{T}_n$ of $I_n(\overline{c_T})$, that is, of $C_n * c_T$, which is T, and $I_n(\overline{c_{T'}})$, that is, of $C_n * c_{T'}$, which is T'. □

Example 2.25. Let $T = \bullet(((\bullet(\bullet\bullet))\bullet)\bullet)$ and $T' = (\bullet(\bullet\bullet))(\bullet(\bullet\bullet))$. Using the method explained in Lemma 3.3 below, one obtains $c_T = a_{11} a_1^2$ and $c_{T'} = a_1 a_\emptyset$. Right-reversing $a_1^{-2} a_{11}^{-1} a_1 a_\emptyset$ leads to $a_{100} a_\emptyset a_0 a_{00} a_\emptyset^{-2}$ (see Figure 17), and we deduce that the least upper bound of T and T' in the Tamari poset is the tree $T * a_{100} a_\emptyset a_0 a_{00}$, namely $(((\bullet(\bullet\bullet))\bullet)\bullet)\bullet$ (which is also $T' * a_\emptyset^2$).

Fig. 17 Computing the right-lcm of c_T and c_T' by right-reversing determines the least upper bound of T and T' in the Tamari lattice, here for $T = \bullet(((\bullet(\bullet\bullet))\bullet)\bullet)$ and $T' = (\bullet(\bullet\bullet))(\bullet(\bullet\bullet))$.

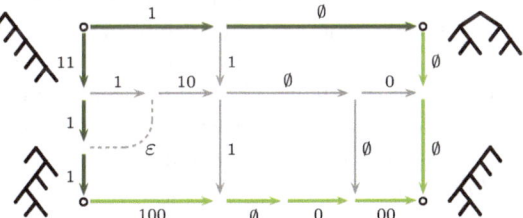

3 The Polish normal form on F

We now develop another approach for determining least common multiples in the monoid F_{sym}^+, whence, equivalently, least upper bounds in the Tamari lattices, namely using what is known as the Polish algorithm. Initially introduced in the case of the self-distributivity law [9, Chapter IX], the latter is easily adapted to our current context, where it provides a unique normal form for the elements of F_{sym}^+ and a method for determining the upper bound of two trees in the Tamari lattice. The main technical tool here is the covering relation of [13], a variant of the weight sequences of [24] – also see [23] and [1].

The section is organized as follows. In Subsection 3.1 we recall the standard Polish encoding of trees and its connection with the Tamari ordering. In Subsection 3.2 we describe an algorithm that, starting with the Polish encoding of two trees, determines a common upper bound of the latter in the Tamari lattice together with a distinguished way of performing the rotations. Then, in Subsection 3.3, we use the covering relation to control the previous algorithm and, in particular, prove that it always determines the least upper bound of the initial trees. Finally, in Subsection 3.4, we deduce a unique normal form for the elements of F that enjoys a sort of weak rationality property.

3.1 The Polish encoding of trees

As is well known, trees or, equivalently, parenthesized expressions can be encoded without parentheses using the Polish notation. Here we consider the right version, and use $\circ$ as the operation symbol.

Definition 3.1. For T a tree, the *(right)-Polish encoding* of T is the word $\langle T \rangle$ recursively defined by $\langle T \rangle = T$ if T has size 0, and $\langle T \rangle = \langle T_0 \rangle \langle T_1 \rangle \circ$ for $T = T_0 {}^\wedge T_1$.

For T of size n, the Polish encoding $\langle T \rangle$ is a word of length $2n + 1$, which we consider as a map of $\{1, \ldots, 2n+1\}$ into $\{\bullet, \circ\}$ (when we restrict to unlabeled trees): thus $\langle T \rangle (k)$ refers to the kth letter of the word $\langle T \rangle$. There exists a natural one-to-one *origin* function from the positions of the letters of $\langle T \rangle$ to the addresses of the nodes of T, recursively defined for $T = T_0 {}^\wedge T_1$ with T_0 of size n_0 by the rule that the origin of k in T is 0α where α is the origin of k in T_0 for $k \leqslant 2n_0 + 1$, it is 1α where α is the origin of $k - 2n_0 - 1$ in T_1 for $2n_0 + 1 < k \leqslant 2n$, and it is $\emptyset$ for $k = 2n + 1$. For instance, the Polish encoding of the tree $\bullet((\bullet\bullet)\bullet)$ of Figure 2 is $\bullet\bullet\bullet\circ\bullet\circ\circ$, and the corresponding origins are $\bullet_{(0)} \bullet_{(100)} \bullet_{(101)} \circ_{(10)} \bullet_{(11)} \circ_{(1)} \circ_{(\emptyset)}$.

For our current purpose, it is important to note the following connection between the Polish encoding and the Tamari order.

Lemma 3.2. *Let $<^{\mathrm{Lex}}$ denote the lexicographical extension of the ordering $\bullet < \circ$ to $\{\bullet, \circ\}$-words. Then, for all trees T, T', the relation $T \leqslant_{\mathscr{T}} T'$ implies $\langle T \rangle \leqslant^{\mathrm{Lex}} \langle T' \rangle$.*

Proof. When translated to the right Polish notation, applying a left-rotation in a tree amounts to replacing some subword of the form $\langle T_0 \rangle \langle T_1 \rangle \langle T_2 \rangle \circ\circ$ with the corresponding word $\langle T_0 \rangle \langle T_1 \rangle \circ \langle T_2 \rangle \circ$. The latter word is $<^{\mathrm{Lex}}$-larger than the former, as the beginning of the word is preserved, until the first letter $\bullet$ associated with $\langle T_2 \rangle$, which is replaced with $\circ$. □

When the initial letter $\bullet$ is erased, the words that are the Polish encoding of a trees identify with Dyck words, defined as those words in the alphabet $\{\bullet, \circ\}$ such that no initial segment has more $\circ$'s than $\bullet$'s, see for instance [26]. Using the standard correspondence between such words and random walks in $\mathbb{N}^2$, we obtain a simple receipe for determining the elements c_T and c'_T of (15) from $\langle T \rangle$.

Lemma 3.3. *(See Figure 18.) Assume that T is a size n tree. For k in $\{1, \ldots, 2n+1\}$ recursively define $v_T(k)$ by $v_T(1) = -1$ and, for $k \geqslant 2$,*

$$v_T(k) = \begin{cases} v_T(k-1)+1 & \text{for } \langle T\rangle(k-1) = \langle T\rangle(k) = \bullet, \\ v_T(k-1)-1 & \text{for } \langle T\rangle(p-1) = \langle T\rangle(p) = \circ, \\ v_T(k-1) & \text{otherwise.} \end{cases} \qquad (21)$$

Then c'_T is obtained from $\langle T\rangle$ by replacing each letter $\langle T\rangle(k)$ with ε if it is $\bullet$ and with a_{1i} with $i = v_T(k)$ if is it $\circ$; the word c_T is obtained similarly after erasing the last block of $\circ$.

Fig. 18 Computing c_T and c'_T from the Polish encoding $\langle T\rangle$ of T: write the kth letter of $\langle T\rangle$ at level $v_T(k)$; then c'_T is read from the levels of the letters $\circ$. Here, for $(\bullet\bullet)((\bullet(\bullet\bullet))\bullet)$, we read $c'_T = a_0 a_{11} a_1^2 a_0$, and, discarding the last two symbols $\circ$, $c_T = a_0 a_{11} a_1$.

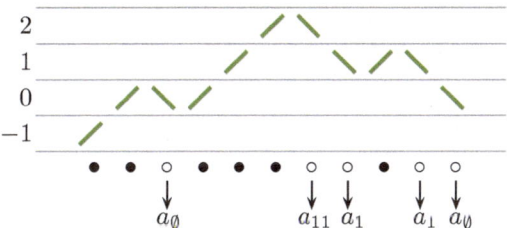

We skip the verification, a comparison of the recursive definitions of $\langle T\rangle$ and c'_T.

3.2 The Polish algorithm

Assume that T, T' are trees of size n and we look for a (minimal) tree T'' that is an upper bound of T and T' in the Tamari order. If T and T' do not coincide, then one of the words $\langle T\rangle$, $\langle T'\rangle$ is lexicographically smaller than the other, say for instance $\langle T\rangle$. This means means that there exists k such that $\langle T\rangle(k)$ is $\bullet$, whereas $\langle T'\rangle(k)$ is $\circ$. In this case, we shall say that T and T' have a *clash at k*. Here is the point.

Lemma 3.4. *Assume that T is a tree and that the kth letter in $\langle T\rangle$ is $\bullet$. Then there exists at most one pair (α, r) such that $T * \hat{a}_{\alpha,r}$ is defined and T and $T * \hat{a}_{\alpha,r}$ have a clash at k. Moreover, if there exists T'' such that T and T'' have a clash at k, there exists exactly one pair as above.*

Proof. As Figure 19 shows, if we have $T' = T * \hat{a}_{\alpha,r}$, then the words $\langle T\rangle$ and $\langle T'\rangle$ coincide up to the first letter coming from the $\alpha 10^{r-1}1$-subtree of T: the latter is $\bullet$ (as is always the first letter of a Polish encoding), whereas, in $\langle T'\rangle$, we have a letter $\circ$ at this position. Thus, the action of $\hat{a}_{\alpha,r}$ on $\langle T\rangle$ is to replace $\bullet$ by $\circ$ at a position whose origin in T has the form $\alpha 10^{r-1}10^i$ for some i.

Consider now the kth letter in $\langle T\rangle$, supposed to be a letter $\bullet$. The origin of k in T is a certain address of leaf in T, say β. By the above argument, a pair (α, r) may result in a clash at k only if we can write $\beta = \alpha 10^{r-1}10^i$ for some $r \geq 1$ and $i \geq 0$. For every β, this happens for at most one pair (α, r), and this happens if and only if β contains at least two digits 1.

Assume now that T and T'' have a clash at k, and consider the value of $v_T(k)$ as defined in (21). By construction (and by the standard properties of Dyck words), we have $v_{T''}(k) \geqslant 0$ as $\langle T'' \rangle(k)$ is $\circ$. By construction, we have $v_T(k) > v_{T''}(k)$ since $\langle T \rangle(k)$ is $\bullet$, whence $v_T(k) \geqslant 1$. This implies (actually an equivalence) that the address β contains at least two digits 1. Hence there exists a pair (α, r) as above. $\qquad\square$

Fig. 19 Action of $\hat{a}_{\alpha,r}$: the Polish encodings coincide up to the first $\bullet$ corresponding to T_2 in $\langle T \rangle$ (black arrow); the latter is replaced with $\circ$ in $\langle T' \rangle$ because, in T', there is one more right-edge after T_1 than in T, and the clash occurs between the marked letters.

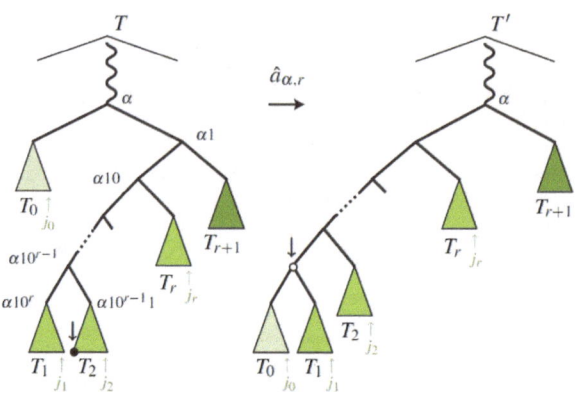

Now the principle of an algorithm should be clear: starting with two trees T, T' such that the Polish encoding $\langle T \rangle$ and $\langle T' \rangle$ coincide up to position $k-1$, we have found a unique way of applying an iterated left-rotation $\hat{a}_{\alpha,r}$ to one of the trees so that the clash is moved further to the right. By iterating the process, we obtain after finitely many steps two trees whose Polish encodings coincide, that is, we obtain a common upper bound for the initial trees T, T'.

Definition 3.5. Assume that T, T' are trees of equal size.

(i) If $\langle T \rangle <^{\mathrm{Lex}} \langle T' \rangle$ holds, we denote by $s(T, T')$ the unique element $\hat{a}_{\alpha,r}$ such that $T * \hat{a}_{\alpha,r}$ and T' have no clash at the position where T and T' have one.

(ii) We denote by $S(T, T')$ the signed $\widehat{A}$-word recursively defined by the rules

$$S(T, T') = \begin{cases} \varepsilon & \text{for } T = T', \\ s(T, T')\, S(T * s(T, T'), T') & \text{for } \langle T \rangle <^{\mathrm{Lex}} \langle T' \rangle, \\ S(T, T' * s(T', T))\, s(T', T)^{-1} & \text{for } \langle T \rangle >^{\mathrm{Lex}} \langle T' \rangle. \end{cases} \qquad (22)$$

Example 3.6. Let us consider the trees of Example 2.25 again, namely $T_0 = \bullet(((\bullet(\bullet\bullet))\bullet)\bullet)$ and $T_0' = (\bullet(\bullet\bullet))(\bullet(\bullet\bullet))$. We find

$\langle T_0 \rangle = \bullet\bullet\bullet\underline{\bullet}\circ\circ\bullet\circ\bullet\circ\circ,$

$\langle T_0' \rangle = \bullet\bullet\bullet\circ\circ\bullet\bullet\bullet\circ\circ\circ.$

Thus we have $\langle T_0 \rangle <^{\mathrm{Lex}} \langle T_0' \rangle$, with a clash at 4 (underlined). The origin of 4 in T_0 is 10011, whence $s(T_0, T_0') = a_{100}$, and $S(T_0, T_0') = a_{100}\, S(T_1, T_1')$ with $T_1 = T_0 * a_{100}$ and $T_1' = T_0'$, corresponding to

$\langle T_1 \rangle = \bullet\bullet\bullet\circ\underline{\bullet}\circ\bullet\circ\bullet\circ\circ,$

$\langle T_1' \rangle = \bullet\bullet\bullet\circ\circ\bullet\bullet\bullet\circ\circ\circ$.

We have now $\langle T_1 \rangle <^{\text{Lex}} \langle T_1' \rangle$, with a clash at 5. The origin of 5 in T_1 is 1001, whence $s(T_1, T_1') = \hat{a}_{\emptyset,3}$, and $S(T_1, T_1') = \hat{a}_{\emptyset,3} \, S(T_2, T_2')$ with $T_2 = T_1 * \hat{a}_{\emptyset,3}$ and $T_2' = T_1'$, hence

$\langle T_2 \rangle = \bullet\bullet\bullet\circ\circ\bullet\circ\bullet\circ\bullet\circ$,

$\langle T_2' \rangle = \bullet\bullet\bullet\circ\circ\bullet\bullet\bullet\circ\circ\circ$.

This time, we have $\langle T_2' \rangle <^{\text{Lex}} \langle T_2 \rangle$, with a clash at 7. The origin of 7 in T_2' is 110, so $s(T_2', T_2)$ is $a_\emptyset$, and $S(T_2, T_2') = S(T_3, T_3') \, a_\emptyset^{-1}$ with $T_3 = T_2$ and $T_3' = T_2' * a_\emptyset$, that is,

$\langle T_3 \rangle = \bullet\bullet\bullet\circ\circ\bullet\circ\bullet\circ\bullet\circ$,

$\langle T_3' \rangle = \bullet\bullet\bullet\circ\circ\bullet\circ\bullet\bullet\circ\circ$.

We find now $\langle T_3' \rangle <^{\text{Lex}} \langle T_3 \rangle$, with a clash at 9. The origin of 9 in T_3' is 11, whence $s(T_3', T_3) = a_\emptyset$, and $S(T_3, T_3') = S(T_4, T_4') \, a_\emptyset^{-1}$ with $T_4 = T_3$ and $T_4' = T_3' * a_\emptyset$, that is,

$\langle T_4 \rangle = \bullet\bullet\bullet\circ\circ\bullet\circ\bullet\circ\bullet\circ$,

$\langle T_4' \rangle = \bullet\bullet\bullet\circ\circ\bullet\circ\bullet\circ\bullet\circ$.

We have $T_4 = T_4'$, so the algorithm halts. The tree T_4 is a common upper bound of T_0 and T_0', and the word $S(T_0, T_0')$ is $a_{100}\hat{a}_{\emptyset,3}a_\emptyset^{-2}$.

Thus, for all equal size trees T, T', we obtained a distinguished signed $\widehat{A}$-word $S(T, T')$, and, by construction, the relation $T' = T * S(T, T')$ is satisfied.

Remark 3.7. As mentioned in the beginning of the section, an entirely similar algorithm can be defined with the self-distributivity law $x(yz) = (xy)(xz)$ replacing the associativity law $x(yz) = (xy)z$. Then tree rotations are replaced with distributions, which consist in replacing subtrees $T_0^\wedge(T_1^\wedge T_2)$ with $(T_0^\wedge T_1)^\wedge(T_0^\wedge T_2)$. In this case, the size of the trees is changed by the transformations, and termination becomes problematic. Actually, in spite of experimental evidence [16] and positive partial results [9], the question, which seems to be extremely difficult, remains open.

3.3 The covering relation

For the moment, we have no connection between the common upper bound of two trees provided by the Polish algorithm of Subsection 3.2 and their least upper bound in the Tamari lattice. In particular, if T, T' are trees satisfying $T \leqslant_{\mathscr{T}} T'$, it is not a priori clear that the Polish algorithm terminates with the pair (T', T'), that is, the clashes always occur on the first of the two current trees. We shall see now that this is actually true. The main tool will be the covering relation, a binary relation that provides a description of the shape of a tree in terms of the addresses of its leaves. We recall that, if T is a size n tree, $T^\#$ denotes the labeled tree obtained by attributing to the leaves of T labels 0 to n from left to right. So, for instance, for $T = \bullet((\bullet\bullet)\bullet)$, we have $T^\# = \bullet_0((\bullet_1\bullet_2)\bullet_3)$, and $\langle T^\# \rangle = \bullet_0\bullet_1\bullet_2\circ\bullet_3\circ\circ$.

Definition 3.8. (See Figure 20.) Assume that T is a size n tree. For $0 \leqslant i \leqslant n$, we define $\text{add}_T(i)$ to be the origin of $\bullet_i$ in $\langle T^\# \rangle$. Then, for $j > i$, we say that j *covers* i in T, written $j \rhd_T i$, if there exists an address γ such that $\text{add}_T(j)$ has the form $\gamma 1^p$ for some positive p and $\text{add}_T(i)$ begins with $\gamma 0$. We write $j \unrhd_T i$ for "$j \rhd_T i$ or $j = i$".

Fig. 20 Covering relation of T: the leaves are numbered 0 to n, and j covers i in T if there exists a subtree T' such that j is the last (rightmost) label in T', whereas i is a non-final label in T'. For instance, in the right-hand tree, 4 covers 1, 2, 3, but does not cover 0, and 3 covers 1 and 2, whereas 2 covers nobody.

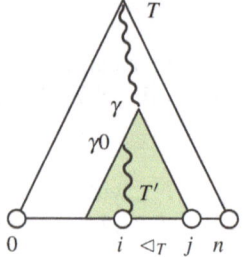 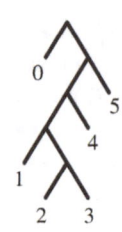

It is easily seen [13] that, for every j occurring in a tree T, the set of all i's covered by j is either empty or is an interval ending in $j-1$: if $j \rhd_T i$ and $j > i' \geqslant i$ hold, then so does $j \rhd_T i'$. Also, the relation $\rhd_T$ is transitive, and it determines T. We shall need the more precise result that every initial fragment of the covering relation determines the corresponding initial fragment of the Polish encoding of T.

Lemma 3.9. *Assume that T is a size n tree. Then, for $1 \leqslant j \leqslant n+1$, the number of symbols $\circ$ following the jth letter $\bullet$ in $\langle T \rangle$ is the number of i's satisfying $j \rhd_T i$ and $k \not\rhd_T i$ for $j > k > i$.*

Proof. Write $j \rhd_T^{\#} i$ if we have $j \rhd_T i$ and $k \not\rhd_T i$ for $j > k > i$. Then, by definition, $j \rhd_T^{\#} i$ holds if and only we have $\mathrm{add}_T(j) = \alpha 1^q$ and $\mathrm{add}_T(i) = \alpha 01^p$ for some α and some $p, q \geqslant 0$. Indeed, if we have $\mathrm{add}_T(i) = \alpha 0\beta$ with β containing at least one 0, say $\beta = 1^p 0\gamma$, then we have $k \rhd_T i$ for k satisfying $\mathrm{add}_T(k) = \alpha 01^p 01^r$.

On the other hand, an induction shows that the jth letter $\bullet$ in $\langle T \rangle$ is followed by r letters $\circ$ if and only if the address $\mathrm{add}_T(j)$ has the form $\gamma 1^r$ for some γ that does not finish with 1, that is, is empty or finishes with 0.

Now, assume that $\mathrm{add}_T(j)$ is $\gamma 1^r$. For $0 \leqslant m < r$, let i_m be the (unique) position whose address in T has the form $\gamma 1^m 01^q$ for some q. By the above characterization, we have $j \rhd_T^{\#} i_m$. So the number of i's satisfying $j \rhd_T^{\#} i$ is at least r.

Conversely, assume that there are r different positions $i_0 < \cdots < i_{r-1}$ satisfying $j \rhd_T^{\#} i_m$. By the above characterization, there exist $\alpha_0, \ldots, \alpha_{r-1}$ satisfying $\mathrm{add}_T(i_m) = \alpha_m 01^*$ and $\mathrm{add}_T(j) = \alpha 1^*$. As the numbers i_m are pairwise distinct, so are the addresses α_m and, therefore, we have $\mathrm{add}_T(j) = \alpha_0 1^{r'}$ with $r' \geqslant r$. $\qquad\square$

It directly follows from Lemma 3.9 that the covering relation of a tree T determines the Polish encoding of T, hence T itself. Actually, the lemma shows more.

Lemma 3.10. *Assume that T, T' are equal size trees, and (as a set of pairs) $\rhd_T$ is properly included in $\rhd_{T'}$. Then $\langle T \rangle <^{\mathrm{Lex}} \langle T' \rangle$ holds.*

Proof. Let j be minimal such that there exists i satisfying $j \rhd_{T'} i$ but not $j \rhd_T i$. For $k < j$, the restriction of the covering relations $\rhd_T$ and $\rhd_{T'}$ to the interval $[1, k]$ coincide and, therefore, by Lemma 3.9, the numbers of symbols $\circ$ following $\bullet_{k-1}$ in $\langle T^{\#} \rangle$ and $\langle T'^{\#} \rangle$ are equal. So, up to $\bullet_{j-1}$, the words $\langle T \rangle$ and $\langle T' \rangle$ coincide.

Consider now $\bullet_{j-1}$. We claim that the number r' of $\circ$ following $\bullet_{j-1}$ in $\langle T'^{\#} \rangle$ is larger than its counterpart r in $\langle T^{\#} \rangle$, resulting in a clash between $\langle T \rangle$ and $\langle T' \rangle$ and in the inequality $\langle T \rangle <^{\mathrm{Lex}} \langle T' \rangle$. To see that $r' > r$ holds, we use Lemma 3.9 again. Using $\rhd_T^{\#}$ as in the proof of Lemma 3.9, we note that $j \rhd_T^{\#} i$ implies $j \rhd_{T'}^{\#} i$ as the restrictions of $\rhd_T$ and $\rhd_{T'}$ to $[1, j-1]$ coincide. By hypothesis, there are r values of i satisfying $j \rhd_T^{\#} i$, and these values also satisfy $j \rhd_{T'}^{\#} i$. Now, by hypothesis, there exists i' satisfying $j \not\rhd_T i'$ and $j \rhd_{T'} i'$. If i' is chosen maximal, we have $j \rhd_{T'}^{\#} i'$. Hence there are strictly more than r values of i satisfying $j \rhd_{T'}^{\#} i$, as expected. □

When a left-rotation transforms a tree T into a tree T', the covering relation of T is included in that of T'. The precise relation is as follows. Hereafter we use $\{0,1\}^*$ (*resp.* $\{1\}^*$) for the set of all addresses (*resp.* all addresses of the form 1^i).

Lemma 3.11. *Assume $T' = T * \hat{a}_{\alpha,r}$. Then $\rhd_{T'}$ is obtained by adding to $\rhd_T$ the pairs (j, i) that satisfy*

$$\exists m \in \{1, \ldots, r\} \ (\mathrm{add}_T(j) \in \alpha 10^{r+1-m} \{1\}^*) \quad and \quad \mathrm{add}_T(i) \in \alpha 0 \{0,1\}^*. \quad (23)$$

Proof. Consider Figure 19 again. Let $j_1, \ldots, j_r$ denote the last variable in the subtrees $T_1, \ldots, T_r$. A direct inspection shows that every covering pair in T is still a covering pair in T', and that the new covering pairs are the pairs (j_m, i) with $1 \leqslant m \leqslant r$ and i occurring in T_0: the action of $\hat{a}_{\alpha,r}$ is to let $j_1, \ldots, j_m$ cover the variables of T_0. Converted into addresses, this gives (23). □

Lemma 3.11 is important for the Polish algorithm as it bounds possible coverings.

Lemma 3.12. *Assume that T, T' are equal size trees satisfying $\langle T \rangle <^{\mathrm{Lex}} \langle T' \rangle$. Then the covering relation of $T * s(T, T')$ is included in the transitive closure of $\rhd_T$ and $\rhd_{T'}$.*

Proof. Assume $s(T, T') = \hat{a}_{\alpha,r}$ and let $T_1 = T * \hat{a}_{\alpha,r}$. We use the notation of Figure 19 once more, calling j_m the rightmost variable occurring in T_m for $0 \leqslant m \leqslant r$. Let I denote the set of all i's occurring in the subtree T_0. By Lemma 3.11, the pairs that belong to $\rhd_{T_1}$ and not to $\rhd_T$ are the pairs $(j_1, i), \ldots, (j_r, i)$ with i in I. The hypothesis that $\hat{a}_{\alpha,r}$ is $s(T, T')$ implies that the number of $\circ$ following $\bullet_{j-1}$ in $\langle T'^{\#} \rangle$ is larger than its counterpart in T, so j_1 must cover strictly more positions in T' than in T. So, necessarily, $j_1 \rhd_{T'} j_0$ holds. On the other hand, $j_0 \unrhd_T i$ holds for every i in I, and $j_m \unrhd_T j_1$ holds for $1 \leqslant m \leqslant r$. It follows that, for all m in $\{1, \ldots, r\}$ and i in I, the pair (j_m, i) belongs to the transitive closure of $\rhd_T$ and $\rhd_{T'}$. □

Lemma 3.13. *Assume that T, T' are equal size trees. Then the Polish algorithm running on (T, T') terminates with a pair (T_∞, T_∞) such that $\rhd_{T_\infty}$ is the transitive closure of $\rhd_T$ and $\rhd_{T'}$.*

Proof. Let (T_t, T_t') denote the pair of trees obtained after t steps of the Polish algorithm running on (T, T'), and N be the total number of steps. By Lemma 3.11, the relations $\rhd_{T_t}$ make a non-decreasing sequence with respect to inclusion, and so do the relations $\rhd_{T_t'}$. So, in particular, the transitive closure of $\rhd_T$ and $\rhd_{T'}$ is included in the transitive closure of $\rhd_{T_N}$ and $\rhd_{T_N'}$. Now, by hypothesis, the latter is $\rhd_{T_\infty}$.

On the other hand, Lemma 3.12 shows that, for every t, the relation $\rhd_{T_{t+1}}$ is included in the transitive closure of $\rhd_{T_t}$ and $\rhd_{T_t'}$, and so is $\rhd_{T_{t+1}'}$. Hence the transitive closure of $\rhd_{T_{t+1}}$ and $\rhd_{T_{t+1}'}$ is the transitive closure of $\rhd_{T_t}$ and $\rhd_{T_t'}$. Hence $\rhd_{T_\infty}$, which is the transitive closure of $\rhd_{T_N}$ and $\rhd_{T_N'}$, is the transitive closure of $\rhd_{T_0}$ and $\rhd_{T_0'}$. □

We are ready to put pieces together and state the main results of this section.

Proposition 3.14. *For T, T', T'' equal size trees, the following are equivalent:*

(i) *The tree T'' is the least upper bound of T and T' in the Tamari lattice;*
(ii) *The Polish algorithm running on (T, T') returns (T'', T'');*
(iii) *The covering relation of T'' is the transitive closure of those of T and T'.*

Proof. Let $T_\vee$ be the least upper bound of T and T' in the Tamari lattice, and T_∞ be the tree such that the Polish algorithm running on (T, T') returns (T_∞, T_∞). By Lemma 3.11, $\rhd_T$ and $\rhd_{T'}$ are included in $\rhd_{T_\vee}$. Hence the transitive closure of $\rhd_T$ and $\rhd_{T'}$, which by Lemma 3.13 is $\rhd_{T_\infty}$, is included in $\rhd_{T_\vee}$.

On the other hand, by definition, we have $T \leqslant_{\mathscr{T}} T_\infty$ and $T' \leqslant_{\mathscr{T}} T_\infty$, whence $T_\vee \leqslant_{\mathscr{T}} T_\infty$. This implies that $\rhd_{T_\vee}$ is included in $\rhd_{T_\infty}$. Hence $\rhd_{T_\vee}$ and $\rhd_{T_\infty}$ coincide, and, therefore, $T_\vee = T_\infty$ holds. So (i) and (ii) are equivalent.

Next, as said above, (ii) implies (iii) by Lemma 3.11. Conversely, if T'' is such that $\rhd_{T''}$ is the transitive closure of $\rhd_T$ and $\rhd_{T'}$, then $\rhd_{T''}$ coincides with $\rhd_{T_\infty}$ and, therefore, we must have $T'' = T_\infty$. So, (ii) and (iii) are equivalent. □

Corollary 3.15. *For T, T' equal size trees, the following are equivalent:*

(i) *We have $T \leqslant_{\mathscr{T}} T'$ in the Tamari order;*
(ii) *There exists f in F_{sym}^+ such that $T' = T * f$ holds;*
(iii) *$S(T, T')$ is a positive **A**-word, that is, the Polish algorithm running on (T, T') finishes with (T', T').*
(iv) *The relation $\rhd_T$ is included in $\rhd_{T'}$.*

Proof. The equivalence of (i) and (ii) has been established in Lemma 2.2.

Next, it is obvious that (iii) implies (ii) as every element $\hat{a}_{\alpha, r}$ belongs to F_{sym}^+. Conversely, if $T \leqslant_{\mathscr{T}} T'$ holds, then the least upper bound of T and T' is T'. Hence, by Proposition 3.14, the Polish algorithm running on (T, T') finishes with (T', T'). This means that the word $S(T, T')$ contains positive letters $\hat{a}_{\alpha, r}$ only. So (ii) implies (iii).

Finally, as observed above, (iii) is equivalent to saying that the Polish algorithm running on (T, T') finishes with (T', T'), whereas (iv) is equivalent to saying that $\rhd_{T'}$ is the transitive closure of $\rhd_T$ and $\rhd_{T'}$. By Proposition 3.14, the latter properties are equivalent and, therefore, (iii) and (iv) are equivalent. □

It should be noted that the equivalence of (i) and (iv) in Corollary 3.15 already appears as [24, Theorem 2.1].

3.4 The Polish normal form

One of the interests of Proposition 3.14 and Corollary 3.15 is that they provide unique distinguished decompositions for every element of F and of F_{sym}^+ in terms of the generators $\hat{a}_{\alpha,r}$. Indeed, we obtained for every pair of equal size trees (T,T') a certain signed $\widehat{A}$-word $S(T,T')$ such that $T * S(T,T')$ is defined and equal to T'. This word $S(T,T')$ does not depend on T.

Lemma 3.16. *Assume that f belongs to F and $T * f$ is defined. Then the signed $\widehat{A}$-word $S(T,T*f)$ is an expression of f, and it does not depend on T.*

Proof. First, we have $T * f = T * S(T,T')$, so, by Proposition 1.7, the word $S(T,T')$ is an expression of f. Next, assume that σ is a substitution, and let us compare the Polish algorithm running on a pair (T,T') and on the pair (T^σ, T'^σ). The word $\langle T^\sigma \rangle$ is obtained from the word $\langle T \rangle$ by replacing every variable $\bullet_i$ with the corresponding word $\langle \sigma(i) \rangle$. As the variables occur in the same order in the words $\langle T \rangle$ and $\langle T' \rangle$, substituting $\bullet_i$ with $\langle \sigma(i) \rangle$ introduces no new clash. Therefore, if (T_t, T_t') are the trees at the tth step of the algorithm running on (T,T'), then $(T_t^\sigma, T_t'^\sigma)$ are the trees at the tth step of the algorithm running on (T^σ, T'^σ), implying $S(T,T') = S(T^\sigma, T'^\sigma)$.

By definition, for every f in F, there exists a unique pair of trees (f_-, f_+) such that every pair $(T, T*f)$ can be expressed as $((f_-^\#)^\sigma, (f_+^\#)^\sigma)$. The above result then shows that $S(T, T*f)$ coincides with $S(f_-, f_+)$, which only depends on f. $\qquad\square$

Definition 3.17. For f in F, the *Polish normal form* of f is the signed $\widehat{A}$-word $S(f_-, f_+)$.

Example 3.18. Let $f = a_0 a_1 = a_{11} a_0 \ (= x_1 x_2 = x_3 x_1)$. Then we have $f_- = \bullet(\bullet(\bullet(\bullet\bullet)))$ and $f_+ = (\bullet\bullet)((\bullet\bullet)\bullet)$. Running the Polish algorithm on these trees returns the (positive) $\widehat{A}$-word $a_0 a_1$: so the latter is the Polish normal form of f. By contrast, $a_{11} a_0$, which is another $\widehat{A}$-expression of f, is not normal. One verifies similarly that the word a_0^2 is normal, whereas the equivalent words $a_1 \hat{a}_{0,2}$ and $a_1 a_0 a_0$ are not.

Corollary 3.15 immediately implies:

Proposition 3.19. *An element of F belongs to the submonoid F_{sym}^+ if and only if its Polish normal form contains no letter $\hat{a}_{\alpha,r}^{-1}$.*

As the family of generators $\widehat{A}$ is infinite, it makes no sense to wonder whether Polish normal words form a rational language or whether the Polish normal form can be connected with an automatic structure. However, let us observe that being Polish normal is a local property that can be characterized in terms of adjacent letters.

Proposition 3.20. *A positive $\widehat{A}$-word $\hat{a}_{\alpha_1,r_1} \cdots \hat{a}_{\alpha_\ell,r_\ell}$ is Polish normal if and only if*

$$\alpha_t 0^{r_t} < \alpha_{t+1} 10^{r_{t+1}-1} 1$$

holds for every $t < \ell$, where $<$ denotes the left-right (partial) ordering of addresses.

Proof. Let $w = \hat{a}_{\alpha_1, r_1} \cdots \hat{a}_{\alpha_\ell, r_\ell}$ and assume that $T * w$ is defined. For $0 \leqslant t \leqslant \ell$, put $T_t = T * \hat{a}_{\alpha_1, r_1} \cdots \hat{a}_{\alpha_t, r_t}$. Then w is normal if, for every $1 \leqslant t \leqslant \ell$, we have $\hat{a}_{\alpha_t, r_t} = s(T_{t-1}, T_\ell)$, that is, $\hat{a}_{\alpha_t, r_t}$ appears at the tth step of the Polish algorithm running on $(T, T * w)$. Now, as shown in Figure 19, the origin in T_t of the letter $\circ$ involved in the clash between T_{t-1} and T_t is $\alpha_t 0^{r_t}$, whereas the origin in T_t of the letter $\bullet$ involved in the clash between T_t and T_{t+1} lies in $\alpha_{t+1} 10^{r_{t+1}-1} 1\{0\}^*$. The normality condition is then that, in $\langle T_t \rangle$, the former letter lies on the left of the latter. By construction of the Polish encoding, this happens if and only if the first address precedes the second in the "left-right-root" linear ordering of addresses. Due to the form of the second address, this is equivalent to $\alpha_t 0^{r_t} < \alpha_{t+1} 10^{r_{t+1}-1} 1$. $\qquad\square$

For instance, the word $a_\emptyset a_\emptyset$ is normal, as we have $\emptyset 0 1^1 = 0 < \emptyset 10^{1-1} 1 = 11$, but $a_1 \hat{a}_{\emptyset, 2}$ is not, as we do not have $10^1 = 10 < \emptyset 10^{2-1} 1 = 101$.

4 Distance in Tamari lattices

We conclude this description of the connections between the Tamari lattice and the Thompson group F with a few observations about distances in $\mathcal{T}_n$. The general principle is that it is easy to obtain upper bounds, but difficult to prove lower bounds and many questions remain open in this area. Our main observation here is that the embedding of the monoid F_{sym}^+ into the group F is not an isometry, and not even a quasi-isometry (Definition 4.9): for every positive constant C, there exist elements of F_{sym}^+ whose length in F is smaller than their length in F_{sym}^+ by a factor at least C. In terms of Tamari lattices, this implies that chains are not geodesic (Corollary 4.16).

The plan of the section is as follows. In Subsection 4.1, we quickly survey the known results about the diameter of Tamari lattices. Then, we show in Subsection 4.2 how to use the syntactic relations of $\boldsymbol{R}$ to obtain (rather weak) distance lower bounds. Finally, in Subsection 4.3, we use the covering relation to establish (stronger) lower bounds.

4.1 The diameter of $\mathcal{T}_n$

Surprisingly, the diameter of the Tamari lattice $\mathcal{T}_n$ is not known for every n.

Definition 4.1. For T, T' in $\mathcal{T}_n$, the *distance* between T and T', denoted by $\mathrm{dist}(T, T')$ is the minimal number of (left and right) rotations needed to transform T into T'. The *diameter* of $\mathcal{T}_n$ is the maximum of $\mathrm{dist}(T, T')$ for T, T' in $\mathcal{T}_n$.

Theorem 4.2 (Sleator, Tarjan, Thurston [25]). *For $n \geqslant 11$, the diameter of $\mathcal{T}_n$ is at most $2n - 6$; for n large enough, it is exactly $2n - 6$.*

The argument uses the fact that the maximal distance between two size n trees is also the maximal number of flips needed to transform two triangulations of an

$(n+2)$-gon one into the other. A lower bound for the latter is obtained by putting the considered triangulations on the two halves of a sphere and bounding the hyperbolic volume of the resulting tiled polyhedron. It is conjectured that the value $2n-6$ is correct for every $n \geqslant 11$. However, due to its geometric nature, the argument of [25] works only for $n \geqslant n_0$, with no estimation of n_0.

By contrast, combinatorial arguments involving the covering relation of Subsection 3.3 lead to (weaker) results that are valid for every n.

Theorem 4.3 ([13]). *For $n = 2p^2$, the diameter of $\mathcal{T}_n$ is at least $2n - 2\sqrt{2n} + 1$ and, for every n, it is at least $2n - \sqrt{70n}$.*

Although no theoretical obstruction seems to exist, the covering arguments have not yet been developed enough to lead to an exact value of the diameter. However some candidates for realizing the maximal distance are known.

Conjecture 4.4 ([13]). *For α an address, let $\langle \alpha \rangle$ denote the tree recursively specified by the rules $\langle \emptyset \rangle = \bullet$, $\langle 0\alpha \rangle = \langle \alpha \rangle^\wedge \bullet$, and $\langle 1\alpha \rangle = \bullet^\wedge \langle \alpha \rangle$. Define*

$$Z_n = \begin{cases} \langle 111(01)^{p-2} \rangle \\ \langle 111(01)^{p-2}0 \rangle \end{cases} \quad Z'_n = \begin{cases} \langle 000(10)^{p-2} \rangle & \text{for } n = 2p+3, \\ \langle 000(10)^{p-2}1 \rangle & \text{for } n = 2p+4, \end{cases}$$

see Figure 21. Then one has $\mathrm{dist}(Z_n, Z'_n) = 2n - 6$ for $n \geqslant 9$.

Fig. 21 The zigzag trees of Conjecture 4.4, here for $n = 15$; the distance is 24, as predicted.

Z_n Z'_n

Conjecture 4.4 has been checked up to size 19 (sizes below 9 are special, because the trees are then too small for the generic scheme to start; by the way, the value $2n - 6$ is valid for $n = 5, 6, 7$, but not for $n \leq 4$ and for $n = 8$).

4.2 Syntactic invariants

A natural way to investigate distances in the Tamari lattices is to use the action of F on trees and to study the length of the elements of F with respect to the generating family $\mathbf{A}$. Indeed, Proposition 1.7 directly implies

Lemma 4.5. *For all trees T, T', we have $\mathrm{dist}(T, T') = \|S(T, T')\|_{\mathbf{A}}$, where $\|f\|_{\mathbf{A}}$ is the $\mathbf{A}$-length of f, that is, the length of the shortest signed $\mathbf{A}$-word representing f.*

In order to establish (lower) bounds on $\|f\|_{\mathbf{A}}$, a natural approach is to use the syntactic properties of the relations of $\mathbf{R}$.

Lemma 4.6. *For w a signed A-word, denote by $|w|_1$ the number of letters $a_{1i}^{\pm 1}$ in w.*

(i) *If u, u' are R-equivalent positive A-words, then $|u|_1 = |u'|_1$ holds.*
(ii) *If w, w' are signed A-words, then $w \curvearrowright_R w'$ implies $|w|_1 \geqslant |w'|_1$ holds.*

Proof. In both cases, it suffices to inspect the relations of R. In the case of the pentagon relations, we have $|a_\alpha^2|_1 = |a_{\alpha 0} a_\alpha a_\alpha a_{\alpha 1}|_1$, both being 2 for α in $\{1\}^*$, and 0 otherwise. Similarly, for (ii), we find $|a_\alpha^{-1} a_{\alpha 1}|_1 = |a_{\alpha \alpha} a_{\alpha 0}^{-1} a_\alpha^{-1}|_1$, both being 2 for α in $\{1\}^*$, and 0 otherwise. The inequality comes from $|a_\alpha^{-1} a_\alpha|_1 = 2 > 0 = |\varepsilon|_1$ for α in $\{1\}^*$. $\qquad\square$

Note that the counterpart of Lemma 4.6(ii) involving left-reversing is false: $a_\emptyset a_0^{-1}$ is left-R-reversible to $a_\emptyset^{-1} a_1 a_\emptyset$, and we have $|a_\emptyset a_0^{-1}|_1 = 1 < 3 = |a_\emptyset^{-1} a_1 a_\emptyset|_1$.

Proposition 4.7. *For every f of F and every A-word w representing f, we have*

$$\|f\|_A \geqslant |D_{\ell r}(w)|_1 + |N_{\ell r}(w)|_1. \tag{24}$$

Proof. Put $\ell = |D_{\ell r}(w)|_1 + |N_{\ell r}(w)|_1$. By definition, the word w is right-R-reversible to the word $N_r(w) D_r(w)^{-1}$, so, by Lemma 4.6(ii), we have

$$|w|_1 \geqslant |N_r(w) D_r(w)^{-1}|_1 = |N_r(w)|_1 + |D_r(w)|_1.$$

Next, it follows from Proposition 2.22 that there exists a positive A-word u satisfying $N_r(w) \equiv_R^+ N_{\ell r}(w) u$ and $D_r(w) \equiv_R^+ D_{\ell r}(w) u$. Then, by Lemma 4.6(i), we deduce $|N_r(w)|_1 + |D_r(w)|_1 \geqslant \ell$, whence $|w| \geqslant |w|_1 \geqslant \ell$.

Now assume $w' \equiv_R w$. By the result above, we have $|w'| \geqslant |N_{\ell r}(w')|_1 + |D_{\ell r}(w')|_1$. By Proposition 2.22, we have $N_{\ell r}(w') \equiv_R^+ N_{\ell r}(w)$ and $D_{\ell r}(w') \equiv_R^+ D_{\ell r}(w)$, whence, by Lemma 4.6(i), $|N_{\ell r}(w')|_1 + |D_{\ell r}(w')|_1 = \ell$. Thus $|w'| \geqslant \ell$ holds for every word w' that represents $\overline{w}$. By definition, this means that $\|f\|_A \geqslant \ell$ is true. $\qquad\square$

Of course, a symmetric criterion involving $\{0\}^*$ instead of $\{1\}^*$ may be stated. *Example* 4.8. Let $f = a_1 a_{11}^{-1} a_\alpha a_{11}^{-1}$. Left-reversing the word $a_1 a_{11}^{-1} a_\alpha a_{11}^{-1}$ yields the word $a_{111}^{-1} a_{1111}^{-1} a_1 a_\emptyset$, which in turn is right-reversible to $a_1 a_\emptyset a_1^{-1} a_{11}^{-1}$. We conclude that $N_{\ell r}(w)$ is $a_1 a_\emptyset$ and $D_{\ell r}(w)$ is $a_{11} a_1$. Then Proposition 4.7 gives $\|f\|_A \geqslant 4$, that is, the word $a_1 a_{11}^{-1} a_\alpha a_{11}^{-1}$ is geodesic.

By construction, the elements c_T involved in proof of Proposition 2.13 are represented by A-words all letters of which are of the form a_α with α in $\{1\}^*$, and, therefore, these words are geodesic. However, elements of this type are quite special, and the criterion of Proposition 4.7 is rarely useful. In particular, it follows from the construction that every element of F can be represented by a word of the form $c_T^{-1} c_{T'}$ but, even when the fraction is irreducible, that is, when the elements represented by c_T and $c_{T'}$ admit no common left-divisor in F_{sym}^+, it need not be geodesic, as shows the example of $a_\emptyset^{-p} a_{1^p} a_{1^{p-1}} \cdots a_1$, an irreducible fraction of length $2p$ which is R-equivalent to the positive-negative word $a_\emptyset a_0^{-p} a_\emptyset^{-1}$ of length $p + 2$.

4.3 The embedding of F_{sym}^+ into F

More powerful results can be obtained using the covering relation of Subsection 3.3. As an example, we shall now establish that the embedding of the monoid F_{sym}^+ into the group F provided by Proposition 2.3 is not an isometry, that is, there exist elements of F_{sym}^+ whose length as elements of F is smaller than their length as elements of F_{sym}^+. This result is slightly surprising: clearly, fractions need not be geodesic in general, but we might expect that, when an element of F belongs to F_{sym}^+, then its length inside F_{sym}^+ equals its length inside F.

Definition 4.9. If $(X,d),(X',d')$ are metric spaces, a map $f : X \to X'$ is a *quasi-isometry* if there exist $C \geqslant 1$ and $C' \geqslant 0$ such that $\frac{1}{C}d(f(x,y)) - C' \leqslant d'(f(x,y)) \leqslant Cd(x,y) + C'$ holds for all x,y in X.

The result we shall prove is as follows.

Proposition 4.10. *For f in F_{sym}^+, let $\|f\|_A^+$ denote the A-length of f in F_{sym}^+, that is, the length of a shortest positive A-word representing f. Then the embedding of F_{sym}^+ into F is not a quasi-isometry of $(F_{sym}^+, \| \ \|_A^+)$ into $(F, \| \ \|_A)$.*

In order to establish Proposition 4.10, it is enough to exhibit a sequence of elements f_p of F_{sym}^+ satisfying $\|f_p\|_A = o(\|f_p\|_A^+)$. This is what the next result provides.

Lemma 4.11. *For every $p \geqslant 1$, let u_p be the $\widehat{A}$-word*

$$\hat{a}_{(10)^p,1}\,\hat{a}_{(10)^{p-1},2} \cdots \hat{a}_{10,p}\,\hat{a}_{\emptyset,p+1}.$$

Then, for every p, we have $\|\overline{u_p}\|_A \leqslant 3p+1$ and $\|\overline{u_p}\|_A^+ = (p+1)(p+2)/2$.

Establishing Lemma 4.11 requires to prove two inequalities, namely an upper bound on $\|\overline{u_p}\|_A$ and a lower bound on $\|\overline{u_p}\|_A^+$. As always, the first task is easier than the second.

Lemma 4.12. *For every $p \geqslant 1$, we have $\|\overline{u_p}\|_A \leqslant 3p+1$.*

Proof. For $p \geqslant 1$, let $w_p = a_{(10)^p}\,a_{(10)^{p-1}1}^{-1}\,a_{(10)^{p-1}}\,a_{(10)^{p-2}1}^{-1}\,a_{(10)^{p-2}} \cdots a_{101}^{-1}\,a_{10}\,a_1^{-1}\,a_{\emptyset}$. Then w_p is a signed A-word of length $2p+1$. An easy induction using the formulas of (13) shows that w_p is right-reversible to the positive-negative word $u_p\hat{a}_{\emptyset,p}^{-1}$, see Figure 22. As the latter word has A-length $3p+1$, the result follows. $\square$

Fig. 22 Right-reversing the signed A-word w_p, here with $p=3$. The numerator is the word u_p of Lemma 4.11, whereas the denominator is the length p word $\hat{a}_{\emptyset,p}$ (as usual, the letters "a" have been skipped).

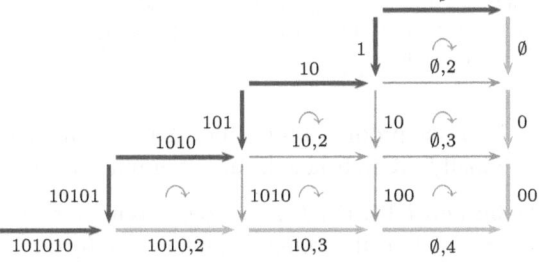

In order to complete the proof of Lemma 4.11, it remains to prove that $\|\overline{u_p}\|_{\mathbf{A}}^+$ is $p(p+3)/2$. As the length of u_p is $p(p+3)/2$, the point is to prove:

Lemma 4.13. *For every $p \geqslant 1$, the word u_p is geodesic in F_{sym}^+.*

Proof. With the notation of Conjecture 4.4, let T_p be the size $2p+2$ tree $\langle(10)^p 1\rangle$. Then $T_p * u_p$ is defined and equal to $T_p' = \langle 0^{p+1} 1^p\rangle$, see Figure 23.

We look at the covering relations that are satisfied in T_p' but not in T_p. First, $2p+1$ covers 0 in T_p' but not in T_p. We deduce that every $\mathbf{A}$-word transforming T_p to T_p' contains at least one step with critical index $2p+1$, the critical index of a_α being defined as the unique j such that applying a_α adds at least one relation $j \rhd i$, that is, the unique j such that $\mathrm{add}_T(j)$ lies in $\alpha 10\{1\}^*$.

Now, here is the point: $2p$ covers 0 and 1 in T_p', but not in T_p. We deduce that every $\mathbf{A}$-word transforming T_p to T_p' contains at least one step with critical index $2p$. But we claim that every such word must actually contain at least *two* such steps, that is, one step cannot be responsible for the two new covering relations. Indeed, $2p$ covers 2 in T_p, but does not cover 1. The only situation when a step adding $2p \rhd 1$ in a tree T can simultaneously add $2p \rhd 0$ is when 1 covers 0 in T. But this cannot happen here, because we are considering positive $\mathbf{A}$-words only, so any possible covering satisfied at an intermediate step must remain in the final tree T_p'. As 1 does not cover 0 in T_p', it is impossible that 1 covers 0 at any intermediate step. Thus two steps are needed to ensure $2p \rhd 1$ and $2p \rhd 0$.

The sequel is similar. In T_p', the number $2p-1$ covers 0, 1, and 2, whereas, in T_p, it covers only 3. Then at least three steps are needed to ensure $2p-1 \rhd 2$, $2p-1 \rhd 1$, and $2p-1 \rhd 0$. Indeed, a step can cause $2p-1$ to simultaneously cover 1 and 2 only if 2 covers 1 at the involved step, which cannot happen as 2 does not cover 1 in T_p'; the argument is then the same for 1 and 0 once we know that 2 does not cover 1.

Similarly, 4 steps are needed to force $2p-2$ to cover 3 to 0, then 5 steps to force $2p-3$ to cover 4 to 0, etc., and $p+1$ steps to force $p+1$ to cover p to 0. We conclude that $1 + 2 + \cdots + (p+1)$, that is, $(p+1)(p+2)/2$, steps at least are needed to go from T_p to T_p'. Hence u_p is geodesic among positive $\mathbf{A}$-words.   □

Fig. 23 The trees of Lemma 4.13 with the leaf labelling involved in the covering relation, here in the case $p = 3$. The tree T_p'' is the intermediate tree $T_p * w_p$ obtained after $2p+1$ steps: applying $\hat{a}_{0,p}$ to T_p'' leads to T_p' in a total of $3p+1$ rotations.

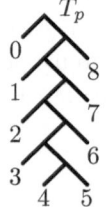 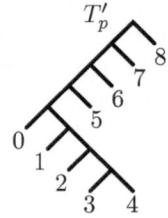 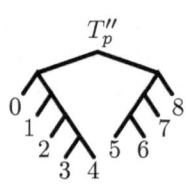

Thus the proof of Proposition 4.10 is complete.

Finally, we translate the above arguments into the language of Tamari lattices.

Definition 4.14. For T, T' in $\mathcal{T}_n$ satisfying $T \leqslant_{\mathcal{T}} T'$, the *positive distance* $\mathrm{dist}^+(T, T')$ from T to T' is the minimal number of left-rotations needed to transform T into T'.

Proposition 4.15. *For every even n, there exist T, T' in $\mathscr{T}_n$ satisfying*

$$\mathrm{dist}(T, T') \leqslant \frac{12}{n} \mathrm{dist}^+(T, T'). \tag{25}$$

Proof. Write $n = 2p + 2$ and let T_p, T'_p be the size n trees of the proof of Lemma 4.13. Then we have $\mathrm{dist}(T_p, T'_p) \leqslant 3p + 1$ and $\mathrm{dist}^+(T_p, T'_p) = (p+1)(p+2)/2$, whence $\mathrm{dist}(T_p, T'_p) \leqslant \frac{12n-16}{n(n+2)} \mathrm{dist}^+(T_p, T'_p)$ in term of n, and, a fortiori, (25). $\square$

Corollary 4.16. *Chains are not geodesic in Tamari lattices; more precisely, for every n, there exists a length ℓ chain of $\mathscr{T}_n$ whose endpoints are at distance less than $12\ell/n$.*

We conclude with a few open questions.

Question 4.17. *For w a positive $\mathbf{A}$-word, is the number $\lambda(w)$ of (10) a least upper bound for the length of the words that are $\mathbf{R}$-equivalent to w?*

The answer is positive in the case of $a_\emptyset^n$. Indeed, we have $\lambda(a_\emptyset^n) = n(n+1)/2$ and $a_\emptyset^n \equiv_{\mathbf{R}} \hat{a}_{1^{n-1},1} \hat{a}_{1^{n-2},2} \cdots \hat{a}_{1,n-1} \hat{a}_{\emptyset,n}$. The general case is not known.

Question 4.18. *Does the Polish normal form of Definition 3.17 satisfy some Fellow Traveler Property, that is, is the distance between the paths associated with the normal forms of elements f and $fa_\alpha^{\pm 1}$ uniformly bounded?*

A positive solution would provide a sort of infinitary automatic structure on F. The question should be connected with a possible closure of Polish normal words under left- or right-$\mathbf{R}$-reversing

Finally, using mapping class groups and cell decompositions, D. Krammer constructed for every size n tree T an exotic lattice structure on $\mathscr{T}_n$ in which T is the bottom element [20].

Question 4.19. *Can the Krammer lattices be associated with submonoids of the Thompson group F?*

More generally, a natural combinatorial description of the Krammer lattices is still missing, but would be highly desirable. Connections with permutations and braids in the line of [2] can be expected. Also, as suggested by Nathan Reading, it would be interesting to investigate a possible connection with maximal cones in type A cluster algebras.

References

1. J.L. Baril and J.M.. Pallo, "Efficient lower and upper bounds of the diagonal-flip distance between triangulations", *Inform. Proc. Letters* **100** (2006) 131–136.
2. A. Björner and M. Wachs, "Shellable nonpure complexes and posets. II", *Trans. Amer. Math. Soc.* **349** (1997) 3945–3975.
3. M. Brin and C. Squier, "Groups of piecewise linear homeomorphisms of the real line", *Invent. Math.* **79** (1985) 485–498.

4. J.W. Cannon and W.J. Floyd, "What is Thompson's group?", *Notices of the AMS* **58-8** (2011) 1112–1113.
5. J.W. Cannon, W.J. Floyd, and W.R. Parry, "Introductory notes on Richard Thompson's groups", *Enseign. Math.* **42** (1996) 215–257.
6. A. Clifford and G. Preston, *The algebraic theory of semigroups, volume 1*, Amer. Math. Soc. Surveys, vol. 7, Amer. Math. Soc., 1961.
7. P. Dehornoy, "The structure group for the associativity identity", *J. Pure Appl. Algebra* **111** (1996) 59–82.
8. _____, "Groups with a complemented presentation", *J. Pure Appl. Algebra* **116** (1997) 115–137.
9. _____, *Braids and Self-Distributivity*, Progress in Math., vol. 192, Birkhäuser, 2000.
10. _____, "Study of an identity", *Algebra Universalis* **48** (2002) 223–248.
11. _____, "Complete positive group presentations", *J. of Algebra* **268** (2003) 156–197.
12. _____, "Geometric presentations of Thompson's groups", *J. Pure Appl. Algebra* **203** (2005) 1–44.
13. _____, "On the rotation distance between binary trees", *Advances in Math.* **223** (2010) 1316–1355.
14. _____, "The word reversing method", *Intern. J. Alg. and Comput.* **21** (2011) 71–118.
15. P. Dehornoy, with F. Digne, E. Godelle, D. Krammer, and J. Michel, "Garside Theory", Book in progress, *http://www.math.unicaen.fr/~garside/Garside.pdf*.
16. O. Deiser, "Notes on the Polish Algorithm", *http://page.mi.fu-berlin.de/deiser/wwwpublic/psfiles/polish.ps*.
17. H. Friedman and D. Tamari, "Problèmes d'associativité: Une structure de treillis finis induite par une loi demi-associative", *J. Combinat. Th.* **2** (1967) 215–242.
18. V.S. Guba, "The Dehn function of Richard Thompson's group *F* is quadratic", *Invent. Math.* **163** (2006) 313–342.
19. S. Huang and D. Tamari, "Problems of associativity: A simple proof for the lattice property of systems ordered by a semi-associative law", *J. Combinat. Th. Series A* **13** (1972) 7–13.
20. D. Krammer, "A class of Garside groupoid structures on the pure braid group", *Trans. Amer. Math. Soc.* **360** (2008) 4029–4061.
21. S. Mac Lane, *Natural associativity and commutativity*, Rice University Studies, vol. 49, 1963.
22. R. McKenzie and R.J. Thompson, "An elementary construction of unsolvable word problems in group theory", in *Word Problems*, Boone and *al*, eds., Studies in Logic, vol. 71, North-Holland, 1973, 457–478.
23. J.M. Pallo, "Enumerating, ranking and unranking binary trees", *The Computer Journal* **29** (1986) 171–175.
24. _____, "An algorithm to compute the Möbius function of the rotation lattice of binary trees", *RAIRO Inform. Théor. Applic.* **27** (1993) 341–348.
25. D. Sleator, R. Tarjan, and W. Thurston, "Rotation distance, triangulations, and hyperbolic geometry", *J. Amer. Math. Soc.* **1** (1988) 647–681.
26. R. Stanley, *Enumerative Combinatorics, vol. 2*, Cambridge Studies in Advances Math., no. 62, Cambridge Univ. Press, 2001.
27. J.D. Stasheff, "Homotopy associativity of *H*-spaces", *Trans. Amer. Math. Soc.* **108** (1963) 275–292.
28. Z. Šunić, "Tamari lattices, forests and Thompson monoids", *Europ. J. Combinat.* **28** (2007) 1216–1238.
29. D. Tamari, "The algebra of bracketings and their enumeration", *Nieuw Archief voor Wiskunde* **10** (1962) 131–146.
30. R.J. Thompson, "Embeddings into finitely generated simple groups which preserve the word problem", in *Word Problems II*, Adian, Boone, and Higman, eds., Studies in Logic, North-Holland, 1980, 401–441.

Parenthetic Remarks

Ross Street

Abstract The Tamari lattice provides an example of a (non-symmetric) operad. We discuss such operads and their associated monads and monoidal categories. Freeness is an important aspect. The free structures are described in various ways using well-formed words (in the spirit of some of Tamari's papers), using string diagrams leading to forests, and in terms of rewrite rules.

1 Introduction

The Tamari Festschrift project[1] has alerted several of us to the papers of Dov Tamari and to our common interests ... except that his work came earlier. My contribution here is a slightly revised version of an incomplete preprint from 1997. While there is little new in what I have to say, except perhaps about the way the topics interact, it is very much in the spirit of Tamari's work. In particular, I move from the example of parenthesising to more general algebraic operations. I invoke a criterion for when an expression involving operations and constants is well formed. By moving from words to cycles, Tamari [23][2] provides added insight to this criterion, aiding enumeration.

My purpose is to provide a gentle introduction to some contemporary mathematical concepts by starting with the parenthesizing example. I emphasise categorical aspects; however, a first course in category theory should be sufficient prerequisite. We look at simple kinds of operads, monads, monoidal categories, free ones, and appropriate string diagrams which lead to forests.

Ross Street
Centre of Australian Category Theory, Macquarie University, NSW 2109, Australia
e-mail: *ross.street@mq.edu.au*

[1] The year of Tamari's birth has another significance for me. My Father, George Street, was born in Sydney on 4 October 1911.

[2] How did I miss this paper, published in an Australian journal, dedicated to Hanna Neumann, and communicated by Mike Newman? At the time I was not ready to absorb what it had to say.

We will not look at the related topic of higher categories. However I must say, soon after the appearance of [20], Jim Stasheff pointed out the similarity between the oriental diagrams therein with associahedra. Also, before any precise definitions had appeared, Todd Trimble [24] proposed the relevance of operads to defining weak higher categories. Michael Batanin in [2, 3] developed a theory of higher operads which achieved one definition of higher category and I recommend [4, 3] for further reading on these matters. For a treatment of polynomial functors more general than in the present paper, see [9].

In Section 2, we look at a basic algebraic structure we call parengebra: it has a single n-ary operation for all $n \geq 2$ formalizing insertion of brackets. Free parengebras lead to associahedra. With this as motivating example, in Section 3 we review the definition of the simplest operads: they are monoids for substitution tensor product of sequences of sets. Sequences of sets provide coefficients for polynomial endofunctors and operads yield monads on the category of sets. These operads also determine monoidal categories (non-symmetric PROPs). Some examples to be used later are explained. Free algebras for operads are constructed in Section 4 in terms of well-formed words. Using this description, some properties of these operads are proved. In Section 5, other constructions for the monoidal category associated with a free operad are provided in terms of string diagrams, forests, and rewrite rules.

2 Parengebras

The bracketing[3] of a string $abcde$ of entities a, b, c, d, e should produce a new entity $(abcde)$. This example shows bracketing as an operation of arity 5. We do not allow $()$ or (a). This leads us to the following kind of universal algebra. A *parengebra* is a set A together with, for all $n \geq 2$, exactly one n-ary operation

$$\beta_n : A^n \to A, \tag{1}$$

subject to no axioms. Let $\mathcal{K}$ denote the category of parengebras and their homomorphisms. The general associativity law implies that every semigroup is canonically a parengebra. In particular, we have the parengebra $\mathbb{N}$ of natural numbers under addition where

$$\beta_n(m_1, \ldots, m_n) = m_1 + \cdots + m_n. \tag{2}$$

Write $F(X)$ for the free parengebra on the set X. The elements of $F(X)$ are built up iteratively as follows:

  i. each $x \in X$ is in $F(X)$;
  ii. if $a_1, \ldots, a_n \in F(X)$ for $n \geq 2$, then $\beta_n(a_1, \ldots, a_n) \in F(X)$.

---

[3] We use the word bracket, interchangeably with parenthesis, for the symbols ")" and "(". The former word conjugates better. For us, the symbols "]" and "[" are square brackets.

Writing $(a_1 \dots a_n)$ for $\beta_n(a_1, \dots, a_n)$, we can imagine elements of $F(X)$ as words in X (that is, elements of the free monoid X^* on X) with brackets meaningfully inserted. This provides the left adjoint functor

$$F : \mathbf{Set} \to \mathscr{K} \tag{3}$$

to the forgetful functor $U : \mathscr{K} \to \mathbf{Set}$ into the category $\mathbf{Set}$ of small sets. Let

$$\mathscr{K} = U \circ F : \mathbf{Set} \to \mathbf{Set} \tag{4}$$

be the monad on $\mathbf{Set}$ generated by the adjunction $F \dashv U$. As with any forgetful functor from a category of universal algebras, the functor $U : \mathscr{K} \to \mathbf{Set}$ is monadic; we have an isomorphism of categories

$$\mathscr{K} \cong \mathbf{Set}^K \tag{5}$$

where $\mathscr{K} \cong \mathbf{Set}^K$ is the category of K-algebras for the monad K in the sense of Eilenberg-Moore [8] (although they used the term "triple" for "monad").

Consider the functor $R : \mathbf{Set} \to \mathbf{Set}$ defined by the power series

$$R(X) = \sum_{n \geq 2} X^n. \tag{6}$$

As with any endofunctor, an R-algebra is an object A of the category ($\mathbf{Set}$ in this case) together with an arrow $\beta : R(A) \to A$, which in this case is the same as a parengebra. Write $\mathbf{Set}^R$ for the category of R-algebras. We have the trivial identification

$$\mathscr{K} = \mathbf{Set}^R. \tag{7}$$

The free monad on an endofunctor is said to exist *pointwise* when the underlying functor from the category of algebras for the endofunctor has a left adjoint. It is an observation of Michael Barr (see [13]) that the free monad is then generated by that adjunction. Therefore, in our case, it follows that K is the pointwise free monad on the endofunctor R. As such, K is expressible as a power series; that is, there is a sequence

$$k_0, k_1, k_2, \dots$$

of sets and a natural bijection

$$K(X) \cong k_0 + k_1 \times X + k_2 \times X^2 + k_3 \times X^3 + \cdots. \tag{8}$$

We can take the set k_n to consist of meaningfully bracketed words of length n. More precisely, we can find the sets k_n using the philosophy of clubs (see [12, 14]). This leads to the expectation that the functor K can be recaptured from the free parengebra $k = K(1)$ on the singleton set 1 augmented by some grounding homomorphism. Let $\gamma : k \to \mathbb{N}$ be the unique parengebra homomorphism taking the element of the generating set 1 to the natural number 1. This gives a grading of k according to the number of occurrences of the generator in the bracketed word; that is,

$$k_n = \gamma^{-1}(n). \tag{9}$$

For any set X, the iterative construction of $K(X)$ given by (i) and (ii) can be interpreted as a bijection

$$K(X) \cong X + K(X)^2 + K(X)^3 + \cdots . \tag{10}$$

These are the components of a natural isomorphism

$$K \cong 1_{\mathbf{Set}} + R \circ K \tag{11}$$

between endofunctors of **Set**. One might hope that there is a category of "virtual sets" which are, to ordinary sets, as integers are to natural numbers; see [10] for virtual species and [17] for Euler characteristic as cardinality. Moreover, we can apply a virtual sets argument [10] to the bijection of the last paragraph to obtain a "formula" for the k_n as mere sets. Put $Y = K(X)$ so that the bijection becomes

$$Y \cong X + Y^2 + Y^3 + \cdots \cong X + Y^2 \times (1-Y)^{-1}. \tag{12}$$

Multiplying on the right by $(1-Y)$ and rearranging, we obtain the quadratic isomorphism

$$2Y^2 - (1+X) \times Y + X \cong 0. \tag{13}$$

Solving by radicals and choosing the meaningful minus sign, we obtain

$$Y \cong \frac{1}{4}(1 + X - (1 - 6X + X^2)^{1/2}). \tag{14}$$

Applying the binomial theorem, we deduce a bijection

$$K(X) \cong X + X^2 + 3X^3 + 11X^4 + 45X^5 + \cdots . \tag{15}$$

If the reader is suspicious of virtual sets, this argument is at least valid in terms of generating functions and so gives the cardinalities of $k_0, k_1, k_2, k_3, k_4, k_5, k_6, \ldots$ as $0, 1, 1, 3, 11, 45, 197, \ldots$ (see [18]), the super Catalan numbers. As an example, we see that for k_4 there are the eleven bracketings occurring as in the Tamari-Stasheff pentagon K_4.

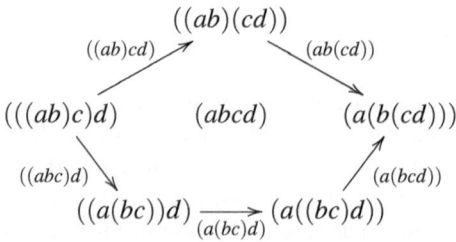

From this cell complex representation, we see that it is natural to define the *dimension* of an element $a \in K(X)$ to be $n - 2$ when n is the largest natural number for which β_n occurs in the iterative construction of a from elements of X. It is also natural to define the boundary $\partial(a)$ of $a \in K(X)$ to be the subset of $K(X)$ consisting of those elements obtained from a by meaningfully inserting precisely one further pair of brackets.

The Stasheff pentagon arose in homotopy theory, specifically, in studying the structure borne by loop spaces [19]. This study was further advanced by the work of Boardman and Vogt [7], May [16] and Kelly [15], out of which arose the concept of operad. We shall see that our sequence $k = (k_n)$ of sets is indeed a planar (or non-permutative) operad.

3 Operads

Given two sequences p and q of sets, we define their substitution product $p \circ q$ to be the sequence of sets

$$(p \circ q)_n = \sum_{m_1 + \cdots + m_r = n} p_r \times q_{m_1} \times \cdots \times q_{m_r} . \tag{16}$$

We regard a sequence p of sets as a functor from the discrete category $\mathbb{N}$, whose objects are natural numbers, to the category **Set**. The substitution product defines a monoidal structure on the category **Set**$^{\mathbb{N}}$ of sequences of sets; the unit for substitution is the sequence u with u_1 a singleton set and $u_n = \varnothing$ for $n \neq 1$. A (planar) operad t is a monoid for the substitution product on **Set**$^{\mathbb{N}}$. Notice that the unit for such a monoid is an element $1 \in t_1$ and the multiplication amounts to a collection of functions

$$t_r \times t_{m_1} \times \cdots \times t_{m_r} \to t_{m_1 + \cdots + m_r} , \tag{17}$$

whose value at $(\tau, \tau_1, \ldots, \tau_r)$ is denoted by $\tau[\tau_1, \ldots, \tau_r]$, called the result of substituting $\tau_1, \ldots, \tau_r$ in τ, such that

$$1[\tau_1] = \tau_1 \tag{18}$$

$$\tau[1, \ldots, 1] = \tau \tag{19}$$

$$\tau[\tau_1[\tau_{11}, \ldots, \tau_{1s_1}], \ldots, \tau_r[\tau_{r1}, \ldots, \tau_{rs_r}]] \tag{20}$$

$$= \tau[\tau_1, \ldots, \tau_r][\tau_{11}, \ldots, \tau_{1s_1}, \ldots, \tau_{r1}, \ldots, \tau_{rs_r}].$$

An operad morphism is just a monoid morphism in **Set**$^{\mathbb{N}}$ with substitution tensor product.

Example 3.1. Let $\mathscr{A}$ be a monoidal category. Each object A of $\mathscr{A}$ gives rise to an operad $t = \mathscr{A}(A)$, where $t_n = \mathscr{A}(A^{\otimes n}, A)$ and substitution

$$\mathscr{A}(A^{\otimes r}, A) \times \mathscr{A}(A^{\otimes m_1}, A) \times \cdots \times \mathscr{A}(A^{\otimes m_r}, A) \to \mathscr{A}(A^{\otimes n}, A) \qquad (21)$$

is given by $(f, g_1, \ldots, g_r) \mapsto f \circ (g_1 \otimes \cdots \otimes g_r)$. The distinguished element of t_1 is of course the identity morphism of A.

This operad t is called the *endomorphism operad* of A in $\mathscr{A}$ and denoted by e_A. For any operad t, an operad morphism $t \longrightarrow e_A$ is called an *action* of t on A. When $\mathscr{A} = \mathbf{Set}$, we call the set A equipped with an action a *t-algebra*. The action can be expressed as a sequence of functions $t_n \times A^{\otimes n} \longrightarrow A$, whose value at $(\tau, a_1, \ldots, a_n)$ is written $\tau[a_1, \ldots, a_n]$, subject to two axioms expressing compatibility with the operad structure.

Example 3.2. Every operad t determines a strict monoidal category $\mathscr{V}_t$ as follows. The objects are the natural numbers and the homsets are given by

$$\mathscr{V}_t(m, n) = \sum_{m_1 + \cdots + m_n = m} t_{m_1} \times \cdots \times t_{m_n}. \qquad (22)$$

Composition

$$\mathscr{V}_t(n, s) \times \mathscr{V}_t(m, n) \to \mathscr{V}_t(m, s) \qquad (23)$$

takes $(\tau_{n_1}, \ldots, \tau_{n_s}, \tau_{m_1}, \ldots, \tau_{m_n}) \in t_{n_1} \times \cdots \times t_{n_s} \times t_{m_1} \times \cdots \times t_{m_n}$ to

$$(\tau_{n_1}[\tau_{m_1}, \ldots, \tau_{m_{n_1}}], \tau_{n_2}[\tau_{m_{n_1}+1}, \ldots, \tau_{m_{n_1}+n_2}], \ldots, \tau_{n_s}[\tau_{m_{n-n_s}+1}, \ldots, \tau_{m_n}]) \qquad (24)$$

where $m = m_1 + \cdots + m_n$ and $n = n_1 + \cdots + n_s$. The tensor product of $\mathscr{V}_t$ is given on objects by addition and on homsets

$$\mathscr{V}_t(m, n) \times \mathscr{V}_t(r, s) \to \mathscr{V}_t(m + r, n + s) \qquad (25)$$

by $((\tau_{m_1}, \ldots, \tau_{m_n}), \ldots, (\tau_{r_1}, \ldots, \tau_{r_s})) \mapsto (\tau_{m_1}, \ldots, \tau_{m_n}, \ldots, \tau_{r_1}, \ldots, \tau_{r_s})$. Clearly we recapture the original operad t from the object $1 \in \mathscr{V}_t$ by the construction of Example 3.1. Furthermore, for any strict monoidal category $\mathscr{A}$ and $A \in \mathscr{A}$, the operad morphisms $\phi : t \to \mathscr{A}(A)$ are in bijection with the strict monoidal functors $M : \mathscr{V}_t \to \mathscr{A}$ with $M(1) = A$; the bijection is defined by putting $M(n) = A^{\otimes n}$ and taking the effect of M on homsets $\mathscr{V}_t(n, 1) \to \mathscr{A}(M(n), M(1))$ to be ϕ_n.

Each operad gives rise to a monad (see [16] for the symmetric case). To see this, we identify each natural number $n \in \mathbb{N}$ with the set $\{0, 1, \ldots, n-1\}$ and so obtain the natural sequence $nat : \mathbb{N} \to \mathbf{Set}$ of sets. Left Kan extension along the functor nat gives a functor

$$Ser : \mathbf{Set}^{\mathbb{N}} \to \mathbf{Set}^{\mathbf{Set}} \qquad (26)$$

defined by $Ser(p) = P$, where

$$P(X) = \sum_{n \geq 0} p_n \times X^n \tag{27}$$

is the power series with coefficient sets p_n, $n \geq 0$. There is a (strict) monoidal structure on the category $\mathbf{Set}^{\mathbf{Set}}$ of endofunctors of $\mathbf{Set}$ defined by composition of functors. A standard calculation with power series shows that Ser is a strong monoidal functor; that is, it coherently preserves the monoidal structures:
- if $P = Ser(p)$ and $Q = Ser(q)$ then $P \circ Q \cong Ser(p \circ q)$; and
- $Ser(u)$ is isomorphic to the identity functor.

It follows that monoids t in $\mathbf{Set}^{\mathbb{N}}$ are taken to monads $T = Ser(t)$ on the category $\mathbf{Set}$. An *algebra for the operad* t is defined to be an Eilenberg-Moore algebra for the monad $T = Ser(t)$. This agrees with the definition in Example 3.1.

Besides the substitution product, there are of course other useful operations on sequences of sets. The *sum* $p + q$ of sequences p and q is given pointwise by

$$(p+q)_n = p_n + q_n \tag{28}$$

(disjoint union of sets); this is the coproduct of p and q in the category $\mathbf{Set}^{\mathbb{N}}$ and, likewise, infinite sums can be considered.

The *convolution product* $p * q$ of p and q is given by

$$(p * q)_n = \sum_{r+s=n} p_r \times q_s. \tag{29}$$

It is easy to see that the functor $Ser : \mathbf{Set}^{\mathbb{N}} \longrightarrow \mathbf{Set}^{\mathbf{Set}}$ takes sum to coproduct and convolution product to product.

We can identify each set z with the sequence z given by $z_0 = z$ and $z_n = 0$ for all $n > 0$. Then each sequence p of sets can be decomposed as a power series

$$p \cong \sum_{n \geq 0} p_n * u^{*n} \tag{30}$$

where u^{*n} is the n-fold convolution power of the unit u for substitution product.

Example 3.3. Let u^* denote the terminal sequence of sets; that is, each set in the sequence is a singleton. This clearly has a unique operad structure. The monad $Ser(u^*)$ on $\mathbf{Set}$ is given by the full geometric series

$$X^* = \sum_{n \geq 0} X^n. \tag{31}$$

The elements of X^* are written as *words* in the *alphabet* X. The algebras for the terminal operad are the Eilenberg-Moore algebras for the geometric series monad, and so are monoids (in $\mathbf{Set}$ with cartesian product as monoidal structure).

Example 3.4. Let z be any set. Consider the sequence $z + u$ of sets given by

$$(z+u)_n = \begin{cases} z & \text{for } n = 0 \\ 1 & \text{for } n = 1 \\ 0 & \text{for } n > 1 . \end{cases} \tag{32}$$

There is a unique operad structure on $z + u$ for which the substitution $(z+u)_1 \times (z+u)_0 \to (z+u)_0$ is the second projection. The monad $Ser(z+u)$ takes X to the monic polynomial $z+X$ of degree 1. The monad structure on $Ser(z+u)$ is induced by the canonical monoid structure on the arbitrary set z with respect to the coproduct as tensor product on **Set**. The algebras for the operad $z + u$ are sets A together with a function $z + A \to A$ whose restriction to A is the identity; that is, the algebras amount to functions $z \to A$ out of z. The category of algebras is the category $z/\textbf{Set}$ of sets under z.

Example 3.5. Any monad T on **Set** admits a distributive law [5] with the monad $Ser(z+u)$ of Example 3.4; that is, there is a natural transformation $\lambda : Ser(z+u) \circ T \to T \circ Ser(z+u)$ satisfying axioms ensuring that T lifts to a monad on $Ser(z+u)$-algebras, and that $T \circ Ser(z+u)$ gains a monad structure whose algebras are the same as the algebras for the lifted monad; in this case, the category of these algebras is just the category $z/\textbf{Set}^T$ of T-algebras under z. The component $\lambda_X : z + T(X) \to T(z+X)$ of λ at X is constructed from the unit of T at z and T of the inclusions of z and X in $z + X$. Suppose now that $T = Ser(t)$ for some operad t. Then $T \circ Ser(z+u) = Ser(t \circ (z+u))$ for the operad $t \circ (z+u)$ calculated as follows:

$$(T \circ Ser(u+z))(X) = \sum_{n \geq 0} t_n \times (z+X)^n = \sum_{n \geq 0} t_n \times \sum_{0 \leq m \leq n} \binom{n}{m} z^{n-m} \times X^m \quad (33)$$

$$= \sum_{\substack{m \geq 0 \\ r \geq 0}} \binom{r+m}{m} t_{r+m} \times z^r \times X^m = Ser(t \circ (z+u))(X) \quad (34)$$

where $(t \circ (z+u))_m = \sum_{r \geq 0} \binom{r+m}{m} t_{r+m} \times z^r$. We shall not bother to explicitly describe the substitution operation of $t \circ (z+u)$ except in the special case of the next example.

Example 3.6. As a particular case of Example 3.5, take $t = u^*$ to be the terminal sequence as discussed in Example 3.3. Then $T \circ Ser(z+u)$ is the monad given by

$$(T \circ Ser(z+u))(X) = (z+X)^*. \quad (35)$$

The operad $t \circ (z+u) = (z+u)^*$ is given by taking $(z+u)^*_m$ to consist of those elements of the free monoid $(z+1)^*$ which are words in elements of z and the symbol 0 with exactly m occurrences of 0. Indeed, we have the natural bijection

$$(z+X)^* \cong \sum_{m \geq 0} (z+u)^*_m \times X^m \quad (36)$$

$$a_0 x_1 a_1 x_2 a_2 \ldots a_{m-1} x_m a_m \leftrightarrow ((a_0 0 a_1 0 a_2 \ldots a_{m-1} 0 a_m), (x_1, x_2, \ldots, x_m)) \quad (37)$$

where the $a_0, \ldots, a_m \in z^*$ and $x_1, \ldots, x_m \in X$. We have the substitution functions

$$(z+u)^*_r \times (z+u)^*_{m_1} \times \cdots \times (z+u)^*_{m_r} \to (z+u)^*_{m_1 + \cdots + m_r} \quad (38)$$

$$(a, b_1, \ldots, b_r) \mapsto a_0 b_1 a_1 \ldots a_{r-1} b_r a_r \tag{39}$$

where $a = a_0 0 a_1 0 \ldots a_{r-1} 0 a_r$. The distinguished element of $(z + u)^*{}_1$ is of course $0 \in 1$.

4 Well-formed words

Proposition 4.1 below will follow from Proposition 4.2 which will follow, in turn, from a discussion of well-formed words. The criterion for when an expression in a universal algebra, written in Polish or Lukasiewicz notation, is well formed I learned from Samuel Eilenberg who caused it, I believe, to be an exercise in Bourbaki. Now I see that this classical result was the starting point for Tamari's paper [23], which went on to provide a criterion (in terms of cyclic words) useful for enumeration.

Proposition 4.1. *Free monads on power series endofunctors of* **Set** *are all of the form* $Ser(t)$ *for an operad* t.

Let P be a power series endofunctor of **Set** with coefficients given by the sequence p of sets. We regard the elements of p_n as n-ary operations. A P-algebra is a set A together with, for each $n \in \mathbb{N}$, and each $\omega \in p_n$, a function $A^n \to A$ which is also denoted by ω. The left adjoint of the underlying functor $\mathbf{Set}^P \to \mathbf{Set}$ can be described iteratively as we did earlier for the case $P = R$. Then the adjunction generates the pointwise free monad on P. We shall give another construction related to Example 3.3.

Consider the free monoid $(P(1) + X)^*$ on the set $P(1) + X$ and regard the set $\mathbb{Z}$ of integers as a monoid under addition. Let $\overline{(\ )} : (P(1) + X)^* \to \mathbb{Z}$ be the monoid morphism given on generators by:

$$\bar{x} = -1 \qquad \text{for all } x \in X \tag{40}$$

and

$$\bar{\omega} = n - 1 \qquad \text{for all } \omega \in p_n. \tag{41}$$

Call $a \in (P(1) + X)^*$ well-formed when

i. $\bar{a} = -1$ and
ii. $a = bc$ implies $\bar{c} < 0$.

Let $W_p(X)$ denote the subset of $(P(1) + X)^*$ consisting of the well-formed words. We can equip $W_p(X)$ with the structure of P-algebra by defining

$$\omega(a_1, \ldots, a_n) = \omega a_1 \ldots a_n \tag{42}$$

for all $\omega \in p_n$ and $a_1, \ldots, a_n \in W_p(X)$. To see that

$$\omega a_1 \ldots a_n$$

is well formed, notice that

$$\overline{\omega a_1 \dots a_n} = \overline{\omega} + \overline{a_1} + \cdots + \overline{a_n} = (n-1) + (-1) + \cdots + (-1) = -1 \quad (43)$$

and if $a_i = bc$, then

$$\overline{ca_{i+1} \dots a_n} = \overline{c} + \overline{a_{i+1}} + \cdots + \overline{a_n} = \overline{c} + (-1) + \cdots + (-1) = \overline{c} + (i-n) < 0. \quad (44)$$

Although we shall not need it here, what Tamari proved in [23] was that, when a word $a \in (P(1)+X)^*$ satisfied the equality (i), there was precisely one cyclic ordering of a which was well formed. This aided enumeration significantly. He proved that the number of well-formed ways of inserting the operations $\omega_1 \dots, \omega_m \in P(1)$ into the words $x_1 \cdots x_n \in X^*$ was

$$\frac{(n+1)(n+2) \cdots (n+m-1)}{\prod\limits_{\omega \in P(1)} (m_\omega)!} \quad (45)$$

where $m_\omega = \#\{r | \omega_r = \omega\}$ (so that $\sum\limits_{\omega \in P(1)} m_\omega = m$).

Proposition 4.2. $W_p(X)$ *is the free P-algebra on the set X.*

Proof. Given a function $f : X \to A$ into a P-algebra A, we must show that there exists a unique extension of f to a P-algebra morphism $g : W_p(X) \to A$. That is, we must show that the equations

$$g(x) = f(x) \quad \text{for} \quad x \in X \text{ and} \quad (46)$$

$$g(\omega a_1 \dots a_n) = \omega(g(a_1), \dots, g(a_n)) \text{ for } \omega \in p_n \text{ and } a_1, \dots, a_n \in W_p(X) \quad (47)$$

uniquely determine g satisfying the equations. To define $g(a)$ for $a \in W_p(X)$, we use induction on the number r of occurrences of elements of $P(1)$ in the word a. By well formedness, if $r = 0$ then $a \in X$ and the definition $g(a) = f(a)$ is forced. Suppose $r > 0$. By well formedness, a must have length > 1; so put $a = tb$ where $t \in X + P(1)$. But $\overline{b} < 0$ and $\overline{t} + \overline{b} = -1$; so $\overline{t} > -1$, whence $t \in p_n$ for some $n = -\overline{b}$ and we can write $b = c_1 \dots c_n$ where $\overline{c_1} = \cdots = \overline{c_n} = -1$. Since a is well formed, it follows that $c_1, \dots, c_n \in W_p(X)$. Define $g(a) = \omega(g(c_1), \dots, g(c_n))$ as we are forced to, and can do since each c_i has fewer than r occurrences of elements of $P(1)$. It remains to prove that g is a P-algebra morphism but that is a direct inductive argument. $\square$

To construct the operad from W_p, notice that $\mathbb{N}$ becomes a P-algebra by taking each n-ary operation to be n-fold addition. Then we obtain a P-algebra morphism

$$\gamma : W_p(1) \to \mathbb{N} \quad (48)$$

which restricts to the function $1 \to \mathbb{N}$ whose value at the one element 0 of 1 is $1 \in \mathbb{N}$. So $\gamma(a)$ is the number of occurrences of the element 0 of 1 in the word a. Let $t_n = \gamma^{-1}(n)$ be the fibre of γ over n. If we put $z = P(1) = p_0 + p_1 + p_2 + \cdots$

we see that W_p is a submonad of the monad $Ser(z+u)^*$ described in Example 3.3. Furthermore, the gradings are respected so that t is a suboperad of $(z+u)^*$.

Corollary 4.3. *The operad t is free on the sequence p of sets. The monad $T = Scr(t)$ is free on the endofunctor P.*

Proof. Suppose o is an operad and $\theta_n : p_n \to o_n$ are functions for all $n \geq 0$. We can make $\sum_{m \geq 0} o_m$ into a P-algebra by defining, for each $\omega \in p_n$, a function

$$\omega : \left(\sum_{m \geq 0} o_m \right)^n = \sum_{m_1,\dots,m_n} o_{m_1} \times \cdots \times o_{m_n} \to \sum_{m \geq 0} o_m \qquad (49)$$

by $\omega(v_1,\dots,v_n) = \theta_n(\omega)[v_1,\dots,v_n]$. By Proposition 4.2, there is a unique P-algebra morphism $\phi : W_p(1) \to \sum_{m \geq 0} o_m$ determined by $\phi(0) = 1 \in u_1$. This clearly respects the gradings and so gives functions $\phi_n : t_n \to o_n$. Since ϕ is a P-algebra morphism, the functions ϕ_n for $n \in \mathbb{N}$ commute with the substitution operations and ϕ_1 preserves the distinguished object. So (ϕ_n) is the unique operad morphism extending (θ_n). The free monad on the endofunctor P on **Set** is the monad generated by the underlying functor **Set**$^P \to$ **Set** and its left adjoint. By Proposition 4.2, this left adjoint is provided by W_p. It is easily seen that the monad structure transports across our isomorphism $W_p \cong Ser(t)$ to the monad structure induced by the operad t. $\qquad\square$

In particular, taking the sequence r of sets given by $r_n = 0$ for $n = 0, 1$ and $r_n = 1$ for $n \geq 2$, so that the power series endofunctor $Ser(r)$ is R, we obtain a parengebra isomorphism

$$W_r(X) \cong F(X), \qquad (50)$$

hence an operad structure on $k = (k_n)$ producing the monad structure on $K = Ser(k)$. A topological version of k is the Stasheff A_∞-operad.

A monad T on a category $\mathscr{C}$ is cartesian (see [6]) when the endofunctor $T : \mathscr{C} \to \mathscr{C}$ preserves pullbacks and the multiplication $\mu : T \circ T \to T$ and unit $\eta : 1_{\mathscr{C}} \to T$ are cartesian natural transformations. A natural transformation $\theta : H \to K : \mathscr{C} \to \mathscr{D}$ is cartesian when, for all morphisms $f : U \to V$ in $\mathscr{C}$, the commutative square

$$\begin{array}{ccc} HU & \xrightarrow{\theta_U} & KU \\ {\scriptstyle Hf} \downarrow & & \downarrow {\scriptstyle Kf} \\ HV & \xrightarrow[\theta_V]{} & KV \end{array} \qquad (51)$$

is a pullback in $\mathscr{D}$.

Corollary 4.4. *For the free operad t on a sequence p of sets, the monad $T = Ser(t)$ is cartesian.*

Proof. It is well known that the free monoid monad $(\ )^*$ on **Set** (see Example 3.3) is cartesian (it is even true also for **Set** replaced by a topos with natural numbers object (see[6]). It is easy to see that the monad $Ser(z+u)$ of Example 3.4 is cartesian. It follows from this that the monad $X \mapsto (z+X)^*$ of Example 3.3 is cartesian. In particular, this is true with $z = P(1)$. The pullbacks proving cartesianness of the monad generated by the adjunction of Proposition 4.2 (that is, W_p and its forgetful right adjoint) are a consequence of those for $X \mapsto (P(1)+X)^*$ and their closure under well-formedness.   □

For recent developments see [9].

5 Free monoidal categories

We shall make a connection with free monoidal categories on certain[4] tensor schemes [11]. A tensor scheme consists of a set whose elements are called objects and a graph whose vertices are words of objects and whose edges are called morphisms. Words $D_1 \dots D_n$ of objects are written $D_1 \otimes \cdots \otimes D_n$. Here we restrict to the case where the target of each morphism is a word of length 1.

Take $p \in \mathbf{Set}^{\mathbb{N}}$. Consider the tensor scheme $\mathscr{D}_p$ with one object D and one morphism

$$\omega : D^{\otimes n} \to D \tag{52}$$

for each natural number n and each $\omega \in p_n$. We represent the morphism ω by a planar string diagram where there are n input strings above the node ω and 1 output string below[5].

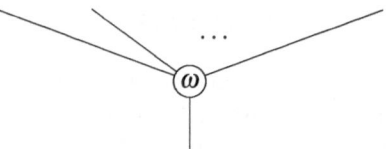

Let $\mathscr{D}_p{}^*$ denote the free strict monoidal category on the tensor scheme $\mathscr{D}_p$: the objects are of the form

$$D^{\otimes n}$$

for $n \in \mathbb{N}$ and the morphisms can be identified with the deformation classes of planar string diagrams generated by those representing morphisms of $\mathscr{D}_p$. Write $\mathrm{Mon}(\mathscr{D}_p{}^*, \mathbf{Set})$ for the category of (strong) monoidal functors from the monoidal category $\mathscr{D}_p{}^*$ to $\mathbf{Set}$ with its cartesian monoidal structure. Clearly there is an equivalence of categories

$$\mathbf{Set}^P \simeq \mathrm{Mon}(\mathscr{D}_p{}^*, \mathbf{Set}) . \tag{53}$$

---

[4] One might call them opetopic [1] tensor schemes or multigraphs.

[5] Here we take the progressive direction to be downward which is opposite to [11] but the same as in [21].

Corollary 5.1. *If t is the free operad on the sequence p of sets, then there is a monoidal isomorphism*

$$\mathscr{D}_p{}^* \cong \mathscr{V}_t \tag{54}$$

taking D to 1 and

$$\omega : D^{\otimes n} \to D$$

to $\omega 0 \ldots 0 \in t_n = \mathscr{V}_t(n, 1)$ for $\omega \in p_n$.

Proof. We must prove that, given a strict monoidal category $\mathscr{A}$, an object A of $\mathscr{A}$, and a morphism $f_\omega : A^{\otimes n} \to A$ for each $\omega \in p_n$, there exists a unique strict monoidal functor $M : \mathscr{V}_t \to \mathscr{A}$ with $M(1) = A$ and $M(\omega 0 \ldots 0) = f_\omega$. Since t is the free operad on p, there is a unique operad morphism $\phi : t \to \mathscr{A}(A)$ (see Example 3.1) whose restriction to p is given by $\omega \mapsto f_\omega$. By Example 3.2, this operad morphism determines a strict monoidal functor M as required. $\square$

The arrows of $\mathscr{D}_p{}^*$ can be identified with p-labelled (planar) forests. To see this, recall from [11] how the general string diagrams are built from the generators in $\mathscr{D}_p$. We take some arrows from $\mathscr{D}_p$ and some identity strings and tensor them; this amounts to placing the representing string diagrams next to each other. Here are two examples.

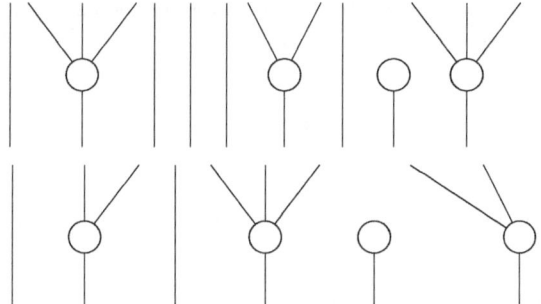

Next we stack such diagrams vertically splicing each lower loose string of one diagram with precisely one upper loose string of the diagram below. The two example diagrams are composable and the composite is represented as follows.

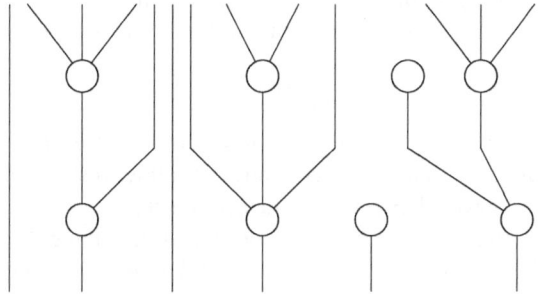

But such a planar diagram can be replaced by a more combinatorial structure. We can represent this last composite by a diagram

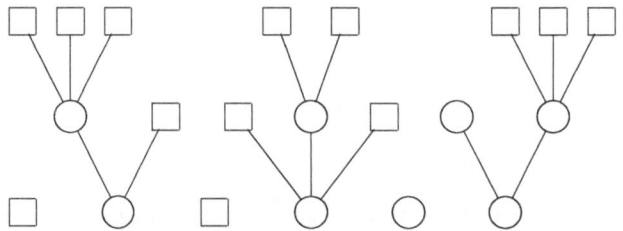

where we have removed the bottom loose strings, put square nodes on the top loose strings, and shortened the strings so that they connect nodes of consecutive height.

We must remember that, although we have omitted it from the diagram, each round node with n strings attached at top is labelled by an element of p_n. The structure that arises in this way is precisely a planar forest; that is, a functor

$$f : [k]^{\mathrm{op}} \to \Delta \tag{55}$$

where Δ is the category whose objects are the linearly ordered sets $[k] = \{0, 1, \dots, k\}$ for $k \geq -1$ and whose morphisms are order-preserving functions. Such a functor gives linearly ordered sets $f(i)$ for $0 \leq i \leq k$, whose elements are called vertices of f of height i, and order-preserving functions $f_i : f(i+1) \to f(i)$ for $0 \leq i < k$. In our example,

$$f(0) = [5], f(1) = [6], f(2) = [7] \tag{56}$$

$$f_0(0) = f_0(1) = 1, f_0(2) = f_0(3) = f_0(4) = 3, f_0(5) = f_0(6) = 5 \tag{57}$$

$$f_1(0) = f_1(1) = f_1(2) = 0, f_1(3) = f_1(4) = 3, f_1(5) = f_1(6) = f_1(7) = 6 . \tag{58}$$

A forest

$$f : [k]^{\mathrm{op}} \to \Delta$$

is called a tree when $f(0) = [0]$. Each forest can be identified with a linearly ordered set of component trees: the number of component trees is n where $f(0) = [n-1]$.

For a forest

$$f : [k]^{\mathrm{op}} \to \Delta$$

of *height* k, a vertex $v \in f(i)$ is called a *leaf* when the fibre of f_i over v is empty. Notice in our last diagram that the square nodes are all leaves; but there are also two round nodes that are leaves. Let p be a sequence of sets. A *labelling* of a forest f in p assigns to each vertex $v \in f(i)$, which is not a leaf, an element $\omega(v) \in p_n$ where $n > 0$ is the cardinality of the fibre of f_i over v, and assigns to some leaves $v \in f(i)$, which are not of height k, an element $\omega(v) \in p_0$. A leaf which is not labelled will be called *fallen*; so all leaves of height k are fallen. A forest together with a labelling in

p will be called a *p-forest*. This leads to our next view of the strict monoidal category $\mathscr{D}_p{}^*$. We see that the morphisms

$$D^{\otimes m} \to D^{\otimes n}$$

can be identified with p-forests having m fallen leaves and n vertices of height 0. The tensor product of two p-forests is given by placing them next to each other in the plane; we shall not describe it combinatorially. Composition is given by *grafting*: we shall just give an example. Let us take the following example of two composable p-forests

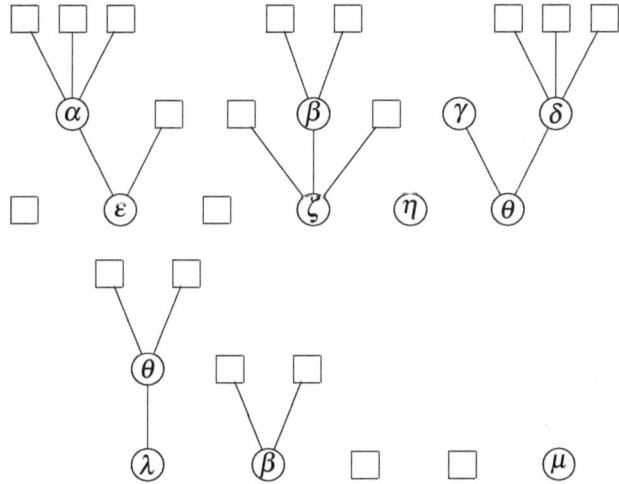

Their composite is the following p-forest in which the component trees of the upper forest are grafted on at the fallen leaves of the lower forest.

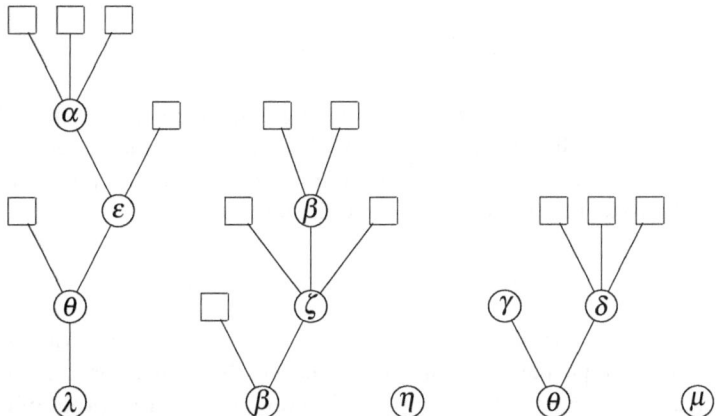

From this, we deduce another view of the free operad t on p in which elements of t_n are identified with p-trees with n fallen leaves.

A description of $\mathscr{D}_p{}^*$ in the spirit of computer science can be obtained using *rewrite systems* (see [21] for example). This seems appropriate to mention here since we see from [22] that Tamari was interested in solubility of word problems. The basic *p*-forest

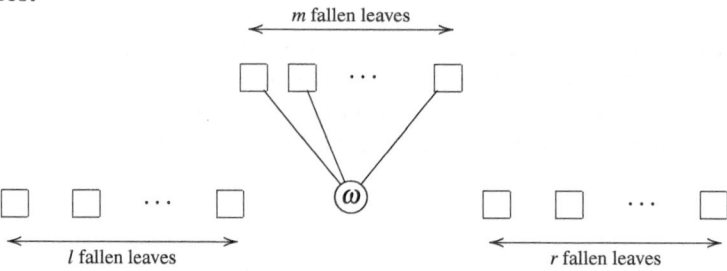

can be identified with triplets (l, ω, r) where l, m, r are natural numbers and $\omega \in p_m$. Consider the directed graph $\mathscr{G}_p$ whose vertices are natural numbers and whose edges are the expressions

$$(\omega, r) : a \to b \tag{59}$$

where a, b, r are natural numbers such that $r < b \le a + 1$ and $\omega \in p_m$ where $m = a - b + 1$. The idea is that each $\omega \in p_m$ is regarded as the name of a (first-order) rewrite rule and

$$(\omega, r) : a \to b$$

represents an allowable application of the rule; the numbers $l = b - 1 - r$ and r are the left and right whiskers, respectively. We introduce some (second-order) rewrite rules on directed paths in the graph $\mathscr{G}_p$.

$$\frac{a \xrightarrow{(\omega,r)} b \xrightarrow{(\omega',r')} c}{a \xrightarrow{(\omega',r'+a-b)} a-b+c \xrightarrow{(\omega,r)} c} \quad \text{for } r < r' \tag{60}$$

It is easy to see that the *p*-forests corresponding to the top and bottom of the rewrite rule are equal. Two directed paths in $\mathscr{G}_p$ with same source and target are said to be equivalent when there is a sequence of applications of the (second-order) rewrite rules which takes one path to the other. A directed path

$$a_0 \xrightarrow{(\omega_1,r_1)} a_1 \xrightarrow{(\omega_2,r_2)} \cdots \xrightarrow{(\omega_n,r_n)} a_n \tag{61}$$

is said to be in *normal form* when $r_i \ge r_{i+1}$ for $1 \le i \le n$. Each directed path is equivalent to a unique path in normal form; indeed, the normal form can be achieved by directed applications of the (second-order) rewrite rules (the proof of confluence and termination is similar to the case considered in [21]. In this way, we obtain a strict monoidal category isomorphic to $\mathscr{D}_p{}^*$. The objects are natural numbers and the arrows are equivalence classes of directed paths in $\mathscr{G}_p$. The composition is induced on equivalence classes by concatenation of paths. The tensor product $a \otimes b$ is given on objects a and b by addition $a + b$ of natural numbers. The functor $c \otimes -$ is given on morphisms by

$$c \otimes [(\omega_n, r_n) \circ \cdots \circ (\omega_1, r_1)] = [(\omega_n, r_n) \circ \cdots \circ (\omega_1, r_1)] \tag{62}$$

and the functor $- \otimes d$ is given on morphisms by

$$[(\omega_n, r_n) \circ \cdots \circ (\omega_1, r_1)] \otimes d = [(\omega_n, r_n + d) \circ \cdots \circ (\omega_1, r_1 + d)] ; \tag{63}$$

the tensor product $\pi \otimes \rho : a \otimes c \to b \otimes d$ of two arrows $\pi : a \to b$ and $\rho : c \to d$ is then given by either route around the following square, the commutativity of which is precisely what we have achieved by our equivalence relation on paths.

$$\begin{array}{ccc} a \otimes c & \xrightarrow{\pi \otimes c} & b \otimes c \\ {\scriptstyle a \otimes \rho} \downarrow & & \downarrow {\scriptstyle b \otimes \rho} \\ a \otimes d & \xrightarrow{\pi \otimes d} & b \otimes d \end{array} \tag{64}$$

The more combinatorial thing to do is, of course, to take the morphisms to be paths in normal form. Then composition and tensoring involve reducing to normal form.

Acknowledgement I am grateful to Jim Stasheff for turning my preprint into something acceptable to the editors and for helpful advice on the content, and to Mark Weber for TeXing the diagrams and for advice on references.

References

1. J.C. Baez and J. Dolan, "Higher-dimensional algebra. III. n-categories and the algebra of opetopes", *Adv. Math.* **135** (1998) 145–206.
2. M.A. Batanin, "On the definition of weak ω-category", Macquarie Math. Report no. 96/207, Macquarie University, NSW Australia (1996).
3. _____, "Monoidal globular categories as a natural environment for the theory of weak n-categories", *Adv. Math.* **136** (1998) 39–103.
4. M.A. Batanin, "Homotopy coherent category theory and A_∞-structures in monoidal categories", *J. Pure Appl. Algebra* **123** (1998) 67–103.
5. J. Beck, "Distributive laws", in *Seminar on Triples and Categorical Homology Theory* (ETH Zürich, 1966/67), B. Eckmann, ed., Lecture Notes in Mathematics, vol. 80, Springer, Berlin, 1969, 119–140. Reprinted in *Reprints in Theory and Applications of Categories*, no. 18 (2008) 95–112, *http://tac.mta.ca/tac/reprints/articles/18/tr18.pdf*.
6. J. Bénabou, "Some remarks on free monoids in a topos", in *Category theory* (Como, Italy, 1990), A. Carboni, M.C. Pedicchio and G. Rosolini, eds., Lecture Notes in Mathematics, vol. 1488, Springer, Berlin, 1991, 20–29.
7. J.M. Boardman and R.M. Vogt, *Homotopy invariant algebraic structures on topological spaces*, Lecture Notes in Mathematics, vol. 347, Springer, Berlin-New York, 1973.
8. S. Eilenberg and J.C. Moore, "Adjoint functors and triples", *Illinois J. Math.* **9** (1965) 381–398.

9. N. Gambino and J. Kock, "Polynomial functors and polynomial monads", *arxiv.org/abs/0906.4931*.

10. A. Joyal, "Foncteurs analytiques et espèces de structures", in *Combinatoire énumérative* (Montreal, Canada, 1985), G. Labelle and P. Leroux, eds., Lecture Notes in Mathematics, vol. 1234, Springer, Berlin, 1986, 126–159.

11. A. Joyal and R. Street, "The geometry of tensor calculus. I", *Adv. Math.* **88** (1991) 55–112.

12. G.M. Kelly, "An abstract approach to coherence", in *Coherence in Categories*, G.M. Kelly, M. Laplaza, G. Lewis and S. Mac Lane, eds., Lecture Notes in Mathematics, vol. 281, Springer, Berlin, 1972, 106–147.

13. _____ , "A unified treatment of transfinite constructions for free algebras, free monoids, colimits, associated sheaves, and so on", *Bull. Austral. Math. Soc.* **22** (1980) 1–83.

14. _____ , "On clubs and data-type constructors", in *Applications of Categories in Computer Science* (Durham, UK, 1991), M.P. Fourman, P.T. Johnstone and A.M. Pitts, eds., London Math. Soc. Lecture Note Ser., vol. 177, Cambridge Univ. Press, Cambridge, 1992, 163–190.

15. _____ , "On the operads of J.P. May", *Reprints in Theory and Applications of Categories*, no. 13 (2005) 1–13, *http://tac.mta.ca/tac/reprints/articles/13/tr13.pdf*.

16. J.P. May, *The Geometry of Iterated Loop Spaces*, Lecture Notes in Mathematics, vol. 271, Springer, 1972.

17. S.H. Schanuel, "Negative sets have Euler characteristic and dimension", in *Category theory* (Como, Italy, 1990), A. Carboni, M.C. Pedicchio and G. Rosolini, eds., Lecture Notes in Mathematics, vol. 1488, Springer, Berlin, 1991, 379–385.

18. N.J.A. Sloane, *A Handbook of Integer Sequences*, Academic Press, New York, 1973.

19. J.D. Stasheff, "Homotopy associativity of *H*-spaces, I", *Trans. Amer. Math. Soc.* **108** (1963) 275–292.

20. R. Street, "The algebra of oriented simplexes", *J. Pure Appl. Algebra* **49** (1987) 283–335.

21. _____ , "Categorical structures", in *Handbook of Algebra*, vol. 1, M. Hazewinkel, ed., Elsevier Science B.V., Amsterdam, 1996, 529–577.

22. D. Tamari, "Le problèmes d'associativité des monoïdes et le problème des mots pour les demi-groupes; algèbres partielles et chaînes élémentaires", *Séminaire Dubreil-Pisot (Algèbre et Théorie des nombres)* **24** (1970/71) 1–15.

23. _____ , "Formulae for well formed formulae and their enumeration", *J. Austral. Math. Soc.* **17** (1974) 154–162 (Collection of articles dedicated to the memory of Hanna Neumann).

24. T. Trimble, "Thoughts on weak n-categories", undated typed notes.

On the Categories of Modules Over the Tamari Posets

Frédéric Chapoton

Abstract One can attach an Abelian category to each Tamari poset, the category of modules over its incidence algebra. This can also be described as the category of modules over the Hasse diagram of the poset, seen as a quiver with relations. The derived category of this category seems to be a very interesting object, with nice properties and many different descriptions. We recall known results and present some conjectures on these derived categories.

Introduction

Let us denote by Y_n the nth Tamari poset, seen as a partial order on the set of planar binary trees with $n + 1$ leaves, n inner vertices and one root. With this convention, the cardinality of Y_n is given by the Catalan number

$$c_n = \frac{1}{n+1}\binom{2n}{n}. \tag{1}$$

In this article, the main object considered is the category of modules over the Tamari poset Y_n and its derived category.

We start by recalling in the first section the general setting for the category of modules over a finite poset.

We then recall briefly a nice formula for the dimension of the incidence algebra of the Tamari poset. This is rather independent of the rest of the article.

In the third section, we formulate the main conjecture: the derived categories of modules over Tamari posets should be fractionally Calabi-Yau categories.

Frédéric Chapoton
Institut Camille Jordan, Université Claude Bernard Lyon 1, 21 Avenue Claude Bernard, 69622 Villeurbanne Cedex, France, e-mail: *chapoton@math.univ-lyon1.fr*

The notion of Calabi-Yau category has its roots in algebraic geometry. Calabi-Yau varieties generalize elliptic curves, Abelian varieties and K3 surfaces, that are very classical objects. They are the subject of a lot of recent activity, in relationship with mirror symmetry. One can translate the geometric properties of Calabi-Yau varieties into algebraic properties of the derived categories of coherent sheaves on them, and this leads to the notion of Calabi-Yau category. This notion has recently been very useful in the algebraic context of cluster categories. Fractionally Calabi-Yau categories, introduced by Kontsevich, satisfy a weaker form of the axioms of Calabi-Yau categories. One can also find more subtle geometric examples of such categories. For instance, there exists varieties whose derived categories of coherent sheaves have some kind of decomposition into pieces, one of which is fractionally Calabi-Yau. Fractionally Calabi-Yau categories also appear in relation with singularity theory. Our main conjecture would provide examples of a more algebraic nature.

In the fourth section, we explain what is known about the main conjecture at the level of Grothendieck groups. In particular, one knows that a periodicity property holds for a natural linear map, which would also follow from the Calabi-Yau property. Moreover, the full spectrum of this periodic map is known.

In the fifth section, we present results of Ladkani who proved that many other posets (including some Cambrian lattices) have the same derived category of modules as the Tamari poset. We also propose a conjecture for yet another poset with this property.

In the last section, we speculate on a possible relation between the derived categories of modules over the Tamari posets and some triangulated categories (of geometric or algebraic nature) associated with well-chosen quasi-homogeneous polynomials.

1 Representations of posets in general

We will work over a fixed ground field k and with finite posets only. Details on the material of this section can be found for example in [17, 16]. For a general introduction to triangulated categories in representation theory, see [9].

To every poset P, one can associate an Abelian category $\mathrm{mod}\,P$, which is called the category of modules over the poset P. One can give two distinct definitions of this category.

The first definition of $\mathrm{mod}\,P$ uses the incidence algebra $I(P)$ of the poset P. This is an associative algebra with a basis indexed by pairs of elements (x,y) such that $x \leq y$ in P. The product of basis elements is given by the rule

$$(x,y)(z,t) = \delta_{y,z}(x,t), \tag{2}$$

where $\delta_{y,z}$ is 1 if $y = z$ and 0 otherwise. The dimension of $I(P)$ is therefore the number of pairs of comparable elements, which are usually called intervals, in P.

The category $\operatorname{mod} P$ can be defined as the category of finite-dimensional modules over the algebra $I(P)$.

One can give an equivalent definition of $\operatorname{mod} P$ using the language of quivers. Recall that the Hasse diagram $H(P)$ of a poset P is an oriented graph, whose vertices are the elements of P, and whose edges are the covering relations in P, which means that there is an edge $x \to y$ if $x > y$ in P and if there is no z such that $x > z > y$. One can consider the Hasse diagram as a quiver, and therefore define a representation of $H(P)$ as the data of a vector space for every vertex and a linear map for every edge. One furthermore imposes on the representations of this quiver the following relations (full commutation): for every two paths with common start and end, the corresponding compositions of linear maps are equal. This category of representations of the Hasse diagram $H(P)$ with relations is the category $\operatorname{mod} P$.

The equivalence of these two definitions is simple and classical. Indeed, the quotient of the path algebra of the quiver $H(P)$ by the ideal generated by the relations of full commutation is exactly the incidence algebra $I(P)$. Conversely one can decompose every module over $I(P)$, using the idempotents (x,x) for $x \in P$, to obtain a representation of the quiver $H(P)$ satisfying the relations of full commutation.

The category $\operatorname{mod} P$ is a k-linear Abelian category with nice homological properties. In particular, it has finite homological dimension, bounded by the cardinality of P.

Let $\mathscr{D}^b \operatorname{mod} P$ be the bounded derived category of $\operatorname{mod} P$. This category can be defined by starting from the category of bounded chain complexes of objects of $\operatorname{mod} P$, and then inverting formally the quasi-isomorphisms, which are the morphisms of complexes that induces isomorphisms in homology, see for instance [12]. The category $\mathscr{D}^b \operatorname{mod} P$ is a triangulated category, and is therefore endowed with a shift functor, denoted by $[1]$, and a collection of distinguished triangles.

The category $\mathscr{D}^b \operatorname{mod} P$ admits a Serre functor S, which is an auto-equivalence such that there is a bi-natural isomorphism

$$\operatorname{Hom}(X,Y)^* \simeq \operatorname{Hom}(Y,SX), \tag{3}$$

where $*$ is the linear dual over the ground field k.

When there exists a Serre functor, one can define another functor $\Theta = S[-1]$. The functor Θ is usually called the Auslander-Reiten translation. It sends every indecomposable projective module to the shift of the corresponding indecomposable injective module.

2 Dimension of the incidence algebra of Y_n

As a side remark, let us note that the incidence algebra of Y_n has the following nice property.

Theorem 2.1. *The dimension of the incidence algebra of* Y_n *is given by*

$$\frac{2(4n+1)!}{(n+1)!(3n+2)!}. \tag{4}$$

This result (or rather the equivalent statement that this formula gives the number of intervals in Y_n) was conjectured by Pallo in [19] and proved in [4] using combinatorial methods and generating series. Since then, a bijection with a family of planar maps known to be counted by the same formula has been obtained by Bernardi and Bonichon in [2].

This formula also gives the sum of the dimensions of indecomposable projective modules in $\mathrm{mod}\,Y_n$.

The formula (4) has been recently extended in [3] to the case of m-Tamari posets (which are new generalisations of the Tamari posets).

3 Main conjecture on the category $\mathscr{D}^b \mathrm{mod}\,Y_n$

Recall that a triangulated category with a Serre functor S is called *Calabi-Yau of dimension d* if there exists a natural equivalence

$$S \simeq [d] \tag{5}$$

between the Serre functor and the dth power of the shift functor. This terminology is motivated by the properties of the bounded derived categories of coherent sheaves on Calabi-Yau manifolds.

This notion has been generalized by Kontsevich as follows. One says that a triangulated category with Serre functor S is *fractionally Calabi-Yau of type* (p,q) if there exists a natural equivalence

$$S^q \simeq [p] \tag{6}$$

between the qth power of the Serre functor and the pth power of the shift functor. It is a common abuse of notation to speak of dimension p/q instead of type (p,q), but we prefer the more precise notation. From the definition, one can see that type (p,q) implies type (Np,Nq) for every integer $N \geq 1$, but one cannot in general simplify the type by division by a common factor of p and q.

Examples are given by the derived categories of modules over quivers of type $\mathbb{A}$, $\mathbb{D}$ or $\mathbb{E}$, which are fractionally Calabi-Yau of type $(h-2,h)$ where h is the Coxeter number.

Some examples of fractionally Calabi-Yau categories also appear in semi-orthogonal decompositions of some derived categories of coherent sheaves on algebraic varieties. This gives a geometric motivation to the study of this notion.

Our interest in the representations of the Tamari posets Y_n is motivated by the following conjecture.

Conjecture 3.1. *The triangulated category $\mathscr{D}^b \mathrm{mod}\,Y_n$ is fractionally Calabi-Yau of type $(n(n-1), 2n+2)$.*

This conjecture holds for $n = 1, 2, 3$, because of the following equivalences

$$\mathscr{D}^b \operatorname{mod} \mathsf{Y}_1 \simeq \mathscr{D}^b \operatorname{mod} \mathbb{A}_1, \tag{7}$$

$$\mathscr{D}^b \operatorname{mod} \mathsf{Y}_2 \simeq \mathscr{D}^b \operatorname{mod} \mathbb{A}_2, \tag{8}$$

$$\mathscr{D}^b \operatorname{mod} \mathsf{Y}_3 \simeq \mathscr{D}^b \operatorname{mod} \mathbb{D}_5. \tag{9}$$

The first two equivalences are in fact equalities and the third is easily proved using the techniques of Ladkani described in [13] for derived equivalences between categories of modules over posets. Let us sketch the proof here. We will use the following theorem from [13].

Theorem 3.2. *Let P be a finite poset. Let P^+ be the poset P with one maximum element added and let P^- be the poset P with one minimum element added. Then the categories $\mathscr{D}^b \operatorname{mod} P^+$ and $\mathscr{D}^b \operatorname{mod} P^-$ are equivalent.*

Applying this theorem when P^+ is the Tamari poset Y_3 (a pentagon), one finds that P^- is an orientation of the Dynkin diagram $\mathbb{D}_5$, Q.E.D.

As the Coxeter numbers of $\mathbb{A}_1$, $\mathbb{A}_2$ and $\mathbb{D}_5$ are 2, 3 and 8, this gives the types $(0, 2)$, $(1, 3)$ and $(6, 8)$, which indeed imply the types given by Conjecture 3.1, namely $(0, 4)$, $(2, 6)$ and $(6, 8)$.

The type $(n(n-1), 2n+2)$ assigned to the Tamari poset Y_n by Conjecture 3.1 is related to the following observation. One can write the Catalan number c_n as

$$\frac{2n}{2} \frac{2n-1}{3} \cdots \frac{n+2}{n}, \tag{10}$$

which is a fraction written as a quotient of products, together with a bijection between factors of the numerator and factors of the denominator, in such a way that the sum of two corresponding factors is $2n+2$ and the sum of all differences between a numerator factor and the associated denominator factor is $n(n-1)$.

When one replaces every factor k in this factorisation by the q-analogue $[k]_q = 1 + q + \cdots + q^{k-1}$, one finds a classical q-analogue of the Catalan number, which is a polynomial in q with positive coefficients [1].

We will refer to this factorisation again in Section 6, in relation with homogeneous singularities.

4 At the level of Grothendieck groups

Let us go back to the general setting of finite posets for a moment.

The Grothendieck group $K_0(\mathscr{A})$ of an Abelian category $\mathscr{A}$ is the quotient of the free Abelian group generated by objects of $\mathscr{A}$, modulo relations $[X] - [Y] + [Z]$ for every short exact sequence

$$0 \to X \to Y \to Z \to 0. \tag{11}$$

Similarly, the Grothendieck group $K_0(\mathscr{C})$ of a triangulated category $\mathscr{C}$ is the quotient of the free Abelian group generated by objects of $\mathscr{C}$, modulo relations $[X] - [Y] + [Z]$ for every distinguished triangle

$$X \to Y \to Z \to X[1]. \tag{12}$$

Let P be a finite poset. The natural inclusion of the category $\bmod P$ into its derived category $\mathscr{D}^b \bmod P$ induces an isomorphism between their Grothendieck groups. One can therefore work with $K_0(\bmod P)$ only. The classes of simples modules give a basis of $K_0(\bmod P)$.

The auto-equivalences $[1]$, S and Θ of the category $\mathscr{D}^b \bmod P$ induce linear maps on the Grothendieck group $K_0(\bmod P)$. The shift $[1]$ corresponds to $-\,\mathrm{Id}$. Let us call θ the linear map corresponding to Θ. This is usually called the Coxeter transformation.

Let C be the matrix defined by $C_{x,y} = 1$ if $x \leq y$ in P and 0 otherwise. Then the matrix of θ in the basis of simple modules is given by the explicit formula $-C(C^{-1})^t$. This is a very concrete object, and the reader may easily check for small Tamari lattices the properties of θ that we present in this article (see the Appendix for the example for Y_4).

Let us now consider the special case of Tamari lattices.

The following statement would be a corollary of the expected Calabi-Yau property of the category $\mathscr{D}^b \bmod \mathsf{Y}_n$ as formulated in Conjecture 3.1.

Theorem 4.1. *The linear map θ satisfies*

$$\theta^{2n+2} = \mathrm{Id}. \tag{13}$$

This has first been proved by relating the square of θ to the anticyclic structure of the Dendriform operad in [5].

More recently, this has been proved again in [7], using a similar method, by relating θ to a ternary anticyclic operad. This new relation has allowed one to obtain the following description of the spectrum of θ.

Let $\lambda(n)$ be the sequence of integers defined by

$$\lambda(n) = (-1)^{\binom{n}{2}} \binom{n-1}{\lfloor \frac{n-1}{2} \rfloor}, \tag{14}$$

and let b_n be the sequence of integers defined by

$$b_n = \frac{1}{n} \sum_{d|n} \mu(d) \lambda(n/d), \tag{15}$$

where μ is the usual Möbius function of number theory, defined by

$$\mu(n) = \begin{cases} (-1)^k & \text{if } n = p_1 p_2 \dots p_k \text{ for distinct prime numbers } p_1, p_2, \dots, p_k, \\ 0 & \text{otherwise.} \end{cases} \tag{16}$$

Because of the division by n in (15), it is not obvious a priori that the b_n are integers. This has been proved in [6] by using properties of symmetric functions.

Theorem 4.2. *The characteristic polynomial of the Coxeter transformation θ of the Tamari poset Y_n is given by*

$$\frac{(x^{2n+2} - 1)^{c_n}}{\left(\prod_{d|2n+2}(x^d - (-1)^{d(n+1)})^{b_d}\right)^{(-1)^{n+1}}}. \tag{17}$$

This formula was conjectured in [6] under an equivalent form. The proof given in [7] uses Koszul duality of cyclic operads and the Legendre transform of symmetric functions.

5 Derived equivalences with other posets

As explained elsewhere in this volume [21], Reading introduced in [20] the Cambrian lattices, and the Tamari lattice Y_n is one of the Cambrian lattices of type $\mathbb{A}_{n-1}$. There is one such lattice $\mathrm{Cam}(Q)$ for every quiver Q with underlying graph the Dynkin diagram of type $\mathbb{A}_{n-1}$. The Tamari lattice is obtained for the equi-oriented quiver

$$\bullet \longrightarrow \bullet \longrightarrow \cdots \longrightarrow \bullet.$$

The following result has been proved (in greater generality) by Ladkani in [14].

Theorem 5.1. *For any two quivers Q and Q' with underlying graph the Dynkin diagram of type $\mathbb{A}_{n-1}$, the categories $\mathscr{D}^b \mathrm{mod}\,\mathrm{Cam}(Q)$ and $\mathscr{D}^b \mathrm{mod}\,\mathrm{Cam}(Q')$ are equivalent.*

In particular, all the categories $\mathscr{D}^b \mathrm{mod}\,\mathrm{Cam}(Q)$, for quivers Q with underlying graph the Dynkin diagram of type $\mathbb{A}_{n-1}$, are derived equivalent to $\mathscr{D}^b \mathrm{mod}\,Y_n$.

In a rather remarkable way, one can also see the Tamari lattice Y_n as a member of a completely different family of posets, see also [23]. Recall that a *tilting module* in the category of modules over a quiver with n vertices is an object T of $\mathrm{mod}\,Q$ which is the direct sum of n pairwise non-isomorphic indecomposable objects and that satisfies $\mathrm{Ext}(T, T) = 0$. For example, the direct sum of all indecomposable projective modules is always a tilting module.

A natural partial order on the set of tilting modules has been introduced and studied for general algebras in [10, 22]. The Tamari lattice can be interpreted as this natural partial order on the set of tilting modules on the equi-oriented quiver of type $\mathbb{A}_n{}^1$. One has a poset $\mathrm{Tilt}(Q)$ for every quiver Q with underlying graph the Dynkin diagram of type $\mathbb{A}_n$.

Ladkani has proved in another work [15] that all these posets have equivalent derived categories of modules.

---

[1] Notice the shift: as a poset of tilting modules, the Tamari lattice Y_n is related to $\mathbb{A}_n$, while as a Cambrian poset, it is related to $\mathbb{A}_{n-1}$.

Theorem 5.2. *For any two quivers Q and Q' with underlying graph the Dynkin diagram of type $\mathbb{A}_n$, the triangulated categories $\mathcal{D}^b \operatorname{mod} \operatorname{Tilt}(Q)$ and $\mathcal{D}^b \operatorname{mod} \operatorname{Tilt}(Q')$ are equivalent.*

In particular, all the categories $\mathcal{D}^b \operatorname{mod} \operatorname{Tilt}(Q)$ are derived equivalent to $\mathcal{D}^b \operatorname{mod} \mathsf{Y}_n$.

From these two theorems, one can conclude that the category $\mathcal{D}^b \operatorname{mod} \mathsf{Y}_n$ appears in many different ways as the derived category of modules over posets. In fact, one expects that at least one more family of posets shares the same derived category.

Recall that a Dyck word of length n is a sequence of $2n$ parentheses, containing n opening parentheses and n closing parentheses, in such a way that any prefix contains more openings than closings. It is a well-known combinatorial fact that the number of Dyck words of length n is the Catalan number c_n. Let DW_n be the set of Dyck words of size n. On can define a partial order on DW_n as follows. Every Dyck

Fig. 1 A pair of comparable Dyck words

word of length n can be drawn as a lattice path from $(0,0)$ to $(2n,0)$ where opening parentheses are mapped to steps $(1,1)$ (Northeast) and closing parentheses to steps $(1,-1)$ (Southeast). Then the partial order on Dyck words is simply the relation "being always below" for the associated lattice paths. This is illustrated in figure 1.

Conjecture 5.3. *Let DW_n be the poset of Dyck words of size n. The triangulated category $\mathcal{D}^b \operatorname{mod} \operatorname{DW}_n$ is equivalent to $\mathcal{D}^b \operatorname{mod} \mathsf{Y}_n$.*

This conjecture is obviously true for $n = 1, 2$ as the poset DW_1 has just one element ( ) and the poset DW_2 is a chain of length 2. The conjecture can be checked easily for $n = 3$ (illustrated in figure 2) using again Theorem 3.2 from [13]. It has also been checked for $n = 4$ by Ladkani using more sophisticated tools[2].

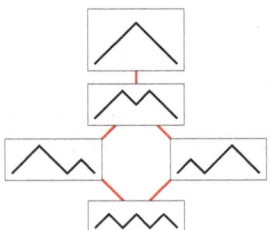

Fig. 2 The poset of Dyck words DW_3

---

[2] Private communication.

The poset of Dyck words DW_n has a natural representation-theoretic meaning, as it can be identified with the distributive lattice of upper ideals in the poset of positive roots of the root system of type $\mathbb{A}_{n-1}$.

6 Quasi-homogeneous isolated singularities

In this section, we want to propose some other categories that should also be related to the Tamari posets. This is more speculative than the rest of the article, due to the lack of expertise of the author in singularity theory.

From now on, one assumes that $n \geq 2$.

Let us introduce $n-1$ variables $x_1, \ldots, x_{n-1}$, where the variable x_i has degree $i+1$. These degrees are the denominator factors in the factorisation (10) of Catalan numbers.

Let now W_n be a generic polynomial in the variables $x_1, \ldots, x_{n-1}$, with coefficients in $\mathbb{C}$, and homogeneous of total degree $2n+2$. This degree is also related to the special factorisation (10).

Let us assume that the generic hypersurface $W_n = 0$ has an isolated singularity at the origin (which we expect to be true for every n).

One would like to compare the category $\mathscr{D}^b \operatorname{mod} Y_n$ to some category built from the data of W_n. For this, one has to use some kind of categorification of the classical theory of isolated singularities. One needs a triangulated category having in particular the property that the rank of its Grothendieck group is the Milnor number of the hypersurface singularity W_n.

There seems to exist two ways to get a hold on such a category. The first option is to use some kind of what is known as the directed Fukaya category associated with W_n, which is a very sophisticated geometric construction using symplectic geometry and Lagrangian submanifolds. The second option, which is expected to be the image of the first one under homological mirror symmetry, is much more algebraic, but one has to use the mirror singularity of W_n instead of W_n. Sometimes this mirror singularity is given by another polynomial V_n, called the Berglund-Hübsch transpose of W_n. When this is the case, the desired algebraic category, called the singularity category of V_n, can then be described in several different ways, either as a quotient of derived categories or using stable categories of maximal Cohen-Macaulay modules or matrix factorisations. The singularity categories are known to be fractionally Calabi-Yau.

For a better informed and much more precise view of this material, the reader may want to consult [11] and [8].

Let us use the notation $\mathscr{D}^b \operatorname{Fuk}^{\rightarrow}(W_n)$ for the appropriate derived category of the directed Fukaya category associated with W_n.

Conjecture 6.1. *The category $\mathscr{D}^b \operatorname{Fuk}^{\rightarrow}(W_n)$ is equivalent to the category $\mathscr{D}^b \operatorname{mod} Y_n$.*

For $n = 2$ and 3, one can reduce generic polynomials to $W_2 = x_1^2$ and $W_3 = x_1^4 + x_1 x_2^2$. It is classical that the singularities associated with these polynomials are related to the Dynkin diagrams $\mathbb{A}_2$ and $\mathbb{D}_5$.

Let us end with a few words on the cases $n = 4$ and $n = 5$.

For $n = 4$, the Catalan number c_4 is 14 and the polynomial W_4 is of degree 10 in 3 variables x_1, x_2, x_3 of degrees $2, 3, 4$. It can be reduced by a linear change of variables to a generic linear combination of the monomials

$$ x_1^5, \quad x_1^2 x_2^2, \quad x_1 x_3^2, \quad x_2^2 x_3. \tag{18} $$

It seems that $\mathscr{D}^b \bmod Y_4$ could be related to the (bimodal) singularity called $S_{1,0}$ (of Milnor number 14) which appears for example in [8, 18, 24]. The mirror singularity is another bimodal singularity, usually called W_{17}.

For $n = 5$, the Catalan number c_5 is 42 and the polynomial W_5 is of degree 12 in 4 variables x_1, x_2, x_3, x_4 of degrees $2, 3, 4, 5$, therefore a generic sum of 10 distinct monomials. It seems that $\mathscr{D}^b \bmod Y_5$ could be equivalent to the tensor product of derived categories of Dynkin type $\mathbb{A}_2$, $\mathbb{A}_3$ and $\mathbb{D}_7$.

Appendix

We present briefly here the concrete example of Y_4, with 14 elements.

Here is the Hasse diagram of Y_4 (edges are oriented implicitly from top to bottom) and the matrix of the Coxeter transformation θ, obtained from the formula $-C(C^{-1})^t$, where C is the matrix of comparability in the partial order Y_4 (see Section 4). To write these linear maps as matrices, one has to choose an arbitrary linear extension of the partial order.

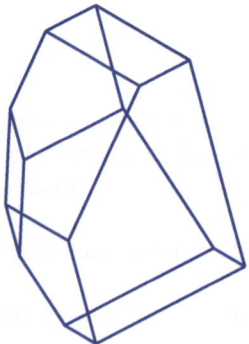

$$
\begin{pmatrix}
0 & 0 & 0 & 0 & 0 & 0 & 0 & 0 & 0 & 0 & 0 & 0 & 0 & -1 \\
0 & 0 & 0 & 0 & 0 & 0 & 0 & 0 & 0 & 0 & 0 & 0 & 1 & -1 \\
0 & 0 & 0 & 0 & 0 & 0 & 0 & 0 & 0 & 0 & 0 & 1 & 0 & -1 \\
0 & 0 & 0 & 0 & 0 & 0 & 0 & 0 & -1 & 1 & 0 & 1 & -1 \\
0 & 0 & 0 & 0 & 0 & 0 & 0 & 0 & -1 & 0 & 1 & 1 & -1 \\
0 & 0 & 0 & 0 & 1 & 0 & 0 & 0 & 0 & 0 & 0 & 0 & 0 & -1 \\
0 & 0 & 0 & -1 & 1 & 0 & 0 & 0 & 0 & 0 & 0 & 1 & 0 & -1 \\
0 & 0 & -1 & 0 & 1 & 0 & 1 & 0 & 0 & 0 & 0 & 0 & 0 & -1 \\
0 & 0 & -1 & 0 & 1 & 0 & 0 & 0 & 0 & 0 & 0 & 0 & 1 & -1 \\
1 & -1 & -1 & 0 & 1 & -1 & 1 & 0 & 1 & 0 & 0 & 0 & 0 & -1 \\
1 & -1 & -1 & 0 & 1 & 0 & 0 & -1 & 1 & 0 & 0 & 0 & 1 & -1 \\
1 & -1 & -1 & 0 & 1 & 0 & 0 & 0 & 0 & -1 & 1 & 0 & 1 & -1 \\
1 & 0 & -1 & -1 & 1 & -1 & 1 & 0 & 0 & 0 & 0 & 1 & 0 & -1 \\
1 & 0 & -1 & -1 & 1 & 0 & 0 & 0 & 0 & -1 & 0 & 1 & 1 & -1
\end{pmatrix}
$$

Fig. 3 The Tamari lattice Y_4

One can check that $\theta^{10} = \mathrm{Id}$ and that the characteristic polynomial is

$$(x+1)^2 \cdot (x^4 + x^3 + x^2 + x + 1) \cdot (x^4 - x^3 + x^2 - x + 1)^2 = \Phi_2^2 \, \Phi_5 \, \Phi_{10}^2, \qquad (19)$$

where Φ_n is the cyclotomic polynomial of order n.

Let us compare this with Theorem 4.2. The first few terms of the sequence b_n defined by (15) are

$$\mathbf{1,\text{-}1}, -1, 1, \mathbf{1}, -1, -3, 4, 8, \mathbf{\text{-}13}, -23, 39, 71, -121, -229, 400, 757, \dots \qquad (20)$$

for $n = 1, 2, \dots$. We have emphasized the coefficients that we will need.

Theorem 4.2 for $n = 4$ therefore gives the following formula

$$\frac{(x^{10} - 1)^{14}}{(x+1)^{-1}(x^2 - 1)^1(x^5 + 1)^{-1}(x^{10} - 1)^{13}} \qquad (21)$$

for the characteristic polynomial of θ, which is indeed equal to (19).

References

1. G.E. Andrews, "On the difference of successive Gaussian polynomials", *J. Statist. Plann. Inference* **34** (1993) 19–22.
2. O. Bernardi and N. Bonichon, "Intervals in Catalan lattices and realizers of triangulations", *J. Combin. Theory Ser. A* **116** (2009) 55–75.
3. M. Bousquet-Mélou, E. Fusy, and L.-F. Préville Ratelle, "The number of intervals in the m-Tamari lattices", *arxiv.org/abs/1106.1498*.
4. F. Chapoton, "Sur le nombre d'intervalles dans les treillis de Tamari", *Sém. Lothar. Combin.* **55** (2005/07) Art. B55f, 18 pp. (electronic).
5. ———, "On the Coxeter transformations for Tamari posets", *Canad. Math. Bull.* **50** (2007) 182–190.
6. ———, "Le module dendriforme sur le groupe cyclique", *Ann. Inst. Fourier (Grenoble)* **58** (2008) 2333–2350.
7. F. Chapoton, "Sur une opérade ternaire liée aux treillis de Tamari", *Annales de la Faculté des Sciences de Toulouse, Sér. 6,* **20** (2011) 843–869.
8. W. Ebeling and D. Ploog, "A geometric construction of Coxeter-Dynkin diagrams of bimodal singularities", 18 p.
9. D. Happel, *Triangulated categories in the representation theory of finite-dimensional algebras*, London Mathematical Society Lecture Note Series, vol. 119, Cambridge University Press, Cambridge, 1988.
10. D. Happel and L. Unger, "On a partial order of tilting modules", *Algebr. Represent. Theory* **8** (2005) 147–156.
11. H. Kajiura, K. Saito, and A. Takahashi, "Triangulated categories of matrix factorizations for regular systems of weights with $\varepsilon = -1$", *arxiv.org/abs/0708.0210*.
12. B. Keller, "Derived categories and tilting", in *Handbook of tilting theory*, London Math. Soc. Lecture Note Ser., vol. 332, Cambridge Univ. Press, Cambridge, 2007, 49–104.
13. S. Ladkani, "Universal derived equivalences of posets", *arxiv.org/abs/0705.0946*.
14. ———, "Universal derived equivalences of posets of cluster tilting objects", *arxiv.org/abs/0710.2860*.
15. ———, "Universal derived equivalences of posets of tilting modules", *arxiv.org/abs/0708.1287*.

16. S. Ladkani, "On derived equivalences of categories of sheaves over finite posets", *J. Pure Appl. Algebra* **212** (2008) 435–451.
17. H. Lenzing, "Coxeter transformations associated with finite-dimensional algebras", in *Computational methods for representations of groups and algebras (Essen, 1997)*, Progr. Math., vol. 173, Birkhäuser, Basel, 1999, 287–308.
18. H. Lenzing and J.A. de la Peña, "Extended canonical algebras and Fuchsian singularities", *Mathematische Zeitschrift* **268** (2011).
19. J. Pallo and C. Germain, "The number of coverings in four Catalan lattices", *Intern. J. Computer Math.* **61** (1996) 19–28.
20. N. Reading, "Cambrian lattices", *Adv. Math.* **205** (2006) 313–353.
21. N. Reading, "From the Tamari lattice to Cambrian lattices and beyond", in *this volume*.
22. C. Riedtmann and A. Schofield, "On a simplicial complex associated with tilting modules", *Comment. Math. Helv.* **66** (1991) 70–78.
23. H. Thomas, "The Tamari lattice as it arises in quiver representations", in *this volume*.
24. T. Urabe, *Dynkin graphs and quadrilateral singularities*, Lecture Notes in Mathematics, vol. 1548, Springer-Verlag, Berlin, 1993.

The Tamari Lattice as it Arises
in Quiver Representations

Hugh Thomas

Abstract In this chapter, we explain how the Tamari lattice arises in the context of the representation theory of quivers, as the poset whose elements are the torsion classes of a directed path quiver, with the order relation given by inclusion.

1 Introduction

In this chapter, we will explain how the Tamari lattice T_n arises in the context of the representation theory of quivers. When studying the representation theory of quivers, one fixes a quiver (quiver being a synonym for "directed graph") and then considers the category of *representations* of that quiver. (Terms which are not defined in this introduction will be defined shortly.) Subcategories of this category with certain natural properties are called *torsion classes*. We show that the Tamari lattice T_n arises as the set of torsion classes, ordered by inclusion, for the quiver consisting of a directed path of length n. It therefore follows that, for any directed graph, we obtain a generalization of the Tamari lattice. At the end of this chapter, we will comment briefly on the lattices that arise in this way, which include the Cambrian lattices discussed in Reading's contribution to this volume [12].

The treatment of quiver representations which we have undertaken is very elementary. In particular, we avoid all use of homological algebra. A reader familiar with quiver representations will have no trouble finding quicker proofs of the results we present here. Introductions to quiver representations from a more algebraically sophisticated point of view may be found in [3, 2].

Hugh Thomas
Department of Mathematics and Statistics, University of New Brunswick, Fredericton, NB, E3B 5A3, Canada. e-mail: *hthomas@unb.ca*

2 Quiver representations

Let Q be a quiver (i.e., a directed graph). Fix a ground field K. A representation of Q is an assignment of a finite-dimensional vector space V_i over K to each vertex i of Q, and a linear map $V_\alpha : V_i \to V_j$ to each arrow $\alpha : i \to j$ of Q.

For a pair of representations V, W of Q, we define a morphism from V to W to consist of a collection of maps $f_i : V_i \to W_i$ for all vertices i, such that for any $\alpha : i \to j$, we have that $W_\alpha \circ f_i = f_j \circ V_\alpha$. We write $\mathrm{Hom}(V, W)$ for the set of morphisms from V to W. It has a natural K-vector space structure. As usual, an isomorphism is a morphism which is invertible. An injection is a morphism all of whose linear maps are injections; surjections are defined similarly.

These definitions make the representations of a quiver into a *category*, which we denote $\mathrm{rep}\,Q$. (The careful reader is encouraged to confirm this.)

Given two representations of Q, their direct sum $V \oplus W$ is defined in the obvious way: setting $(V \oplus W)_i = V_i \oplus W_i$, and $(V \oplus W)_\alpha = V_\alpha \oplus W_\alpha$.

A representation is called *indecomposable* if it is not isomorphic to the direct sum of two non-zero representations.

3 Subrepresentations, quotient representations, and extensions

If Y is a representation of Q, a *subrepresentation* of Y is a representation X such that for each i, X_i is a subspace of Y_i, and for $\alpha : i \to j$, we have that X_α is induced from the inclusions of X_i and X_j into Y_i and Y_j respectively. The inclusions of X_i into Y_i define an injective morphism from X to Y.

If Y is a representation of Q, and $x \in Y_i$, the subrepresentation of Y generated by x is the representation X such that X_j is spanned by all images of x under linear maps corresponding to (directed) walks from i to j in Q.

If Y is a representation of Q, and X is a subrepresentation of Y, then it is also possible to form the *quotient representation* Y/X. By definition $(Y/X)_i = Y_i/X_i$, and the maps of Y/X are induced from the maps of Y. The quotient maps from Y_i to $(Y/X)_i$ define a surjective morphism from Y to Y/X.

Suppose X, Y, Z are representations of Q. Then Y is said to be an *extension* of Z by X if there is a subrepresentation of Y which is isomorphic to X, such that the corresponding quotient representation is isomorphic to Z. The extension is called *trivial* if there is a morphism s from Y to X which is the identity on X. Such a morphism is said to split the inclusion of X into Y.

Lemma 3.1. *If Y is a trivial extension of Z by X, then Y is isomorphic to $X \oplus Z$.*

Proof. Let s be the map which splits the inclusion of X into Y. Write g for the quotient map from Y to Z. Then $s \oplus g$ is a morphism from Y to $X \oplus Z$, which is an isomorphism over each vertex. It follows that it is an isomorphism of representations. $\qquad\square$

The following discussion is not necessary for our present considerations, but may be of interest, in that it connects our discussion to notions of homological algebra. It is possible to define a notion of equivalence on extensions, as follows: two extensions Y, Y' of X by Z are said to be equivalent if there is an isomorphism from Y to Y' which induces the identity maps on X and Z. We write $\mathrm{Ext}(Z, X)$ for the set of extensions of Z by X up to equivalence. This turns out to have a natural K-vector space structure. $\mathrm{Ext}(Z, X)$ can then be identified as $\mathrm{Ext}^1(Z, X)$ in the usual sense of homological algebra. See [2, Appendix A.5].

4 Pullbacks of extensions

Lemma 4.1. *Let Y be an extension of Z by X. Suppose we have a surjective map $h : Z' \to Z$. Then there is a representation Y' which is an extension of Z' by X, and such that Y' admits a surjection to Y.*

The extension which we will exhibit in order to prove this lemma is called the *pullback* of the extension Y along the surjection h.

Proof. Let i be a vertex of Q. We are given surjective maps $g : Y_i \to Z_i$ and $h : Z'_i \to Z_i$. Define Y'_i to be the pullback of these two maps, that is to say,

$$Y'_i = \{(y, z') \mid y \in Y, z' \in Z', \text{ and } g(y) = h(z')\}.$$

For $\alpha : i \to j$ an arrow of Q, define $Y'_\alpha = Y_\alpha \times Z'_\alpha$. One verifies that this defines a map from Y'_i to Y'_j. It follows that Y' is a representation of Q.

Write $f_i : X_i \to Y_i$ for the given injection from X_i to Y_i. There is an injective morphism f' from X to Y', defined on X_i by sending x to $(f_i(x), 0)$. One checks that $(f_i(x), 0) \in Y'_i$, and that these maps define a morphism from X to Y'.

It is also easy to see that there is a surjective morphism from Y' to Z' defined on Y'_i by sending (y, z') to $z' \in Z'_i$. The elements of Y'_i which are sent to zero by this map are those of the form $(y, 0)$, and $(y, 0) \in Y'_i$ iff $g(y) = 0$ iff $(y, 0) \in f'_i(X_i)$. So Z' is isomorphic to Y'/X, as desired.

Finally, one defines a map from Y'_i to Y_i by sending $(y, z') \to y$. One checks that this defines a morphism from Y' to Y, which is surjective since h is. $\qquad\square$

5 Indecomposable representations of the quiver A_n

Consider the quiver which consists of an oriented path: the vertices are numbered 1 to n, and for $1 \leq i \leq n-1$, there is a unique arrow α_i whose tail is at vertex i, and whose head is at vertex $i+1$. We will refer to this quiver as A_n.

For $1 \leq i \leq j \leq n$, define a representation E^{ij} by putting one-dimensional vector spaces at all vertices p with $i \leq p \leq j$, with identity maps between successive one-dimensional vector spaces, and zero vector spaces and maps elsewhere.

Proposition 5.1. *The representations E^{ij} are indecomposable, and any indecomposable representation of A_n is isomorphic to some E^{ij}.*

Proof. Suppose $E^{ij} \cong X \oplus Y$. Since for E^{ij} the vector spaces at each vertex are at most one-dimensional, for a given vertex p, at most one of X_p and Y_p is non-zero. If neither X nor Y is zero, there must be some $i \leq p < j$ such that either X_p is zero and Y_{p+1} is zero, or vice versa. In either case, $(X \oplus Y)_{\alpha_p} = 0$. However, $(E^{ij})_{\alpha_p} \neq 0$, and it follows that E^{ij} is not isomorphic to $X \oplus Y$.

Let V be an indecomposable representation of A_n. Write p_j for V_{α_j}. Choose i minimal such that $V_i \neq 0$, and choose a non-zero $t \in V_i$. Let T be the subrepresentation of V generated by t, which admits a natural injection into V. We have natural inclusions of the vector space at vertex k for T into the vector space at vertex k for V, and we denote this inclusion by $f_k : T_k \to V_k$.

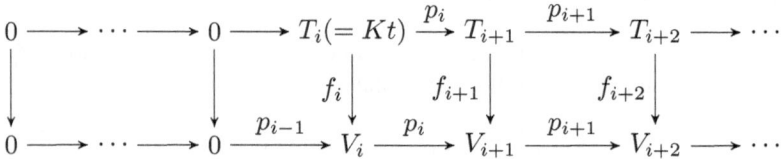

Let j be maximal so that $p_{j-1} \ldots p_i(t) \neq 0$. Observe that T_k is one-dimensional for $i \leq k \leq j$, and zero otherwise. Define a map s_j which splits the inclusion f_j, that is to say, a map such that $s_j \circ f_j$ is the identity. Now inductively define $s_{j-1}, s_{j-2}, \ldots, s_i$ so that, when constructing s_k, we have that s_k splits f_k, and $p_k \circ s_k = s_{k+1} \circ p_k$. For k not between i and j, define $s_k = 0$.

We claim the maps s_k define a morphism from V to T. The only conditions which we did not explicitly build into the construction of s_k are the commutativity conditions $p_{i-1} \circ s_{i-1} = s_i \circ p_{i-1}$ (if $i > 1$) and $p_j \circ s_j = s_{j+1} \circ p_j$ (if $j < n$). The first is satisfied by our assumption that i is minimal such that $V_i \neq 0$, and the second is satisfied by our assumption that $p_j \ldots p_i(t) = 0$, which implies that $p_j|_{T_j} = 0$.

By Lemma 3.1, it follows that V is isomorphic to the direct sum of T and V/T. Since V is indecomposable by assumption, and T is non-zero, V/T must be zero, so V is isomorphic to T, which is isomorphic to E^{ij}, proving the proposition. $\square$

6 Morphisms and extensions between indecomposable representations of A_n

Proposition 6.1. *The space of morphisms from E^{ij} to E^{kl} is either 0-dimensional or 1-dimensional. It is one-dimensional iff $k \leq i \leq l \leq j$.*

Proof. E^{ij} is generated by $(E^{ij})_i$, so a morphism $f : E^{ij} \to E^{kl}$ is determined by its restriction to $(E^{ij})_i$. The space of maps from $(E^{ij})_i$ to $(E^{kl})_i$ is one-dimensional if $k \leq i \leq l$, and zero otherwise. If $l > j$, then the commutativity condition corresponding to α_j cannot be satisfied for a non-zero morphism; if on the other hand $l \leq j$, then we see that non-zero morphisms do exist. $\square$

Proposition 6.2. *The only circumstance in which there is a non-trivial extension of E^{ij} by E^{kl} is if $i+1 \leq k \leq j+1 \leq l$. In this case, any non-trivial extension is isomorphic to $E^{il} \oplus E^{kj}$. (If $k = j+1$, we interpret E^{kj} as zero.)*

Proof. Let Y be an extension of E^{ij} by E^{kl}. Let t be an element of Y_i which maps to a non-zero element of $(E^{ij})_i$. Let T be the subrepresentation of Y which is generated by t. The representation T is definitely non-zero at the vertices p with $i \leq p \leq j$, since the image of t in $(E^{ij})_p$ is non-zero for such p. If $T_{j+1} = 0$, then T is isomorphic to E^{ij}, and the projection from Y to E^{ij} splits the inclusion of T into Y, so Y is isomorphic to $E^{ij} \oplus E^{kl}$ by Lemma 3.1.

Therefore, in order for there to exist a non-trivial extension, we must have $k \leq j+1 \leq l$. Consider now the case that $k \leq i$, which we must also exclude. Let v be the image of t in Y_{j+1}, and suppose it is non-zero. Since E^{ij} is not supported over $j+1$, we must have that v lies in the image of E^{kl}. By our assumption that $k \leq i$, there is an element x of $(E^{kl})_i$ such that its image in $(E^{kl})_{j+1}$ coincides with v. Now set $t' = t - x$, and repeat the above analysis with t'. By construction the image of t' in Y_{j+1} is zero, so E^{ij} is a direct summand of Y and the extension is trivial.

Finally, suppose that $i+1 \leq k \leq j+1 \leq l$, and that we have some t in Y_i whose image in Y_{j+1} is non-zero. It follows necessarily that the subrepresentation T of Y generated by t must be isomorphic to E^{il}. As in the proof of Proposition 5.1, we see that T is a direct summand of Y, so Y is isomorphic to $E^{il} \oplus Z$ for some Z, and it is clear that we must have $Z \cong E^{kj}$ (or $Z = 0$ if $k = j+1$). $\square$

The only nontrivial extension of indecomposable representations of A_2 is an extension of E^{11} by E^{22}, isomorphic to E^{12}. There are several examples of nontrivial extensions among representations of A_3, such as, for example, the extension of E^{12} by E^{23} which is isomorphic to $E^{13} \oplus E^{22}$. Note that this latter example is obviously non-trivial even though the extension is a direct sum, since the summands in the direct sum are not the same as the two indecomposables from which the extension was built.

7 Subcategories of $\operatorname{rep} Q$

By definition, a subcategory of a category is a category whose objects and morphisms belong to those of the original category, and such that the identity maps in the subcategory and category coincide. This notion is not strong enough for our purposes. A full additive subcategory $\mathscr{B}$ of $\operatorname{rep} Q$ is a subcategory satisfying the following conditions:

- For X, Y objects of $\mathscr{B}$, we have that $\operatorname{Hom}_{\mathscr{B}}(X, Y) = \operatorname{Hom}(X, Y)$.
- There is some set of indecomposable objects of $\operatorname{rep} Q$ such that the objects of $\mathscr{B}$ consist of all finite direct sums of indecomposable objects from this set.

From now on, when we speak of subcategories, we always mean full additive subcategories.

In this chapter, we are particularly interested in *torsion classes* in $\operatorname{rep} Q$. A torsion class in an abelian category $\mathscr{A}$ is a full additive subcategory closed under quotients and extensions. That is to say, if $Y \in \mathscr{A}$, and there is a surjection from Y to Z, then $Z \in \mathscr{A}$, and if X, Z are in $\mathscr{A}$, and Y is an extension of Z by X, then $Y \in \mathscr{A}$. Torsion classes play an important role in tilting theory, which it is beyond the scope of this chapter to review. See [3, 2] for more information on this subject.

8 Quotient-closed subcategories

As a prelude to classifying the torsion classes of $\operatorname{rep} A_n$, we consider the subcategories of $\operatorname{rep} A_n$ which are closed under quotients. This class of subcategories of $\operatorname{rep} Q$ was not studied classically but has received some recent attention (see [11] or, considering the equivalent dual case, [14]).

Let $\mathbf{M}$ be the set of n-tuples $(a_1, \ldots, a_n)$ with $0 \le a_i \le n + 1 - i$. For $\mathbf{a}$ in $\mathbf{M}$, define

$$\mathscr{F}_\mathbf{a} = \{(i, j) \mid i \le j < i + a_i\}$$

and let $\mathscr{C}_\mathbf{a}$ be the full subcategory consisting of direct sums of indecomposables E^{ij} with $(i, j) \in \mathscr{F}_\mathbf{a}$.

Proposition 8.1. *The quotient-closed subcategories of* $\operatorname{rep} A_n$ *are exactly those categories of the form* $\mathscr{C}_\mathbf{a}$ *for* $\mathbf{a} \in \mathbf{M}$.

In order to prove this proposition, we need a lemma:

Lemma 8.2. *Suppose X is a representation of A_n which admits a surjection to E^{kl}. Then any expression for X as a direct sum of indecomposables includes an indecomposable E^{kj} with $j \ge l$.*

(Note that the statement of the lemma avoids assuming any uniqueness of the decomposition of X as a direct sum of indecomposable representations. In fact, for $X \in \operatorname{rep} Q$, and Q any quiver, the collection of summands appearing in an expression for X as a direct sum of indecomposable representations is unique up to permutation and isomorphism. This is called the Krull-Schmidt property, and it is established, for example, in [2, Section 1.4]. However, in the interests of self-containedness, we have preferred to avoid the use of this.)

Proof. Consider an expression of X as a direct sum of indecomposables. By Proposition 6.1, E^{kl} does not admit any morphisms from E^{ij} with $i < k$, so we may assume X contains no such summands. On the other hand, when we consider summands of X of the form E^{ij} with $i > k$, we see that the map from such summands to E^{kl} cannot be surjective at vertex k. Therefore X must have some summand of the form E^{kj} which admits a morphism to E^{kl}; using Proposition 6.1 again, we see that $l \leq j$. □

Now we prove the proposition:

Proof. Clearly there are surjections:

$$E^{ii} \leftarrow E^{i(i+1)} \leftarrow \cdots \leftarrow E^{in}$$

It follows that a quotient-closed subcategory which contains E^{ij} necessarily contains E^{ip} for all $i \leq p \leq j$, and thus that any quotient-closed subcategory is of the form $\mathscr{C}_{\mathbf{a}}$ for some $\mathbf{a} \in \mathbf{M}$.

Next, we verify that any such subcategory is quotient-closed. Suppose $X \in \mathscr{C}_{\mathbf{a}}$, and X admits a surjection to Y. Assume for the sake of contradiction that Y is not in $\mathscr{C}_{\mathbf{a}}$. So Y has some indecomposable direct summand E^{ij} which is not in $\mathscr{C}_{\mathbf{a}}$, and E^{ij}, in particular, admits a surjection from X. By Lemma 8.2, it follows that X has some direct summand of the form E^{ik} with $k \geq j$, so $(i,k) \in \mathscr{F}_{\mathbf{a}}$, so $(i,j) \in \mathscr{F}_{\mathbf{a}}$, contradicting the assumption that $Y \notin \mathscr{C}_{\mathbf{a}}$. □

9 Subcategories ordered by inclusion

We consider the obvious order on $\mathbf{M}$, the order it inherits as a Cartesian product, and we write that $\mathbf{a} = (a_1, \ldots, a_n) \leq \mathbf{b} = (b_1, \ldots, b_n)$ iff $a_i \leq b_i$ for all i.

Proposition 9.1. *For* $\mathbf{a}, \mathbf{b} \in \mathbf{M}$, $\mathscr{C}_{\mathbf{a}} \subseteq \mathscr{C}_{\mathbf{b}}$ *iff* $\mathbf{a} \leq \mathbf{b}$.

Proof. Clearly if $\mathbf{a} \leq \mathbf{b}$, then $\mathscr{F}_{\mathbf{a}} \subseteq \mathscr{F}_{\mathbf{b}}$, and therefore $\mathscr{C}_{\mathbf{a}} \subseteq \mathscr{C}_{\mathbf{b}}$. Conversely, if $\mathscr{C}_{\mathbf{a}} \subseteq \mathscr{C}_{\mathbf{b}}$, then in particular the indecomposable objects of $\mathscr{C}_{\mathbf{a}}$ are contained among those of $\mathscr{C}_{\mathbf{b}}$. Since the objects of $\mathscr{C}_{\mathbf{a}}$ are direct sums of objects E^{ij} with $(i,j) \in \mathscr{F}_{\mathbf{a}}$, the indecomposable objects of $\mathscr{C}_{\mathbf{a}}$ are exactly those E^{ij} with $(i,j) \in \mathscr{F}_{\mathbf{a}}$. It follows that $\mathscr{F}_{\mathbf{a}} \subseteq \mathscr{F}_{\mathbf{b}}$, and thus $\mathbf{a} \leq \mathbf{b}$. □

10 Torsion classes in $\operatorname{rep} A_n$

We now define a subset of $\mathbf{M}$. We say that an n-tuple $(a_1,\ldots,a_n) \in \mathbf{M}$ is a *bracket vector* if, for all $1 \leq i \leq n$ and $j \leq a_i$, we have that $j + a_{i+j} \leq a_i$. The well-formed bracket strings of length $2n+2$ correspond bijectively to bracket vectors of length n: for each open-parenthesis, find the corresponding close-parenthesis, and then record the number of open-parentheses strictly between them. Reading these numbers from left to right, and skipping the last one (which is necessarily zero), we obtain a bracket vector. Thus, for example, $()(())$ is encoded by the bracket vector 01, while $(()())$ is encoded by the bracket vector 20. The notion of bracket vector goes back to Huang and Tamari [8]. They show in addition that the poset structure induced on bracket vectors from their inclusion into $\mathbf{M}$ is isomorphic to the Tamari lattice.

The main result of this section is the following theorem:

Theorem 10.1. *The torsion classes of* $\operatorname{rep} A_n$ *are exactly the subcategories* $\mathscr{C}_\mathbf{a}$ *for* $\mathbf{a}$ *a bracket vector.*

Before we begin the proof of this theorem, we will first state and prove the corollary which is the main result of this chapter.

Corollary 10.2. *The torsion classes in* $\operatorname{rep} A_n$, *ordered by inclusion, form a poset isomorphic to the Tamari lattice.*

Proof. We have already observed that $\mathscr{C}_\mathbf{a} \subseteq \mathscr{C}_\mathbf{b}$ iff $\mathbf{a} \leq \mathbf{b}$. It follows that the torsion classes for $\operatorname{rep} A_n$, ordered by inclusion, form a poset isomorphic to the poset structure induced on bracket vectors from their inclusion into $\mathbf{M}$, which, as we have already remarked, is shown in [8] to be isomorphic to the Tamari lattice. $\qquad\square$

Next, we need to establish some terminology and prove a lemma.

For $\mathbf{a}$ a bracket vector, let

$$\mathscr{G}_\mathbf{a} = \{(i, i+a_i-1) \mid 1 \leq i \leq n, a_i \geq 1\}.$$

Let $\mathscr{D}_\mathbf{a}$ be the full subcategory consisting of direct sums of E^{ij} for $(i,j) \in \mathscr{G}_\mathbf{a}$. Observe that $\mathscr{D}_\mathbf{a}$ is a subcategory of $\mathscr{C}_\mathbf{a}$, and any object in $\mathscr{C}_\mathbf{a}$ is a quotient of some object in $\mathscr{D}_\mathbf{a}$.

Lemma 10.3. *Let* $\mathbf{a}$ *be a bracket vector. If* $X \in \mathscr{C}_\mathbf{a}$, *and* $Z \in \mathscr{D}_\mathbf{a}$, *then any extension of* Z *by* X *is trivial.*

Proof. Write $Z = Z^1 \oplus \cdots \oplus Z^m$. Observe that any extension of Z by X can be realized by first forming an extension of Z^1 by X, call it Y^1, then forming an extension of Z^2 by Y^1, call it Y^2, and so on. If the extension at each step is trivial, then the total extension is trivial, so it suffices to consider the case that Z is indecomposable. Suppose therefore that $Z \cong E^{i(i+a_i-1)}$, and let Y be an extension of Z by X, for some $X \in \mathscr{C}_\mathbf{a}$.

Let t be an element of Y_i which maps to a nonzero generator of Z. Let T be the subrepresentation of Y generated by t. Let v be the image of t in Y_{i+a_i}.

If v is non-zero then, since Z is not supported over $i + a_i$, we must have that $v \in X_{i+a_i}$. Let E^{kl} be a direct summand of X in which v is non-zero. So we know that $k \leq i + a_i \leq l$. By the assumption that $(k, l) \in \mathscr{F}_\mathbf{a}$, it follows that $k \leq i$. Therefore, it follows that, as in the proof of Proposition 6.2, we can find an element x of X_i whose image in X_{i+a_i} equals v. Now let $t' = t - x$. The image of t' in Y_{i+a_i} is zero, so T is isomorphic to Z. Therefore, the projection from Y to Z splits the inclusion of T into Y, and the extension of Z by X is trivial.

$\square$

Proof of Theorem 10.1. First, we show that if $\mathbf{a} = (a_1, \ldots, a_n) \in \mathbf{M}$ is not a bracket vector, then $\mathscr{C}_\mathbf{a}$ is not a torsion class. So suppose we have some i, j such that $1 \leq i \leq n$, $j \leq a_i$, and $j + a_{i+j} > a_i$. We know that $\mathscr{C}_\mathbf{a}$ contains $E^{(i,i+a_i-1)}$ and $E^{(i+j,i+j+a_{i+j}-1)}$, and from our assumptions, $i + 1 \leq i + j \leq (i + a_i - 1) + 1 \leq i + j + a_{i+j} - 1$. By Proposition 6.2, it follows that $E^{(i,i+j+a_{i+j}-1)} \oplus E^{(i+j,i+a_i-1)}$ is an extension of $E^{(i+j,i+j+a_{i+j}-1)}$ by $E^{(i,i+a_i-1)}$, and since $i + j + a_{i+j} > i + a_i$, we know that $E^{(i,i+j+a_{i+j}-1)}$ is not contained in $\mathscr{C}_\mathbf{a}$. Thus $\mathscr{C}_\mathbf{a}$ is not closed under extensions, so it is not a torsion class.

Now, we show that if $\mathbf{a}$ is a bracket vector, then $\mathscr{C}_\mathbf{a}$ is a torsion class. We have already shown that $\mathscr{C}_\mathbf{a}$ is quotient-closed, so all that remains is to show that it is closed under extensions.

Let X and Z be representations in $\mathscr{C}_\mathbf{a}$. If we could assume that X and Z were indecomposable, our lives would be much easier – an argument very similar to the converse direction would suffice. However, there is no reason that we can assume that.

Choose an object $Z' \in \mathscr{D}_n$ such that Z' has a surjection onto Z. Let Y be the extension of Z by X, and let Y' be the pullback along $Z' \to Z$ of this extension. By Lemma 10.3, the extension of Z' by X is trivial, so $Y' \in \mathscr{C}_\mathbf{a}$. By Lemma 4.1, Y' admits a surjective map to Y. Thus Y is a quotient of an object of $\mathscr{C}_\mathbf{a}$, and thus lies in $\mathscr{C}_\mathbf{a}$. Therefore, $\mathscr{C}_\mathbf{a}$ is closed under extensions. $\square$

11 Related posets

As was already mentioned, for arbitrary Q, we obtain a poset of torsion classes ordered by inclusion. In fact, it is easy to check from the definition that the intersection of an arbitrary set of torsion classes is again a torsion class. Thus, this poset is closed under arbitrary meets, and it has a maximum element, so it is a lattice.

For Q a connected quiver, the number of indecomposable representations of Q is finite if and only if Q is an orientation of a simply-laced Dynkin diagram. (This is part of the celebrated theorem of Gabriel, see, for example, [2, Theorem VII.5.10].) For such Q (and only such Q), the lattice of torsion classes is a (finite) Cambrian lattice, for a Coxeter element chosen based on the orientation of Q. See Reading's chapter in this volume [12] for more on Cambrian lattices, and [9, 1] for this result. For further work related to the lattice of torsion classes, see also [10].

The poset of torsion classes was not classically studied in representation theory. However, a closely related poset does appear. For the remainder of this section, suppose that Q is a quiver with no oriented cycles.

A representation T of Q is called a *tilting object* if the only extension of T with itself is the trivial extension, and T has n pairwise non-isomorphic direct summands (n being the number of vertices of Q). For $T \in \text{rep} \, Q$, write $\text{Gen} \, T$ for the subcategory of $\text{rep} \, Q$ consisting of all quotients of direct sums of copies of T. A poset was defined by Riedtmann and Schofield [13] on the tilting objects of $\text{rep} \, Q$, by $T \geq V$ iff $\text{Gen} \, T \supseteq \text{Gen} \, V$. This poset was studied further by Happel and Unger [5, 6, 7].

This poset structure is related to the one discussed in this paper, because if T is a tilting object, then $\text{Gen} \, T$ is a torsion class. If we suppose further that Q is an orientation of a simply-laced Dynkin diagram, then the torsion classes arising in this way can be described easily: they are just the torsion classes which include all the injective representations of Q, see [2, Corollary VI.6.6]. Thus, the torsion classes arising in this way form an interval in the poset of all torsion classes, whose minimal element is the torsion class consisting only of injective representations, and whose maximal element is the torsion class consisting of all representations.

The Riedtmann-Schofield order on tilting objects was first analyzed for $\text{rep} \, A_n$ in [4]. The torsion class $\mathscr{C}_\mathbf{a}$ contains all the injective representations iff $a_1 = n$, since the injective indecomposable representations are those of the form E^{1j} for $1 \leq j \leq n$. There is a bijection from bracket vectors with $a_1 = n$ to bracket vectors of length $n - 1$, by removing a_1. (A bracketing corresponding to a bracket vector with $a_1 = n$ has its first open parenthesis closed by the final close parenthesis of the bracketing.) It therefore follows that the Riedtmann-Schofield order on tilting objects for $\text{rep} \, A_n$ is isomorphic to the Tamari lattice T_{n-1}.

Acknowledgement The author was partially supported by an NSERC Discovery Grant. He thanks the editors of the Tamari Festschrift for the invitation to contribute to this volume, and for their helpful editorial suggestions. He also thanks Sefi Ladkani and Vic Reiner for useful comments, and Charles Paquette for a careful reading of the manuscript which led to numerous corrections and improvements.

References

1. C. Amiot, O. Iyama, I. Reiten, and G. Todorov, "Preprojective algebras and c-sortable words", *arxiv.org/abs/1002.4131*.
2. I. Assem, D. Simson, and A. Skowronski, *Elements of the representation theory of associative algebras. Vol. 1. Techniques of representation theory*, London Mathematical Society Student Texts, vol. 65, Cambridge University Press, Cambridge, 2006.
3. M. Auslander, I. Reiten, and S. Smalø, *Representation theory of Artin algebras*, Cambridge Studies in Advanced Mathematics, vol. 36, Cambridge University Press, Cambridge, 1995.
4. A.B. Buan and H. Krause, "Tilting and cotilting for quivers of type $\widetilde{A}_n$", *J. Pure App. Algebra* **190** (2004) 1–21.
5. D. Happel and L. Unger, "On a partial order of tilting modules", *Algebr. Represent. Theory* **8** (2005) 147–156.
6. _____ , "On the quiver of tilting modules", *J. Algebra* **284** (2005) 857–868.

7. _____ , "Reconstruction of path algebras from their posets of tilting modules", *Trans. Amer. Math. Soc.* **361** (2009) 3633–3660.
8. S. Huang and D. Tamari, "Problems of associativity: A simple proof for the lattice property of systems ordered by a semi-associative law", *J. Combinatorial Theory Ser. A* **13** (1972) 7–13.
9. C. Ingalls and H. Thomas, "Noncrossing partitions and representations of quivers", *Compos. Math.* **145** (2009) 1533–1562.
10. S. Ladkani, "Universal derived equivalences of posets of cluster tilting objects", *arxiv.org/abs/0710.2860.*
11. S. Oppermann, I. Reiten, and H. Thomas, "Quotient closed subcategories of quiver representations", in preparation.
12. N. Reading, "From the Tamari lattice to Cambrian lattices and beyond", in *this volume*.
13. C. Riedtmann and A. Schofield, "On a simplicial complex associated with tilting modules", *Comment. Math. Helv.* **66** (1991) 70–78.
14. C.M. Ringel, "Minimal infinite submodule-closed subcategories", *arxiv.org/abs/1009.0864.*

From the Tamari Lattice to
Cambrian Lattices and Beyond

Nathan Reading

Abstract We trace the path from the Tamari lattice, via lattice congruences of the weak order, to the definition of Cambrian lattices in the context of finite Coxeter groups, and onward to the construction of Cambrian fans. We then present sortable elements, the key combinatorial tool for studying Cambrian lattices and fans. The chapter concludes with a brief description of the applications of Cambrian lattices and sortable elements to Coxeter-Catalan combinatorics and to cluster algebras.

1 A map from permutations to triangulations

The road from the Tamari lattice to Cambrian lattices starts with a simple map from the set S_{n+1} of permutations of $\{1, \ldots, n+1\}$ to the set of triangulations of a convex polygon with $n+3$ vertices. This map connects the Tamari lattice to the weak order on permutations, and opens the door to understanding the Tamari lattice in a broader lattice-theoretic context.

One of the many realizations of the Tamari lattice is as a partial order on triangulations of a convex polygon. Specifically, take Q to be a convex $(n+3)$-gon in the plane and identify the vertices of Q with the numbers $0, 1, \ldots, n+1, n+2$. We require that the vertices 0 and $n+2$ be on a horizontal line, with 0 to the left and with all other vertices below that line. Furthermore, we require that the vertices 1 through $n+1$ be placed so that, for all i from 0 to $n+1$, the vertex i is strictly further left than the vertex $i+1$.

A *triangulation* of Q is a tiling of Q by triangles whose vertices are contained in the vertex set of Q. The triangulation is specified by the collection of n diagonals of Q appearing as edges of the triangles. A *diagonal flip* on a triangulation of Q is the operation of removing a diagonal of the triangulation to create a quadrilateral from

Nathan Reading
Department of Mathematics, North Carolina State University, Raleigh, NC, USA
e-mail: *nathan_reading@ncsu.edu*

two triangles, and then inserting the other diagonal of the quadrilateral to create a
new triangulation. The Tamari lattice is a partial order on triangulations of Q whose
cover relations are given by diagonal flips. The two triangulations in the cover differ
by exactly one diagonal of Q, and the higher triangulation in the cover relation is the
one in which this diagonal has larger slope. The Tamari lattice, for $n = 3$, is shown
in Figure 1.a.

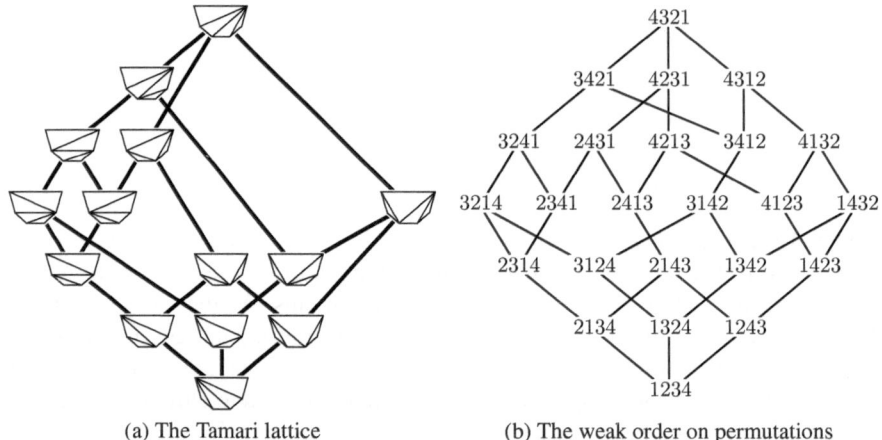

(a) The Tamari lattice   (b) The weak order on permutations

Fig. 1 The Tamari lattice and the weak order on permutations

This definition of the Tamari lattice highlights its connection to the associahedron.
Since the vertices of the associahedron can be labeled by triangulations of a fixed
convex polygon such that edges are given by diagonal flips, the Hasse diagram of the
Tamari lattice is isomorphic to the 1-skeleton of the associahedron.

To define the weak order, we first write permutations in one-line notation, meaning
that we represent a permutation x of $\{1,\ldots,n+1\}$ by the sequence $x_1 x_2 \cdots x_{n+1}$,
where x_i means $x(i)$. There is a cover relation $x \lessdot y$ in the weak order whenever the
one-line notations of x and y differ only by swapping a pair of adjacent entries. The
permutation x is the one in which the two entries appear in numerical order, and y is
the permutation in which the two entries appear out of order. For example, the weak
order on S_4 is shown in Figure 1.b.

We now define a map η from S_{n+1} to the set of triangulations of Q. Start with a
path along the bottom edges of Q, as shown in the first frame of Figure 2. Given a
permutation $x \in S_{n+1}$, read from left to right in the one-line notation for x. For each
entry, create a new path by deleting the corresponding vertex from the old path. The
triangulation $\eta(x)$ is defined by the union of the sequence of paths, as illustrated in
Figure 2 for the permutation with one-line notation 3246175. Figure 3.a shows the
result of applying η to every permutation in S_4. The shaded edges indicate covering
pairs in the weak order which map to the same triangulation.

This map and similar maps have appeared in many papers, including [6, 7, 29,
30, 36, 50]. The map can be seen in a broader context in the chapter by Rambau and

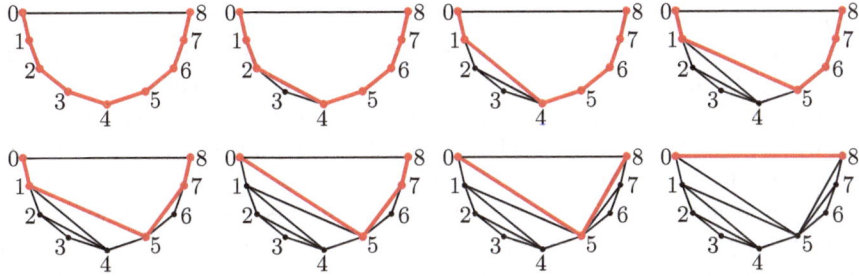

Fig. 2 The triangulation $\eta(3246175)$

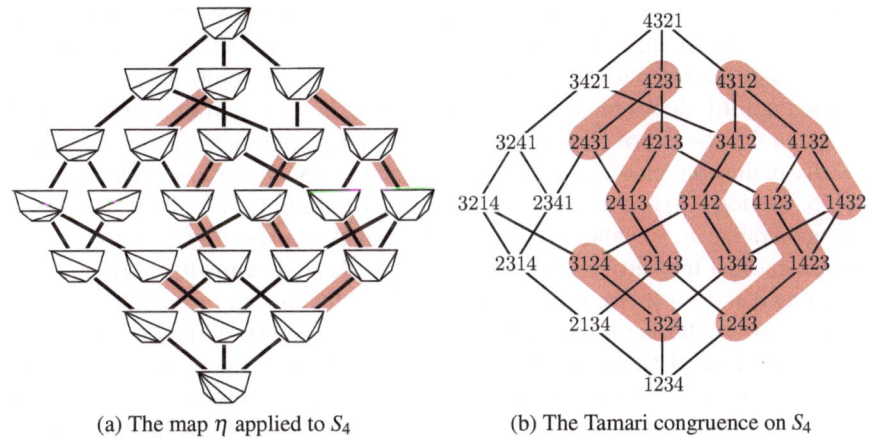

(a) The map η applied to S_4 (b) The Tamari congruence on S_4

Fig. 3 The map η applied to every permutation in S_4 and the Tamari congruence on S_4

Reiner [31] in this volume, specifically by giving some thought to [31, Theorem 9] and the accompanying figure.

Björner and Wachs [7, Section 9] studied a map τ from permutations to binary trees that is, up to a standard bijection from triangulations to binary trees, identical to η. We describe their results in terms of the map η. First, the fiber $\eta^{-1}(\Delta)$ of each triangulation Δ is a non-empty interval in the weak order on S_{n+1}. A permutation is the minimal element in its η-fiber if and only if it avoids the pattern 312. That is, a permutation x is minimal in its fiber if and only if there is no sequence of three (not-necessarily adjacent) entries in the one-line notation for x such that the largest of the three is first, followed by the smallest of the three, and finally the median-valued. Similarly, a permutation is the maximal element in its η-fiber if and only if it avoids the pattern 132. For example, comparing Figures 1.b and 3.a, we see that the permutation 4213 is not the minimal element of its η-fiber, and indeed, the sequence 413 (or the sequence 423) is an instance of the pattern 312 in the permutation 4213. However, 4213 is the maximal element in its η-fiber because it avoids the pattern 132.

Björner and Wachs also showed that the weak order and the Tamari lattice are closely related. Specifically, the restriction of the weak order to 312-avoiding permutations is a sublattice of the weak order, and the restriction of η to this sublattice is an isomorphism from the sublattice to the Tamari lattice. This is readily seen in the case of S_4 by inspection of Figures 1 and 3.a. The sublattice of the weak order consisting of 312-avoiding permutations (and thus the Tamari lattice) is also a quotient of the weak order in an order-theoretic sense. Indeed, the results of [7] go most of the way to establishing something stronger: As we will see in Section 2, the Tamari lattice is a lattice quotient (i.e., a lattice-homomorphic image) of the weak order, because the map η is a lattice homomorphism. This is the key insight that leads to the notion of Cambrian lattices.

Before we shift the discussion to lattice theory, we give a generalization, in a more combinatorial direction, of the map η. We will see in Section 3 that this generalization is also an essential step towards Cambrian lattices. The generalization, which was exploited in [36], draws on the description in [45, Section 4.3] of a similar family of maps in the context of signed permutations. These families of maps arise quite naturally in the context of (equivariant) iterated fiber polytopes, as explained in [6] and [45, Section 4.3] and as summarized in [36, Sections 4, 6].

To generalize η, we alter the construction of the polygon Q by removing the requirement that the vertices 1 through $n+1$ be located below the horizontal line containing 0 and $n+2$. We keep the requirement that, for all i from 0 to $n+1$, the vertex i is strictly further left than the vertex $i+1$. Again we start with a path along the bottom edges of Q, and read the one-line notation of a permutation from left to right. When we read an entry whose corresponding vertex is on the bottom of Q, we *remove* that vertex from the path, as before. When we read an entry whose corresponding vertex is on the top of Q, we *insert* that vertex into the path. Figure 4.a shows this new permutations-to-triangulations map applied to all of the permutations in S_4, in the case where the vertices 1, 2, and 4 are on the bottom of Q and 3 is on the top. To avoid a profusion of notation, we use the symbol η to refer to any of the permutations-to-triangulations maps, tacitly assuming a choice of Q. We use the phrase "the Tamari case" to distinguish the original definition of Q and of η.

As another example, consider the case where all of the vertices 1 through n are *above* the line containing 0 and $n+2$. In this case, the symmetries of the problem imply that η has the same pattern-avoidance properties as described above, except that "312" is replaced by "231" throughout the description, and "132" is replaced by "213" throughout.

When some vertices are on top of Q and others are on bottom, as in the example of Figure 4.a, the behavior of the map is a mixture of the "231-behavior" and the "312-behavior," as we now explain. The locations, top or bottom, of the vertices are recorded by upper- or lower-barring the symbols from 1 to $n+1$. Thus, for example, we write $\overline{3}$ to indicate that the vertex 3 is on top of the polygon Q or we write $\underline{3}$ to indicate that 3 is on the bottom of Q. In [36, Proposition 5.7], it is shown that a permutation is a minimal element in its η-fiber if and only if it avoids the patterns $31\underline{2}$ and $\overline{2}31$. That is, the permutation may contain no 312-pattern such that the "2"

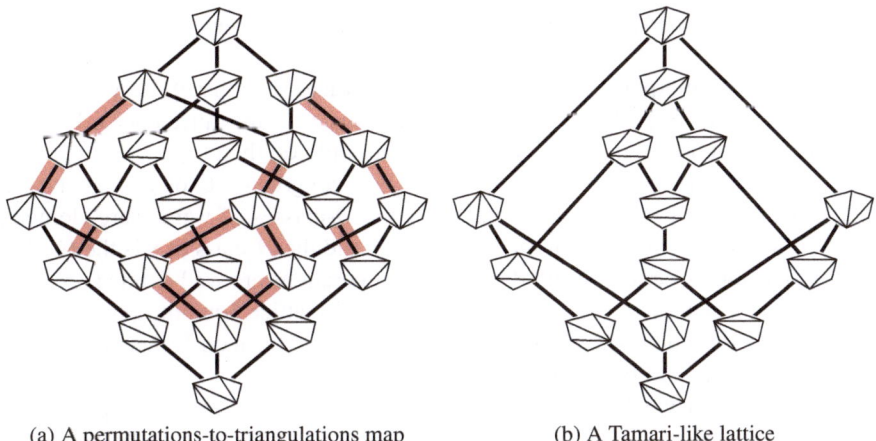

(a) A permutations-to-triangulations map   (b) A Tamari-like lattice

Fig. 4 A non-Tamari permutations-to-triangulations map applied to every permutation in S_4, and a Tamari-like (Cambrian) lattice.

in the pattern is a lower-barred number and may contain no 231-pattern such that the "2" in the pattern is an upper-barred number.

The subposet consisting of permutations avoiding $31\underline{2}$ and $\overline{2}31$ is a sublattice, isomorphic to a "Tamari-like" lattice on triangulations [36, Theorem 6.5]. As in the original Tamari lattice, cover relations are diagonal flips and moving up in the order means increasing the slope. In particular, just as in the Tamari lattice, the undirected Hasse diagram of the lattice is isomorphic to the 1-skeleton of the associahedron.

Figure 4.b shows this Tamari-like lattice in the case where the vertices 1, 2, and 4 are on the bottom of Q and 3 is on the top. These Tamari-like lattices are the first examples of Cambrian lattices beyond the Tamari lattices. The next several sections develop Cambrian lattices in general.

2 Bringing lattice congruences into the picture

A *lattice* is a partially ordered set such that any pair x, y of elements has a unique maximal lower bound (the *meet* of x and y, written $x \wedge y$) and a unique minimal upper bound (the *join* of x and y, written $x \vee y$). The Tamari lattice is, as its name suggests, a lattice, and so is the weak order.

The meet and join operation in a lattice satisfy certain properties, and indeed the definition of a lattice can be rephrased in universal-algebraic language: A lattice is a set with two operations $\wedge$ and $\vee$ satisfying a certain list of axioms. The exact form of the axioms is not important for our present purposes, but the idea that a lattice is an algebraic object is fundamental. From the algebraic viewpoint, it becomes natural to consider *lattice homomorphisms* and *lattice congruences*. In this section, we point out that the permutations-to-triangulations map η is a lattice homomorphism and characterize its associated congruence in a way that can be generalized to all finite Coxeter groups.

A lattice homomorphism is a map η between lattices such that $\eta(x \wedge y) = \eta(x) \wedge \eta(y)$ and $\eta(x \vee y) = \eta(x) \vee \eta(y)$. A lattice congruence is an equivalence relation Θ on a lattice L such that, if $x_1 \equiv y_1$ and $x_2 \equiv y_2$ modulo Θ, then $(x_1 \wedge x_2) \equiv (y_1 \wedge y_2)$ and $(x_1 \vee x_2) \equiv (y_1 \vee y_2)$ modulo Θ. Given a lattice congruence Θ on L, we construct the quotient lattice L/Θ, whose elements are the Θ-classes. If x and y are elements of the lattice L, then the meet of the Θ-class of x and the Θ-class of y is the Θ-class of $x \wedge y$. The join is defined similarly, and these operations are well defined precisely because Θ is a congruence. The nonempty fibers of a lattice homomorphism are the classes of a lattice congruence, and, given a lattice congruence on L, there is a natural lattice homomorphism from L to the quotient lattice.

Some readers may only be familiar with the notion of congruences and quotients in the context of groups or rings, and perhaps without reference to the term "congruence." In a group, the congruence classes are the cosets of a normal subgroup; in a ring, they are the additive cosets of an ideal. In a lattice, congruences are more complicated, but are still often amenable to study.

There is much to be gained, for the study of lattice congruences as for other lattice-theoretic topics, by passing back and forth between a universal-algebraic and an order-theoretic point of view. This is especially true for finite lattices. When L is a finite lattice, congruences in L have a particularly simple order-theoretic characterization. It is straightforward to verify that an equivalence relation Θ on a finite lattice L is a lattice congruence if and only if it satisfies the following three properties:

(i)   Every equivalence class is an interval.
(ii) The projection $\pi_\downarrow : L \to L$, mapping each element a of L to the minimal element in its Θ-class, is order-preserving.
(iii) The projection $\pi^\uparrow : L \to L$, mapping each element a of L to the maximal element in its Θ-class, is order-preserving.

This simple observation and its generalizations appear to be the subject of many independent rediscoveries, including in [9, 11, 25, 32].

The order-theoretic rephrasing of the definition of a lattice congruence shows, in particular, how to recognize, combinatorially, when a map of lattices might be a lattice homomorphism. For example, what we have already learned about the map η, in the Tamari case, leads us to suspect that the fibers of η constitute a lattice congruence of the weak order, and thus to suspect that η is a lattice homomorphism from the weak order to the Tamari lattice. Indeed, our suspicions are correct, as is easily proven using the tools already developed in [7]; the fibers of η, in the Tamari case, define a lattice congruence called the *Tamari congruence*. The Tamari congruence on S_4 is shown in Figure 3.b. As a special case of the Cambrian construction, we will see that η is a lattice homomorphism for the other choices of Q as well.

The permutations-to-triangulations setup does more than simply find a lattice homomorphism between two given lattices. Suppose we are given a surjective map η from a lattice L to a *set* U, and suppose that the fibers of η satisfy conditions (i)–(iii).

The fibers therefore constitute a lattice congruence Θ on L, so there is a quotient lattice L/Θ and the map η factors through a bijection from L/Θ to U. We can use this bijection to *define* a lattice structure on U isomorphic to L/Θ, and when we do so, the map η is a lattice homomorphism from L to U. In particular, the map η from the weak order on permutations to the *set* of triangulations can be thought of as *defining* the Tamari lattice. In [36, Sections 4–5], the Tamari lattice is constructed from the ground up by this method, together with the other Tamari-like lattices.

Indeed, to define the Tamari lattice, we don't even need the map η. We can define the Tamari lattice to be the quotient of the weak order modulo the Tamari congruence. In doing so, we lose the combinatorial realization of the Tamari lattice as a partial order on the set of triangulations. But we gain a point of view on the Tamari lattice that leads us to a broad generalization. To find this generalization, we need to understand what makes the Tamari congruence special among the vast number of lattice congruences of the weak order.

First, we describe some general features of congruences of a finite lattice L. Much of this description does not generalize to infinite lattices, so we keep strictly to the finite case. Suppose Θ is a congruence on L. If $x \lessdot y$ (i.e., x is covered by y) in L and $x \equiv y$ modulo L, then we say Θ *contracts* the edge $x \lessdot y$. Since congruence classes are intervals, we can describe Θ completely just by recording which edges in the Hasse diagram are contracted by Θ. Indeed, this is exactly how congruences are indicated in Figures 3 and 4.a.

As one would expect, edges cannot be contracted independently to form a congruence; rather, contracting one edge may force other edges to be contracted. As a simple example of this *edge-forcing*, consider the hexagonal poset shown at the left of Figure 5. We call the two edges incident to the minimal element the *bottom edges*,

Fig. 5 Local forcing requirements

the two edges incident to the top element the *top edges*, and the other two edges *side edges*. The following facts are easily checked using either the definition or the order-theoretic characterization of a lattice congruence: For either side edge, there is a congruence contracting only that edge. For any bottom or top edge, a congruence contracting that edge must also contract the opposite edge and both side edges. Here the opposite edge means the edge related by a half-turn of the diagram.

In a general finite lattice, edge-forcing can be much more complicated. In the weak order on S_{n+1}, however, edge-forcing can be understood in terms of local forcing requirements of the kind illustrated in the hexagon example. These intervals, and the local forcing requirements, are illustrated in Figure 5. In these pictures, shaded edges represent edges in the lattice that are contracted by a given congruence. The pictures show, given a contracted edge, which other edges must be contracted. By symmetry (turning intervals upside-down or reflecting them in a vertical line), each requirement pictured describes four requirements. In words, the local forcing requirements are that whenever a bottom or top edge of the interval is contracted, the opposite bottom or top edge is also contracted, as well as all side edges (if side edges are present). A collection of edges is *closed under local forcing* if the local forcing requirements are satisfied on every hexagonal and quadrilateral interval. As a consequence of [33, Theorem 25], we have the following characterization of lattice congruences of the weak order on permutations: Given a collection $\mathscr{E}$ of edges in the weak order on S_{n+1}, there exists a lattice congruence Θ contracting exactly the edges in $\mathscr{E}$ if and only if $\mathscr{E}$ is closed under local forcing.

Using the local forcing requirements, the reader can readily verify that the Tamari congruence on S_4, shown in Figure 3.b, is the unique finest (in the sense of refinement of set partitions) lattice congruence contracting the edges $1324 < 3124$ and $1243 < 1423$. This characterization generalizes to larger values of n: In general, the Tamari congruence is the finest congruence contracting the edges

$$1\cdots(j-2)(j-1)(j+1)j(j+2)\cdots(n+1)$$
$$< 1\cdots(j-2)(j+1)(j-1)j(j+2)\cdots(n+1)$$

for $j = 2,\ldots,n$. If we vary the polygon Q, to alter the map η, the corresponding congruence Θ retains a similar characterization [36, Theorem 6.2]. It is the finest congruence contracting the edges

$$1\cdots(j-2)(j-1)(j+1)j(j+2)\cdots(n+1)$$
$$< 1\cdots(j-2)(j+1)(j-1)j(j+2)\cdots(n+1)$$

for $j = 2,\ldots,n$ such that j is lower-barred and the edges

$$1\cdots(j-2)j(j-1)(j+1)(j+2)\cdots(n+1)$$
$$< 1\cdots(j-2)j(j+1)(j-1)(j+2)\cdots(n+1)$$

for $j = 2,\ldots,n$ such that j is upper-barred.

Although it is the above characterization that leads to the Cambrian lattices, it is worth mentioning another characterization of the Tamari congruence and its cousins, found recently by Santocanale and Wehrung [46]. The set of all congruences on a lattice is itself a finite (distributive) lattice under the refinement partial order. The Tamari-like congruences are exactly the minimal *meet-irreducible congruences* in the lattice of congruences of the weak order on permutations. The natural generalization (into the context of finite Coxeter groups) suggested by the Santocanale-Wehrung characterization leads, not to the Cambrian lattices, but to a different object that has not yet been studied.

3 Cambrian lattices

The symmetric group S_{n+1} is a classical example of a finite *Coxeter group*. Coxeter groups are given by a simple combinatorial presentation, which we review below, but there is also a much more geometric point of view: A finite group can be realized as a Coxeter group if and only if it can be realized in Euclidean space as a group of transformations generated by reflections.

Generalized associahedra were introduced, in combinatorial terms, by Fomin and Zelevinsky [13] and were realized as polytopes by those authors and Chapoton [10]. There is a generalized associahedron for each finite Coxeter group W, called the W-associahedron. More precisely, there is a generalized associahedron for each *root system*, but we gloss over this distinction for the purpose of brevity. In fact, we skip the definition of the generalized associahedron entirely. The key point, for our purposes, is that there is a W-associahedron for each finite Coxeter group W and that the W-associahedron for $W = S_{n+1}$ is the usual n-dimensional associahedron. Since we also constructed the usual associahedron from a lattice congruence of the weak order on the symmetric group, and since there is a weak order on every Coxeter group, it is natural to wonder whether we can generalize the lattice-theoretic construction. That is, we would like a construction that, starting with an arbitrary Coxeter group W, produces a lattice congruence of the weak order on W such that the Hasse diagram of the quotient is the 1-skeleton of the generalized associahedron for W.

In this section, after filling in some background material on Coxeter groups, we describe the construction of the desired congruence, which we call a *Cambrian congruence*. We then discuss some properties of the Cambrian congruences and their quotients, the *Cambrian lattices*. Our coverage of Coxeter groups is necessarily no more than a sketch. A more in-depth, but still gentle, introduction, including the definition of the generalized associahedron, is found in [12]. Hohlweg's chapter [21] in this volume is another accessible description of Coxeter groups, culminating in a different construction of generalized associahedra.

To define a Coxeter group, choose a finite generating set $S = \{s_1, \ldots, s_n\}$ and for every $i < j$, choose an integer $m(i, j) \geq 2$, or $m(i, j) = \infty$. Define W to be the group with the following presentation:

$$W = \langle S \mid s_i^2 = 1, \forall i \text{ and } (s_i s_j)^{m(i,j)} = 1, \forall i < j \rangle.$$

By convention, the "relation" $(s_i s_j)^\infty = 1$ is interpreted as the absence of a relation of the form $(s_i s_j)^m = 1$. This abstract definition captures the essence of groups of transformation generated by reflections by taking a generating set of involutions and requiring that the product of two generators resembles the composition of two reflections (a rotation).

We mentioned above that the symmetric group is a Coxeter group. Specifically, we take $S = \{s_1, \cdots s_n\}$, where s_i is the transposition $(i \ i+1)$. We easily compute the order of each $s_i s_j$ and conclude that we must set

$$m(i,j) = \begin{cases} 3 \text{ if } j = i+1, \text{ or} \\ 2 \text{ if } j > i+1. \end{cases}$$

One can check that S_{n+1} is indeed isomorphic to the abstract Coxeter group with this choice of S and $m(i,j)$.

Besides S_{n+1}, we will follow another example of a Coxeter group: The dihedral group of order 8, which we call B_2. This Coxeter group has $S = \{s_1, s_2\}$ and $m(1,2) = 4$. Thus B_2 is $\langle \{s_1, s_2\} \mid s_1^2 = s_2^2 = (s_1 s_2)^4 = 1 \rangle$, and its elements are

$$1, s_1, s_2, s_1 s_2, s_2 s_1, s_1 s_2 s_1, s_2 s_1 s_2, s_1 s_2 s_1 s_2 = s_2 s_1 s_2 s_1.$$

The *Coxeter diagram* of W is a graph with vertex set $\{1, \ldots, n\}$ and edges $i - j$ whenever $m(i,j) \geq 3$. Each edge is labeled by $m(i,j)$, except that, by convention, we omit edge labels "3." Note that an edge between i and j is absent if and only if $m(i,j) = 2$, which happens if and only if s_i and s_j commute. For example, the dihedral group of order 8 has a diagram with two vertices connected by an edge labeled 4, and the diagram for S_{n+1} is

$$1 \text{——} 2 \text{——} 3 \text{——} \quad \cdots \quad \text{——} n$$

Coxeter diagrams provide a convenient way of encoding the defining data of a Coxeter group. This encoding makes it easy to describe one of the fundamental results on Coxeter groups: The classification of finite Coxeter groups. If the Coxeter diagram of W is not connected as a graph, then it is easy to see that W is a direct product of the Coxeter groups encoded by the connected components of the diagram. If the Coxeter diagram of W is connected, then W is called irreducible. Figure 6 lists the Coxeter diagrams of finite irreducible Coxeter groups and shows their standard names. In particular, the name B_2 for our earlier example is part of this naming convention, and the Coxeter group S_{n+1} appears as A_n in the classification.

Since W is generated by S, each element w of W can be written (in many ways) as a word in the "alphabet" S. A word of minimal length, among words for w, is called a *reduced word* for w. The *length* $\ell(w)$ of w is the length of a reduced word for w. The *weak order* on a Coxeter group W sets $u \leq w$ if and only if a reduced word for u occurs as a prefix of some reduced word for w. The cover relations in the weak order are $w \lessdot ws$ for $w \in W$ and $s \in S$ with $\ell(w) < \ell(ws)$. It is easy to check that this definition of the weak order reduces, in the case where $W = S_{n+1}$, to the earlier definition of the weak order on permutations. (See Figure 1.b.) The weak order on B_2 is pictured in Figure 7.a. To see that this picture is correct, keep in mind that the element $s_1 s_2 s_1 s_2$ at the top of the picture is also equal to $s_2 s_1 s_2 s_1$. The weak order is a meet semilattice (meaning that meets exist, but not necessarily joins) in general, and a lattice when W is finite. (What we have defined is sometimes called the *right weak order*. There is also an isomorphic, but not identical, *left weak order*. See, for example, Forcey's chapter [18] in this volume.)

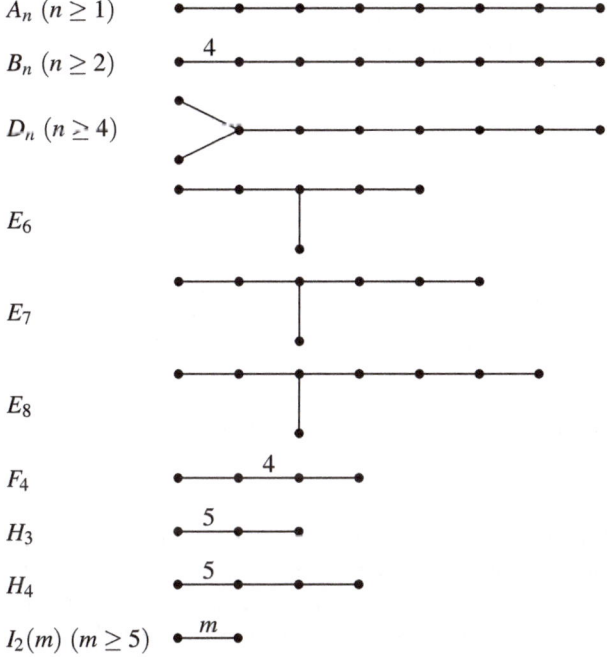

Fig. 6 Coxeter diagrams of finite irreducible Coxeter systems

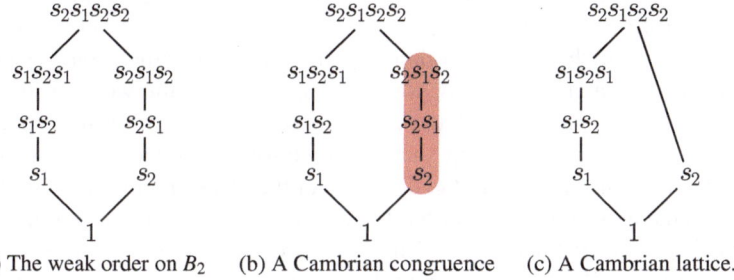

(a) The weak order on B_2 (b) A Cambrian congruence   (c) A Cambrian lattice.

Fig. 7 The weak order on B_2, a Cambrian congruence, and the corresponding Cambrian lattice

Having defined Coxeter diagrams and the weak order, we are prepared to state the general definition of Cambrian lattices. Informally: we orient the Coxeter diagram, find a copy of the Coxeter diagram within the weak order, and contract edges according to the orientation. More formally, notice that for each pair s_i, s_j of generators in S, the datum $m(i, j)$ is easily read off from the weak order by finding the join (least upper bound) of s_i and s_j. The join $s_i \vee s_j$ is of the form $s_i s_j s_i s_j \cdots$ and has length $m(i, j)$. The interval in the weak order below $s_i \vee s_j$ has the form of a polygon with $2m(i, j)$ sides. The union of all of these intervals, as i and j vary, is essentially the Coxeter diagram. To make the resemblance more perfect, we leave out those intervals where $m(i, j) = 2$, or in other words, those intervals that form 4-gons. For example, Figure 8 shows this union of intervals in $A_3 = S_4$ and in B_3.

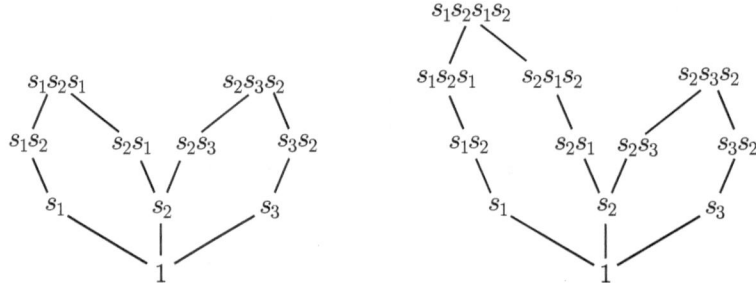

Fig. 8 The Coxeter diagram inside the weak order. On the left, the Coxeter group is $A_3 = S_4$ and on the right is B_3.

Now recall the characterization, from Section 2, of the Tamari congruence on S_{n+1} as the smallest congruence contracting a certain set of edges. These edges are $s_2 \lessdot s_2 s_1$, $s_3 \lessdot s_3 s_2$, ..., $s_n \lessdot s_n s_{n-1}$. More generally, as we let the polygon Q vary in the permutations-to-triangulations map, there is a choice of $s_1 \lessdot s_1 s_2$ or $s_2 \lessdot s_2 s_1$, $s_2 \lessdot s_2 s_3$ or $s_3 \lessdot s_3 s_2$, and so forth, such that the fibers of η constitute the smallest congruence contracting the chosen edges.

Accordingly, we define the *Cambrian congruences* on a general finite Coxeter group W as follows: For each edge i——j in the Coxeter diagram of W, declare one of i or j to be "before" the other. We may as well name i and j so that i is before j. There is a chain $s_j \lessdot s_j s_i \lessdot \cdots \lessdot (s_j s_i s_j s_i \cdots)$ in the weak order with the top element having $m(i, j) - 1$ letters. We contract all covers in this chain. After *orienting* the diagram by making such a choice for each edge of the diagram, the Cambrian congruence is the finest congruence of the weak order on W contracting all of the chosen edges in the weak order. In S_{n+1}, edges in the diagram connect $(i - 1)$ to i for $i = 2, \ldots, n$. We declare $i - 1$ to be before i if i is lower-barred and declare i to be before $i - 1$ if i is upper-barred. Figure 7.b shows one of the two Cambrian congruences on B_2. The contracted edges are indicated by shading.

To understand larger examples beyond the S_{n+1} case, we need to know how edge-forcing works for the weak order on a general finite Coxeter group. In general, we must consider all *polygonal intervals* (intervals whose Hasse diagrams are cycles) in the weak order. The weak order on permutations has the property that every polygonal interval is a quadrilateral or a hexagon. In a general finite Coxeter group, larger polygonal intervals may be present. Edge-forcing for the weak order on a general finite Coxeter group is completely analogous to edge-forcing for the weak order on permutations. There are local forcing requirements for every polygonal interval, with the same description: Whenever a bottom or top edge of the interval is contracted, the opposite bottom or top edge is also contracted, as well as all side edges (if side edges are present). The only difference is that, in general, there might be more than two side edges. We define *closure under local forcing* in the same way, with reference to all polygonal intervals, and again, [33, Theorem 25] implies that a collection $\mathscr{E}$ of edges in the weak order on W is the set of edges contracted by some congruence if and only if $\mathscr{E}$ is closed under local forcing. Figure 9.a shows the weak order on the Coxeter group B_3, with certain edges shaded. Using the local forcing criterion and comparing to Figure 8, the reader can readily verify that the congruence indicated by the shaded edges is the Cambrian congruence given by orienting 1 before 2 and 3 before 2.

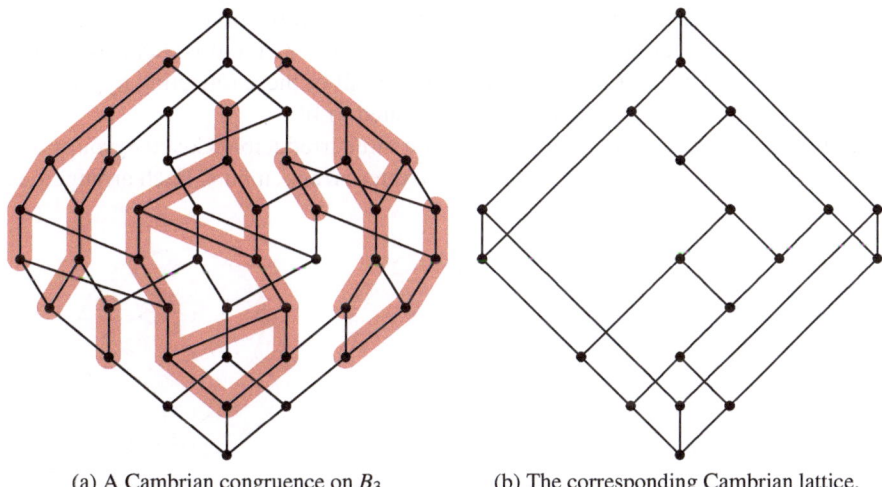

(a) A Cambrian congruence on B_3 (b) The corresponding Cambrian lattice.

Fig. 9 A Cambrian congruence and Cambrian lattice for the Coxeter group B_3

A *Cambrian lattice* is the quotient of the weak order on a finite Coxeter group W modulo some Cambrian congruence on W. A general fact about quotients of finite lattices says that this quotient is isomorphic to the subposet induced by the bottom elements of congruence classes. Above, we described how to define a Cambrian congruence on S_{n+1} starting from a choice of the polygon Q. The Cambrian lattice obtained from this orientation is isomorphic to the Tamari-like lattice on triangulations of Q. Thus two Cambrian lattices associated to S_4 are shown in Figures 1.a and 4.b. As further examples, Figure 7.c depicts the Cambrian lattice arising from the Cambrian congruence shown in Figure 7.b, while Figure 9.b shows the Cambrian lattice arising from the Cambrian congruence shown in Figure 9.a.

It should seem unlikely, *a priori*, that this generalization of the Tamari lattice along lattice-theoretic lines should yield anything useful. But amazingly, the Cambrian lattices have the same relationship to the generalized associahedra that the Tamari lattice has to the usual associahedron. The following theorem was conjectured in [36] and proved in [37, 38, 41].

Theorem 3.1. *The Hasse diagram of any Cambrian lattice associated to a finite Coxeter group W is isomorphic to the 1-skeleton of the generalized associahedron for W.*

4 Cambrian fans

Theorem 3.1 makes the connection, in general, between Cambrian lattices and generalized associahedra. In this section, we expand on that connection by describing how a Cambrian congruence determines a polyhedral structure closely related to the generalized associahedron: a Cambrian fan.

Consider a finite Coxeter group W, represented concretely as a group generated by the reflections S in a real vector space of dimension $|S|$. Typically, there are additional elements of W, besides the generators S, that act as reflections. Let T be the set of elements of W that act as reflections. For each $t \in T$, there is a corresponding reflecting hyperplane. Let $\mathscr{A}$ be the collection of all of these reflecting hyperplanes. This is called the *Coxeter arrangement* associated to W.

Referring to Figure 6, we see that there are three irreducible Coxeter groups with $|S| = 3$. Figure 10 pictures the associated Coxeter arrangements. Each arrangement

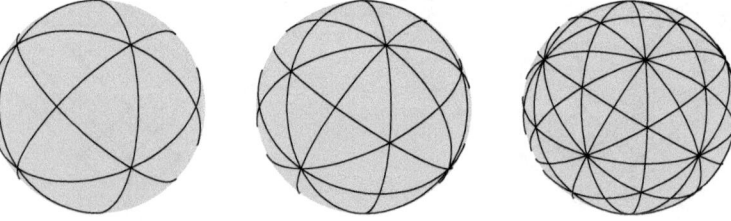

Fig. 10 The Coxeter arrangements for A_3, B_3, and H_3

is a collection of planes, through the origin, in $\mathbb{R}^3$. The intersection of these planes with the unit sphere about the origin is an arrangement of great circles. The figure shows these great circles on the sphere. The sphere is opaque, so that we only see each great circle as it intersects the near side of the sphere.

The complement $\mathbb{R}^n \setminus (\bigcup_{H \in \mathscr{A}} H)$ of the arrangement $\mathscr{A}$ is a collection of unbounded, n-dimensional, disjoint, open sets. The closures of these sets are called *regions*. The regions are the maximal cones of a *fan*, meaning that any two regions intersect in a *face* of each. We call this fan the *Coxeter fan* $\mathscr{F}$ associated to W. In Figure 10, the regions appear as spherical triangles, each representing an unbounded cone with triangular cross section. In general, the regions are unbounded cones whose cross sections are $(n-1)$-dimensional simplices, and therefore $\mathscr{F}$ is a *simplicial fan*.

The Coxeter arrangement $\mathscr{A}$ and the Coxeter fan $\mathscr{F}$ it defines are closely related to the combinatorics of the Coxeter group W. We now highlight two well-known aspects of this close relation. First and most importantly, the regions are in bijection with the elements of W. There is a region D whose facets are contained in the reflecting hyperplanes for the reflections S. (For some Coxeter groups W, it is possible to choose a representation of W such that no such D exists, but for any W, there is a standard way of producing a representation of W as a group generated by reflections, and in this standard representation, D exists.) Then each element w in W is mapped to the region wD obtained by acting on D by the transformation w.

As an example, consider the case where W is B_2. The Coxeter fan for B_2 is shown in Figure 11.a. Each region is labeled by the corresponding element of W. As another

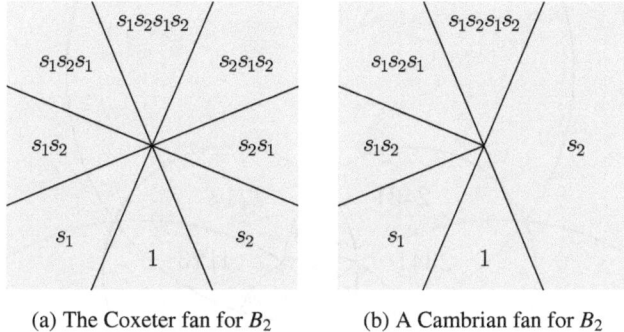

(a) The Coxeter fan for B_2 (b) A Cambrian fan for B_2

Fig. 11 The Coxeter fan and a Cambrian fan for B_2

example, we give a different depiction of the Coxeter arrangement for A_3, which appeared as the left picture in Figure 10. As before, we first pass from a collection of six planes through the origin in $\mathbb{R}^3$ to a collection of six great circles on the unit sphere. But in Figure 12, instead of showing only the near side of an opaque sphere, we show a stereographic projection of the sphere to the plane. This allows us to label the 24 regions (including the region outside all of the circles) by the 24 permutations in S_4.

The second aspect we wish to highlight is that cover relations in the weak order correspond to pairs Q, R of adjacent regions. The pair Q, R is separated by a unique hyperplane $H \in \mathscr{A}$. Taking, without loss of generality, Q to be on the same side of H as D, the cover relation is $Q \lessdot R$. This is readily verified in our running examples by comparing Figure 11.a to Figure 7.a and comparing Figure 12 to Figure 1.b.

Lattice congruences on the weak order respect the geometry of $\mathscr{F}$. Recall that every congruence class is an interval. Using the geometric characterization of the weak order, it is not hard to show that, given any interval in the weak order, the union of the corresponding regions in $\mathscr{F}$ is a convex cone. Thus the lattice congruence defines a collection of convex full-dimensional cones, and [35, Theorem 1.1] says that these are the maximal cones of a fan. When the lattice congruence is a Cambrian congruence, the resulting fan is called a *Cambrian fan*. Figure 11.b illustrates the Cambrian fan associated to the Cambrian congruence on B_2 shown in Figure 7.b. In the figure, each maximal cone is labeled by the bottom element of the corresponding Cambrian congruence class. A Cambrian fan associated to S_4 is shown in Figure 13. This is the fan associated to the Cambrian congruence (or Tamari congruence) pictured in Figure 3.b. The fan is drawn in stereographic projection as explained in connection with Figure 12, and each maximal cone of the fan is labeled with the bottom element of the corresponding congruence class.

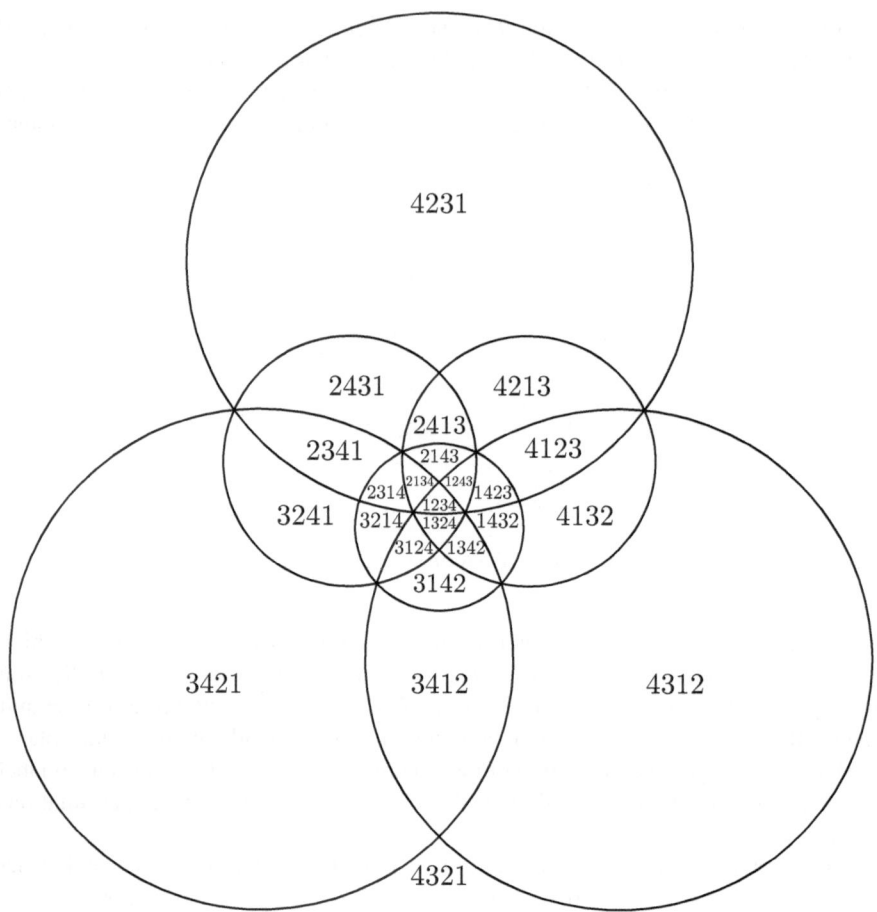

Fig. 12 Regions in the Coxeter arrangement for S_4, labeled by permutations.

General considerations about fans constructed from lattice congruences of the
weak order imply the following connection between Cambrian fans and Cambrian
lattices: Two maximal cones in the Cambrian fan are adjacent (share a codimension-1
face) if and only if the corresponding elements of the Cambrian lattice are related by
a cover. In particular, by Theorem 3.1, the adjacency graph on maximal cones in the
Cambrian fan is isomorphic to the 1-skeleton of the generalized associahedron. In
fact, more is true, as conjectured in [36] and proved in [37, 38, 41]:

Theorem 4.1. *Each Cambrian fan associated to W is combinatorially isomorphic
to the normal fan of the W-associahedron.*

Rather than actually defining the normal fan of a polytope, let us informally
describe the relationship between a polytope P and its normal fan $\mathscr{F}$. First, there
is a one-to-one correspondence between facets (maximal proper faces) of P and
rays in the fan $\mathscr{F}$ such that each ray ρ consists of outward-facing normal vectors

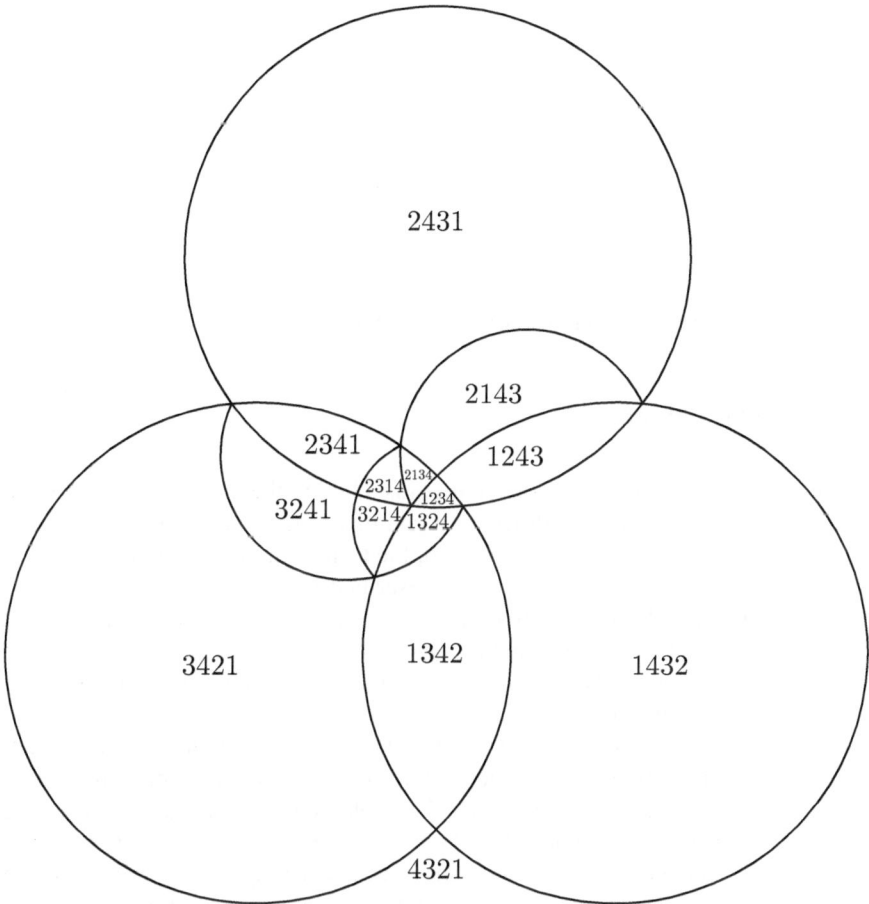

Fig. 13 The $s_1 s_2 s_3$-Cambrian fan associated to S_4

to the corresponding facet F. Then, there is a one-to-one correspondence between codimension-2 faces of P and two-dimensional cones in $\mathscr{F}$ such that a ray is contained in a two-dimensional cone if and only if the corresponding facet contains the corresponding codimension-2 face. Continuing, similar correspondences and reversals of containment are required in each dimension. We illustrate in Figure 14 by picturing a realization of the B_2-associahedron whose normal fan is the Cambrian fan pictured in Figure 11.b.

The combinatorial isomorphism with the normal fan of the W-associahedron is not the ultimate result on Cambrian fans. Indeed, there is a natural polytopal realization of the W-associahedron whose normal fan actually coincides with the Cambrian fan. Hohlweg's chapter [21] in this volume describes this realization.

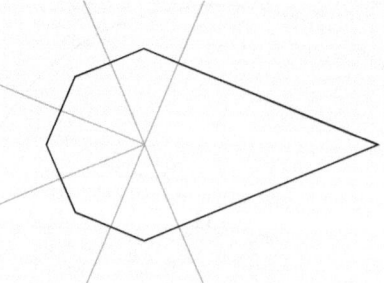

Fig. 14 A B_2-associahedron and its normal fan (a Cambrian fan)

5 Sortable elements

Although the definition of Cambrian lattices is a lattice-theoretic one, most of what has been proved about them follows from combinatorial models. In the paper [36], where Cambrian lattices were first defined, the Cambrian lattices associated to the Coxeter groups $S_{n+1} = A_n$ and B_n were described in terms of the combinatorics of triangulations, with the help of some results from the theory of fiber polytopes [5, 6, 45]. A general model for Cambrian lattices was later provided by the *Coxeter-sortable elements* (or simply *sortable elements*). These were introduced in [37], and the connection to Cambrian lattices was made in [38].

In defining a Cambrian lattice, our initial datum is an orientation of the Coxeter diagram of W. To define sortable elements, it is useful to write this initial datum in a different form. A *Coxeter element* is an element c of W that has a reduced word of the form $s_1 \cdots s_n$, where $S = \{s_1, \ldots, s_n\}$ and $|S| = n$. That is, a Coxeter element is the product of the generators in S, each occurring exactly once. Choosing a Coxeter element c in a finite Coxeter group W is equivalent to choosing, for each non-commuting pair of generators s_i and s_j, which of the two is before the other in a word for c. Since s_i and s_j are non-commuting if and only if i—j is an edge in the Coxeter diagram, Coxeter elements are equivalent to orientations of the diagram. For this reason, we use the term *c-Cambrian congruence* to describe the Cambrian congruence arising from the diagram orientation corresponding to c. As an example, if we orient the diagram s_1——s_2——s_3 of S_4 with s_1 before s_2 and s_3 before s_2, we define the Coxeter element $c = s_1 s_3 s_2 = s_3 s_1 s_2$.

In the case of the symmetric group, the Coxeter element c can be read off directly from the polygon Q. Specifically, c is the $(n+1)$-cycle in S_{n+1} obtained by reading the vertices of Q (excluding 0 and $n+2$) in counter-clockwise order.

To define sortable elements, it is useful to think about how one might write down a reduced word for an element w of W. We assume that we can easily determine the length of a given element of W. (Although this assumption seems to ignore the fact that the length of w is defined in terms of reduced words, it is nevertheless a useful point of view.) If w is the identity, then the only reduced word for w is the empty word. Otherwise, we write down a reduced word $a_1 \cdots a_k$ for w from left to right as follows. First, try each element $s_i \in S$ in some order until we find one such that $\ell(s_i w) < \ell(w)$. We will eventually find such an s_i because w has some nonempty reduced word, and we can take s_i to be the first letter of that word. Set a_1 equal to s_i, and write $w' = a_1 w$. If w' is the identity, then $k = 1$ and a_1 is the desired reduced word. Otherwise, find $s_{i'}$ such that $\ell(s_{i'} w') < \ell(w')$, set a_2 equal to $s_{i'}$, and define $w'' = a_2 w'$. If w'' is the identity, then $k = 2$ and $a_1 a_2$ is the desired word. Otherwise, continuing in this manner, we eventually find a reduced word $a_1 \cdots a_k$. The output depends, of course, on the order in which we try the elements of S in each step.

The usefulness of this method of writing down a reduced word is that it lends itself to a global choice of a canonical reduced word for each element. For example, one might, at each step, try the elements of S in the order $s_1, \ldots, s_n$. This would have the effect of representing each element by its lexicographically first reduced word, in the sense of lexicographic order on subscripts.

The definition of sortable elements depends on a more subtle choice of a canonical word. Fix a total order $(s_1, \ldots, s_n)$ on S. Follow the method described above, trying the letters in S cyclically in the order $(s_1, \ldots, s_n)$. This method differs from the lexicographic method described above, in that we do not start again at s_1 at every step. Instead, if we choose s_i at some step, in the next step we try the letters in the order $s_{i+1}, \ldots, s_n, s_1, \ldots, s_i$. The resulting reduced word is called the $(s_1, \ldots, s_n)$-*sorting word* for w. In effect, we take repeated "passes" through the sequence $(s_1, \ldots, s_n)$, adding whatever letters we can to the word. For convenience, between passes, we insert the symbol "$\mid$" as a "divider."

An $(s_1, \ldots, s_n)$-sorting word can be described by a sequence of subsets of S: The ith subset in the sequence is the set of letters that are added to the $(s_1, \ldots, s_n)$-sorting word in the ith pass. Equivalently, these are the sets of letters occurring between dividers. Let c be the Coxeter element $s_1 \cdots s_n$. We say that w is c-*sortable* if this sequence of subsets is weakly decreasing under containment. (We say "c-sortable" instead of "$(s_1, \ldots, s_n)$-sortable" because it is easy to show that the notion depends only on c, not on the reduced word $s_1 \cdots s_n$ for c.)

The (s_1, s_2)-sorting words for elements of B_2 are the empty word, s_1, $s_1 s_2$, $s_1 s_2 \mid s_1$, $s_1 s_2 \mid s_1 s_2$, s_2, $s_2 \mid s_1$, and $s_2 \mid s_1 s_2$. Six of the elements of B_2 are c-sortable for $c = s_1 s_2$. The exceptions are $s_2 \mid s_1$ (because $\{s_2\} \not\supseteq \{s_1\}$) and $s_2 \mid s_1 s_2$ (because $\{s_2\} \not\supseteq \{s_1, s_2\}$). The (s_1, s_2, s_3)-sorting words for elements of S_4 are shown in Figure 15.a, in the same positions as the permutations shown in Figure 1.b. Figure 15.b shows the subposet of the weak order induced by the $s_1 s_2 s_3$-sortable elements.

Comparing Figure 15.a with Figure 3.b, we see an example of the following theorem, which is the concatenation of [38, Theorem 1.1] and [38, Theorem 1.4].

Theorem 5.1. *An element w of W is the bottom element in its c-Cambrian congruence class if and only if it is c-sortable.*

As mentioned earlier, the quotient of a finite lattice modulo a congruence Θ is isomorphic to the subposet induced by the bottom elements of Θ-classes. Thus, we have the following corollary, which is exemplified by Figures 1.a and 15.b.

Corollary 5.2. *The c-Cambrian lattice is isomorphic to the subposet of the weak order on W induced by the c-sortable elements.*

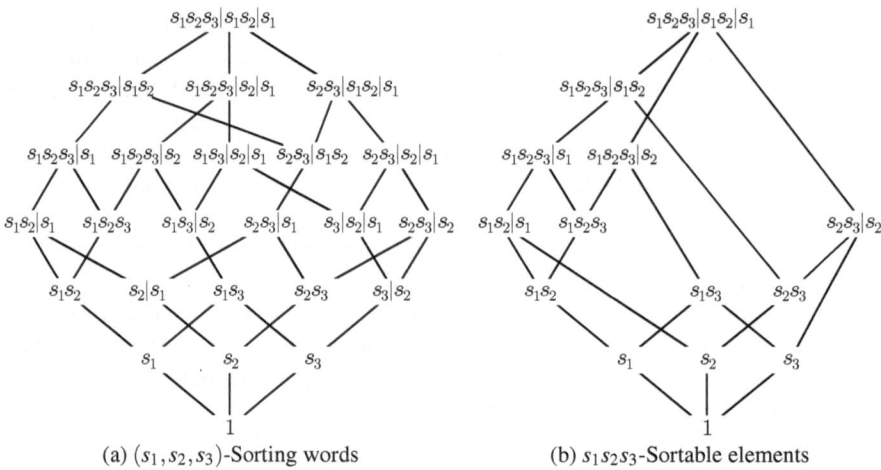

(a) (s_1, s_2, s_3)-Sorting words   (b) $s_1 s_2 s_3$-Sortable elements

Fig. 15 (s_1, s_2, s_3)-Sorting words for elements of S_4, and the subposet of the weak order induced by $s_1 s_2 s_3$-sortable elements

Typically, the bottom elements of a lattice congruence on a finite lattice L are not a sublattice of L. But [38, Theorem 1.2] shows that the Cambrian congruence is an exception:

Theorem 5.3. *The c-sortable elements in W constitute a sublattice of the weak order on W.*

The realization of the Cambrian lattice in terms of sortable elements provides considerable combinatorial traction. Most importantly, the sortable elements allow for *uniform* proofs. A non-uniform (or *case-by-case* or *type-by-type*) proof of a result on finite Coxeter groups is a proof obtained by checking each case of the classification of finite Coxeter groups (Figure 6). Typically, the result is argued separately for each of the infinite families (in order of difficulty) $I_2(m)$, A_n, B_n and D_n, and then (perhaps by computer) for the exceptional cases E_6, E_7, E_8, F_4, H_3, and H_4. A uniform proof is a single argument that is valid for an arbitrary finite Coxeter group.

We now describe in detail three specific types of combinatorial traction provided by sortable elements. The first is a natural recursive structure that lends itself well to arguments by induction on two parameters: the length of elements of W and the rank of W. (The rank of a Coxeter group is the size of its defining generating set S.) The recursive structure is encapsulated in two easy lemmas [37, Lemmas 2.4, 2.5], which are given below as Lemmas 5.4 and 5.5.

Suppose $s \in S$ and let c be a Coxeter element of W. Then s is *initial* in c if there is a reduced word for c having s as its first letter. Equivalently, in the diagram orientation corresponding to c, no other generator is declared to be "before" s. The notation $W_{\langle s \rangle}$ stands for the subgroup of W generated by the set $S \setminus \{s\}$. This is an example of a *standard parabolic subgroup* of W, and it is a Coxeter group in its own right, with generating set $S \setminus \{s\}$.

Lemma 5.4. *Let s be initial in c and suppose $w \not\geq s$. Then w is c-sortable if and only if it is an sc-sortable element of $W_{\langle s \rangle}$.*

To see why this lemma is true, write $c = s_1 \cdots s_n$ with $s_1 = s$. The assumption that $w \not\geq s$ means that there is no reduced word for w having the word s as a prefix, or in other words, there is no reduced word for w having s as its first letter. In particular, the $(s_1, \ldots, s_n)$-sorting word for w does not start with s_1. If w is c-sortable, then s_1 does not appear in its $(s_1, \ldots, s_n)$-sorting word, so w is in $W_{\langle s \rangle}$, and its $(s_2, \ldots, s_n)$-sorting word is the same sequence of letters as its $(s_1, \ldots, s_n)$-sorting word. Now $sc = s_2 \cdots s_n$, and the fact that w is c-sortable immediately implies that w is sc-sortable, as an element of $W_{\langle s \rangle}$. For the converse, we need a well-known fact about standard parabolic subgroups: If w is contained in $W_{\langle s \rangle}$, then no reduced word for w (as an element of W) contains the letter s. Thus if w is in $W_{\langle s \rangle}$ and is sc-sortable as an element of $W_{\langle s \rangle}$, then its $(s_1, \ldots, s_n)$-sorting word must coincide with its $(s_2, \ldots, s_n)$-sorting word. Now the c-sortability of w follows from the sc-sortability of w.

Lemma 5.5. *Let s be initial in c and suppose $w \geq s$. Then w is c-sortable if and only if sw is scs-sortable.*

To explain this lemma, again write $c = s_1 \cdots s_n$ with $s_1 = s$ and note that scs is the Coxeter element $s_2 \cdots s_n s_1$. To make the $(s_1, \ldots, s_n)$-sorting word for w, we first need to try the letter s_1. Under the hypotheses of the lemma, (with $s = s_1$), we take s_1 as the first letter of the c-sorting word, and then continue trying the letters in the cyclic order given by $(s_2, \ldots, s_n, s_1)$. The sequence of letters obtained after s_1 is exactly the $(s_2, \ldots, s_n, s_1)$-sorting word for sw and the criterion for w to be c-sortable is exactly the criterion for sw to be scs-sortable.

The second type of combinatorial traction provided by sortable elements is the natural *search tree* structure on sortable elements. Each c-sortable element has a naturally-defined predecessor among c-sortable elements, obtained by deleting the last letter of the $(s_1, \ldots, s_n)$-sorting word. Starting with the Hasse diagram of the c-Cambrian lattice (realized as a partial order on c-sortable elements), delete all edges except those connecting an element to its predecessor. The result is a spanning tree of the Hasse diagram, rooted at the identity element. (This tree may depend on $(s_1, \ldots, s_n)$, not only on c.) Furthermore, given any c-sortable element w, there is a simple algorithm for determining all c-sortable elements whose predecessor is w. Under these circumstances, it is a simple matter to efficiently traverse the set of c-sortable elements. For a general description of efficient traversal, see [48, Section 4C]. Figure 16 shows the search tree structure on $s_1 s_2 s_3$-sortable elements in S_4.

The third type of combinatorial traction provided by sortable elements is the extra combinatorial information available in the sorting word for a c-sortable element. We explain by giving an important example of information that can be read off from the sorting word. (Another example appears in Hohlweg's chapter [21, Section 3.3].) To explain the example, we need to define the *simple roots* associated to our chosen reflection representation of W. For each generator $s_i \in S$, there is a corresponding vector α_i called a simple root. Recall that D is the region identified with the identity element, and that its facets are contained in the reflecting hyperplanes for the generators S. The full definition of a *root system* specifies the length of each α_i up to a

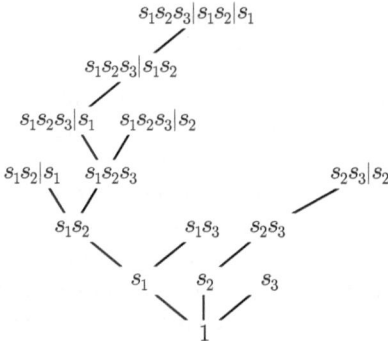

Fig. 16 The search tree structure on $s_1s_2s_3$-sortable elements in S_4

global scaling, but for our purposes, all we need to say is that α_i is a nonzero normal vector to the reflecting hyperplane for s_i, pointing in the direction of the interior of D.

Let $c = s_1 \cdots s_n$ and let v be a c-sortable element of W with $(s_1, \ldots, s_n)$-sorting word $a_1 \cdots a_k$. Fix some $s_i \in S$. In the process of building the $(s_1, \ldots, s_n)$-sorting word, there was some first time that s_i was tested but *not* included in the $(s_1, \ldots, s_n)$-sorting word. Let $a_1 \cdots a_j$ be the part of the $(s_1, \ldots, s_n)$-sorting word that was already determined before s_i was first tested but not included. We say $a_1 \cdots a_k$ *skips* s_i after position j. Define $C_c^{s_i}(v)$ to be the vector $a_1 \cdots a_j \cdot \alpha_i$. That is, act on the simple root α_i by the reflection a_j, then the reflection a_{j-1}, on so forth until acting by a_1. As indicated by the notation, the vector $C_c^{s_i}$ depends only on c, not on $(s_1, \ldots, s_n)$. Define $C_c(v) = \{C_c^{s_i}(v) : s_i \in S\}$. The set $C_c(v)$ can also be defined recursively by induction on the length of v and the rank of W. Recall now that c-sortable elements are the bottom elements of c-Cambrian congruence classes, and that each c-Cambrian congruence class defines a cone. Surprisingly, the cone for v is completely described by the data $C_c(v)$. Specifically, by [42, Theorem 6.3], the cone has n facets, and the vectors in $C_c(v)$ are the normal vectors to these facets, pointing into the interior of the cone. In particular, each Cambrian fan, previously defined in terms of a Cambrian congruence, has an explicit combinatorial description in terms of sortable elements.

In the example of B_2, we refer to the Coxeter fan and s_1s_2-Cambrian fan pictured in Figure 11. The region D is, as usual, identified with the identity element 1. Let α_1 and α_2 be the inward facing normal vectors to D, with α_1 orthogonal to the line separating D from the region identified with s_1. Then s_1 is the reflection orthogonal to α_1 and s_2 is the reflection orthogonal to α_2. Let us calculate $C_{s_1s_2}(s_2)$. The letter s_1 is skipped in the first step of constructing the (s_1, s_2)-sorting word for s_2, so $C_{s_1s_2}^{s_1}(s_2)$ is α_1. In the next step, s_2 becomes the first letter of the (s_1, s_2)-sorting word. The next time s_2 is tried, the sorting word is already complete, so s_2 is skipped. Thus $C_{s_1s_2}^{s_2}(s_2)$ is $s_2\alpha_2 = -\alpha_2$. We see in Figure 11.b. that α_1 and $-\alpha_2$ are indeed inward-facing normal vectors to the cone associated to s_2. Similarly, we calculate $C_{s_1s_2}(s_1s_2)$. The letter s_1 is skipped after both s_1 and s_2 have already been tried and placed in the sorting word, so $C_{s_1s_2}^{s_1}(s_1s_2) = s_1s_2\alpha_1$. For the same reason, $C_{s_1s_2}^{s_2}(s_1s_2) = s_1s_2\alpha_2$, and we verify that the cone associated to s_1s_2 is indeed defined by inward-facing normal vectors $s_1s_2\alpha_1$ and $s_1s_2\alpha_2$.

6 Applications

In this section, we briefly mention several mathematical applications of Cambrian lattices and sortable elements.

6.1 Coxeter-Catalan combinatorics

The combinatorics of objects counted by the Catalan numbers has a rich history. Recently, several of these *Catalan objects* have been shown to be special cases (the case $W = S_{n+1}$) of combinatorial constructions whose input is a finite Coxeter group W. The more general objects are counted by a generalization of the Catalan number. For more background, including references, we refer the reader to [3, 12].

One of the Coxeter-Catalan constructions, the generalized associahedron, has already been mentioned. Its vertices are indexed by a generalization of triangulations. Another construction generalizes the classical noncrossing partitions. Initially, the W-noncrossing partitions were noticed to be equinumerous with the vertices of the W-associahedron, for any W, but no uniform explanation was known. Sortable elements were used in [37] to give the first uniform explanation, by uniformly defining a bijection from sortable elements to noncrossing partitions and a bijection to the vertices of the generalized associahedron. (Although the bijections were uniformly *defined* in [37], the proofs of some key lemmas were non-uniform. These lemmas were later proved uniformly in [42]. Meanwhile, another uniform bijection between noncrossing partitions and vertices of the generalized associahedron was given in [4], using ideas from [8].)

Recall the discussion at the end of Section 5 of three types of combinatorial traction provided by sortable elements. The bijection to vertices of the generalized associahedron is defined in terms of sorting words for sortable elements, and both bijections are proved using induction on length and rank. The bijections also make the natural search tree structure on sortable elements available in the context of noncrossing partitions and in the context of the generalized associahedron.

6.2 Sortable elements in quiver theory

Sortable elements and Cambrian lattices have figured in the study of quiver representations and various topics. Details are far beyond the scope of this chapter, but we refer the reader to the research papers [1, 2, 23, 24] and to Thomas' chapter [49] in this volume.

6.3 Combinatorial Hopf algebras

Here we depart from the theme of the section to mention, rather than an application of Cambrian lattices, another profitable generalization of the Tamari lattices. Both generalizations arise from the same obxervation about the permutations-to-triangulations map. In [29, 30], Loday and Ronco constructed a combinatorial Hopf algebra whose operations are closely tied to the combinatorics of the Tamari lattice. A permutations-to-triangulations map (in an equivalent formulation using planar binary trees rather than triangulations) was also relevant to the Loday-Ronco Hopf algebra. An infinite family of Hopf algebras, containing the Loday-Ronco Hopf algebra, is constructed in [35], building on the observation that the permutations-to-triangulations map is a lattice homomorphism. The construction in [35] yields Hopf algebras built on congruence classes of permutations, and suggests the project of instead realizing the Hopf algebras in terms of natural combinatorial objects. The goal is a description of the product and coproduct in terms of natural operations on the combinatorial objects, and a description of the quotient lattice as a partial order on the combinatorial objects, with cover relations given by combinatorial moves analogous to diagonal flips on triangulations. Besides the motivating examples of the Loday-Ronco Hopf algebra and the Hopf algebra of noncommutative symmetric functions [19], two Hopf algebras built on *rectangulations* (tilings of a rectangle by rectangles) have already been studied [20, 28, 40], as well as a Hopf algebra of *sashes* [27], certain tilings counted by the Pell numbers.

6.4 Cluster algebras

Cluster algebras were defined by Fomin and Zelevinsky in [14] and have since become the subject of intense research, in part because of their unexpected connections to a variety of mathematical areas. Sortable elements and Cambrian lattices/fans have found applications in the structural theory of cluster algebras. These applications are surprising *a priori*, but less surprising once the connection between Cambrian lattices and generalized associahedra is known, because generalized associahedra are the combinatorial structure underlying cluster algebras of finite type. To give the reader an idea of these applications, we start with the gentlest possible introduction to cluster algebras, giving only the ideas behind the definition. The lecture notes [12], while still gentle, do contain some actual definitions. Many other expository works are listed in the Cluster Algebras Portal, maintained by Sergey Fomin and easily found online.

The setting for a cluster algebra is a field of rational functions in n variables. We start with elements $x_1, \ldots x_n$ of the field and some "combinatorial data." This pair

$$\{\text{Combinatorial Data}\}, (x_1, \ldots x_n)$$

is called the initial *seed*. The rational functions $x_1, \ldots x_n$ are the *cluster variables*. There is an operation called *mutation*, which takes a seed and gives a new seed. The combinatorial data tells how to do mutation. Mutation can take place any of n "directions," and it does two things: It switches out one cluster variable, replacing it with a new one, and it alters the combinatorial data. The result is a new seed. The mutation operation is involutive, in the sense that mutating in the kth direction, and then mutating the result in the kth direction returns the original seed. The basic setup is represented schematically in Figure 17 for $n = 3$, with dangling edges to indicate that the picture continues outward indefinitely. Mutation in the kth direction is indicated by the symbol μ_k.

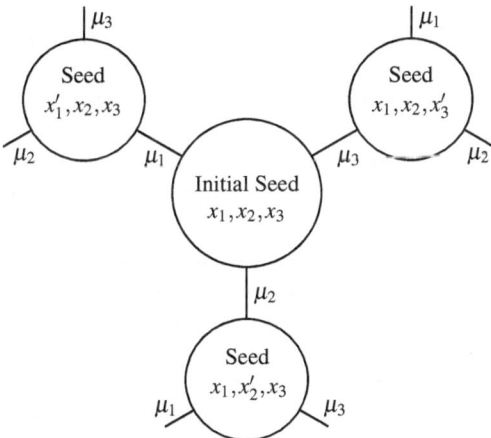

Fig. 17 A schematic representation of seed mutations

To construct the cluster algebra, we do all possible sequences of mutations, and collect all the cluster variables that appear. The *cluster algebra* for the given initial seed is the algebra of rational functions generated by all cluster variables. As a matter of bookkeeping, we place the seeds on the vertices of an infinite tree, with each vertex incident to n edges labeled μ_1 through μ_n to indicate the direction of mutation. Once the initial seed is placed on some vertex of the tree, since each other vertex is connected to the initial vertex by a unique path, each vertex indicates a seed: the seed obtained from the initial seed by the sequence of mutations labeling the edges on this unique path.

However, depending on the initial combinatorial data, seeds at different vertices on the tree may coincide. In this case, we obtain a smaller graph by identifying vertices of the infinite tree whenever they have the same seed. The smaller graph is called the *exchange graph*. The cluster algebra is said to be of *finite type* when the exchange graph is a finite graph. For example, when $n = 2$, the infinite tree is an infinite path. There are four cases where the initial combinatorial data causes this infinite path to collapse to a cycle (of length 4, 5, 6 or 8). More generally, the main result of [15] identifies the exchange graphs of cluster algebras of finite type as the 1-skeleta of generalized associahedra.

To see how Cambrian lattices and sortable elements contribute, it is first useful to know a bit more about the combinatorial data that describes the initial seed. This data takes the form of an $n \times n$ matrix B, called the *exchange matrix*, and some other data that should be thought of as a choice of coefficients. The key point is that choosing B is equivalent to choosing a Coxeter group W, a specific representation of W as a group generated by reflections, and an orientation of the Coxeter diagram of W. (Strictly speaking, we must choose a *crystallographic* Coxeter group, thus allowing the representation of W to be given by integer matrices.) When W is finite, the orientation of its diagram defines a Cambrian lattice, and having a specific representation of W then implies a specific realization of the corresponding Cambrian fan. The Hasse diagram of this Cambrian lattice is isomorphic to the exchange graph of the cluster algebra defined by B, with the initial seed corresponding to the bottom element of the Cambrian lattice (the identity element in W). But in fact, even more structural information is contained in the sortable elements and Cambrian fan. The cluster variables appearing in the various seeds can be identified by various statistics known as *denominator vectors* and **g**-*vectors*. The denominator vectors are easily read off from the sorting words of sortable elements. The **g**-vectors are encoded geometrically in the Cambrian fan, which coincides with what one might call the **g**-*vector fan* of the cluster algebra. Thus we have arrived, via the lattice theory and combinatorics of Coxeter groups, at the fundamental combinatorial and polyhedral structure underlying cluster algebras. In particular, Figures 11.b and 13 are illustrations of this underlying structure.

The definition of the weak order and the definition of sortable elements are both valid even when W is not of finite type. The weak order becomes a semilattice (i.e., meets exist) but not a lattice, so the lattice-theoretic definition of the Cambrian lattice is no longer available. However, the weak order restricted to sortable elements *is* still available and useful. This *Cambrian semilattice* serves as a model for *part of* the exchange graph, and in that part, we can still read off **g**-vectors. We can also define a Cambrian fan using the sets $C_c(v)$, and this Cambrian fan coincides with part of the **g**-vector fan. An example of an infinite Cambrian fan is shown in Figure 18, drawn in the same stereographic projection as previous fan pictures. The cones of the Cambrian fan are outlined in black, and the reflecting hyperplanes are drawn as thin blue curves. As one moves right from the center of the figure, one encounters infinitely many smaller and smaller Cambrian cones.

In an important infinite case, the Cambrian model still recovers the entire **g**-vector fan. When W is of *affine* type, the **g**-vector fan is the union of a Cambrian fan with the antipodal opposite of a Cambrian fan for the opposite orientation of the diagram. Such a *doubled Cambrian fan*, built on the fan from Figure 18, is shown in Figure 19. The second Cambrian fan is indicated by the red arcs, some of which are obscured by the original black arcs.

The Cambrian model not only recovers the combinatorics and polyhedral geometry of cluster algebras, but also leads to proofs of some key structural conjectures from [16, 17], in the cases where the underlying Coxeter group is of finite or affine type.

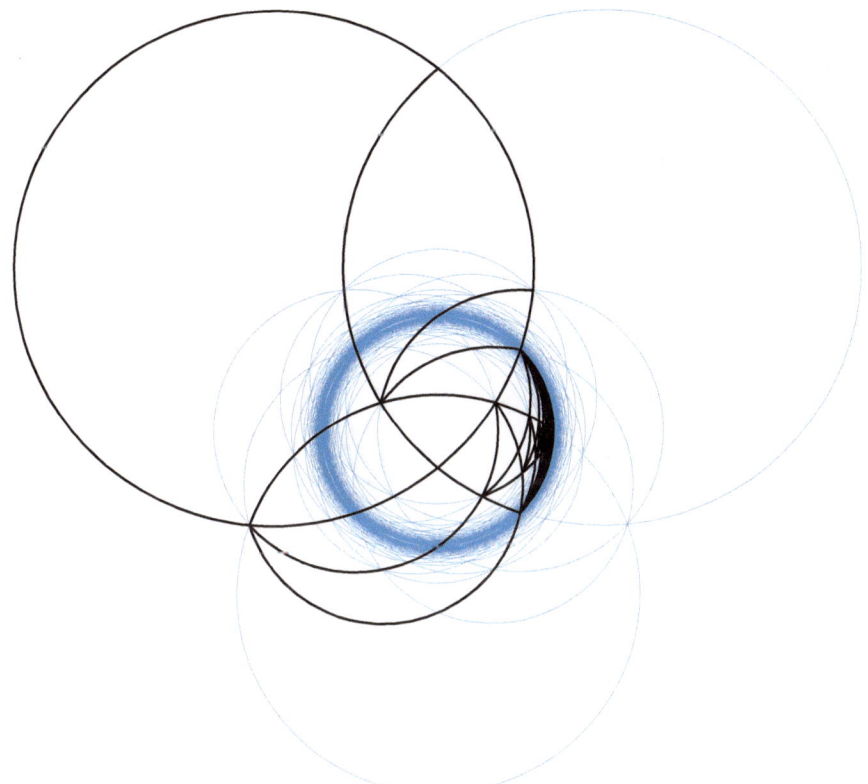

Fig. 18 A Cambrian fan for an infinite Coxeter group

For some of these cases, the Cambrian model is currently the only known approach to proving the structural results.

Of necessity, we have given almost no details on cluster algebras and how they interact with Cambrian lattices and sortable elements. Readers interested in filling in the details can find the necessary cluster algebras definitions in [17], details about **g**-vectors and Cambrian fans in [41, 43, 51], and more about the Cambrian model of cluster combinatorics in [43, 44].

References

1. C. Amiot, "A derived equivalence between cluster equivalent algebras", *J. Algebra* **351** (2012) 107–129.
2. C. Amiot, O. Iyama, I. Reiten and G. Todorov, "Preprojective algebras and c-sortable words", *Proc. London Math. Soc.* (3) **104** (2012) 513–539.
3. D. Armstrong, *Generalized Noncrossing Partitions and Combinatorics of Coxeter Groups*, Mem. Amer. Math. Soc., vol. 202, no. 949, Amer. Math. Soc., Providence, RI, 2009.

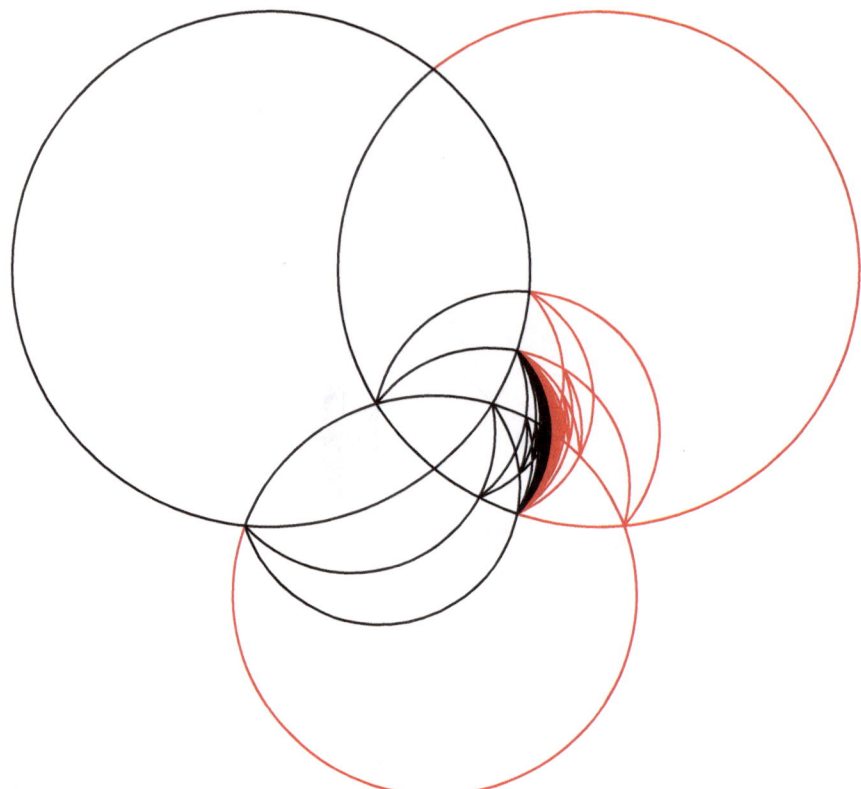

Fig. 19 A doubled Cambrian fan for an affine Coxeter group

4. C.A. Athanasiadis, T. Brady, J. McCammond and C. Watt, "*h*-vectors of generalized associahe-dra and noncrossing partitions", *Int. Math. Res. Not.* **2006**, Art. ID 69705, 28 pp.
5. L. Billera and B. Sturmfels, "Fiber polytopes", *Ann. of Math.* (2) **135** (1992) 527–549.
6. _____ , "Iterated Fiber polytopes", *Mathematika* **41** (1994) 348–363.
7. A. Björner and M. Wachs, "Shellable nonpure complexes and posets. II.", *Trans. Amer. Math. Soc.* **349** (1997) 3945–3975.
8. T. Brady and C. Watt, "Non-crossing partition lattices in finite real reflection groups", *Trans. Amer. Math. Soc.* **360** (2008) 1983–2005.
9. I. Chajda and V. Snášel, "Congruences in Ordered Sets", *Math. Bohem.* **123** (1998) 95–100.
10. F. Chapoton, S. Fomin and A. Zelevinsky, "Polytopal realizations of generalized associahedra", *Canad. Math. Bull.* **45** (2002) 537–566.
11. G. Dorfer, "Lattice-extensions by means of convex sublattices", in *Contributions to general algebra* (Linz, 1994), vol. 9, G. Pilz, ed., Hölder-Pichler-Tempsky, Vienna, 1995, 127–132.
12. S. Fomin and N. Reading, "Root systems and generalized associahedra", in *Geometric Combinatorics*, E. Miller, V. Reiner and B. Sturmfels, eds., IAS/Park City Math. Ser., vol. 13, Amer. Math. Soc., Providence, RI, 2007, 63–131.
13. S. Fomin and A. Zelevinsky, "Y-systems and generalized associahedra", *Ann. of Math.* **158** (2003) 977–1018.
14. _____ , "Cluster algebras. I. Foundations", *J. Amer. Math. Soc.* **15** (2002) 497–529.
15. _____ , "Cluster Algebras II: Finite Type Classification", *Inventiones Mathematicae* **154** (2003) 63–121.

16. _____, "Cluster algebras: notes for the CDM-03 conference", in *Current Developments in Mathematics*, A.J. de Jong, D. Jerison, G. Lustig, B. Mazur, W. Schmid and S.T. Yau, eds., Int. Press, Somerville, MA, 2003, 1–34.
17. _____, "Cluster Algebras IV: Coefficients", *Compositio Mathematica* **143** (2007) 112–164.
18. S. Forcey, "Extending the Tamari lattice to some compositions of species", in *this volume*.
19. I.M. Gelfand, D. Krob, A. Lascoux, B. Leclerc, V.S. Retakh and J.-Y. Thibon, "Noncommutative symmetric functions", *Adv. Math.* **112** (1995) 218–348.
20. S. Giraudo, "Algebraic and combinatorial structures on Baxter permutations", *arxiv.org/abs/1011.4288*.
21. C. Hohlweg, "Generalized associahedra from the geometry of finite reflection groups", in *this volume*.
22. C. Hohlweg, C. Lange and H. Thomas, "Permutahedra and generalized associahedra", *Adv. Math.* **226** (2011) 608–640.
23. C. Ingalls and H. Thomas, "Noncrossing partitions and representations of quivers", *Compositio Mathematica* **145** (2009) 1533–1562.
24. M. Kleiner and A. Pelley, "Admissible sequences, preprojective representations of quivers, and reduced words in the Weyl group of a Kac-Moody algebra", *Int. Math. Res. Not.* **2007**, no. 4, Art. ID rnm013, 28 pp.
25. M. Kolibiar, "Congruence relations and direct decompositions of ordered sets", *Acta Sci. Math. (Szeged)* **51** (1987) 129–135.
26. G. Kreweras, "Sur les partitions non croisées d'un cycle", *Discrete Math.* **1** (1972) 333–350.
27. S. Law, "The Hopf algebra of sashes", in preparation (2012).
28. S. Law and N. Reading, "The Hopf algebra of diagonal rectangulations", *J. Combin. Theory Ser. A* **119** (2012) 788–824.
29. J.-L. Loday and M. Ronco, "Hopf algebra of planar binary trees", *Adv. Math.* **139** (1998) 293–309.
30. _____, "Order structure on the algebra of permutations and of planar binary trees", *J. Algebraic Combin.* **15** (2002) 253–270.
31. J. Rambau and V. Reiner, "A survey of the higher Stasheff-Tamari orders", in *this volume*.
32. N. Reading, "Order Dimension, Strong Bruhat Order and Lattice Properties for Posets", *Order* **19** (2002) 73–100.
33. _____, "Lattice and Order Properties of the Poset of Regions in a Hyperplane Arrangement", *Algebra Universalis* **50** (2003) 179–205.
34. _____, "Lattice congruences of the weak order", *Order* **21** (2004) 315–344.
35. _____, "Lattice congruences, fans and Hopf algebras", *J. Combin. Theory Ser. A* **110** (2005) 237–273.
36. _____, "Cambrian Lattices", *Adv. Math.* **205** (2006) 313–353.
37. _____, "Clusters, Coxeter-sortable elements and noncrossing partitions", *Trans. Amer. Math. Soc.* **359** (2007) 5931–5958.
38. _____, "Sortable elements and Cambrian lattices", *Algebra Universalis* **56** (2007) 411–437.
39. _____, "Noncrossing partitions and the shard intersection order", *J. Algebraic Combin.* **33** (2011) 483–530.
40. _____, "Generic rectangulations", *European J. Combin.* **33** (2012) 610–623.
41. N. Reading and D. Speyer, "Cambrian Fans", *J. Eur. Math. Soc.* **11** (2009) 407–447.
42. _____, "Sortable elements in infinite Coxeter groups", *Trans. Amer. Math. Soc.* **363** (2011) 699–761.
43. _____, "Combinatorial frameworks for cluster algebras", *arxiv.org/abs/1111.2652*.
44. _____, "Cambrian frameworks for cluster algebras of affine Cartan type", in preparation (2012).
45. V. Reiner, "Equivariant fiber polytopes", *Doc. Math.* **7** (2002) 113–132.
46. L. Santocanale and F. Wehrung, "Sublattices of associahedra and permutohedra", *arxiv.org/abs/1103.3488*.
47. R.P. Stanley, *Enumerative Combinatorics*, vol. 1, Cambridge University Press, Cambridge, 1997.

48. J.R. Stembridge, "Computational aspects of root systems, Coxeter groups, and Weyl characters", in *Interaction of combinatorics and representation theory*, J.R. Stembridge, J.-Y. Thibon and M.A.A van Leeuwen, eds., MSJ Memoirs., vol. 11, Math. Soc. Japan, 2001, 1–38.

49. H. Thomas, "The Tamari lattice as it arises in quiver representations", in *this volume*.

50. A. Tonks, "Relating the associahedron and the permutohedron", in *Operads: Proceedings of Renaissance Conferences* (Hartford, CT/Luminy, 1995), J.-L. Loday, J.D. Stasheff and A.A. Voronov, eds., Contemp. Math., vol. 202, Amer. Math. Soc., Providence, RI, 1997, 33–36.

51. S. Yang and A. Zelevinsky, "Cluster algebras of finite type via Coxeter elements and principal minors", *Transform. Groups* **13** (2008) 855–895.

Catalan Lattices on Series Parallel Interval Orders

Filippo Disanto, Luca Ferrari, Renzo Pinzani, Simone Rinaldi

Abstract Using the notion of series parallel interval order, we propose a unified setting to describe Dyck lattices and Tamari lattices (two well-known lattice structures on Catalan objects) in terms of basic notions of the theory of posets. As a consequence of our approach, we find an extremely simple proof of the fact that the Dyck order is a refinement of the Tamari one. Moreover, we provide a description of both the weak and the strong Bruhat order on 312-avoiding permutations, by recovering the proof of the fact that they are isomorphic to the Tamari and the Dyck order, respectively; our proof, which simplifies the existing ones, relies on our results on series parallel interval orders.

Key words: Catalan numbers, Tamari lattice, Dyck lattice, series parallel interval orders (MSC2010: 05A19,06A07)

Filippo Disanto
Institut für Genetik, Universität Köln, Zülpicher Str. 47a, 50674 Köln, Germany, e-mail: *disafili@yahoo.it*

Luca Ferrari
Dipartimento di Sistemi e Informatica, University of Firenze, viale Morgagni 65, 50134, Firenze, Italy, e-mail: *ferrari@dsi.unifi.it*

Renzo Pinzani
Dipartimento di Sistemi e Informatica, University of Firenze, viale Morgagni 65, 50134, Firenze, Italy, e-mail: *pinzani@dsi.unifi.it*

Simone Rinaldi
Dipartimento di Scienze Matematiche e Informatiche, University of Siena, Pian dei Mantellini 44, 53100 Siena, Italy, e-mail: *rinaldi@unisi.it*

323

1 Introduction

A *Dyck path* is a lattice path starting from the origin of a fixed Cartesian coordinate system, ending on the x-axis, never falling below the x-axis and using only two kind of steps, namely up steps $(1, 1)$ and down steps $(1, -1)$ (see Figure 1).

Fig. 1 A Dyck path.

The set of all Dyck paths of fixed length can be ordered by setting $P \leq Q$ when P lies weakly below Q (in the usual two-dimensional drawings of Dyck paths). In Figure 2 a pair of comparable Dyck paths is shown: the path in Figure 2 (a) is smaller than the path in Figure 2 (b), as shown in Figure 2 (c).

This partial order is in fact a lattice, called *Dyck lattice* (and sometimes also *Stanley lattice* [4, 13]).

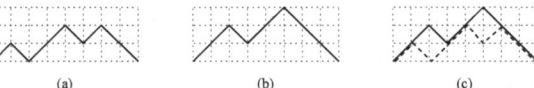

Fig. 2 Two comparable Dyck paths.   (a)   (b)   (c)

The Hasse diagram of the Dyck lattice on the set of Dyck paths of length 6 is represented in Figure 3 (a).

It is well known that Dyck paths of length $2n$ are counted by the n-th *Catalan* number $C_n = \frac{1}{n+1}\binom{2n}{n}$. The different incarnations of the Catalan family give rise to several further lattices beside Dyck's. Among them, the Tamari lattice [11] is indeed one of the more widely known, and appears naturally in the study of binary trees and of the Stasheff polytope. See Figure 3 (b) for the Hasse diagram of the Tamari lattice with five elements.

Using suitable bijections between Dyck paths, binary trees and planar trees, the two mentioned Catalan lattices can be defined on the set of planar trees of size n in such a way that the Dyck lattice with n elements is an extension of the Tamari lattice with n elements (see [13]).

In this paper, we will consider yet another occurrence of Catalan structures, namely *series parallel interval orders*. Our aim is to define the above two Catalan lattices on the set of series parallel interval orders with the aid of a special kind of linear extension. We propose this unified interpretation since, in our opinion, it allows to better understand the connections between Dyck and Tamari lattices. To obtain our characterization of the Dyck and Tamari lattices, we will make use of some basic notions of the theory of posets. The known relationship between the two lattices will be quite simple to achieve in our setting.

Thanks to this approach we will also be able to provide a link between the Dyck (respectively, Tamari) lattice and the strong (respectively, weak) Bruhat order, when

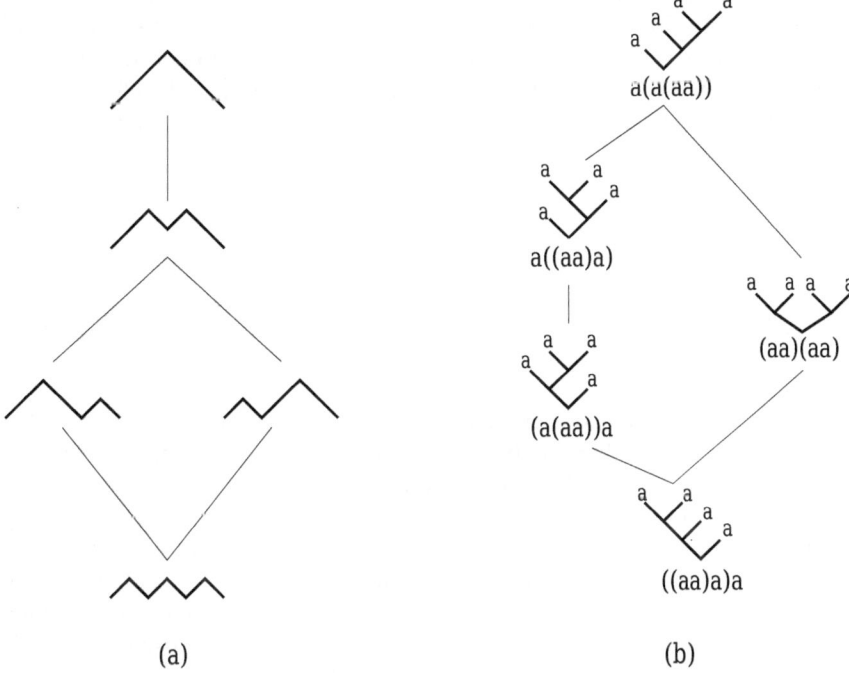

Fig. 3 (a) The Dyck lattice with five elements; (b) The Tamari lattice with five elements.

the latter is considered on the class of permutations avoiding the pattern 312. Recall that a permutation π is said to *avoid* a pattern ρ whenever there does not exist a subsequence of entries of π that have the same relative order as ρ. Here the pattern 312 comes out in a natural way, as we hope to make clear in Section 4.

As already recalled, the main combinatorial objects in this approach are series parallel interval orders, which are the intersection of two important classes of partially ordered sets, namely series parallel orders [16] and interval orders [10] (see also [5] for recent work in enumerative combinatorics involving interval orders) . For the purposes of the present paper, a *series parallel order* is a poset having no induced subposet isomorphic to the fence of order four, and an *interval order* is a poset having no induced subposet isomorphic to the poset $2+2$. Our approach will express important features of series parallel interval orders and so their use in this unified version of the two Catalan lattices seems to be relevant on its own.

2 Series-parallel interval orders

In this section we will focus on those posets having no induced subposet isomorphic either to the poset $2+2$ or to the fence of order four, shown in Figure 4. These partial orders are called *series parallel interval orders*. We will denote by $\mathcal{O}$ the class of such posets, also writing $\mathcal{O}(n)$ for those having precisely n elements. We should stress that, due to technical reasons, in what follows we will rather deal with the strict order relation associated with a series parallel interval order. Therefore we will always use the expression $R \in \mathcal{O}$ to mean that R is the strict order relation associated with a series parallel interval order.

Fig. 4 (a) The poset $2 + 2$; (b) The fence of order four.

(a)   (b)

These posets have been considered recently in [7], where the authors show that they are counted by Catalan numbers with respect to their size (i.e., number of elements); some bijections with other structures enumerated by Catalan numbers are also established. For our purposes, we need to recall here a bijection ρ (stated in [7]) between planar trees with $n + 1$ nodes and $\mathcal{O}(n)$. Given any planar tree T, we define the binary relation $R = \rho(T)$ on the set of its nodes other than the root, by setting xRy whenever x and y cannot be joined by a directed path in T (in the directed graph canonically determined by T) and x lies to the left of y in T. The resulting poset is indeed in $\mathcal{O}$. In Figure 5 we can see an instance of the bijection ρ. In what follows, we will always represent rooted trees with the root at the bottom.

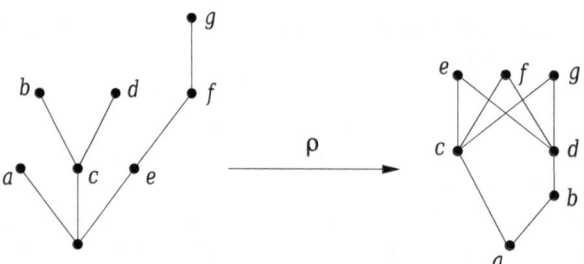

Fig. 5 The bijection ρ: a planar tree and the associated poset in $\mathcal{O}$.

2.1 Preorder linear extensions of series parallel interval orders

In this section we will define a particular type of linear extension for the posets in $\mathcal{O}$ corresponding to the preorder traversal in the associated planar tree; for this reason we will call it the *preorder linear extension*. Recall that a *preorder traversal* of a planar tree is a recursive way of visiting all the nodes of the tree consisting of the following operations:

1. visit the root;
2. traverse the subtrees of the root, in the order given by the planar tree structure.

In order to define the preorder linear extension of $R \in \mathcal{O}$, we need to define an auxiliary binary relation $Z(R) = Z$ on the support of R. Given a binary relation B, we set $\bar{B} = B \cup B^{-1}$ and we use the notation B^c to indicate the complement of B. Observe that, for a strict partial order relation B, $x\bar{B}y$ means that x and y are (distinct and) comparable in the partial order. Now define $Z = ((\bar{R})^c \circ \bar{R}) \setminus \bar{R}$. Recall that, for any two binary relations X and Y defined on the same set, the *composition* $X \circ Y$ is defined by setting $x(X \circ Y)y$ when there exists an element z such that xXz and zYy (see, for instance, [15] p. 48). Thus, we can rephrase the above definition by saying that xZy whenever $x\bar{R}y$ and there exists z such that $z\bar{R}y$ and $z\bar{R}x$. Given $R \in \mathcal{O}$, if x, y, z are such that zRy and x is incomparable with both z and y, then Z can be described as illustrated in Figure 6.

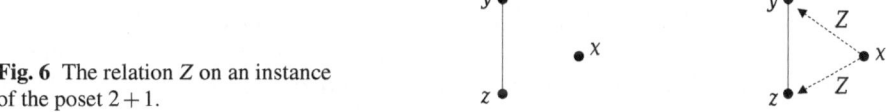

Fig. 6 The relation Z on an instance of the poset $2 + 1$.

We say that a linear extension λ of R is a *preorder linear extension* of R when xZy implies $x\lambda y$. Figure 7 depicts a poset $R \in \mathcal{O}$, the relation $Z(R)$ and an associated preorder linear extension.

Fig. 7 (a) An element $R \in \mathcal{O}$; (b) The relation Z associated with R, where pairs of Z are joined with an arrow. The labels of the nodes are given according to a preorder linear extension.

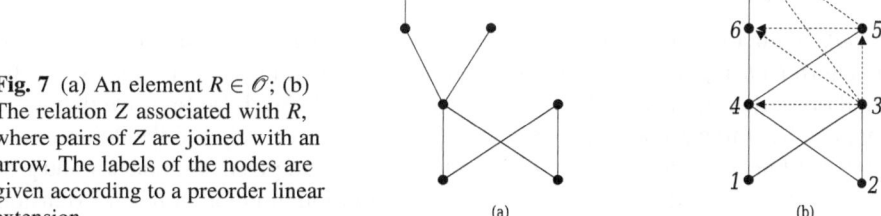

(a)   (b)

The next proposition shows that, for any $R \in \mathcal{O}$, there exists at most one preorder linear extension of R up to (order) automorphisms. The proposition needs a preceding lemma. In what follows, we will say that two elements of a poset are *order-equivalent* when there exists an order automorphism mapping one of them into the other.

Lemma 2.1. *Let $R \in \mathcal{O}$ and suppose that λ_1 and λ_2 are two preorder linear extensions of R. For any x, y in the support of R, if $\lambda_1(x) > \lambda_1(y)$ and $\lambda_2(x) < \lambda_2(y)$, then x and y must be order-equivalent in R.*

Proof. If $\lambda_1(x) > \lambda_1(y)$ then $(x, y) \notin R \cup Z$ and, by $\lambda_2(x) < \lambda_2(y)$, it follows that $(y, x) \notin R \cup Z$. From the definition of Z, this implies that, for every z, $z\overline{R}y$ or $z\overline{R}x$, and also that, for every z, $z\overline{R}y$ or $z\overline{R}x$. Equivalently, for every z, we get that either $z\overline{R}y$ and $z\overline{R}x$, or $z\overline{R}y$ and $z\overline{R}x$. Since $x\overline{R}y$, it is now easy to show that, for every $a \neq x, y$, aRx if and only if aRy and xRa if and only if yRa. $\square$

Proposition 2.2. *Let $R \in \mathcal{O}$ and suppose that λ_1 and λ_2 are two preorder linear extensions of R. If $\lambda_1(x) = \lambda_2(y)$, then the two elements x and y must be order-equivalent in R.*

Proof. If $\lambda_1(x) > \lambda_1(y)$ and $\lambda_2(x) < \lambda_2(y)$, then the assertion follows from the above lemma. Otherwise, without loss of generality, suppose that $\lambda_1(x) > \lambda_1(y)$ and $\lambda_2(x) > \lambda_2(y)$. We claim that there exists an element $\tilde{z}$ such that x is equivalent to $\tilde{z}$ which in turn is equivalent to y, and this will be enough to conclude.

Indeed, denoting by X the support of R, we have that $|\{z \in X \setminus \{x, y\} : \lambda_2(z) > \lambda_2(y) = \lambda_1(x)\}| + 1 = |\{z \in X \setminus \{x, y\} : \lambda_1(z) > \lambda_2(y) = \lambda_1(x)\}|$. Then there exists an element $\tilde{z} \in X \setminus \{x, y\}$ such that $\lambda_1(\tilde{z}) > \lambda_1(x) = \lambda_2(y)$ and $\lambda_2(\tilde{z}) < \lambda_1(x) = \lambda_2(y)$. Since $\lambda_1(\tilde{z}) > \lambda_1(x)$ and $\lambda_2(\tilde{z}) < \lambda_2(y) < \lambda_2(x)$, then $\tilde{z}$ must be order-equivalent to x in R (once again thanks to the above lemma). Analogously, since $\lambda_1(\tilde{z}) > \lambda_1(x) > \lambda_1(y)$ and $\lambda_2(\tilde{z}) < \lambda_2(y)$, $\tilde{z}$ and y are order-equivalent. $\square$

Thanks to the above proposition, we can assert that, for any $R \in \mathcal{O}$, there is at most one preorder linear extension of R. The next proposition shows that indeed a (the) preorder linear extension exists, and also suggests how to find it.

Proposition 2.3. *Let T be a planar tree and ρ be the bijection described in Section 2. Suppose that the nodes of T are labelled according to the preorder traversal (with the root labelled 0). Then the induced labelling on $\rho(T)$ determines a preorder linear extension of $\rho(T)$.*

Proof. With a slight abuse of notation, in this proof we will denote the elements of T and $\rho(T)$ using their labels in the appropriate linear extensions. Denote by $<_t$ the total order determined by the preorder traversal on T. We have to prove that $<_t$ is mapped by ρ to a preorder linear extension of $\rho(T)$. According to the definition of preorder linear extension, what we have to show is that, given two nodes x and y of T such that $x <_t y$, the pair (x, y) satisfies the definition of preorder linear extension. Due to our abuse of notations, this means that we have to prove that, if xZy in $\rho(T)$, then necessarily $x <_t y$. Indeed, xZy immediately implies that $x\overline{R}y$. This means that, in T, either x is a descendant of y, or y is a descendant of x. Suppose that the former case holds. The fact that xZy also implies that there exists an element z in $\rho(T)$ such that $z\overline{R}y$ and $z\overline{R}x$. In particular, this would mean that, in T, x should be a descendant of z and, at the same time, neither z could be a descendant of y nor y could be a descendant of z, which is plainly impossible (since T is a tree). Therefore we must have that y is a descendant of x in T, and so $x <_t y$. $\square$

We remark that the definition of preorder linear extension has a meaning only for elements in $\mathscr{O}$: just observe that it is not possible to construct a linear extension with the required properties neither of $2+2$ nor of the fence of order 4 (see again Figure 4).

In Figure 8 an example of the correspondence between preorder traversal and preorder linear extension is shown.

Fig. 8 A planar tree with the labelling of its nodes according to the preorder traversal and the Hasse diagram of the associated series parallel interval order with its preorder linear extension.

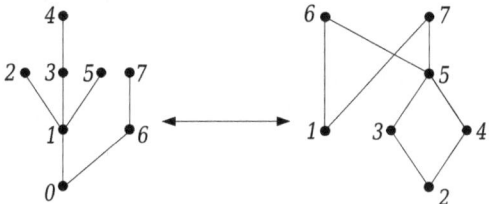

3 Catalan lattices on series parallel interval orders

In this section we will define the Dyck lattice and the Tamari lattice on series parallel interval orders whose support is equipped with a preorder linear extension.

In the sequel we will refer to a node of a planar tree using its label in the preorder traversal. Similarly, we will tacitly assume that posets in $\mathscr{O}$ are equipped with their preorder linear extension, and we will refer to their elements using the corresponding labels. Moreover, if x and y are labels, we will write $x < y$ referring to the usual order on natural numbers.

Given a planar tree T, let $\rho(T)$ denote the poset obtained through the bijection ρ defined above. In particular, we will sometimes refer to $\rho(T)$ as a binary relation. Given $x \in X$, for a binary relation B on X we consider the set $B(x) = \{y \in X : xBy\}$. If B is a partial order, then the set $B(x)$ is the *principal (order) filter* generated by the element x, whereas $B^{-1}(x) = \{y \in X : yBx\}$ is the *principal (order) ideal* generated by x. In the sequel, we will always use the terms "filter" and "ideal" in place of "order filter" and "order ideal". Moreover, we warn the reader that, according to our definition, a principal filter (respectively, ideal) will be considered without its minimum (respectively, maximum). In some sources, these are also called "open filter" and "open ideal", respectively. Finally, observe that the terms "up-sets" and "down-sets" are sometimes used in place of what are called "filters" and "ideals" here. We refer the reader to the example illustrated in Figure 12, which shows some filters in a specific case.

3.1 The Dyck lattice

According to [13], we start by recalling the definition of the Dyck lattice for the set of planar rooted trees with a fixed number of nodes. If T is a planar tree and k is a node of T, then define $h_T(k)$ as the set of ancestors of k in the tree T. Given two planar trees T_1 and T_2 having n nodes, T_1 is less than or equal to T_2 in the Dyck order, written as $T_1 \leq_D T_2$, whenever, for every node k, $|h_{T_1}(k)| \leq |h_{T_2}(k)|$.

The above definition allows us to give a characterization of the Dyck order in terms of series parallel interval orders.

Proposition 3.1. *Let T_1, T_2 be two planar trees having n nodes and let $\rho(T_1) = R_1$ and $\rho(T_2) = R_2$. Then the following conditions are equivalent:*

a) *$T_1 \leq_D T_2$;*
b) *$|R_1^{-1}(k)| \geq |R_2^{-1}(k)|$, for every k.*

Proof. By definition, $T_1 \leq_D T_2$ if and only if $|h_{T_1}(k)| \leq |h_{T_2}(k)|$ for all k. This is equivalent to:

$$|\{x \in T_1 : x \notin h_{T_1}(k)\}| \geq |\{x \in T_2 : x \notin h_{T_2}(k)\}|.$$

Consider now the series parallel interval orders R_1 and R_2 associated with T_1 and T_2 respectively. The previous condition may be expressed by saying that, for all k:

$$|\overline{R_1}(k) \cup \{x \in T_1 : x > k\}| \geq |\overline{R_2}(k) \cup \{x \in T_2 : x > k\}|. \tag{1}$$

To show that (1) is equivalent to $b)$ observe that, for a generic element k, the inequality

$$|\overline{R_1}(k) \cup \{k+1, \ldots, n\}| \geq |\overline{R_2}(k) \cup \{k+1, \ldots, n\}|$$

holds if and only if

$$|R_1^{-1}(k) \cup \{k+1, \ldots, n\}| \geq |R_2^{-1}(k) \cup \{k+1, \ldots, n\}|,$$

since $R_i(k) \subseteq \{k+1, \ldots n\}$, for $i = 1, 2$. Thus, since also $R_i^{-1}(k) \cap \{k+1 \ldots, n\} = \emptyset$, we immediately get

$$|R_1^{-1}(k)| + |\{k+1, \ldots, n\}| \geq |R_2^{-1}(k)| + |\{k+1, \ldots, n\}|,$$

which is precisely $b)$. □

In Figure 9 an application of this proposition is shown. The figure depicts two comparable elements in the Dyck order, and the trees on the left correspond to the posets on the right through ρ. The line connecting the two elements of the poset reads that the element placed above is greater than the one below, according to the Dyck order.

As a corollary, we find that the Dyck order on series parallel interval orders can be defined in terms of the cardinalities of the principal ideals of these posets.

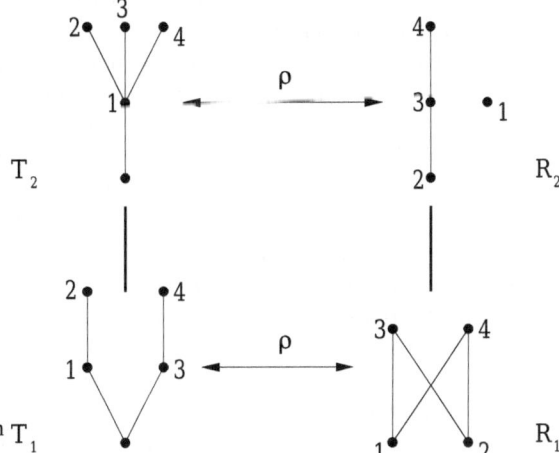

Fig. 9 Two comparable elements in the Dyck order.

Corollary 3.2. *Given* $R_1, R_2 \in \mathcal{O}(n)$, $R_1 \leq_D R_2$ *if and only if* $|R_1^{-1}(k)| \geq |R_2^{-1}(k)|$, *for every* $k \leq n$.

See also Figure 11 (a) for an example.

3.2 The Tamari lattice

According to [13], we start by recalling the definition of the Tamari lattice for the set of planar rooted trees with a fixed number of nodes. If T is a planar tree and k is a node of T, then define $u_T(k)$ as the set of descendants of k in the tree T. Given two planar trees T_1 and T_2 having n nodes, T_1 is less than or equal to T_2 in the Tamari order, written as $T_1 \leq_T T_2$, whenever, for every node k, $|u_{T_1}(k)| \leq |u_{T_2}(k)|$.

The following proposition provides equivalent conditions to define the Tamari order on the set of planar rooted trees, and on series parallel interval orders.

Proposition 3.3. *Let* T_1, T_2 *be two planar trees having n nodes and let* $\rho(T_1) = R_1$ *and* $\rho(T_2) = R_2$. *Then the following conditions are equivalent:*

a) $T_1 \leq_T T_2$;
b) $u_{T_1}(k) \subseteq u_{T_2}(k)$, *for every node k;*
c) $R_1(k) \supseteq R_2(k)$, *for every node k;*
d) $|R_1(k)| \geq |R_2(k)|$, *for every node k.*

Proof. We start showing that a) is equivalent to b). This is trivial by observing that, for any given node k in a planar tree T, if $|u_T(k)| = j$, then $u_T(k) = \{k+1, k+2, \ldots, k+j-1, k+j\}$.

Now we prove that b) is equivalent to c). Given a node k in the planar tree T, consider the set $d(k) = u(k) \cup h(k)$, i.e., the set of ancestors and descendants of k.

We first want to prove that $u_{T_1}(k) \subseteq u_{T_2}(k)$ (for every node k) if and only if $d_{T_1}(k) \subseteq d_{T_2}(k)$ (for every node k).

In fact, if $u_{T_1}(k) \subseteq u_{T_2}(k)$ for all k, then $h_{T_1}(k) \subseteq h_{T_2}(k)$ for all k as well. Indeed, if $x \in h_{T_1}(k)$, then $k \in u_{T_1}(x)$, hence $k \in u_{T_2}(x)$ and so $x \in h_{T_2}(k)$. Therefore $d_{T_1}(k) = u_{T_1}(k) \cup h_{T_1}(k) \subseteq u_{T_2}(k) \cup h_{T_2}(k) = d_{T_2}(k)$, for every node k.

Now suppose that, for every node k, $d_{T_1}(k) \subseteq d_{T_2}(k)$. If $x \in u_{T_1}(k)$ then $x \in d_{T_1}(k)$ and then $x \in d_{T_2}(k)$ with $x > k$. So $x \in u_{T_2}(k) \cup h_{T_2}(k)$, that is $x \in u_{T_2}(k)$, since $x > k$.

From the previous statement we have that condition b) holds if and only if, for every k, $d_{T_1}(k) \subseteq d_{T_2}(k)$. Now $d_{T_1}(k) \subseteq d_{T_2}(k)$ if and only if, for every y, kR_2y implies kR_1y, that is, for every k, $R_1(k) \supseteq R_2(k)$. Indeed, suppose that, for every k, $R_1(k) \supseteq R_2(k)$. If $y \notin d_{T_2}(k)$ then $k\overline{R_2}y$, whence $k\overline{R_1}y$, and so $y \notin d_{T_1}(k)$. Vice versa, suppose that, for every k, $d_{T_1}(k) \subseteq d_{T_2}(k)$. If $y\overline{R_2}k$, then $y\overline{R_1}k$. Now, if kR_2y, then obviously $k < y$, which cannot hold together with $y R_1 k$ and so kR_1y. Thus we can conclude that $y R_2 k$ implies $y R_1 k$, as desired.

The equivalence between c) and d) is trivial, and it is left to the reader. □

Figure 10 shows an application of this proposition. The figure depicts two comparable elements in the Tamari order, and the trees on the left correspond to the posets on the right through ρ. Recalling how the bijection ρ works, we have that, for instance, node 1 in T_2 becomes an isolated point in R_2 since it can be joined to any other node of T_2 by means of a directed path; moreover, the presence of the chain $2 < 3 < 4$ in R_2 is motivated by the fact that, in T_2, 2 lies to the left of 3, 3 lies to the left of 4 and there is no directed path joining any two of these nodes.

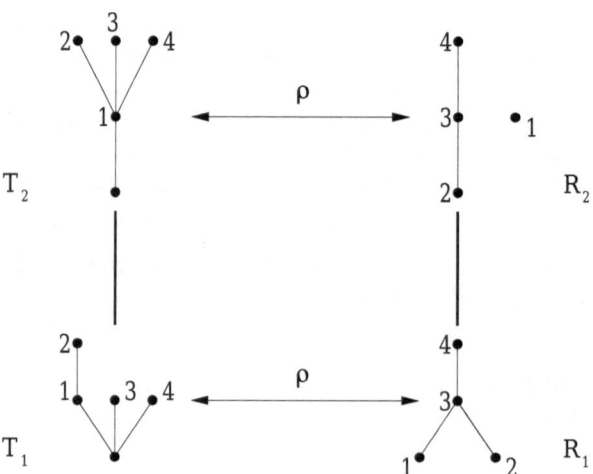

Fig. 10 Two comparable elements in the Tamari order.

In Figure 11 (b) the Tamari lattice on the five elements belonging to $\mathcal{O}(3)$ is depicted.

Remark 3.4. We know from [13] that the Dyck and the Tamari orders are related by the following refinement property: given two Catalan structures T_1, T_2 of the same size, if $T_1 \leq_T T_2$ then $T_1 \leq_D T_2$. This fact is an obvious consequence of Proposition 3.3. Our approach seems to be particularly interesting since it is now possible to prove such a refinement property in a very neat way. Indeed, if $R_1(k) \supseteq R_2(k)$ holds for any k, then we also have that, for all k, $R_1^{-1}(k) \supseteq R_2^{-1}(k)$, and so, for all k, $|R_1^{-1}(k)| \geq |R_2^{-1}(k)|$.

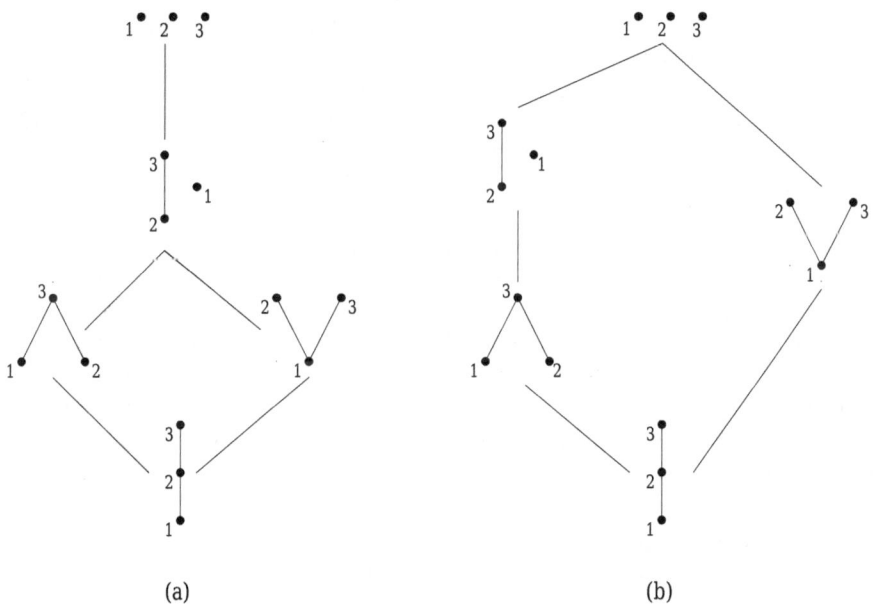

(a)   (b)

Fig. 11 The two Catalan lattices defined on the class of series parallel interval orders of size three: (a) Dyck lattice (b) Tamari lattice.

4 Series parallel interval orders and pattern avoiding permutations

The *strong Bruhat* order ($\leq_B$) and the *weak Bruhat* order ($\leq_b$) are two well-known partial orders defined on the set of permutations having fixed length [16]. We briefly recall here their definitions.

Given a permutation $\pi = a_1 a_2 \ldots a_n$, a *reduction* of π is a permutation obtained from π by interchanging some a_i with some a_j, provided that $i < j$ and $a_i > a_j$. We say that $\pi_1 <_B \pi_2$ whenever π_1 is obtained from π_2 through a sequence of reductions. Define a *simple reduction* of $\pi = a_1 a_2 \ldots a_n$ as a permutation obtained from π by interchanging some a_i with some a_{i+1}, provided that $a_i > a_{i+1}$. We say that $\pi_1 <_b \pi_2$ whenever π_1 is obtained from π_2 through a sequence of simple reductions.

In this section we will consider another well-known Catalan structure, namely the class of permutations avoiding the pattern 312, and we will prove, using a characterization given in [1], that the strong Bruhat order, when restricted to such a class of pattern avoiding permutations, is isomorphic to the Dyck order. Moreover we will show that an analogous isomorphism also exists between the Tamari lattice and the weak Bruhat order on the same class of permutations. We remark that these two results have been already obtained independently in [1] (for the Dyck case) and in [6] (for the Tamari case; see also [8]). Here our main aim is to find a common language for these two results.

We start by describing a bijection between series parallel interval orders on n elements and permutations of length n avoiding the pattern 312, denoted by $Av_n(312)$. Our approach can be compared with the one used in [5] to enumerate posets avoiding $2+2$.

First of all recall that the set of principal filters of a poset avoiding $2+2$ is linearly ordered by inclusion. The interested reader can find a proof of this fact in [9], where it is also proved that this condition completely characterizes such a class of posets.

Given a poset $R \in \mathcal{O}(n)$, consider the labelling of its elements determined by its preorder linear extension and denote its principal filters by $R(1), R(2), \ldots, R(n)$. Define a permutation $\pi(R)$ of length n as follows: k precedes j in $\pi(R)$ precisely when either $R(k) \supset R(j)$ or $R(k) = R(j)$ and $k > j$. It is easy to show that, for each $R \in \mathcal{O}$, $\pi(R)$ does not contain the pattern 312, and the function $R \mapsto \pi(R)$ is a bijection between $\mathcal{O}(n)$ and $Av_n(312)$.

Remark 4.1. Observe that our bijection cannot be described in terms of principal ideals (instead of principal filters), due to our choice of taking the preorder linear extension of a poset.

For instance, the permutation associated with the poset depicted in Figure 12 is 2146753. Indeed the filters of such a poset are $R(1) = R(2) = \{3,4,5,6,7\}, R(3) = R(5) = R(7) = \emptyset, R(4) = \{5,6,7\}, R(6) = \{7\}$ and then they are listed as follows:

$$R(2) = R(1) \supset R(4) \supset R(6) \supset R(7) = R(5) = R(3).$$

Remark 4.2. Given a permutation $\pi = a_1 \cdots a_n$ of length n and the partial order relation $R \in \mathcal{O}(n)$ associated with it, it is not difficult to observe that $R(a_j)$ is the set of the elements of π greater than a_j and following a_j in π. Analogously, $R^{-1}(a_j)$ is the set of the elements of π lesser than a_j and preceding a_j in π (see again Figure 12). In what follows, we will use the notations $f_\pi(a_j)$ and $i_\pi(a_j)$ in place of $R(a_j)$ and $R^{-1}(a_j)$ (respectively) when dealing with permutations rather than partial order relations.

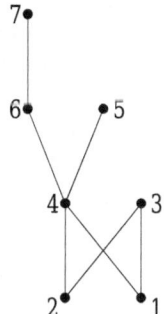

Fig. 12 The Hasse diagram of a
series parallel interval order, together
with its preorder extension.

4.1 The Tamari lattice and the weak Bruhat order on Av(312)

According to [12], it is possible to characterize the weak Bruhat order on permutations using inversions. Recall that an inversion of $\pi = a_1 a_2 \ldots a_n$ is a pair (a_i, a_j) such that $i < j$ and $a_i > a_j$. Given two permutations of the same length π_1 and π_2, it is $\pi_1 \leq_b \pi_2$ if and only if the set $E(\pi_1)$ of inversions of π_1 is a subset of the set $E(\pi_2)$ of inversion of π_2. The following simple proposition provides the key ingredient to prove that the Tamari lattice is isomorphic to the weak Bruhat order on $Av(312)$.

Proposition 4.3. *Let π_1 and π_2 be two permutations of length n. Then $E(\pi_1) \subseteq E(\pi_2)$ if and only if $f_{\pi_1}(k) \supseteq f_{\pi_2}(k)$, for every $1 \leq k \leq n$.*

Proof. Suppose that, for every k, $f_{\pi_1}(k) \supseteq f_{\pi_2}(k)$. If $(i,j) \in E(\pi_1)$, with $i > j$, then $i \notin f_{\pi_1}(j)$ and then $i \notin f_{\pi_2}(j)$. This implies that $(i,j) \in E(\pi_2)$.

Vice versa, suppose that $E(\pi_1) \subseteq E(\pi_2)$. If $i \in f_{\pi_2}(j)$, then $(i,j) \notin E(\pi_2)$, whence $(i,j) \notin E(\pi_1)$. This implies that $i \in f_{\pi_1}(j)$. $\square$

Corollary 4.4. *The Tamari order is isomorphic to the weak Bruhat order restricted to $Av_n(312)$.*

4.2 The Dyck lattice and the strong Bruhat order on Av(312)

For a given permutation π of length n, define the vector $\max_\pi$ as follows: $\max_\pi(k) = \max\{\pi(i) : i \leq k\}$. According to [1], we recall that, given two 312-avoiding permutations π_1 and π_2 of length n, $\pi_1 \leq_B \pi_2$ if and only if, for all $1 \leq k \leq n$, $\max_{\pi_1}(k) \leq \max_{\pi_2}(k)$. For instance, considering the two permutations $\pi_1 = 468753921, \pi_2 = 768543921 \in Av_9(312)$, we have

$$\max_{\pi_1} = (4,6,8,8,8,8,9,9,9) \qquad \max_{\pi_2} = (7,7,8,8,8,8,9,9,9),$$

whence $\pi_1 \leq_B \pi_2$. Indeed, starting from π_2, we obtain π_1 by the following reductions: $\pi_2 = 768543921 \rightarrow 768453921 \rightarrow 468753921 = \pi_1$. Observe that, in the above sequence of reductions, the permutation 768453921 is not 312-avoiding.

Given a permutation π of length n, consider the set of its *consecutive noninversions*, i.e., the set of all $m \in \{1,\ldots,n\}$ such that either $m = n$ or the pair $(m, m+1)$ is a noninversion of π (that is m appears before $m+1$ in π). The following lemma provides a characterization of consecutive noninversions in permutations avoiding 312.

Lemma 4.5. *Let $\pi \in Av_n(312)$, then the following properties hold:*

  i) *m is a consecutive noninversion of π if and only if either $m = n$ or $|i_\pi(m)| \neq |i_\pi(m+1)|$;*
  ii) *if $j < k$, then $|i_\pi(j)| \leq |i_\pi(k)|$;*
  iii) *if $k = |i_\pi(m)|$ and m is a consecutive noninversion of π, then $\pi(k+1) = m$;*
  iv) *the consecutive noninversions of π are those elements of π preceded only by lesser entries;*
  v) *the set of all consecutive noninversions of π coincides with the set of components of $\max_\pi$. Moreover, the index of the consecutive noninversion m in π coincides with the index of the first occurrence of m in $\max_\pi$.*

Proof. i) If $m \neq n$ is a consecutive noninversion, then obviously $|i_\pi(m+1)| \geq |i_\pi(m)| - 1$, whence $|i_\pi(m)| \neq |i_\pi(m+1)|$. Vice versa, suppose that $m \neq n$ is such that $|i_\pi(m)| \neq |i_\pi(m+1)|$; if $(m+1, m)$ were an inversion of π, then there should be an entry $j < m$ between $m+1$ and m, which is impossible since $\pi \in Av_n(312)$.

ii) This is an immediate consequence of the fact that $\pi \in Av_n(312)$.

iii) This is obvious when $m = n$. Otherwise, suppose that m is an element of π having more than i elements on its left; then there would be at least one element j such that j precedes m and $j > m$, and the three elements $j, m, m+1$ would show a 312-pattern in π, a contradiction.

iv) Observe that, if m is a consecutive noninversion of π, then, by iii), all elements of π preceding m are less than m. Vice versa, if m is an entry of π preceded only by lesser elements, one cannot have $|i_\pi(m)| = |i_\pi(m+1)|$, since in this case $m+1$ would precede m. Then $m = n$ or $|i_\pi(m)| < |i_\pi(m+1)|$.

v) This is a direct consequence of iv).   □

Proposition 4.6. *Given $\pi_1, \pi_2 \in Av_n(312)$, the following conditions are equivalent:*

  a) *$|i_{\pi_1}(k)| \geq |i_{\pi_2}(k)|$, for every $1 \leq k \leq n$;*
  b) *$\max_{\pi_1}(k) \leq \max_{\pi_2}(k)$, for every $1 \leq k \leq n$.*

Proof. a) $\Rightarrow$ b) Let $j' = \pi_1(j)$ be a consecutive noninversion of π_1. Thanks to the above lemma, item iii), all the elements of π_1 preceding j' are smaller than j'. We claim that there exists $k = \pi_2(h)$ such that $k > j'$, $h \leq j$ and all the elements of π_2 before k are smaller than k. Indeed, we have $j - 1 = |i_{\pi_1}(j')| \geq |i_{\pi_2}(j')| = |i_{\pi_2}(k)| = t$, where k is the first consecutive noninversion of π_2 on the left of j' with $k \geq j'$ (such a k does indeed exist, as the reader can immediately check). Then $k = \pi_2(t+1)$ is the desired element of π_2. As a consequence, we have that, if j' is a consecutive noninversion of π_1, then $\max_{\pi_1}(j') \leq \max_{\pi_2}(j')$. From this we can immediately deduce the same inequality for any $j' \leq n$.

b) $\Rightarrow$ a) Set $t = |i_{\pi_1}(j)|$, consider the consecutive noninversion j' of π_1 such that $|i_{\pi_1}(j')| = t$ with $j' \geq j$ (this is the first consecutive noninversion of π_1 on the left of j). From the previous lemma we have $j' = \max_{\pi_1}(t+1) \leq \max_{\pi_2}(t+1) = \max_{\pi_2}(h)$, where $h \leq t+1$ is the index of the first component of $\max_{\pi_2}$, which is equal to $\max_{\pi_2}(t+1)$. Again from the previous lemma, we have that $\max_{\pi_2}(h) = k$ is a consecutive noninversion of π_2 and $|i_{\pi_2}(k)| = h - 1$. Finally, using in particular item ii) of the above lemma in the first two inequalities, we have:

$$|i_{\pi_2}(j)| \leq |i_{\pi_2}(j')| \leq |i_{\pi_2}(k)| = h - 1 \leq t = |i_{\pi_1}(j)|. \qquad \square$$

Corollary 4.7. *The Dyck order is isomorphic to the strong Bruhat order restricted to* $Av_n(312)$.

5 Further works

In the present work we have considered two well-known Catalan posets and we have proposed a unifying language to describe them based on the notion of series parallel interval order. There are of course several other poset structures which can be considered on the objects of the Catalan family. Maybe the most famous one is the *Kreweras order* [14], which is naturally defined on noncrossing partitions of a set of given cardinality by refining the classical partial order on set partitions. Other less classical posets have been defined by Baril and Pallo on Dyck words (the *phagocyte lattice* [2]) and on binary trees (the *pruning-grafting lattice* [3]). It seems natural to ask if series parallel interval orders can be used also to describe these (and maybe other) Catalan posets. Unfortunately, we have not been able to find an answer to such a question yet.

References

1. E. Barcucci, A. Bernini, L. Ferrari, M. Poneti, "A distributive lattice structure connecting Dyck paths, noncrossing partitions and 312-avoiding permutations", *Order* **22** (2005) 311–328.
2. J.L. Baril, J.M. Pallo, "The phagocyte lattice of Dyck words", *Order* **23** (2006) 97–107.
3. J.L. Baril, J.M. Pallo, "The pruning-grafting lattice of binary trees", *Theoret. Comput. Sci.* **409** (2008) 382–393.
4. O. Bernardi, N. Bonichon, "Intervals in Catalan lattices and realizers of triangulations", *J. Combin. Theory Ser. A* **116** (2009) 55–75.
5. M. Bousquet-Melou, A. Claesson, M. Dukes, S. Kitaev, "Unlabeled $(2+2)$-free posets, ascent sequences and pattern avoiding permutations", *J. Combin. Theory Ser. A* **117** (2010) 884–909.
6. A. Björner, M. Wachs, "Shellable nonpure complexes and posets 2", *Trans. Amer. Math. Soc.* **349** (1997) 3945–3975.
7. F. Disanto, L. Ferrari, R. Pinzani, S. Rinaldi, "Catalan pairs: A relational-theoretic approach to Catalan numbers", *Adv. Appl. Math.* **45** (2010) 505–517.

8. B. Drake, "The weak order on pattern-avoiding permutations", in *FPSAC* 17*th International Conference on Formal Power Series and Algebraic Combinatorics*, L. Carini, J.Y. Thibon, eds., 2005.
9. M.H. El-Zahar, "Enumeration of ordered sets", in Algorithms and order, I. Rival, ed., Kluwer Acad. Publ., Dordrecht (1989) 327–352.
10. P.C. Fishburn, "Intransitive indifference with unequal indifference intervals", *J. Math. Psych.* **7** (1970) 144–149.
11. S. Huang, D. Tamari, "Problems of associativity: A simple proof of the lattice property of systems ordered by a semi-associative law", *J. Combin. Theory Ser. A* **13** (1972) 7–13.
12. D. Knuth, *The art of computer programming*, Vol. 3, Addison Wesley, 1998.
13. D. Knuth, *The art of computer programming*, Vol. 4, Addison Wesley, 2006.
14. G. Kreweras, "Sur les partitions non croisées d'un cycle", *Discrete Math.* **1** (1972) 333–350.
15. J.P.S. Kung, G.-C. Rota, C.H. Yan, *Combinatorics: The Rota Way*, Cambridge University Press, Cambridge, 2009.
16. R.P. Stanley, *Enumerative Combinatorics*, Vol. 1, Cambridge University Press, 1997.
17. R.P. Stanley, *Enumerative Combinatorics*, Vol. 2, Cambridge University Press, 1999.

Generalized Tamari Order

María Ronco

Abstract In [2], M. Carr and S. Devadoss introduced the notion of *tubing* on a finite simple graph G. When G is the linear graph L_n, with n vertices, the polytope $\mathcal{K} L_n$ is the Stasheff polytope or associahedron. Our goal is to describe a partial order on the set of tubings of a simple graph, which generalizes the Tamari order on the set of vertices of the associahedron. For certain families of graphs, this order induces an associative product on the vector space spanned by the tubings of all the graphs.

1 Introduction

The graph associahedra, introduced by M. Carr and S. Devadoss, associate to any simple finite graph G the set of its tubings, which describe the faces of a polytope $\mathcal{K} G_n$ in $\mathbb{R}^{n-1}$, where n is the number of vertices of G. When G is the linear graph L_n, with n vertices, the polytope $\mathcal{K} L_n$ is the Stasheff polytope or associahedron, while the polytope $\mathcal{K} K_n$ associated to the complete graph K_n is a permutohedron. In [5], S. Forcey and D. Springfield proved that for any simple graph G with n vertices, the polytope KG may be obtained from the permutohedron of dimension $n-1$ by contracting some faces, each contraction corresponds to the elimination of an edge of the complete graph K_n.

Moreover, there exist bijective maps from the set of vertices of the associahedron of dimension $n-1$ and the set Y_n of planar binary rooted trees with $n+1$ leaves on one hand, and between the set of vertices of the permutohedron of dimension $n-1$ and the set S_n of permutations of n elements. The vector space spanned by all the permutations has a natural structure of graded Hopf algebra, described in [10], which is denoted $\mathscr{S}Sym$; the vector space spanned by the set of all planar binary trees has also a natural structure of Hopf algebra described in [6], which is denoted $\mathscr{Y}Sym$.

María Ronco

MOR: Instituto de Matemáticas y Física, Universidad de Talca, Avda. Lircay s/n, Talca, Chile, e-mail: *maria.ronco@inst-mat.utalca.cl*

S. Forcey and D. Springfield showed in [5] how to understand the product and coproduct of the Hopf algebras $\mathscr{S}Sym$ and $\mathscr{Y}Sym$ in terms of tubings. Furthermore, they introduced new algebraic structures on the spaces spanned by maximal tubings of the cyclohedron and the standard simplices.

In the present work, we extend the definition of the associative products $/$ and $\backslash$, defined in [7] on the space spanned by the vertices of the associahedron, to the spaces spanned by the tubings of any family $\{G_n^i \mid 1 \le i \le r_n\}_{n \ge 1}$ of simple connected graphs satisfying $|\mathrm{Vert}(G_n^i)| = n$.

There exists a natural way to understand the associative products of the algebras $\mathscr{S}Sym$ and $\mathscr{Y}Sym$ in terms of restrictions on maximal tubings of the families $\{K_n\}_{n \ge 1}$ of complete graphs and $\{L_n\}_{n \ge 1}$ of linear graphs, respectively. It is therefore possible to define a binary operation $*$ on the spaces spanned by the set of tubings of a family $\mathscr{G} = \{G_n^i \mid 1 \le i \le r_n\}_{n \ge 1}$ of simple finite graphs, and we show that this product is associative if all the graphs are connected and $\mathscr{G}$ satisfies some extra condition.

Finally, for any simple finite graph G, we define a partial order on the set $Tub(G)$ of all tubings of G, and prove that the binary operation $*$ can be described in terms of this order and the associative products $/$ and $\backslash$. In the case of the maximal tubings of the cyclohedron, our order coincides with the one introduced by S. Forcey in [4].

All the vector spaces considered in the present work are over a field $\mathbb{K}$. For any set X, we denote by $\mathbb{K}[X]$ the vector space spanned by X.

2 Graph associahedra

We give a brief description of graph associahedra introduced by M. Carr and S. Devadoss in [2], for a complete description of their construction and further details we refer to their work.

Let G be a graph, we denote by $\mathrm{Vert}(G)$ the set of its vertices and by $\mathrm{Edg}(G)$ the set of its edges.

Definition 2.1. Let G be a graph. For any subset S of $\mathrm{Vert}(G)$, let G_S be the subgraph of G whose set of vertices is S and whose set of edges is $\mathrm{Edg}(G_S) := \{e \in \mathrm{Edg}(G) \mid$ both vertices of e belong to $S\}$. The graph G_S is called the *subgraph induced by S*.

Definition 2.2. Let G be a simple finite graph. A *tube* in G is a proper subset S of $\mathrm{Vert}(G)$ such that the induced subgraph G_S is connected.

Definition 2.3. Let t and t' be two different tubes in a simple finite graph G. We say that t and t' are *compatible* if they satisfy one of the following conditions:

1. either $t \subset t'$, or $t' \subset t$,
2. $t \cap t' = \emptyset$, and there does not exist any edge $e \in \mathrm{Edg}(G)$ such that one of the vertices of e is in t and the other vertex belongs to t'.

For a finite simple graph G, we identify the set of vertices of G with the ordered set $\mathrm{Vert}(G) = \{1 < \cdots < n\}$.

Definition 2.4. Let G be a simple finite graph. A *tubing* of a simple finite graph G is a family $T = \{t^i\}_{1 \le i \le k}$ of tubes of G such that every pair of tubes in T is compatible. Let $Tub(G)$ denotes the set of all tubings of G.

Note that a tubing T of G has at most $n-1$ tubes if G has n vertices. A tubing with k tubes is called a *k-tubing* of G, for $0 \le k \le n-1$.

The set $Tub(G)$ of all tubings of a graph G is partially ordered by the relation $T \prec T'$ if T is obtained from T' by adding tubes. A *maximal tubing* of a graph G with n vertices is a tubing with $n-1$ tubes, we denote by $MTub(G)$ the set of maximal tubings of G.

For any tubing $T \in Tub(G)$, let $\overline{T} := T \cup \mathrm{Vert}(G)$. In other words $\overline{T}$ is the collection of tubes $t \in T$ plus the improper tube of all vertices of G. For a fixed graph G, we denote by $\overline{Tub(G)}$ the set of all $\overline{T}$, with $T \in Tub(G)$.

In [2], M. Carr and S. Devadoss proved that for any simple finite graph G, with n vertices, the geometric realization of the poset $(Tub(G), \prec)$ is the barycentric division of a simple, convex polytope $\mathcal{K}G$ of dimension $n-1$, whose faces of dimension r are indexed by the $n-1-r$ tubings of G, for $0 \le r \le n-1$.

For $n \ge 1$, they consider, as particular examples, four types of graphs with n vertices $\{1 < \ldots, n\}$:

1. the graph S_n with $\mathrm{Edg}(S_n) = \emptyset$,
2. the linear graph L_n with $\mathrm{Edg}(L_n) = \{(i, i+1) \mid 1 \le i \le n-1\}$,
3. the cyclic graph C_n with $\mathrm{Edg}(C_n) = \{(i, i+1) \mid 1 \le i \le n-1\} \cup \{(1, n)\}$,
4. the complete graph K_n with $\mathrm{Edg}(K_n) = \{(i, j) \mid 1 \le i < j \le n\}$,

and proved that:

1. $\mathcal{K}S_n$ is the standard $n-1$-simplex

$$\Delta^{n-1} = \{(t_0, \ldots, t_{n-1}) \in \mathbb{R}^n \mid \sum_{i=0}^{n-1} t_i = 1 \text{ and } 0 \le t_i, 0 \le i \le n-1\}.$$

2. $\mathcal{K}L_n$ is the Stasheff polytope of dimension $n-1$, whose faces are indexed by all planar rooted trees with $n+1$ leaves (see [13]).
3. $\mathcal{K}C_n$ is the cyclohedron of dimension $n-1$ (see [11]).
4. $\mathcal{K}K_n$ is the permutohedron of dimension $n-1$, whose faces of dimension k are indexed by the set of surjective maps $\{1, \ldots, n\} \longrightarrow \{1, \ldots, k\}$ (see [13]).

We refer to [2] and [5] for a nice and detailed description of the polytopes $\mathcal{K}G$, when G is one of the graphs enumerated above. We just draw two examples.

a) For $L_3 = \;{}_1\!\bullet\!\!-\!\!\bullet_2\!\!-\!\!\bullet_3$, $\mathcal{K}L_3$ is given by

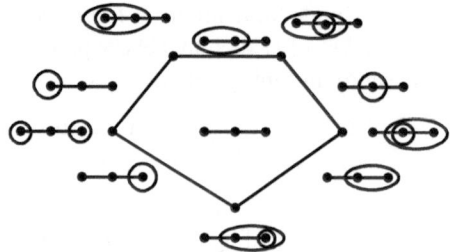

b) For $G_3 = {}_1 \bullet\!\!-\!\!\bullet {}_2 \quad {}_3$, $\mathcal{K} G_3$ is given by

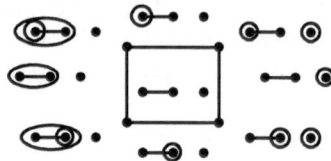

Note that the vertices of the polytope $\mathcal{K} G$ coincide with the subset $MTub(G)$ of maximal tubings, while the empty tubing corresponds to the $(n-1)$-cell of $\mathcal{K} G$.

3 Binary operations on tubings

Let Σ_n denote the set of permutations of n elements. In [10], C. Malvenuto and C. Reutenauer defined an autodual Hopf algebra structure on the graded $\mathbb{K}$-vector space spanned by $\bigcup_{n \geq 0} \Sigma_n$, denoted $\mathcal{S} Sym$.

In a quite similar way, in a joint paper with J.-L. Loday (see [6]), we describe a Hopf algebra structure on the graded vector space spanned by the set of all planar binary rooted trees, which is denoted $\mathcal{Y} Sym$, and proved that $\mathcal{Y} Sym$ is a sub-Hopf algebra of $\mathcal{S} Sym$.

In [5], S. Forcey and D. Springfield described these Hopf algebraic structures in terms of natural operations on tubings of the families $\{K_n\}_{n \geq 1}$ of complete graphs and $\{L_n\}_{n \geq 1}$ of linear graphs. They also show that the injective Hopf algebra homomorphism $\mathcal{Y} Sym \hookrightarrow \mathcal{S} Sym$ may be also described in terms of restrictions of tubings. The algebra and coalgebra structures of $\mathcal{S} Sym$ induce a product and a coproduct on the vector spaces spanned by the set of all tubings of some families of graphs, like the family of cyclic graphs and $\{S_n\}_{n \geq 1}$. As S. Forcey and D. Springfield pointed out in [5], these structures are not always coassociative. However, they introduced new algebraic structures on the space spanned by all the 0-faces of cyclohedra, on one hand; and on the space spanned by the 0-faces of standard simplices.

Let us point out that the Hopf algebra structure $\mathcal{Y} Sym$ was extended, in two different ways (see [3] and [8]), to the vector space spanned by all the faces of permutohedra. In fact, F. Chapoton's Hopf algebra structure is the associated graded structure of the Hopf algebra defined in [7].

From now on all the graphs considered in this work are finite and connected.

Definition 3.1. Let G be a simple finite graph, and let $S \subseteq \mathrm{Vert}(G)$. Suppose that H is a finite graph such that $\mathrm{Vert}(H) = S$ and $\mathrm{Edg}(H) \subseteq \mathrm{Edg}(G_S)$. For any tube $t = \{i_1, \ldots, i_l\}$ of G, the set of vertices $S \cap \{i_1, \ldots, i_l\}$ determines either a tubing of H or is the whole set of vertices of H, we denote it by $t|_H$.

Remark 3.2. For any tubing $T \in Tub(G)$, any subset S of $\mathrm{Vert}(G)$, and any graph H satisfying the conditions of the Definition 3.1, the element $T|_H := \{t|_H\}_{t \in T}$ is the element of $Tub(H) \bigcup \overline{Tub(H)}$ obtained by restricting each tube of T to H and keeping only one copy of each tube.

Example 3.3. Let us show how restriction works from G to H in the following example

Definition 3.4. Suppose that, for any $n \geq 1$, there exists a positive integer s_n and a collection of simple graphs $\{G_n^i\}_{1 \leq i \leq s_n}$ such that $\mathrm{Vert}(G_n^i) = \{1, \ldots, n\}$. Let $\mathscr{G} = \{\{G_n^i\}_{1 \leq i \leq s_n}\}_{n \geq 1}$ be the infinite collection of simple graphs. We say that $\mathscr{G}$ is *admissible* if it satisfies the following conditions:

1. there exists a collection of maps

$$\alpha(n,m) : \{1, \ldots, s_n\} \times \{1, \ldots, s_m\} \longrightarrow \{1, \ldots, s_{n+m}\}, \quad \text{for } n, m \geq 1,$$

   which is associative, that is:

$$\alpha(n,m,r) := \alpha(n+m,r) \circ (\alpha(n,m) \times Id)$$
$$= \alpha(n,m+r) \circ (Id \times \alpha(m,r)).$$

2. $(G_{n+m}^{\alpha(n,m)(i,j)})_{\{1,\ldots,n\}}$ is obtained from G_n^i by eliminating some edges, for all $1 \leq i \leq s_n$,

3. $(G_{n+m}^{\alpha(n,m)(i,j)})_{\{n+1,\ldots,n+m\}}$ is obtained from $G_m^j + n$ by eliminating some edges, where $G_m^j + n$ is the graph G_m^j with the vertex k replaced by $k+n$, for $1 \leq k \leq m$.

Example 3.5.

1. The families of linear graphs $\{L_n\}_{n \geq 1}$, of cyclic graphs $\{C_n\}_{n \geq 1}$ and of complete graphs $\{K_n\}_{n \geq 1}$ are admissible, as well as the family $\{S_n\}_{n \geq 1}$ of finite graphs without edges. Note that in these examples $s_n = 1$ for all $n \geq 1$, and $\alpha(n,m)$ is the unique map from $\{1\} \times \{1\}$ to $\{1\}$.

2. Let $\mathscr{G} = \{G^r\}_{1 \leq r \leq N}$ be a family of finite simple graphs, with $\mathrm{Vert}(G^r) = \{1, \ldots, n_r\}$, such that there exists a positive integer m satisfying $\{1, \ldots, m\} \subseteq \{n_1, \ldots, n_N\}$. There exist two easy ways to construct admissible families of simple finite graphs from it:

a. Consider $\mathcal{G}_1 := \mathcal{G} \cup \{G^{r,s} \mid 1 \leq r,s \leq N\}$, where $G^{r,s}$ is the graph obtained by taking the disjoint union of G^r and $G^s + n_r$ and linking the vertices n_r and $n_r + 1$ with a new edge.

Define recursively $\mathcal{G}_k := (\mathcal{G}_{k-1})_1$, for $k \geq 2$, and $\tilde{\mathcal{G}} := \bigcup_{k \geq 1} \mathcal{G}_k$.

The family of simple graphs $\tilde{\mathcal{G}}$ is admissible and has the following property. For any $n \in \mathbb{N}$, there exists a finite number of graphs in $\tilde{\mathcal{G}}$ with n vertices.

Note that the family of linear graphs $\{L_n\}_{n \geq 1}$ coincides with $\widetilde{\{\bullet\}}$, where $\{\bullet\}$ is the family of a unique graph with one vertex and no edges.

b. Consider $\mathcal{G}_1 := \mathcal{G} \cup \{G^{r,s} \mid 1 \leq r,s \leq N\}$, where $G^{r,s}$ is the graph obtained by taking the disjoint union of G^r and $G^s + n_r$ and adding a new edge for any pair of vertices $(i,j) \in \{1,\ldots,n_r\} \times \{n_r + 1,\ldots,n_r + n_s\}$.

Applying a recursive argument, define $\mathcal{G}_k := (\mathcal{G}_{k-1})_1$, for $k \geq 2$, and $\tilde{\mathcal{G}} := \bigcup_{k \geq 1} \mathcal{G}_k$. As in the previous case the family $\tilde{\mathcal{G}}$ has the property that, for any $n \in \mathbb{N}$, there exists a finite number of graphs in $\tilde{\mathcal{G}}$ with n vertices.

Note that in this case, the family $\{K_n\}_{n \geq 1}$ of complete graphs coincides with $\widetilde{\{\bullet\}}$.

Definition 3.6. Let $\mathcal{G}$ be an admissible family of simple graphs. We describe three binary operations on the graded vector space $\mathbb{K}[Tub(\mathcal{G})]$, spanned by the set of all tubings $\bigcup_{n \geq 1} \bigcup_{1 \leq i \leq s_n} Tub(G_n^i)$. For a pair of positive integers (n,m) and $1 \leq i \leq s_n$ and $1 \leq j \leq s_m$, define maps $*, /, \backslash : Tub(G_n^i) \times Tub(G_m^j) \longrightarrow \mathbb{K}[Tub(G_{n+m}^{\alpha(n,m)(i,j)})]$ as follows:

- $T * W := \sum U$, where the sum is taken over all tubings U of $G_{n+m}^{\alpha(n,m)(i,j)}$ satisfying $U|_{\{1,\ldots,n\}} \in \{T, \overline{T}\}$ and $U|_{\{n+1,\ldots,n+m\}} \in \{W+n, \overline{W}+n\}$,
- $T/W := T \cup \{1,\ldots,n\} \cup \{(w+n) \cup \{1,\ldots,n\}\}_{w \in W}$,
- $T\backslash W := \{t \cup \{n+1,\ldots,n+m\}\}_{t \in T} \cup \{n+1,\ldots,n+m\} \cup (W+n)$,

for any pair of tubings $T \in Tub(G_n^i)$ and $W \in Tub(G_m^j)$, where the vertices of G_n^i are identified with the vertices $\{1,\ldots n\}$ of $G_{n+m}^{\alpha(n,m)(i,j)}$ and the vertices of G_m^j are identified with $\{n+1,\ldots,n+m\} \subseteq \mathrm{Vert}(G_{n+m}^{\alpha(n,m)(i,j)})$. These maps are extended by bilinearity to give binary operations on $\mathbb{K}[Tub(\mathcal{G})]$.

Remark 3.7.

1. Note that T/W and $T\backslash W$ are addends of the sum of $T * W$. Moreover, if T is a k-tubing and W is an h-tubing, then T/W and $T\backslash W$ are both $(k+h+1)$-tubings. So, we may restrict

$$/, \backslash : MTub(\mathcal{G}) \otimes MTub(\mathcal{G}) \longrightarrow MTub(\mathcal{G}),$$

and get products on the space spanned by the vertices of the polytopes $\mathcal{K}G_m^j$.

2. Since $U \in Tub(G_{n+m}^{\alpha(n,m)(i,j)})$ satisfies $U|_{\{1,\ldots,n\}} \prec T$, we have $U|_{\{1,\ldots,n\}} = T$ or $U|_{\{1,\ldots,n\}} = \overline{T}$. In an analogous way, $U|_{\{n+1,\ldots,n+m\}} \prec W + n$ implies that either $U|_{\{n+1,\ldots,n+m\}} = W+n$, or $U|_{\{n+1,\ldots,n+m\}} = \overline{W}+n$.

3. The product $*$ induces a product

$$*_M : \mathbb{K}[MTub(\mathcal{G})] \otimes \mathbb{K}[MTub(\mathcal{G})] \longrightarrow \mathbb{K}[MTub(\mathcal{G})],$$

which is obtained as $*$ but defining $T *_M W$ to be the sum over all maximal tubings U of $Tub(G_{n+m}^{\alpha(n,m)(i,j)})$ whose restrictions to $\{1,\dots,n\}$ and to $\{n+1,\dots,n+m\}$ contain T and W, respectively.

Example 3.8. Consider $G_3 \times G_1$ in G_4 in the following example, where the vertex i of G_3 maps to the vertex i of G_4, while the vertex of G_1 maps to the vertex 4 of G_4. We compute the product of a tubing in G_3 and the empty tubing of G_1, which is the unique tubing of the graph.

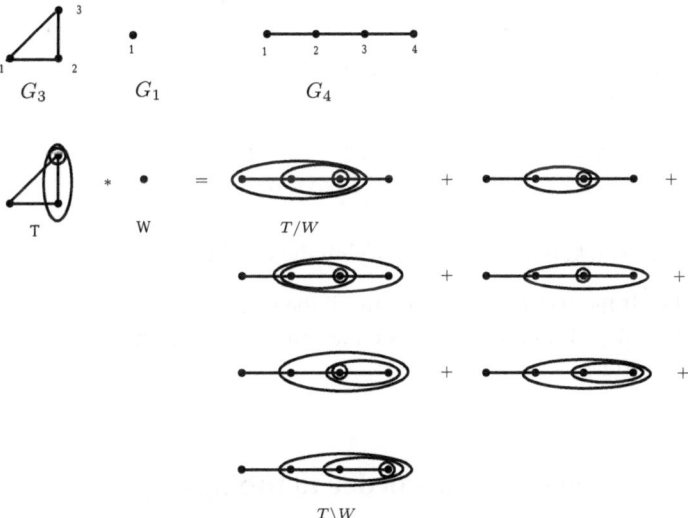

The proof of the following Lemma is straightforward.

Lemma 3.9. *For any admissible family $\mathcal{G}$ of simple finite graphs, the products $/$ and $\backslash$ are associative.*

For the operation $*$ we have the following result:

Theorem 3.10. *Let $\mathcal{G}$ be an admissible family of simple finite graphs. If all the graphs $G_n^i \in \mathcal{G}$ are connected and, for all $n \geq 1$, $1 \leq k \leq s_n$ and $1 \leq i < n$, they satisfy:*

(I) *the restrictions $G_n^k|_{\{1,\dots,i\}} = G_i^h$ and $G_n^k|_{\{i+1,\dots,n\}} = G_{n-i}^l + i$, whenever $\alpha(i, n - i)(h, l) = k$,*

then the product $* : \mathbb{K}[Tub(\mathcal{G})] \otimes \mathbb{K}[Tub(\mathcal{G})] \longrightarrow \mathbb{K}[Tub(\mathcal{G})]$ is associative.

Proof. For any pair of tubings $T \in Tub(G_n^i)$ and $W \in Tub(G_m^j)$, both sets $\{1,\dots,n\}$ and $\{n+1,\dots n+m\}$ are tubes in $G_{n+m}^{\alpha(n,m)(i,j)}$ because all the graphs are connected.

In general, the fact that $G_{n+m}^{\alpha(n,m)(i,j)}|_{\{1,\dots,n\}} = G_n^i$ and $G_{n+m}^{\alpha(n,m)(i,j)}|_{\{n+1,\dots,n+m\}} = G_m^j$, implies that:

- if $t = \{i_1,\dots,i_l\}$ is a tube of G_n^i, then t is a tube in $G_{n+m}^{\alpha(n,m)(i,j)}$, contained in $\{1,\dots,n\}$,
- if $t = \{i_1,\dots,i_l\}$ is a tube of G_m^j, then $t+n$ is a tube in $G_{n+m}^{\alpha(n,m)(i,j)}$, contained in $\{n+1,\dots,n+m\}$.

Let $T \in Tub(G_n^i)$, $W \in Tub(G_m^j)$ and $V \in Tub(G_r^k)$ be three tubings. If U is a tubing in $G_{n+m+r}^{\alpha(n,m,r)(i,j,k)}$ such that:

(1) $U|_{\{1,\dots,n\}}$ is equal to T or to $\overline{T}$,

(2) $U|_{\{n+1,\dots,n+m\}}$ is equal to W or to $\overline{W}$,

(3) $U|_{\{n+m+1,\dots,n+m+r\}}$ is equal to V or to $\overline{V}$,

then the condition (I) implies that the tubings $U|_{\{1,\dots,n+m\}}$ and $U|_{\{n+1,\dots,n+m+r\}}$ are uniquely determined. So, we easily obtain:

$$(T*W)*V = \sum_{U \in Tub(G_{n+m+r}^{\alpha(n,m,r)(i,j,k)})} U = T*(W*V),$$

where the sum is taken over all U satisfying the conditions (1), (2) and (3). $\qquad\square$

Remark 3.11. If the family $\mathcal{G}$ does not fulfill the condition (I), then the tubings which appear in $(T*W)*V$ and in $T*(W*V)$ are the same, but the coefficients of some tubings may differ.

4 Generalization of Tamari order to tubings

In the present section we define a partial order on the set of tubings $Tub(\mathcal{G})$ of a simple finite graph G, which restricts naturally to a partial order $\leq$ on $MTub(\mathcal{G})$. When G is the linear graph L_n, this order is the Tamari order.

Let t be a tube on a graph G, with $Vert(G) = \{1,\dots,n\}$. The smaller vertex of t is denoted $m(t)$, and the largest one is denoted $M(t)$. If $t \subset t'$ are two tubes such that there does not exist another tube w with $t \subset w \subset t'$ we denote it $t \ll t'$.

Definition 4.1. For a simple finite graph G with set of vertices $\{1,\dots,n\}$ and any tubing $T \in Tub(\mathcal{G})$, let $t = \{i_1 < i_1+1 < \cdots < i_1+k < i_2 < \cdots < i_p < i_p+1 < \cdots < i_p+h\}$ be a tube of G, with $i_1+k+1 < i_2$ and $i_{p-1}+1 < i_p$. We define a covering relation on $Tub(\mathcal{G})$, as follows:

1. $T \setminus \{t\}$ covers T if the tube $t \in T$ satisfies one of the following conditions:

   a. $t \ll t'$ for some t' in T, $i_1 = m(t')$ and for any vertex $j \in \{i_2,\dots,i_p\}$ there exists a tube $w \subset t$ in T such that $j \in w$.

   b. t is not contained in any other tube of T, $i_1 = 1$, and for any vertex $j \in \{i_2,\dots,i_p+h\}$ there exists a tube $w \subset t$ in T such that $j \in w$,

2. $T \cup t$ covers T if $T \cup t$ is a tubing, and the tube $t \notin T$ satisfies one of the following conditions:

   a. there exists a tube $t' \in T$ such that $t \subset t'$, $i_p + h = M(t')$, and for any vertex $j \in \{i_1, \ldots, i_{p-1}\}$ there exists a tube $w \subset t'$ in T such that $j \in w$,
   b. t is not contained in any tube of T, $i_p + h = n$, for any vertex $j \in \{i_1, \ldots, i_{p-1}\}$ there exists a tube $w \subset t'$ in T such that $j \in w$.

Definition 4.2. Let G be a simple finite graph with set of vertices $\{1, \ldots, n\}$, the relation $\leq$ is the transitive relation on the set of tubings $Tub(\mathcal{G})$ spanned by the covering relation defined above.

Lemma 4.3. *The relation $\leq$ defines a partial order on $Tub(\mathcal{G})$.*

Proof. We have to check that for any sequence $T_1, \ldots, T_r$, with $r \geq 3$, where T_{i+1} covers T_i, we have that $T_1 \neq T_r$.

Note that if T_{i+1} covers T_i, then the number of tubes of T_{i+1} differs from the number of tubes of T_i. On the other hand, if T_j has the same number of tubes than T_1, then there exists a vertex i_0 such that the numbers of tubes of T_1 containing i_0 is smaller than the number of tubes of T_j containing i_0, which proves our statement. $\square$

Example 4.4.
1) Consider the cyclic graph

The following sequence of tubings is ordered, where the dotted arrow means $<$.

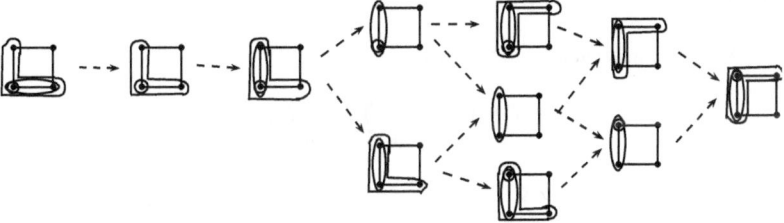

2) For the linear graph $1 \bullet\!\!-\!\!\bullet_2\!\!-\!\!\bullet_3$ we get

The order considered on the set of tubings of the linear graph L_n, $n \geq 1$, has been previously defined in [9] and in [12].

Clearly, for any simple finite graph G, the partial order $<$ on $Tub(\mathcal{G})$, may be restricted to the set maximal tubings $MTub(\mathcal{G})$.

There exists a natural bijection between the set $MTub(\mathcal{L}_n)$ of maximal tubings of the linear graph L_n and the set Y_n of planar binary rooted trees with n leaves, defined recursively in the following way:

1. The unique tubing of the graph L_1 which has one vertex and no edges maps to the tree $\curlyvee$.

2. Suppose that we have defined a bijection $\varphi_j : MTub(\mathcal{L}_j) \longrightarrow Y_j$, for $1 \leq j < n$. Given an element $T \in MTub(\mathcal{L}_n)$, there exists a unique vertex $r \in \text{Vert}(L_n)$ which does not belong to any tube of T. We have $T|_{\{1,\ldots,r-1\}} = \overline{T}_1$ and $T|_{\{r+1,\ldots,n\}} = \overline{T}_2 + r$, for a pair of maximal tubings $T_1 \in MTub(\mathcal{L}_{r-1})$ and $T_2 \in MTub(\mathcal{L}_{n-r})$. We define:

$$\varphi_n(T) = \varphi_{r-1}(T_1) \vee \varphi_{n-r}(T_2),$$

where, for any pair of planar binary rooted trees $w_1 \in Y_{r-1}$ and $w_2 \in Y_{n-r}$, the tree $w_1 \vee w_2 \in Y_n$ is obtained by joining the roots of w_1 and w_2 to a new root, keeping w_1 on the left side and w_2 on the right side.

For instance, we have:

$$\varphi_3(\text{●—⬭●}) = \curlywedge\curlyvee \qquad \varphi_3(\text{◉—●—◉}) = \curlyvee\curlyvee$$

Remark 4.5. The order $<$ on the set $MTub(\mathcal{L}_n)$ of maximal tubings of the linear graph coincides with the Tamari order defined on the set of planar binary rooted trees with $n+1$ leaves.

The following result is an extension of the result proved for the Tamari order and the Hopf algebra of planar binary rooted trees in [7].

Theorem 4.6. *Let $\mathcal{G}$ be an admissible family of simple finite graphs. If all the graphs $G_n^i \in \mathcal{G}$ are connected and, for all $n \geq 1$, $1 \leq k \leq s_n$ and $1 \leq i < n$, they satisfy:*

(I) *the restrictions $G_n^k|_{\{1,\ldots,i\}} = G_i^h$ and $G_n^k|_{\{i+1,\ldots,n\}} = G_{n-i}^l + i$, whenever $\alpha(i, n-i)(h, l) = k$*

then the product $*$ satisfies:

$$T * W = \sum_{T/W \leq U \leq T \backslash W} U,$$

for all tubings $T, W \in Tub(\mathcal{G})$.

Proof. From the definitions of T/W and $T \backslash W$, it is clear that, for any tubing $T/W \leq U \leq T \backslash W$, $U|_{\{1,\ldots,n\}}$ coincides with T or $\overline{T}$, and $U|_{\{n+1,\ldots,n+m\}}$ coincides with $W + n$ or with $\overline{W} + n$.

Suppose now that a tubing U is such that

- $U|_{\{1,\ldots,n\}} = T$ or $U|_{\{1,\ldots,n\}} = \overline{T}$,
- $U|_{\{n+1,\ldots,n+m\}} = W + n$ or $U|_{\{n+1,\ldots,n+m\}} = \overline{W} + n$.

Given a tube $u \in U$, we may assert that

1. If $u \cap \{1,\ldots,n\} \neq \emptyset$, then either $u = t \cup w$ or $u = \{1,\ldots,n\} \cup (w+n)$, with $t \in T$ and $w \in W \cup \{1,\ldots,m\}$ or $w = \emptyset$.
2. If $u \cap \{1,\ldots,n\} = \emptyset$, then either $u = \{n+1,\ldots,n+m\}$ or $u \in W + n$.

For $t \in T$, there exist $u_t \in U$ such that $u_t = t$ or $u_t = t \cup \{j_1, \ldots, j_s\}$ with $\{j_1, \ldots, j_s\} \subseteq \{n, n+1, \ldots, n+m\}$. Moreover, for any $t \subset t'$ in T, we have $u_t \subset u_{t'}$, which implies $T/W \leq U$.

On the other hand, for any $w \in W \mid n$, we have that there exists $u_w \in U$ with $u_w \cap \{n+1, \ldots, n+m\} = w$. So, for any $w \subset w'$ in $W + n$, we get $u_w \subset u_{w'}$, which implies $U \leq T \backslash W$.

The result now follows from the fact that, for a family of simple finite connected graphs $\mathscr{G}$ satisfying condition (1), the coefficient of a tubing U in $T * W$ must be 1 or 0. $\qquad \square$

Example 4.7. Let us compute the product:

in the tubings of the graph

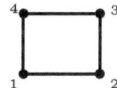

We have the interval of tubings:

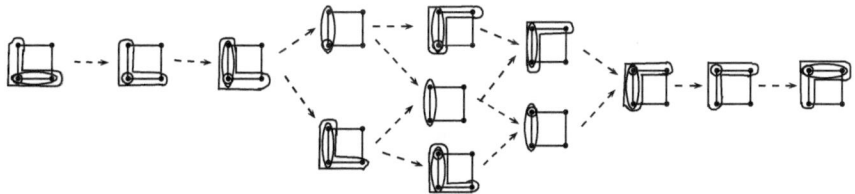

The sum of all tubings in the interval above gives the product.

Acknowledgement The author was partially supported by the Project FONDECYT Regular 1100380 and the MathAmSud Project math OPECSHA. I want to thank S. Forcey for introducing me into Carr's and Devadoss's work, for revising the first version of the present work, and for all the discussions we had about the existence of algebraic structures on the spaces of tubings.

References

1. M. Aguiar and F. Sottile, "Structure of the Loday-Ronco algebra of trees", *J. Algebra* **295** (2006) 473–511.
2. M. Carr and S. Devadoss, "Coxeter complexes and graph associahedra", *Topology and its Applications* **153** (1-2) (2006) 2155–2168.
3. F. Chapoton, "Bigèbres différentielles graduées associées aux permutoèdres, associaèdres et hypercubes", *Ann. Inst. Fourier (Grenoble)* **50** (2000) 1127–1153.
4. S. Forcey, "Extending the Tamari lattice to some compositions of species", in *this volume*.
5. S. Forcey, D. Springfield, "Geometric combinatorial algebras: cyclohedron and simplex", *J. Algebraic Comb.* **32** (2010) 597–627.

6. J.-L. Loday and M. Ronco, "Hopf algebra of the planar binary trees", *Adv. Math.* **139** (2) (1998) 293–309.

7. J.-L. Loday and M. Ronco, "Order structure and the algebra of permutations and of planar binary trees", *J. of Algebraic Combinatorics* **15** N* 3 (2002), 253–270.

8. J.-L. Loday and M. Ronco, "Trialgebras and families of polytopes", in *Homotopy theory: Relations with Algebraic Geometry, Group Cohomology and Algebraic K-Theory*, Contemporary Mathematics, vol. 346, AMS, Providence, 2004, 369–398.

9. J.-C. Novelli and J.-Y. Thibon, "Parking functions and descent algebras", *Annals of Combinatorics* **11** (2007) 59–68.

10. C. Malvenuto and C. Reutenauer, "Duality between quasi-symmetric functions and the Solomon descent algebra", *J. Algebra* **177** (3) (1995) 967–982.

11. M. Markl, "Simplex, associahedron, and cyclohedron", in *Higher Homotopy Structures in Topology and Mathematical Physics*, Contemp. Math., vol. 227, AMS, Providence, 1999, 235–265.

12. P. Palacios and M. Ronco, "Weak Bruhat order on the set of faces of the permutahedra", *J. Algebra* **299** (2006) 648–678.

13. A. Tonks, "Relating the associahedron and the permutohedron", in *Operads: Proceedings of Renaissance Conferences, Hartford, CT/Luminy 1995*, Contemp. Math., vol. 202, AMS, Providence, 1997, 33–36.

A Survey of the Higher Stasheff-Tamari Orders

Jörg Rambau and Victor Reiner

Abstract The Tamari lattice, thought as a poset on the set of triangulations of a convex polygon with n vertices, generalizes to the higher Stasheff-Tamari orders on the set of triangulations of a cyclic d-dimensional polytope having n vertices. This survey discusses what is known about these orders, and what one would like to know about them.

1 Introduction

One often thinks of the Tamari order as a partial order on parenthesizations, or on binary trees. But it can also be taken as an order on triangulations of any n-gon whose vertices lie in convex position.

Choosing the vertices of the n-gon to lie on a parabola, or 2-dimensional *moment curve*, lends itself to a beautiful geometric interpretation for the order. This interpretation generalizes to give two closely related orders on the set of triangulations of a *cyclic polytope* $\mathbf{C}(n,d)$, which is the convex hull of any n points on the d-dimensional moment curve.

These orders, called the *higher Stasheff-Tamari orders* $\mathrm{HST}_1(n,d)$ and $\mathrm{HST}_2(n,d)$, first appeared roughly 20 years ago in the work of Kapranov and Voevodsky [24, Defn. 3.3], and are somewhat mysterious. Nevertheless, they share many beautiful properties with the Tamari order. Here we survey the work on them by Edelman and Reiner [15], Rambau [31], Reiner [36, §6], Edelman, Rambau and Reiner [14], Thomas [43, 44], and most recently, Oppermann and Thomas [26]. We also discuss work on the closely related *Baues problem* for subdivisions of cyclic polytopes and

Jörg Rambau
University of Bayreuth, Germany, e-mail: *joerg.rambau@uni-bayreuth.de*

Victor Reiner
University of Minnesota, Minneapolis, USA, e-mail: *reiner@math.umn.edu*

zonotopes, as studied by Rambau and Santos [33], Athanasiadis, Rambau and Santos [3], and Athanasiadis [2].

Along the way, we indicate which questions about them remain open.

2 Cyclic polytopes

One way to realize the vertices of an n-gon in convex position is to pick the vertices as n points with distinct x-coordinates on the parametrized parabola $\{(t,t^2) : t \in \mathbb{R}\}$ within $\mathbb{R}^2$. More generally, one can define (see [47, Example 0.6]) the d-dimensional *moment curve* in R^d as the image of the parametrization

$$\mathbb{R} \overset{v_d}{\to} \mathbb{R}^d \tag{1}$$
$$t \mapsto (t,t^2,\ldots,t^d).$$

Definition 2.1. The *d-dimensional cyclic polytope with n vertices* $\mathbf{C}(n,d)$ is the convex hull of any n points $v_d(t_1),\ldots,v_d(t_n)$ with distinct x_1-coordinates

$$t_1 < t_2 < \cdots < t_n. \tag{2}$$

We adopt the convention when $d = 0$ that these n points are copies of the unique point of $\mathbb{R}^0$.

An exercise in Vandermonde determinants and polynomial algebra and inequalities [47, Example 0.6, Theorem 0.7, Exercise 0.8] shows that, no matter how one chooses the x_1-coordinates in (2), the polytope $\mathbf{C}(n,d)$ has these combinatorial properties:

- $\mathbf{C}(n,d)$ is a *simplicial polytope*, meaning that its boundary faces are all simplices,
- $\mathbf{C}(n,d)$ has the same subsets of indices $\{i_0,i_1,\ldots,i_k\}$ indexing boundary faces $\text{conv}\{v_d(t_{i_0}), v_d(t_{i_1}),\ldots,v_d(t_{i_k})\}$, dictated by *Gale's evenness criterion*, and in particular,
- $\mathbf{C}(n,d)$ is $\lfloor \frac{d}{2} \rfloor$-*neighborly*, meaning that every vertex subset of size at most $\frac{d}{2}$ spans a simplex on the boundary.

In light of these properties, it is fair to talk about $\mathbf{C}(n,d)$ and its boundary faces indexed by sets of subscripts $\{i_0,i_1,\ldots,i_k\}$, without reference to the choice of x_1-coordinates in (2). In the terminology of oriented matroid theory, the *affine* point configuration given by the points with homogeneous coordinates $\{(1,t_i,t_i^2,\ldots,t_i^d)\}_{i=1,2,\ldots,n}$ realizes the *alternating oriented matroid* [9, Cor. 8.2.10], regardless of the choice in (2).

Note also that if one fixes this choice (2), but varies the dimension d, then one has canonical projection maps $\pi : C(n,d') \to C(n,d)$ for $d' \geq d$ by forgetting the last $d'-d$ coordinates. Figure 1 shows the cyclic polytopes $C(7,d)$ for $d = 0,1,2,3$, along with these projection maps[1].

---

[1] The astute reader will notice that the point configurations $\mathbf{C}(7,1)$ and $\mathbf{C}(7,0)$ are not really determined by the polytope which is their convex hull. We will tacitly use the term "polytope", even

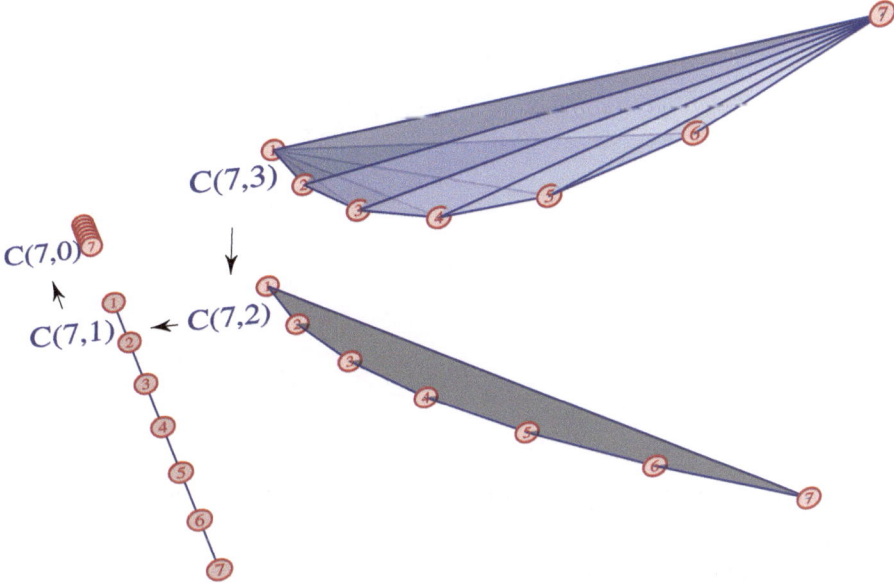

Fig. 1 Cyclic polytopes $\mathbf{C}(7,3)$, $\mathbf{C}(7,2)$, $\mathbf{C}(7,1)$, and $\mathbf{C}(7,0)$ (seven repeated points at the origin) together with the canonical projections forgetting the last coordinate. The bottom triangulation $\hat{0}_{7,2}$ of $\mathbf{C}(7,2)$, discussed in Section 3, is faintly visible as the (obscured) lower facets of $\mathbf{C}(7,3)$.

Because the oriented matroid data for the affine point configuration $\{v_d(t_i)\}_{i=1,2,\ldots,n}$ is independent of the choice (2), it is also well defined to say when a collection $\mathscr{T}$ of $(d+1)$-element subsets $\{i_1, i_2, \ldots, i_{d+1}\}$ indexes the maximal simplices $\mathrm{conv}\{v_d(t_{i_1}), \ldots, v_d(t_{i_{d+1}})\}$ in a triangulation of the cyclic polytope $\mathbf{C}(n,d)$. For complete discussions of the motivations and technicalities here, see Rambau [31, §2] and De Loera, Rambau, and Santos [11, Chap. 2].

We will say more about how one encodes or characterizes the collections $\mathscr{T}$ of $(d+1)$-subsets that index triangulations of $\mathbf{C}(n,d)$ in Section 4.

3 The two orders

The two Stasheff-Tamari orders come from thinking about how a triangulation $\mathscr{T}$ of $\mathbf{C}(n,d)$ induces a section

$$\mathbf{C}(n,d) \overset{s_{\mathscr{T}}}{\rightarrow} \mathbf{C}(n,d+1)$$

---

though in certain situations, there is a point configuration in the background which is really part of the data. This becomes even more apparent in the case of *cyclic zonotopes* discussed in Section 8. We elaborate no further on this here, but refer to [11, Chp. 2] for a technically satisfying setup.

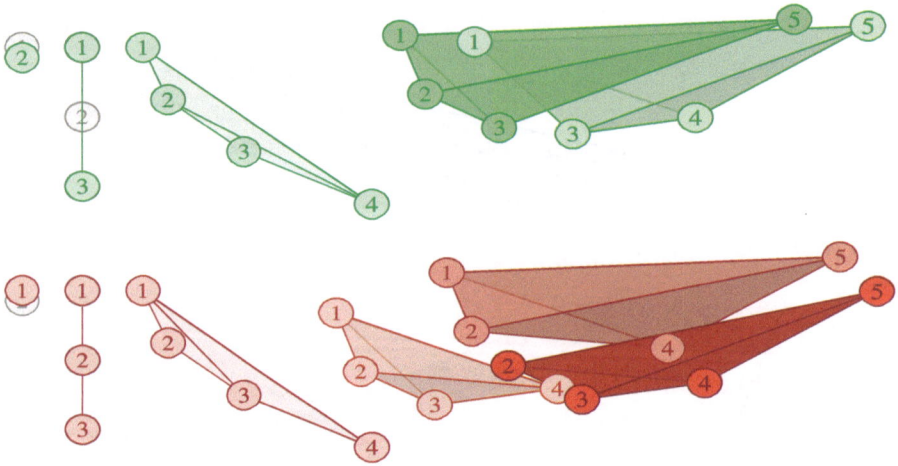

Fig. 2 The two triangulations (green and red) of $\mathbf{C}(d+2,d)$ for small d, specifically,

$d = 0$: $\{1\}$ versus $\{2\}$

$d = 1$: $\{1,3\}$ versus $\{1,2\},\{2,3\}$

$d = 2$: $\{1,2,4\},\{2,3,4\}$ versus $\{1,2,3\},\{1,3,4\}$

$d = 3$: $\{1,2,3,4\},\{1,2,4,5\},\{2,3,4,5\}$ versus $\{1,2,3,5\},\{1,3,4,5\}$, depicted here in an exploded view: the 3-simplices are moved slightly apart to clarify how they assemble.

of the projection map $\mathbf{C}(n,d+1) \xrightarrow{\pi} \mathbf{C}(n,d)$, defined uniquely by insisting that $s_{\mathscr{T}}$ sends $v_d(t_i) \mapsto v_{d+1}(t_i)$, and then extending $s_{\mathscr{T}}$ piecewise-linearly over each simplex in the triangulation $\mathscr{T}$.

From this point of view (and after staring at $\mathbf{C}(n,3)$ in Figure 1 for a bit), one realizes that the top and bottom elements in the usual Tamari poset correspond to the two canonical triangulations of $\mathbf{C}(n,2)$ that come from the "upper" and "lower" facets of $\mathbf{C}(n,3)$. In general, one obtains a canonical *upper* (resp. *lower*) triangulation of $\mathbf{C}(n,d)$ by projecting via $\pi : \mathbf{C}(n,d+1) \to \mathbf{C}(n,d)$ the boundary facets of $\mathbf{C}(n,d+1)$ visible from points with large (resp. small) x_{d+1} coordinate. It is not hard to see that when $n = d+2$, these are the only two triangulations of a cyclic polytope $\mathbf{C}(d+2,d)$; for $d = 0, 1, 2, 3$, they are pictured in Figure 2. See also Figure 10 for the $d = 3$ case. Explicit descriptions of these canonical upper and lower triangulations for general d may be found in [15, Lemma 2.3].

Definition 3.1. Given two triangulations $\mathscr{T}, \mathscr{T}'$ of the cyclic polytyope $\mathbf{C}(n,d)$, say that they are related as $\mathscr{T} \leq_2 \mathscr{T}'$ in the *second higher Stasheff-Tamari order* $\mathrm{HST}_2(n,d)$ if $s_{\mathscr{T}}(x)_{d+1} \leq s_{\mathscr{T}'}(x)_{d+1}$ for every point x of $\mathbf{C}(n,d)$, that is, the section $s_{\mathscr{T}}$ lies weakly below the section $s_{\mathscr{T}'}$ with respect to their x_{d+1}-coordinates.

Definition 3.2. To define the *first higher Stasheff-Tamari order* $\mathrm{HST}_1(n,d)$ on triangulations of $\mathbf{C}(n,d)$, first define when $\mathscr{T}'$ is obtained from $\mathscr{T}$ by an *upward flip*: this means that there exists a $(d+2)$-subset $i_1 < i_2 < \cdots < i_{d+2}$ whose convex hull gives a subpolytope $\mathbf{C}(d+2,d)$ of $\mathbf{C}(n,d)$ with the property that $\mathscr{T}, \mathscr{T}'$ restrict to

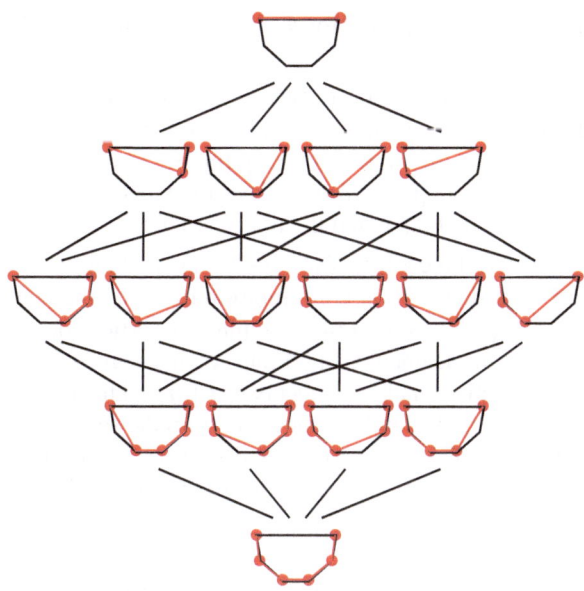

Fig. 3 The (lower!) Stasheff-Tamari orders $\mathrm{HST}_2(6,1) = \mathrm{HST}_1(6,1)$ on the set of triangulations $\mathscr{T}$ of the line segment $\mathbf{C}(6,1)$. Instead of the triangulation $\mathscr{T}$, its image under the piecewise linear section $s_{\mathscr{T}} : \mathbf{C}(6,1) \to \mathbf{C}(6,2)$ is depicted in red.

the lower, upper triangulations of this $\mathbf{C}(d+2,d)$, and otherwise $\mathscr{T}, \mathscr{T}'$ agree on all of their other simplices not lying in this $\mathbf{C}(d+2,d)$.

Then define $\mathscr{T} \leq_1 \mathscr{T}'$ in $\mathrm{HST}_1(n,d)$, if there is a sequence of upward flips starting with $\mathscr{T}$ and ending with $\mathscr{T}'$. That is, $\mathrm{HST}_1(n,d)$ is the transitive closure of the upward flip relation.

Figure 3 illustrates $\mathrm{HST}_2(6,1)$. It should be clear from the definitions and the above discussion that $\leq_1$ is a weaker partial order than $\leq_2$, and that the lower and upper triangulations of $\mathbf{C}(n,d)$ give the unique minimal $\hat{0}_{n,d}$ and maximal $\hat{1}_{n,d}$ elements of $\mathrm{HST}_2(n,d)$. It was left open in [15], and resolved by Rambau affirmatively in [31], that these two triangulations also give unique minimal and maximal elements of $\mathrm{HST}_1(n,d)$. In particular, this resolves the question of *bistellar connectivity* for triangulations of $\mathbf{C}(n,d)$: any pair of triangulations can be related by a sequence of bistellar flips (see Section 6). It is also closely related to the *Generalized Baues Problem* for cyclic polytopes, discussed in Section 7 below.

It was shown in [15] that the two orders $\mathrm{HST}_1(n,d)$ and $\mathrm{HST}_2(n,d)$ are the same for $d = 0,1,2,3$, and this is also not hard to check that they are the same when $n - d = 1,2,3$. This raises the following question that remains open.

Open Problem 3.3. *Are* $\mathrm{HST}_1(n,d)$ *and* $\mathrm{HST}_2(n,d)$ *the* **same** *orders?*

Historically, the order $\mathrm{HST}_1(n,d)$ is the one introduced, in the different terminology of *pasting schemes*, by Kapranov and Voevodsky [24, Def. 3.3]; the second order $\mathrm{HST}_2(n,d)$ was defined in [15, p. 132].

The higher Stasheff-Tamari posets for $d = 0, 1, 2$ are familiar objects, as we next explain.

Example 3.4. For $d = 0$, the cyclic polytope $\mathbf{C}(n, 0)$ is the unique point of $\mathbb{R}^0$, however, it is viewed as a point configuration in which there are n different possible labels i in $\{1, 2, \ldots, n\}$ for this point. A triangulation $\mathscr{T}$ of $\mathbf{C}(n, 0)$ is a choice of one of these labels i, and an upward flip replaces the label i by the label $i + 1$. Thus $\mathrm{HST}_1(n, d)$ and $\mathrm{HST}_2(n, d)$ both equal the linear order $1 < 2 < \cdots < n$.

Example 3.5. For $d = 1$, the cyclic polytope $\mathbf{C}(n, 1)$ is a line segment $[t_1, t_n]$ inside $\mathbb{R}^1$, however, it is viewed as a point configuration in which there are $n - 2$ interior vertices $\{t_2, t_3, \ldots, t_{n-1}\}$. Any subset of these interior vertices determines a unique triangulation $\mathscr{T}$ of the line segment $\mathbf{C}(n, 1)$ into smaller segments. A typical upward flip replaces two consecutive smaller segments $[t_i, t_j], [t_j, t_k]$ having $i < j < k$ with the single segment $[t_i, t_k]$, or equivalently, removes t_j from the subset of interior vertices used in the triangulation. Thus $\mathrm{HST}_1(n, d)$ and $\mathrm{HST}_2(n, d)$ are both isomorphic to the Boolean algebra $2^{\{t_2, t_3, \ldots, t_{n-1}\}}$. This was illustrated for $n = 5$ already in Figure 3, depicting $\mathrm{HST}_2(6, 1) = \mathrm{HST}_1(6, 1)$, which is isomorphic to the Boolean algebra $2^{\{t_2, t_3, t_4, t_5\}}$.

Example 3.6. For $d = 2$, as mentioned above, the cyclic polytope $\mathbf{C}(n, 2)$ is a convex n-gon. A typical upward flip starts with a triangulated sub-quadrilateral $\mathbf{C}(4, 2)$ with four vertices $i < j < k < \ell$ which is triangulated via the two triangles $\{ijk, ik\ell\}$, and replaces it with the same triangulation except for using the two triangles $\{ij\ell, jk\ell\}$ instead. Thus $\mathrm{HST}_1(n, 2)$ is equivalent to one of the usual definitions of the Tamari order. It is not completely obvious that $\mathrm{HST}_1(n, 2) = \mathrm{HST}_2(n, 2)$; a proof appears in [15, Theorem 3.8].

Example 3.7. Figures 4 through 6 show pictures of $\mathrm{HST}_1(6, 2)$, $\mathrm{HST}_1(6, 3)$, and $\mathrm{HST}_1(7, 3)$, respectively, all supported by TOPCOM [32].

The following property, suggested by the previous examples and scrutiny of the accompanying figures, is easily deduced from the definitions.

Proposition 3.8 ([15, Prop. 2.11]). *In both posets* $\mathrm{HST}_1(n, d), \mathrm{HST}_2(n, d)$, *reversal of the labelling, that is, the relabelling* $1 \mapsto n, 2 \mapsto n - 1, \ldots, n \mapsto 1$

- *induces a non-trivial poset automorphism for d odd, and*
- *induces a poset anti-automorphism for d even.*

Scrutiny of the examples and figures also suggests the following properties, which are not as obvious, but deduced by Rambau in [31, Cor. 12.(i)].

Proposition 3.9. *Given a triangulation* $\mathscr{T}$ *of* $\mathbf{C}(n, d)$, *let* $|\mathscr{T}|$ *denote its number of maximal simplices.*

- *For d even,* $|\mathscr{T}|$ *is constant, independent of* $\mathscr{T}$, *equal to* $\binom{n-e-1}{e}$ *if* $d = 2e$.
- *For d odd,* $|\mathscr{T}|$ *takes on all values in the range* $\left[\binom{n-e-1}{e-1}, \binom{n-e}{e} \right]$ *if* $d = 2e - 1$. *In fact,* $\mathrm{HST}_1(n, d)$ *is a ranked poset in which* $\mathscr{T}$ *has rank* $\binom{n-e}{e} - |\mathscr{T}|$.

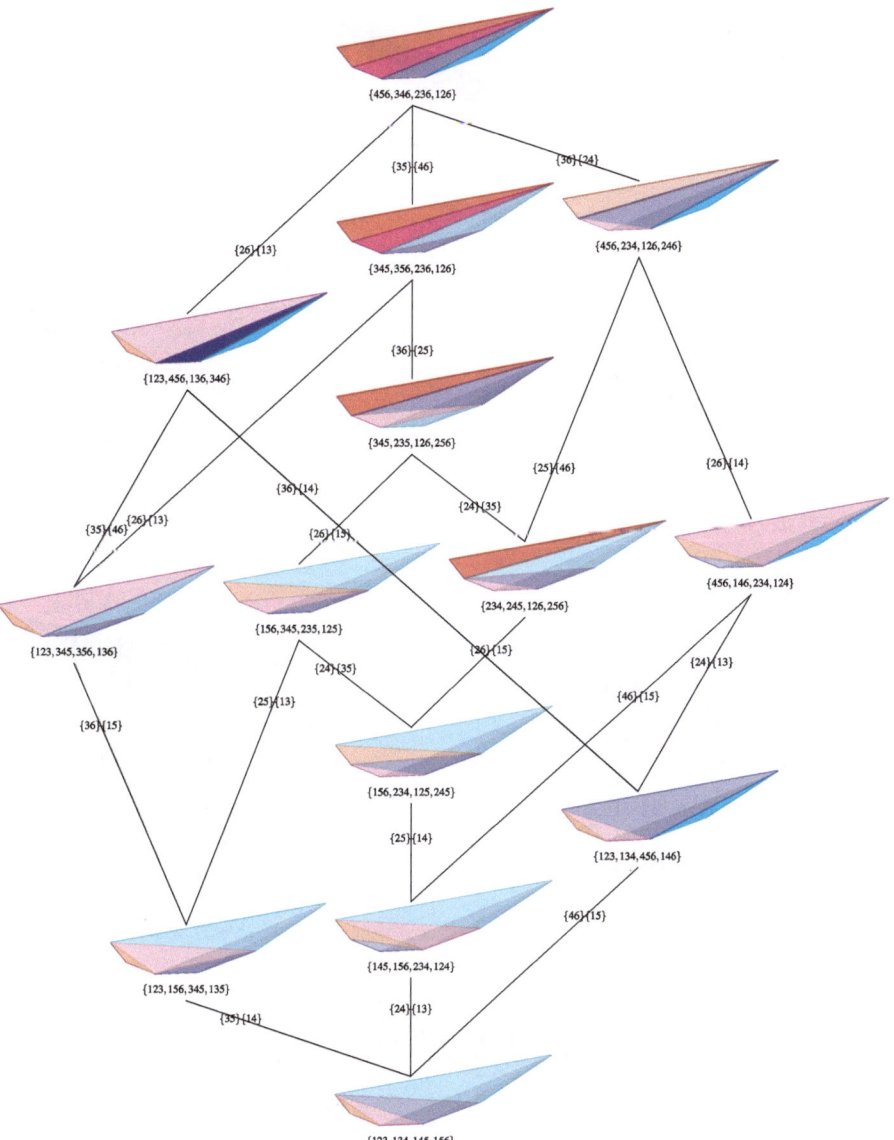

Fig. 4 A picture of $\mathrm{HST}_1(6,2) = \mathrm{HST}_2(6,2)$, similar to [15, Fig. 4(a)]. Triangulations $\mathscr{T}$ of $\mathbf{C}(6,2)$ are depicted as the images of their corresponding sections $s_{\mathscr{T}} : \mathbf{C}(6,2) \to \mathbf{C}(6,3)$, viewed from above $\mathbf{C}(6,3)$. Labels $\{j\ell, ik\}$ on covering relations indicate supports of the corresponding flips as follows: the 3-simplex $\{i, j, k, \ell\}$ with $i < j < k < \ell$ supporting the flip has lower facets $\{ijk, ik\ell\}$, and upper facets $\{ij\ell, jk\ell\}$.

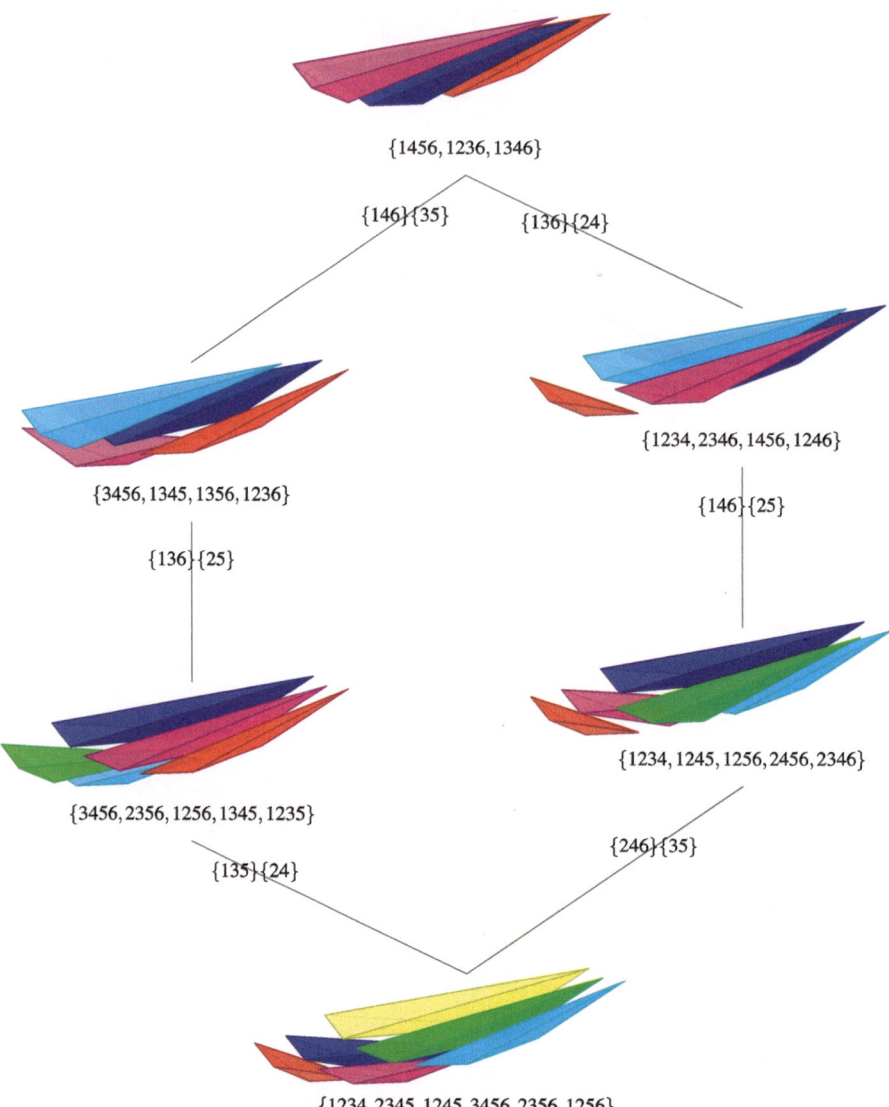

Fig. 5 A picture of $\mathrm{HST}_1(6,3)$; the labels of the covering relations indicate the support of the corresponding flip. After reading Theorem 6.6 below, the interested reader may want to find, for each of the 6 triangulations in this figure, at least one maximal chain in Figure 4 which induces it.

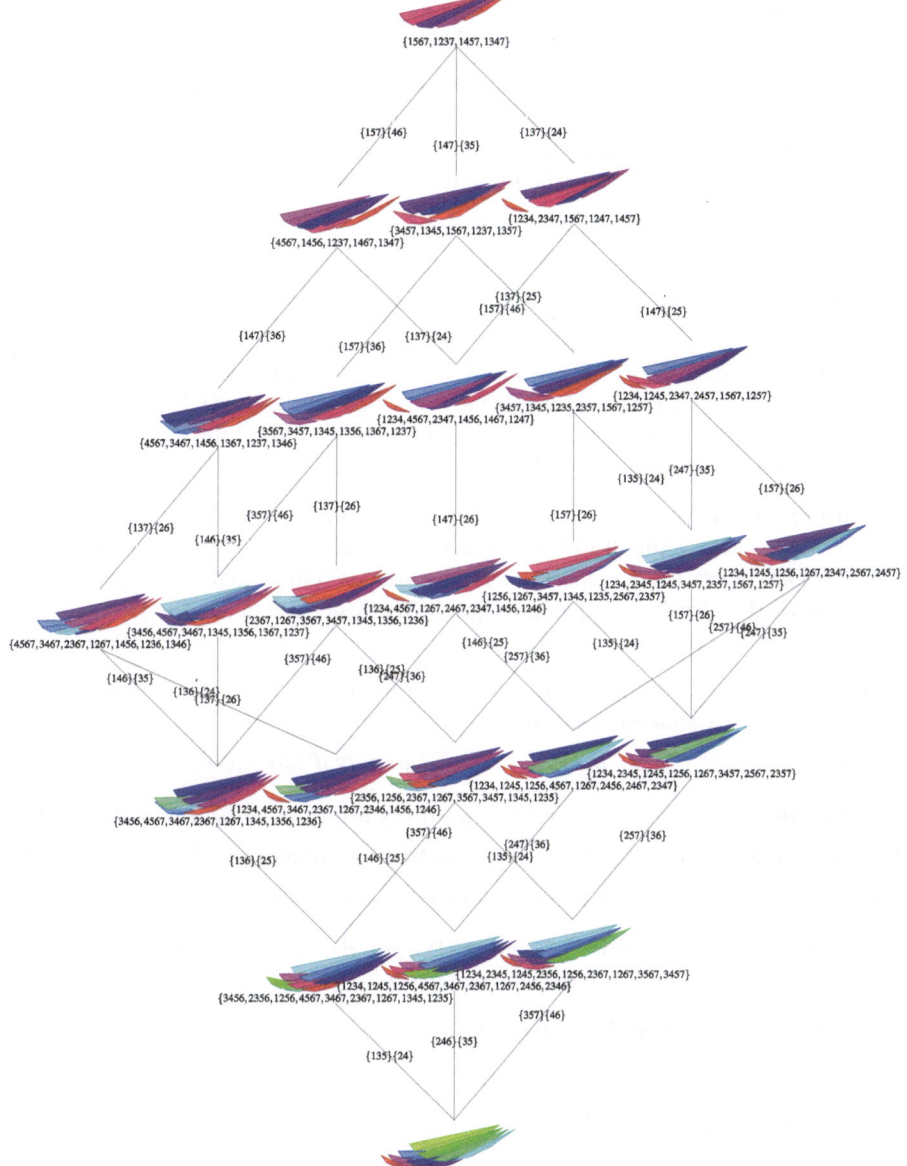

Fig. 6 A picture of $HST_1(7,3)$ (data generated by TOPCOM [32]), similar to [15, Fig. 4(b)].

4 Encodings

Just as it is sometimes useful to encode elements of the Tamari lattice by other means, such as Huang and Tamari's *bracketing vectors* [22], it has also proven useful to encode triangulations of $\mathbf{C}(n,d)$ and the Stasheff-Tamari orders in various ways. We discuss three such encodings in the literature, as they appeared in historical order.

4.1 Submersion sets

For $k \geq 0$, a subset $\{i_1,\ldots,i_{k+1}\}$ of $\{1,2,\ldots,n\}$, is identified with a k-simplex $\sigma = \mathrm{conv}\{v_d(t_{i_1}),\ldots,v_d(t_{k+1})\}$ inside the cyclic polytope $\mathbf{C}(n,d)$. Denote by s_σ the unique map $\sigma \to \mathbb{R}^{d+1}$ that maps its vertices $v_d(t_{i_j}) \mapsto v_{d+1}(t_{i_j})$ for $j = 1,2,\ldots,k+1$ and then extends piecewise-linearly over σ.

Definition 4.1. Given a triangulation $\mathscr{T}$ of $\mathbf{C}(n,d)$, say that σ is *submerged* by $\mathscr{T}$ if $s_\sigma(x)_{d+1} \leq s_{\mathscr{T}}(x)_{d+1}$ for all x in σ. In other words, when one lifts σ into $\mathbf{C}(n,d+1)$, it lies weakly below (with respect to x_{d+1}-coordinates) the image of the section $s_{\mathscr{T}}$.

Define the *k-submersion set* $\mathrm{sub}_k(\mathscr{T})$ to be the collection of subsets $\{i_1,\ldots,i_{k+1}\}$ indexing k-simplices σ submerged by $\mathscr{T}$.

Proposition 4.2 ([15, Prop. 2.15]). *A triangulation $\mathscr{T}$ of $\mathbf{C}(n,d)$ can be recovered uniquely from its submersion set $\mathrm{sub}_{\lceil \frac{d}{2} \rceil}(\mathscr{T})$.*

Furthermore, $\mathscr{T} \leq_2 \mathscr{T}'$ in $\mathrm{HST}_2(n,d)$ if and only if $\mathrm{sub}_{\lceil \frac{d}{2} \rceil}(\mathscr{T}) \subseteq \mathrm{sub}_{\lceil \frac{d}{2} \rceil}(\mathscr{T}')$.

This encoding of $\mathscr{T}$ via $\mathrm{sub}_{\lceil \frac{d}{2} \rceil}(\mathscr{T})$ is used mainly in [15] for $d \leq 3$. There it is explained how to read off $\mathrm{sub}_{\lceil \frac{d}{2} \rceil}(\mathscr{T})$ from the d-simplices of $\mathscr{T}$, and the subsets which can appear as $\mathrm{sub}_{\lceil \frac{d}{2} \rceil}(\mathscr{T})$ are characterized as follows.

Proposition 4.3 ([15, Props. 3.3 and 4.2]). *For $d = 2$, a collection $I = \{ij\} \subset \{1,2,\ldots,n\}$ has $I = \mathrm{sub}_1(\mathscr{T})$ for some triangulation of $\mathbf{C}(n,2)$ if and only if*

- *I contains every boundary edge of $\mathbf{C}(n,2)$.*
- *Assume ik is in I.*
  If $i < j < k$, then ij is also in I.
- *If $ik, j\ell$ are both in I, with $i < j < k < \ell$, then $i\ell$ is also in I.*

For $d = 3$, a collection $I = \{ijk\} \subset \{1,2,\ldots,n\}$ has $I = \mathrm{sub}_2(\mathscr{T})$ for some triangulation of $\mathbf{C}(n,3)$ if and only if

- *I contains every boundary triangle of $\mathbf{C}(n,3)$.*
- *Assume ijk is in I.*
  If $j < k' < k$, then ijk' is also in I.
  If $i < i' < j$, then $i'jk$ is also in I.
- *If ijk, abc are in I, with $a < i < b < j < c < k$, then abk, ajk are also in I.*

Note that these conditions characterizing the sets $\mathrm{sub}_1(\mathscr{T})$ for $d = 2$ and $\mathrm{sub}_2(\mathscr{T})$ for $d = 3$ are *closure* conditions, and hence they are preserved when one intersects sets. This immediately implies that the second Stasheff-Tamari order $\mathrm{HST}_2(n,d)$ is a meet semilattice for $d \leq 3$, with meet operation given by intersecting these sets. Since it has the unique maximal element $\hat{1}_{n,d}$, one immediately deduces the following.

Theorem 4.4 ([15, Thms. 3.6 and 4.9]). *For $d \leq 3$, the higher Stasheff-Tamari order $\mathrm{HST}_2(n,d)$ is a lattice.*

With some work, these encodings can also be used to show the following previously mentioned result.

Proposition 4.5 ([15, Thms. 3.8 and 4.10]). *For $d \leq 3$, the two Stasheff-Tamari orders $\mathrm{HST}_1(n,d), \mathrm{HST}_2(n,d)$ are the same.*

These encoding also have consequences for the homotopy types of intervals and Möbius functions $\mu(x,y)$ in the orders for $d \leq 3$, to be discussed in Section 5 below.

4.2 Snug rectangles

In [43], Thomas presents an amazingly simple encoding of the triangulations $\mathscr{T}$ of $\mathbf{C}(n,d)$, and an accompanying reformulation of the order $\mathrm{HST}_2(n,d)$.

Definition 4.6. Let $L(n,d)$ denote the set of all strictly increasing integer sequences $(a_1, a_2, \ldots, a_d)$ of length d with $1 \leq a_i \leq n$. For each d-simplex, indexed by the $(d+1)$-subset $i_1 < i_2 < \cdots < i_{d+1}$, appearing in $\mathscr{T}$, associate the subset of $L(n,d)$ (called a *snug rectangle*) which is the following d-fold Cartesian product:

$$[i_1, i_2 - 1] \times [i_2, i_3 - 1] \times \cdots \times [i_d, i_{d+1} - 1].$$

Given the triangulation $\mathscr{T}$ and its various snug rectangles, let

$$
\begin{aligned}
U_{\mathscr{T}} := \{ &(a_1, a_2, \ldots, a_{d-1}, a_d) \in L(n,d) : \\
&(a_1, a_2, \ldots, a_{d-1}, a_d), (a_1, a_2, \ldots, a_{d-1}, a_d - 1) \text{ lie in} \\
&\text{the same snug rectangle of } \mathscr{T} \}.
\end{aligned}
$$

Theorem 4.7 ([43, Theorems 1.1, 1.2]). *The map sending a triangulation $\mathscr{T}$ of $\mathbf{C}(n,d)$ to its collection of snug rectangles is a bijection between all triangulations of $\mathbf{C}(n,d)$ and all decompositions of $L(n,d)$ into snug rectangles.*
Furthermore, $\mathscr{T} \leq_2 \mathscr{T}'$ in $\mathrm{HST}_2(n,d)$ if and only if $U_{\mathscr{T}} \subseteq U_{\mathscr{T}'}$.

Thomas goes on to exhibit a natural poset embedding

$$\mathrm{HST}_2(n,d) \hookrightarrow \prod_{j=d}^{n-1} \mathrm{HST}_2(j, d-2)$$

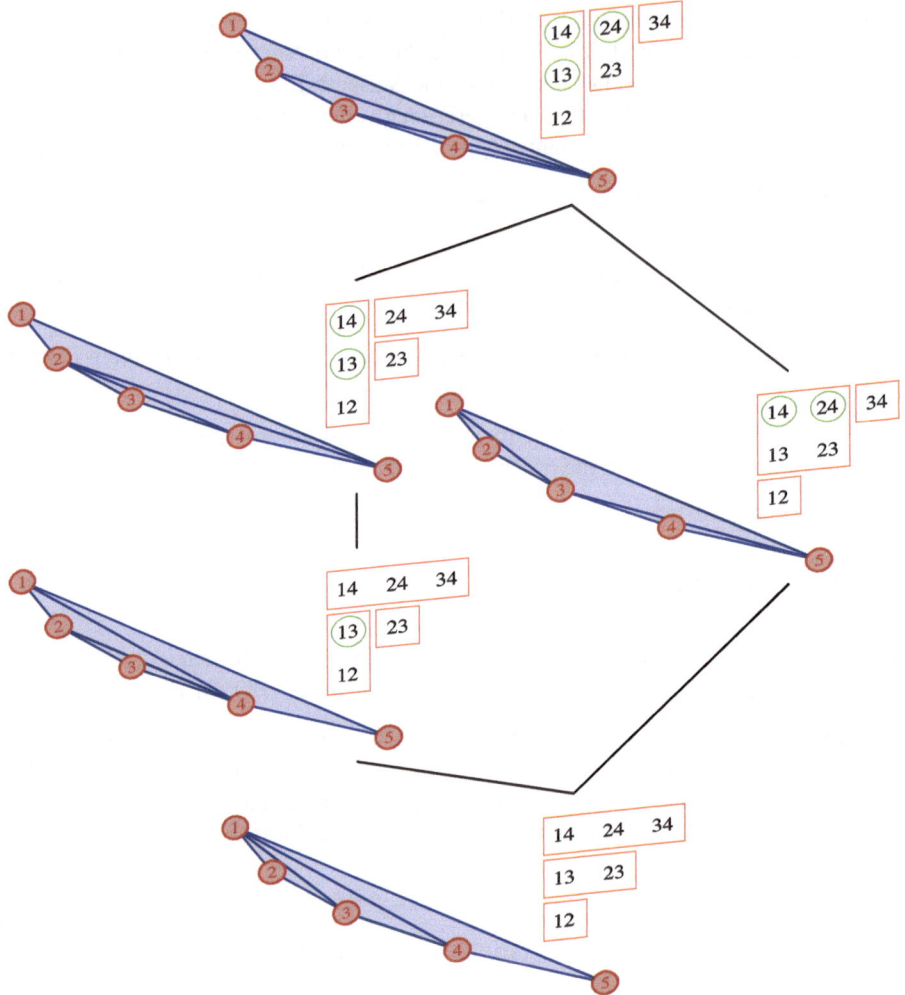

Fig. 7 Triangulations $\mathscr{T}$ of $\mathbf{C}(5,2)$ with their snug rectangle encodings (red), and the points of $U_{\mathscr{T}}$ circled green within the rectangles.

which for $d = 2$ turns into Huang and Tamari's *bracket vector* encoding [22]. By iterating this poset embedding, he improves the upper bounds that had been given in [15] on the *order dimension* $\dim \mathrm{HST}_2(n,d)$, that is, the smallest N for which $\mathrm{HST}_2(n,d)$ has a poset embedding into a product of N linear orders. Furthermore, he gets an *exact* calculation for the *2-dimension* $\dim_2 \mathrm{HST}_2(n,d)$, that is, the smallest N for which $\mathrm{HST}_2(n,d)$ has a poset embedding into a Boolean algebra 2^N of rank N.

Theorem 4.8 ([43, Theorem 6.1]). *The higher Stasheff-Tamari order* $\mathrm{HST}_2(n,d)$ *has*

- *for $d = 2e + 1$ odd,*

$$\dim \mathrm{HST}_2(n,d) \leq \dim_2 \mathrm{HST}_2(n,d) = \binom{n-e-2}{e+1}, \; and$$

- *for $d = 2e$ even,*

$$\dim \mathrm{HST}_2(n,d) \leq \binom{n-e-2}{e},$$

$$\dim_2 \mathrm{HST}_2(n,d) = \binom{n-e-1}{e+1}.$$

4.3 Non-interlacing separated $\frac{d}{2}$-faces

Oppermann and Thomas [26] recently uncovered a fascinating connection between the representation theory of certain algebras and triangulations of the even-dimensional cyclic polytopes $\mathbf{C}(m,d)$ with $d = 2e$. We will not do justice to their results here and refer the interested reader to their paper for more details and precise statements.

Very roughly, they give two generalizations to all even d of the following algebraic results for $d = 2$: when one considers the *path algebra* of the linearly oriented type A_n quiver, the set of *indecomposables* in the module category which are not simultaneously *projective* and *injective* (resp. the set of all indecomposables in the *cluster category*) can be identified with the internal diagonals of $\mathbf{C}(n+2,2)$ (resp. of $\mathbf{C}(n+3,2)$). Furthermore, this can be done in such a way that *basic tilting modules* correspond to triangulations, and *mutations* correspond to diagonal flips. For more on this, see the references in Oppermann and Thomas [26], as well as in Thomas's survey [45] in this volume.

In their work, the role played by the internal diagonals of a triangulation $\mathscr{T}$ of $\mathbf{C}(m,2)$ is played by the nonboundary e-dimensional faces contained in a triangulation $\mathscr{T}$ of $\mathbf{C}(m,2e)$. They begin with an old observation of Dey [13] that for any d, a triangulation $\mathscr{T}$ of any d-dimensional polytope is completely determined by the $\lfloor \frac{d}{2} \rfloor$-dimensional faces that it contains.

In the special case of cyclic polytopes $\mathbf{C}(m,2e)$, these collections of e-faces have a convenient characterization.

Definition 4.9. Say that an e-face with vertex set $i_1 < i_2 < \cdots < i_{e+1}$ is *separated* if $i_{\ell+1} - i_\ell > 1$ for all $0 \leq \ell \leq e$. Say that two such separated e-faces

$$i_1 < i_2 < \cdots < i_{e+1} \text{ and } j_1 < j_2 < \cdots < j_{e+1}$$

intertwine if either

$$i_1 < j_1 < i_2 < j_2 < \cdots < i_{e+1} < j_{e+1}$$

or if the same holds reversing the roles of i's and j's.

When $d = 2$ and $e = 1$, it is not hard to see that the separated e-faces of $\mathbf{C}(m,2)$ correspond to the internal diagonals as well as the "upper" boundary edge $\{1,m\}$, and that the collections of $m - 2$ separated e-simplices which are pairwise non-intertwining are exactly the sets of internal diagonals of triangulations $\mathbf{C}(m,2)$, combined with $\{1,m\}$. Oppermann and Thomas generalize this as follows.

Theorem 4.10 ([26, Theorems 2.3 and 2.4]). *Given a triangulation $\mathcal{T}$ of $\mathbf{C}(m,2e)$, consider the collections of all of its separated e-simplices, or equivalently, all of its e-simplices that do not lie within the lower boundary of $\mathbf{C}(m,2e)$.*

Then these are exactly the collections of $\binom{m-e-1}{e}$ separated and pairwise non-intertwining e-simplices inside $\mathbf{C}(m,2e)$.

They go on to use this characterization in their study of certain categories derived from the module category of the $(e-1)$-fold *higher Auslander algebra* A_n^e of the linearly oriented type A_n quiver. For $e = 1$, this algebra A_n^e is simply the path algebra of the quiver discussed earlier in this section. They obtain generalizations of the above algebraic statements for $e = 1$, by identifying [26, Theorems 1.1, 1.2] the internal e-simplices in $\mathbf{C}(n+2e,2e)$ (resp. $\mathbf{C}(n+2e+1,2e)$) with certain kinds of indecomposable objects in two different categories constructed from A_n^e-modules. Furthermore, they do this in such a way that, in each case, basic tilting modules correspond to triangulations, and the appropriate analogues of mutation correspond to bistellar flips [26, Theorems 4.4, 6.4].

As crucial tools in their proofs, not only do they use the above encoding of triangulations, but also the result of Rambau [31] mentioned in Section 3: all triangulations of $\mathbf{C}(m,d)$ are connected by a sequence of bistellar flips.

5 Lattice property, homotopy types and Möbius function

Theorem 4.4 showed that for $d \leq 3$, the two higher Stasheff-Tamari orders $\mathrm{HST}_1(n,d)$ and $\mathrm{HST}_2(n,d)$ coincide, and both are lattices. Although it was conjectured there that they remain lattices for all d, counterexamples were later found by computer search showing that the lattice property fails, at least for $\mathrm{HST}_2(n,d)$, when $(n,d) = (9,4)$ and $(10,5)$; see [14, §7].

On the other hand, for $d \leq 3$, the two coinciding Stasheff-Tamari lattices $\mathrm{HST}_1(n,d)$ and $\mathrm{HST}_2(n,d)$ enjoy the following pleasant property, which is checked easily for $d \leq 1$, proven for $d = 2$ by Pallo [28, Lemma 4.1], and proven for $d = 3$ in [15, Theorem 4.11].

Theorem 5.1. *Let $d \leq 3$. For an interval $[x,y]$ in the lattices $\mathrm{HST}_1(n,d) = \mathrm{HST}_2(n,d)$, let $\{z_1,z_2,\ldots,z_c\}$ be its set of coatoms, that is the elements $z_i \geq x$ which are covered by y. Then distinct subsets of $\{z_1,z_2,\ldots,z_c\}$ have distinct meets.*

In particular,

- *if $z_1 \wedge \cdots \wedge z_c = x$, then the open interval (x,y) is homotopy equivalent to a $(c-2)$-dimensional sphere, and the Möbius function $\mu(x,y) = (-1)^c$,*

- *if $z_1 \wedge \cdots \wedge z_c > x$, then (x,y) is contractible and $\mu(x,y) = 0$.*

For dimensions $d > 3$, this homotopy type issue is not yet resolved for all intervals, but it is known for the improper open interval $(\hat{0}_{n,d}, \hat{1}_{n,d})$, that is, the proper part of the posets, and the answer for all intervals is conjectured, as we next discuss.

Firstly, the useful tool of Rambau's *suspension lemma for bounded posets* [30], developed to handle the homotopy of the proper parts of the higher Bruhat orders, similarly allowed Edelman, Rambau and Reiner [14, Theorem 1.1] to prove the following.

Theorem 5.2 ([14, Theorem 1.1]). *For $n > d + 1$, the proper parts of both posets* $\mathrm{HST}_1(n,d)$ *and* $\mathrm{HST}_2(n,d)$ *are homotopy equivalent to* $(n - d - 3)$*-dimensional spheres*[2].

Next, when considering intervals $[x,y]$ in $\mathrm{HST}_2(n,d)$, an exact conjecture on their homotopy type was formulated in [14]. For this one must introduce the notion of *polyhedral subdivisions* $\mathscr{S}$ of $\mathbf{C}(n,d)$ (or any point configuration), which are more general than triangulations; we will return to this notion when discussing *secondary polytopes* in Sections 6.2 and 7.1.

Informally, such a subdivision $\mathscr{S}$ is a decomposition of $\mathbf{C}(n,d)$ into subpolytopes $\{P_i\}_{i \in I}$, with these properties:

- each subpolytope P_i has vertex set which is a subset of the vertices of $\mathbf{C}(n,d)$, and
- each pair P_i, P_j of subpolytopes has pairwise intersection $P_i \cap P_j$ equal to a face (possibly empty) common to both.

There is an obvious notion for when one such subdivision *refines* another. Having fixed a particular subdivision $\mathscr{S}$ of $\mathbf{C}(n,d)$, when one considers the collection of all triangulations $\mathscr{T}$ that refine it, it is not hard to see that they form an interval $[x_{\mathscr{S}}, y_{\mathscr{S}}]$ in $\mathrm{HST}_2(n,d)$ (or in $\mathrm{HST}_1(n,d)$). Specifically, if $\mathscr{S}$ has the subpolytope P_i isomorphic to $\mathbf{C}(n_i,d)$, then $x_{\mathscr{S}}$ (resp. $y_{\mathscr{S}}$) triangulates P_i according to the triangulation $\hat{0}_{n_i,d}$ (resp. $\hat{1}_{n_i,d}$) of P_i. It is also not hard to see that the *closed* interval $[x_{\mathscr{S}}, y_{\mathscr{S}}]$ will be poset-isomorphic to the Cartesian product $\prod_{i \in I} \mathrm{HST}_2(n_i,d)$. Hence its proper part, the *open* interval $(x_{\mathscr{S}}, y_{\mathscr{S}})$ will have the homotopy type of a sphere of dimension $-2 + \sum_{i \in I}(n_i - d_i - 1)$, combining Theorem 5.2 with a lemma of Walker [8, eqn (9.8)]: the proper part of the Cartesian product $P_1 \times P_2$ of two bounded posets P_1, P_2 is homeomorphic to the suspension of the join of their proper parts.

Open Problem 5.3 ([14, Conjecture 7.1]). *Prove that the noncontractible open intervals in* $\mathrm{HST}_2(n,d)$ *are exactly the* $(x_{\mathscr{S}}, y_{\mathscr{S}})$ *coming from subdivisions* $\mathscr{S}$ *of* $\mathbf{C}(n,d)$. *In particular, the Möbius function of* $\mathrm{HST}_2(n,d)$ *only takes on values in* $\{0, \pm 1\}$.

These assertions are well known for $d \leq 2$; in the case $d = 2$, they assert that the noncontractible intervals in the Tamari lattice $\mathrm{HST}_2(n,2)$ are exactly the coatomic or

---

[2] For $n = d + 2$, we are using a standard combinatorial convention: the proper part of a poset having only two elements $\{x,y\}$ with $x < y$ is the simplicial complex $\{\varnothing\}$ having only the (-1)-dimensional empty face and no other faces, and considered to triangulate a (-1)-dimensional sphere.

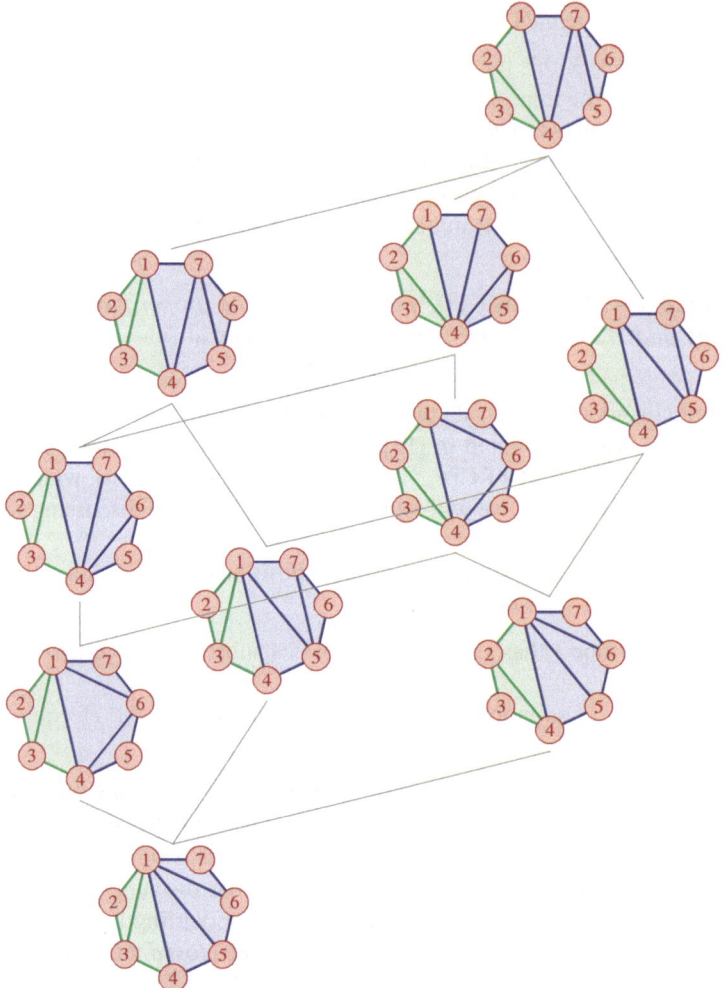

Fig. 8 A subdivision $\mathscr{S}$ of $\mathbf{C}(7,2)$ into a green quadrangle and a blue pentagon, along with its facial interval $[x_{\mathscr{S}}, y_{\mathscr{S}}] \cong \mathrm{HST}_2(4,2) \times \mathrm{HST}_2(5,2)$ in $\mathrm{HST}_2(7,2)$. The open interval $(x_{\mathscr{S}}, y_{\mathscr{S}})$ is homotopy equivalent to a 1-sphere (circle). The heptagon $\mathbf{C}(7,2)$ is depicted with respect to coordinates on the Caratheodory curve, rather than the moment curve, for better visibility of triangles.

facial intervals $[x, y]$, that is, those in which $x = x_{\mathscr{S}}, y = y_{\mathscr{S}}$ are the minimum and maximum elements lying on a particular face of the associahedron, indexed by a polygonal subdivision $\mathscr{S}$ of the n-gon $\mathbf{C}(n,2)$; see Huguet and Tamari [23], and Pallo [28]. Figure 8 shows an example of such an interval $[x_{\mathscr{S}}, y_{\mathscr{S}}]$ within $\mathrm{HST}_2(7,2)$, with in this case an isomorphism $[x_{\mathscr{S}}, y_{\mathscr{S}}] \cong \mathrm{HST}_2(4,2) \times \mathrm{HST}_2(5,2)$.

6 Connection to flip graph connectivity

The Hasse diagram for the higher Stasheff-Tamari order $HST_1(n,d)$, considered as an *undirected* graph, is a special case of an important concept from discrete and computational geometry, which we discuss here: the *flip graph* of all triangulations and *(bistellar) flips* for an arbitrary affine point configuration $\mathbf{A}$ in $\mathbb{R}^d$.

6.1 Bistellar flips

Recall that an edge in the Hasse diagram for $HST_1(n,d)$ corresponds to two triangulations $\mathcal{T}, \mathcal{T}'$ of $\mathbf{A} = \mathbf{C}(n,d)$ that share almost all of the same simplices except that they restrict to the two different possible triangulations (*upper* and *lower*) of the convex hull of a certain subset $\mathbf{A}' = \mathbf{C}(d+2,d)$ of cardinality $d+2$.

It remains true generally that for $d+2$ points $\mathbf{A}'$ in general position in $\mathbb{R}^d$, there will be exactly two triangulations of their convex hull, using only vertices in $\mathbf{A}'$. It even remains true that these two triangulations will again be the set of "upper" and "lower" facets for *some* lifting of the points $\mathbf{A}'$ in $\mathbb{R}^d$ to the vertices of a $(d+1)$-simplex in $\mathbb{R}^{d+1}$, but the combinatorics of these two triangulations will depend upon the signs in the unique affine dependence (up to scaling) among these points, or the *oriented matroid* of the affine point configuration $\mathbf{A}'$; see again [11, §2.4].

Definition 6.1. Two triangulations $\mathcal{T}, \mathcal{T}'$ of the convex hull of an affine point configuration $\mathbf{A}$ in $\mathbb{R}^d$ using only vertices in $\mathbf{A}$, are said to differ by a *(d-dimensional) bistellar flip* if they share almost all of the same simplices, but restrict to the two possible triangulations of the convex hull of some $d+2$ element subset $\mathbf{A}' \subset \mathbf{A}$.

More generally than the d-dimensional bistellar flips, one also allows lower-dimensional *bistellar flips* between two triangulations $\mathcal{T}, \mathcal{T}'$, involving a subset $\mathbf{A}' \subset \mathbf{A}$ of cardinality $e+2$ whose affine span is e-dimensional; see again [11, §2.4] for the precise definitions. Figure 9 illustrates some of the variety of flips possible already for points $\mathbf{A}$ in $\mathbb{R}^2$, with the rightmost example being lower-dimensional flip. Although the variety of possible types of flips grows in higher dimensions (see Figure 10 for one example), when $\mathbf{A}$ in $\mathbb{R}^d$ is in *general position* (no $d+1$ of its points lie on an affine hyperplane of $\mathbb{R}^d$), the flips are *local* modifications, that affect at most $d+1$ simplices on $d+2$ points in a triangulation. Thus, flips are important in computational geometry ($d=2$ or $d=3$, mostly!) as a means to improve triangulations by local modifications (see [17] for just one example or [16] and [10, Chps. 3 and 9] for the low-dimensional viewpoint of Computational Geometry). In non-general position, flips can become quite large modifications. (See also [11, Chp. 8] for a more detailed discussion on algorithmic issues in general dimension).

We should warn the reader that there is a closely related notion of bistellar flip in the literature, which is *not quite* the same: bistellar equivalences for triangulations of *PL-manifolds*, as in the work of Pachner [27]. There one does not insist that the manifolds have a fixed embedding into space nor that the vertices in the triangulation

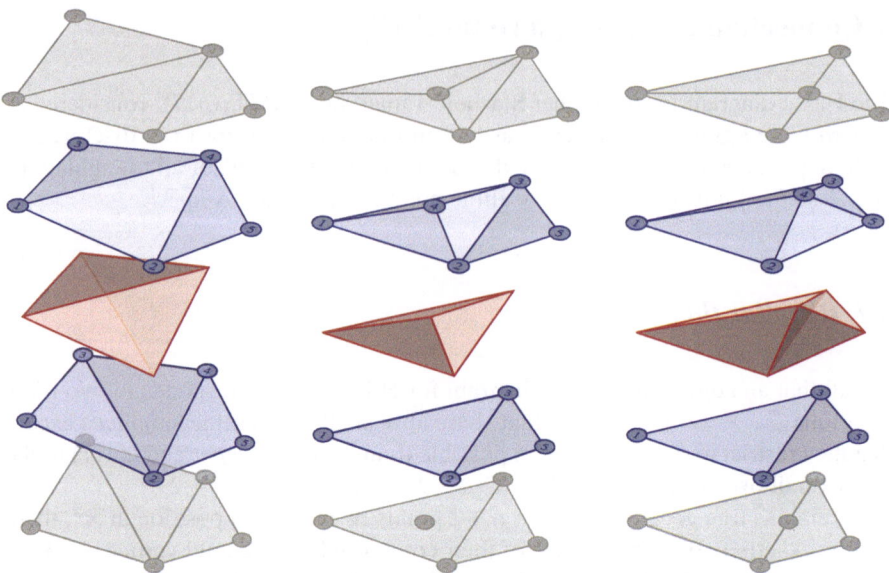

Fig. 9 An edge flip and a vertex flip in dimension two (grey), whose combinatorics can be represented topologically by pushing a surface in dimension three (blue) through a tetrahedron (red) all the way from the lower facets to the upper facets. The rightmost figure is a lower-dimensional flip, adding vertex 4 in the middle of edge 23 (grey): its combinatorics can represented topologically by pushing a surface in dimension three (blue) through a vertical triangle (= 2-simplex!) linked to two vertices (red).

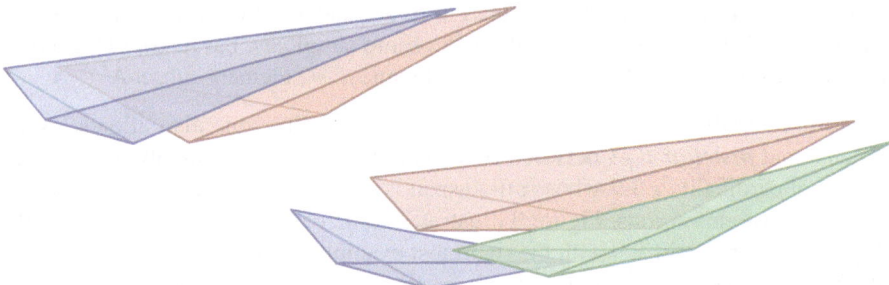

Fig. 10 In dimension three, general position flips will change the number of simplices, as in $\mathbf{C}(5,3)$ depicted here, which has exactly these two triangulations (exploded view). Compare with the discussion of $(2,3)$-*Pachner moves* in the survey by Stasheff in this volume [41, §4.2].

have fixed coordinates. In contrast, triangulations in our context have vertices coming from the point set **A**, with fixed coordinates in $\mathbb{R}^d$.

6.2 The flip connectivity question

In discrete and computational geometry, one would like to use bistellar flips to explore the set of all triangulations of **A**, or to get to any triangulation (for example, a special desired one) from any other triangulation (for example, an obvious one, such as the popular *Delaunay triangulation* [11, § 2.2.2]). This motivates the following definition and question.

Definition 6.2. Given an affine point configuration **A** in $\mathbb{R}^d$, its *flip graph* $\mathscr{G}_{\text{tri}}(\mathbf{A})$ has vertex set indexed by the triangulations $\mathscr{T}$ of the convex hull of **A** using only vertices in **A**, and edges between pairs of triangulations $\mathscr{T}, \mathscr{T}'$ whenever they differ by a bistellar flip.

Question 6.3. *Given an affine point configuration* **A** *in* $\mathbb{R}^d$, *is* $\mathscr{G}_{\text{tri}}(\mathbf{A})$ *connected?*

When either $d \leq 2$, or $|\mathbf{A}| - d \leq 3$, it is not hard to prove that the answer is "Yes". For higher dimensions d and point configurations **A**, this question tantalized researchers for quite some time until resolved negatively by Santos, first in [38], where he found a counter-example with $d = 6$, double-checked by computer-calculations with TOPCOM [32]. Later Santos [39] produced another counter-example $d = 5$ and in general position, which can be turned into convex-position examples by a standard construction, the *Lawrence construction* [11, §5.5].

Theorem 6.4 ([39, Theorem 1]). *There is a 5-dimensional polytope with vertex set* **A** *of cardinality 26 for which the flip graph* $\mathscr{G}_{\text{tri}}(\mathbf{A})$ *is disconnected.*

This should be compared with the positive results of Gelfand, Kapranov and Zelevinsky on *secondary polytopes* [19]. They distinguish a particularly well-behaved subgraph of $\mathscr{G}_{\text{tri}}(\mathbf{A})$, which is not only connected, but even $(|A| - d - 1)$-vertex-connected in the graph-theoretic sense, because it gives the 1-skeleton (vertices and edges) of the $(|A| - d - 1)$-dimensional *secondary polytope*. This subgraph consists of the *regular triangulations* or *coherent triangulations* (and the *regular flips* or *coherent flips* between them), namely those that arise as projections of lower facets of a lifting of the point configuration.

6.3 The flip graph of a cyclic polytope

Returning to cyclic polytopes $\mathbf{C}(n, d)$, it is known and not hard to see that for $d = 2$, *all* triangulations are regular/coherent. This corresponds to the fact that the Hasse diagram of the Tamari order is the 1-skeleton of the *Stasheff polytope* or *associahedron*, which is the secondary polytope for the point configuration $\mathbf{C}(n, 2)$. However, for any fixed $d \geq 3$, one can show that, asymptotically in n, most triangulations of $\mathbf{C}(n, d)$ are *not* regular/coherent, [11, §6.1], which raises that question of connectivity for their flip graphs.

Theorem 6.5 ([31, Thm. 1.1, Cor. 1.2]). *The first higher Stasheff-Tamari order* $\text{HST}_1(n, d)$ *is bounded, with the same bottom* $\hat{0}_{n,d}$ *and top* $\hat{1}_{n,d}$ *triangulations as the*

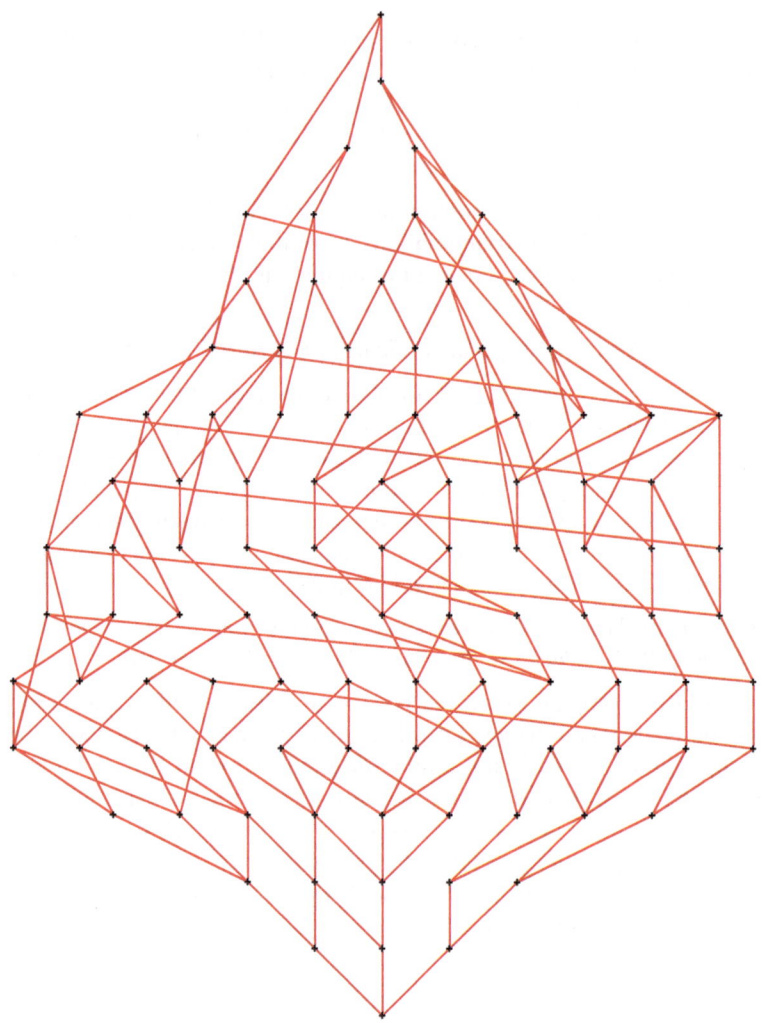

Fig. 11 The Hasse-diagram of $HST_1(10,6)$ generated by an unpublished maple package of the first author and the Stembridge posets package [42].

second higher Stasheff-Tamari order $HST_2(n,d)$. *In particular, the Hasse diagram for* $HST_1(n,d)$, *which is the flip graph* $\mathcal{G}_{tri}(\mathbf{C}(n,d))$, *is connected.*

Figure 11 shows the Hasse-diagram of $HST_1(10,6)$, a non-trivial case for which boundedness was unknown before.

6.4 Diameter

Since the flip graph $\mathcal{G}_{\mathrm{tri}}(\mathbf{C}(n,d))$ is connected, it makes sense to ask for its *diameter*, that is, how many flips are required to reach a triangulation from any other, in the worst case. We explain here how the following structural result on $\mathrm{HST}_1(n,d)$ leads to the exact diameter when d is odd, and diameter bounds when d is even.

Theorem 6.6 ([31, Thm. 1.1]). *There is a one-to-one correspondence between equivalence classes of maximal chains in $\mathrm{HST}_1(n,d)$ and triangulations of $\mathbf{C}(n,d+1)$. Two chains are equivalent if their covering relations are flips on identical sets of $d+1$-simplices. This correspondence is induced by mapping each flip in a maximal chain in $\mathrm{HST}_1(n,d)$ to the corresponding $(d+1)$-simplex in $\mathbf{C}(n,d+1)$.*

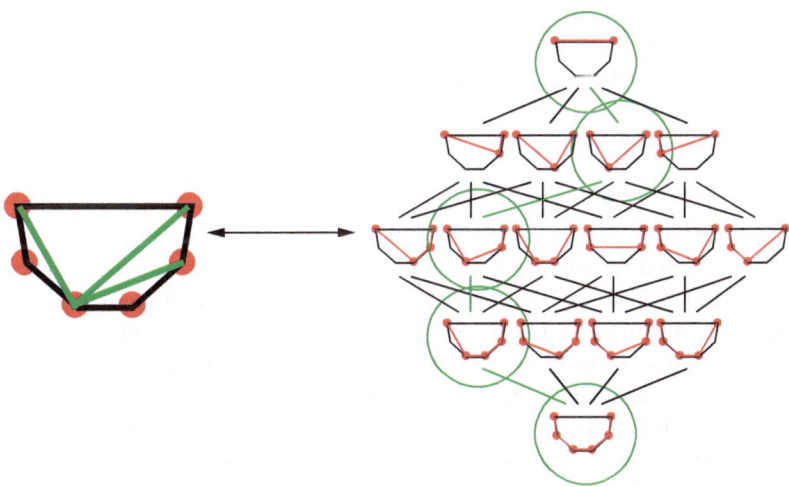

Fig. 12 The connection between a chain in $\mathrm{HST}_1(6,1)$ (represented by characteristic sections) and an element of $\mathrm{HST}_1(6,2)$ (figures from [11, Chp. 5]).

When d is odd, so that $\mathrm{HST}_1(n,d)$ is both ranked and bounded, this determines the diameter of $\mathcal{G}_{\mathrm{tri}}(\mathbf{C}(n,d))$ exactly, combining the previous result, Proposition 3.9, and the following well-known fact.

Proposition 6.7. *A bounded ranked poset of rank r has Hasse diagram diameter r.*

Proof. Every element lies in a maximal chain of length r, and hence any pair of elements are contained in a closed cyclic path of $2r$ edges that concatenates two such maximal chains; thus they lie at distance at most r. On the other hand, the unique bottom and top elements are at distance at least r. $\qquad\square$

Corollary 6.8 ([31, Cor. 1.2]). *For odd $d = 2e - 1$, the diameter of the flip graph of* $\mathbf{C}(n,d)$ *is* $\binom{n-e-1}{e}$.

Since a triangulation of $\mathbf{C}(n, d+1)$ for d even has no more simplices than there are lower facets of $\mathbf{C}(n, d+2)$ and no fewer simplices than there are upper facets of $\mathbf{C}(n, d+2)$, the same argument at least gives these bounds for the diameter.

Corollary 6.9 ([31, Cor. 1.2]). *For even $d = 2e$, the diameter of the flip graph of* $\mathbf{C}(n,d)$ *is bounded between* $\binom{n-e-2}{e}$ *and* $2\binom{n-e-2}{e}$.

6.5 The case $d = 2$: the rotation graph of binary trees

In the case where $d = 2$, the above diameter bounds show that the diameter of $\mathscr{G}_{\text{tri}}(\mathbf{C}(n,2))$ is between $n - 3$ and $2n - 6$. However, this case has been extremely well studied under the guise of the *rotation graph on binary trees*, e.g., in the work of Pallo; see the survey by Dehornoy [12] in this volume for references, and for the close connection with Thompson's group. In particular, the above diameter bound is superseded by the following celebrated result of Sleator, Thurston, and Tarjan.

Theorem 6.10 ([40, Thm. 2]). *The diameter of $\mathscr{G}_{\text{tri}}(\mathbf{C}(n,2))$ is, for sufficiently large values of n, exactly $2n - 10$.*

The proof that the diameter is at least $2n - 10$ for sufficiently large n employs the three-dimensional interpretation of flips sketched above: flipping can be seen as shifting a surface from the lower facets of a (not necessarily straight-line) tetrahedron through the tetrahedron all the way to the upper facets of the tetrahedron.

Moreover, a sequence of flips can be seen as moving a surface all the way through a three-dimensional triangulation, consisting of one tetrahedron per flip and having one triangulation as the bottom and the other triangulation as the top surface. If one could show that there are triangulations of an n-gon so that the three-dimensional space between them needs at least $2n - 10$ tetrahedra to be triangulated, then the claim would follow. And indeed: by embedding the situation in hyperbolic geometry (where volumes of simplices are bounded!), Sleator, Tarjan, and Thurston established the lower bound along these lines. Along their way, they had to master a wealth of technical difficulties, though. No combinatorial or more intuitive proof has been given of this lower bound to date.

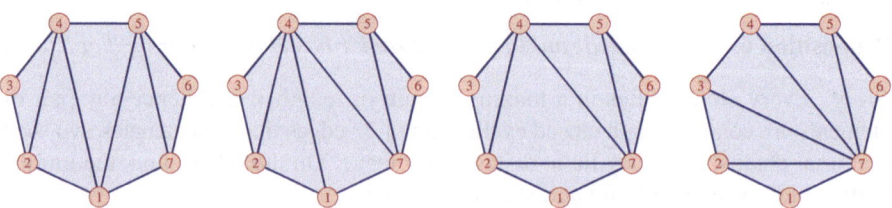

Fig. 13 Flipping (from left to right) to the standard triangulation with respect to vertex 7.

On the other hand, their argument for the diameter upper bound of $2n - 10$ is easy enough to reproduce here. Pick an arbitrary vertex $\mathbf{p}$ of an n-gon with $n > 12$ and an arbitrary triangulation $\mathscr{T}$. Unless $\mathbf{p}$ lies in all possible interior edges, that is, its degree $\deg_{\mathscr{T}}(\mathbf{p})$ in the *interior edge graph* of $\mathscr{T}$ is $n - 3$, we can find a flip that increases the degree of $\mathbf{p}$ by one. (In that case, not all adjacent triangles in the star of $\mathbf{p}$ in $\mathscr{T}$ can form a non-convex quadrilateral.) Thus, we need at most $n - 3 - \deg_{\mathscr{T}}(\mathbf{p})$ flips to transform $\mathscr{T}$ into the unique triangulation with $\deg_{\mathscr{T}}(\mathbf{p}) = n - 3$, the *standard triangulation with respect to* $\mathbf{p}$. The same holds for any other triangulation $\mathscr{T}'$, so that the flip distance $\mathrm{dist}(\mathscr{T}, \mathscr{T}')$ between $\mathscr{T}$ and $\mathscr{T}'$ is at most

$$\mathrm{dist}(\mathscr{T}, \mathscr{T}') \leq \min_{\mathbf{p}} 2n - 6 - \deg_{\mathscr{T}}(\mathbf{p}) - \deg_{\mathscr{T}'}(\mathbf{p}) \qquad (3)$$

If one uses the worst case of this relation as an upper bound, one can not get past $2n - 6$. However, symmetry comes to our aid: Since every triangulation of an n-gon has $n - 3$ interior edges, the average interior-edge degree of a vertex is $(2n - 6)/n = 2 - 6/n$. Summarized:

$$\mathrm{dist}(\mathscr{T}, \mathscr{T}') \leq 2n - 6 - 2 + 6/n - 2 + 6/n = 2n - 10 + 12/n. \qquad (4)$$

Since $n > 12$ and the distance is integral, the claim follows.

7 Subdivisions and the Baues problem

We have already seen, in the discussion of Möbius functions for $\mathrm{HST}_2(n, d)$ in Section 5, the relevance of polytopal subdivisions $\mathscr{S}$ of $\mathbf{C}(n, d)$ which are coarser than triangulations, and the importance of the refinement ordering on them.

The flip graph $\mathscr{G}_{\mathrm{tri}}(\mathbf{A})$ is a one-dimensional object built from these triangulations and bistellar flips relating them. It turns out that bistellar flips can also be thought of as subdivisions which are only slightly coarser than triangulations, namely those that have exactly two refinements, both triangulations. They form part of a larger structure, the *Baues poset*, built from *all* subdivisions. The connectivity question for $\mathscr{G}_{\mathrm{tri}}(\mathbf{A})$ is closely related to the question of homotopy type for this Baues poset. We discuss this somewhat informally here – see [36] for further discussion and references.

7.1 Subdvisions and secondary polytopes

Polytopal subdivisions of the convex hull of a point configuration $\mathbf{A}$, using only vertices in $\mathbf{A}$, already appeared naturally in the work of Gelfand, Kapranov, and Zelevinsky [19, 20] on the *secondary polytope* of $\mathbf{A}$ that was discussed in Section 6.2: the face poset of the second polytope is exactly the poset of all *regular* polytopal subdivisions $\mathscr{S}$ of the convex hull of $\mathbf{A}$, ordered by refinement. See Figure 14 for the example of a pentagon (isomorphic to $\mathbf{C}(5,2)$). See also [11, Chp. 5] for a more elementary introduction to this theory.

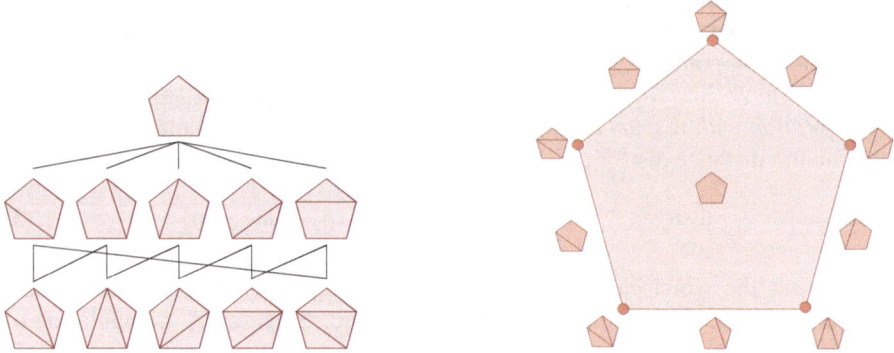

Fig. 14 The refinement poset of a five-gon is isomorphic to the face lattice of its secondary polytope (in this case also a five-gon); figures from [11, Chp. 5].

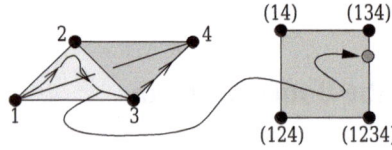

Fig. 15 A path in a tetrahedron and the corresponding cell in the square (figure from [29]).

7.2 Baues's original problem

Meanwhile, a conjecture of Baues in the model theory of loop spaces [5] motivated Billera, Kapranov, and Sturmfels [6] to generalize this subdivision poset. We give here a rough idea of Baues's goal, before explaining their generalization.

The *loop space* ΩX of a base-pointed topological space (X,x) has elements which are closed paths γ in X starting and ending at x, equipped with a certain topology. If X happens to come from a simplicial complex, that is, it is glued from simplices, then one might hope to model ΩX via some type of cell complex; this idea goes back to J.F. Adams [1] who applied it to compute the homology of ΩX.

To this end, consider a piece of a closed path γ inside a d-simplex, with vertices numbered $\{0,1,2,\ldots,d\}$, with γ entering each visited (open) face at its minimal vertex and exiting at its maximal vertex d. Moreover, we require that it enters the simplex at vertex 0 and exits at vertex d. The various substantially distinct options for how this piece of γ can traverse the simplex (in terms of visited open faces) can be modeled by a $(d-1)$-cube: the extreme possibilities are edge paths with increasing vertex labels in the simplex, which biject with vertices of a cube: the vertices 1 through $d-1$ of the simplex that are visited by γ determine the ones in the coordinates of the vertex of the cube. All intermediate options where γ can wander specify in a rather obvious way faces of the cube, where a path meeting the interior of the simplex corresponds to the improper face of the cube, that is, the whole cube.

Thus, one might think that the loop space of a simplicial complex can be modeled by a cubical complex. As always, there are technical subtleties, one of which is that a certain structure must have the homotopy type of a sphere for things to work. Baues conjectured that this structure actually always does have the homotopy type of a sphere.

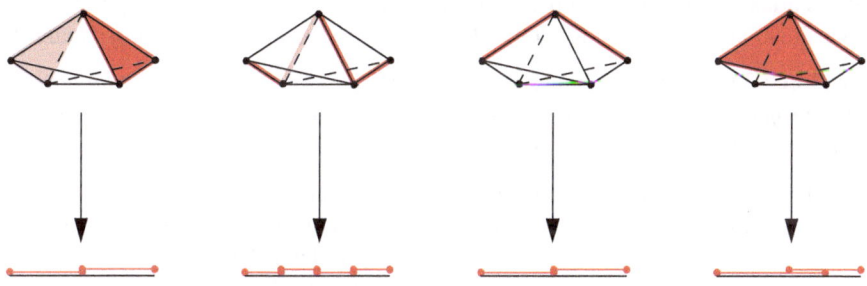

Fig. 16 How cellular strings in the bipyramid project to compatible subdivisions of the line; the rightmost set of faces is *not* a cellular string, because the projections of those faces overlap (figure derived from a figure in [29, Chap. 1]).

7.3 Cellular strings and the generalized Baues problem

Billera, Kapranov, and Sturmfels [6] discovered that the structure Baues was after is an example of the following construction.

Definition 7.1. Consider a d'-dimensional polytope P and linear functional $\mathbb{R}^{d'} \xrightarrow{\pi} \mathbb{R}^1$ taking distinct values $\pi(v) \neq \pi(v')$ whenever v, v' are vertices lying on an edge of P.

Say that a subdivision of the line segment $\pi(P)$ in $\mathbb{R}^1$ into consecutive intervals $[v_0, v_1], [v_1, v_2], \ldots, [v_{\ell-1}, v_\ell]$ is π-*compatible*[3] if, for each $i = 1, 2, \ldots, \ell$, one can

---

[3] The original term "π-induced" in [7, 6] was modified in [11] to "π-compatible" because, in general, there are many subdivisions that are projections of faces under π, induced by the corresponding cellular strings and π, not π alone.

assign a face F_i of P for which $\pi(F_i) = [v_i, v_{i+1}]$. In fact, identify the subdivision with the sequence of faces $(F_1, \ldots, F_\ell)$ in P. Call this sequence a π-*cellular string* in P.

For example, among the π-cellular strings one finds all π-monotone edge paths from the π-minimizing vertex to the π-maximizing vertex of P, but one also has π-cellular strings that take steps through faces which are higher dimensional than edges; see Figure 16.

One defines a refinement ordering on all such π-cellular strings in P via containment of faces, which gives a poset that was baptized the *Baues poset* of P and π. The result that triggered a whole line of research was this.

Theorem 7.2 ([6]). *For any d'-dimensional polytope and linear functional $\mathbb{R}^{d'} \xrightarrow{\pi} \mathbb{R}^1$ as above, the Baues poset is homotopy equivalent to a sphere of dimension $d' - 2$.*

Billera, Kapranov, and Sturmfels also defined a Baues poset of π-compatible subdivisions for any linear projection π of a d'-dimensional polytope P to a d-dimensional polytope $\pi(P)$ for some $d < d'$. The following question arose naturally.

Question 7.3 (Generalized Baues Problem (GBP)). *For a d'-polytope P and for any linear projection π to $\mathbb{R}^d$, does the (Generalized) Baues poset of P and π have the homotopy type of a $d' - d - 1$-sphere?*

At the time when this question was phrased it had almost the status of a conjecture. This thinking was fueled by the work of Billera and Sturmfels on the theory of *fiber polytopes* [7], generalizing Gelfand, Kapranov and Zelevinsky's secondary polytopes. The fiber polytope of the projection π out of P distinguishes geometrically a certain subposet of the π-compatible subdivisions $\mathscr{S}$ of the image polytope $\pi(P) =: Q$, namely those subdivisions which are π-*coherent*: one requires that the collection of faces $\{F_i\}$ of P projecting to the subdivision $\mathscr{S}$ does not "wrap around P", in the sense that there exists a single linear functional g on the $(d' - d)$-dimensional real space $\ker(\pi)$ so that the union $\cup_i F_i$ is exactly the union over all the points q in $\pi(P)$ of the g-maximizing subfaces of the $(d' - d)$-dimensional polytopal fibers $\pi^{-1}(q) \cap P$.

Denoting the subposet of π-coherent subdivisions the *coherent (generalized) Baues poset* of P and π, one has the following striking result.

Theorem 7.4 ([7]). *The coherent generalized Baues poset of P and π is always isomorphic to the face lattice of a polytope, the* fiber polytope *of P and π. In particular, this subposet is homeomorphic to a $d' - d - 1$-sphere.*

In the example of Figure 17 we see that the poset of coherent compatible subdivisions (solid covering relations) is indeed a proper sub-poset of the Baues poset; this sub-poset is isomorphic to a hexagon whereas the whole poset is only homotopy equivalent to a 1-sphere. By Theorem 7.4, the subdivisions connected by dashed covering relations cannot be compatible, because they lie only in chains that are too long to appear in the face lattice of a 2-dimensional polytope. Note that though the bipyramid is isomorphic to $\mathbf{C}(5, 3)$, the indicated projection is *not* the canonical projection between cyclic polytopes, since the induced order of vertices is $2, 1, 3, 5, 4$, as opposed to $1, 2, 3, 4, 5$ in the canonical projection.

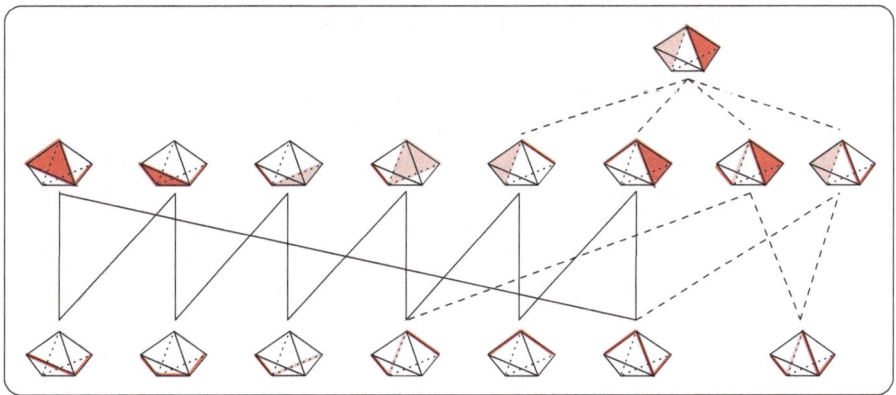

Fig. 17 The (proper part of the) Baues poset of coherent and incoherent cellular strings on a bipyramid, projecting down to compatible subdivisions of a line (example and figure from [47, Chp. 9]; figure also in [29, Chp. 1]).

As time went by with no affirmative answer to Question 7.3, hope diminished, and finally a surprisingly small counterexample was constructed by Rambau and Ziegler [34].

Theorem 7.5 ([34, Thm. 1.5]). *There is a generic projection π of a 5-polytope P with 10 vertices to the plane, having a disconnected generalized Baues poset. In particular, its generalized Baues poset is not homotopy equivalent to a 2-sphere.*

This yields counterexamples for any set of larger parameters d, d' with $d > 2$ and $d' - d > 2$ by standard constructions. For the only missing parameters $d' - d \leq 2$, an affirmative answer could be given, with an involved proof, though:

Theorem 7.6 ([34, Thm. 1.4]). *For $d' = d + 2$, the generalized Baues poset always has the homotopy type of a 1-sphere.*

Thus one has a recurring dichotomy: geometrically distinguished subdivisions form friendly structures, whereas the general subdivisions do not.

On the other hand, as in the case of triangulations, there is a family of particularly friendly polytopes where everything is nice, and it is again – the cyclic polytopes. Work of several authors showed[4] that the canonical projections between cyclic polytopes have indeed well-behaved generalized Baues posets.

Theorem 7.7 ([33, 3]). *For all $d' > d$, the generalized Baues poset of the canonical projection from $\mathbf{C}(n, d')$ to $\mathbf{C}(n, d)$ has the homotopy type of a $d' - d - 1$-sphere.*

---

[4] For some of the history on the progress toward this result, see [36, §4].

8 Connection to the higher Bruhat orders

We discuss here the *higher Bruhat orders* $B(n,k)$ of Manin and Schechtman [25] generalizing the *weak Bruhat order* $B(n,1)$ on the set $\mathfrak{S}_n$ of all permutations of n letters. Their intimate connection to the higher Stasheff-Tamari orders appears already in the original paper of Kapranov and Voevodsky, who discuss [24, §4] a poset map $B(n,k) \to \mathrm{HST}_1(n+2,k+1)$ generalizing the classical poset surjection from the weak Bruhat order on $\mathfrak{S}_n$ to the Tamari order on triangulations of $\mathbf{C}(n+2,2)$; see the survey by Reading [35] and by Hohlweg [21] for more perspectives and different generalizations of this map. Further discussion of higher Bruhat orders, higher Stasheff-Tamari orders, and the poset map between them appears in [36, §6].

8.1 Definition of higher Bruhat orders

One can think of the higher Bruhat orders $B(n,k)$ as orders on orders on orders ... of subsets. When defined for general k, they can seem a bit technical. Here we choose instead to work our way up from $k = 0, 1, 2, \ldots$

Example 8.1. When $k = 0$, the (lower!) Bruhat order $B(n,0)$ is the Boolean algebra $2^{\{1,2,\ldots,n\}}$. Thus it is isomorphic to the two lower Stasheff-Tamari orders on the set of triangulations of $\mathbf{C}(n+2,1)$, as described in Example 3.4.

Note that this isomorphism is most natural if one renumbers the vertices on the line segment $\mathbf{C}(n+2,1)$ as $0, 1, 2, \ldots, n+1$, rather than our usual numbering $1, 2, \ldots, n+2$, so that the internal vertices are labelled $\{1, 2, \ldots, n\}$.

Also, note that one can think of $B(n,0) = 2^{\{1,2,\ldots\}}$ in two ways:

- It is the transitive closure of the relation $S < T$ whenever $S \subset T$ and $|T| = |S| + 1$. This is analogous to $\mathrm{HST}_1(n+2,1)$.
- One has $S \leq T$ for two subsets S, T in $2^{\{1,2,\ldots,n\}}$ if $S \subseteq T$. This is analogous to $\mathrm{HST}_2(n+2,1)$. When we wish to emphasize this analogy, we will borrow Ziegler's notation from [46] where he denotes this *inclusion* order $B_{\subseteq}(n,0)$, to distinguish it from the definition via *single-step inclusion*.

Example 8.2. When $k = 1$, the poset $B(n,1)$ is actually the *weak Bruhat order* on the symmetric group $\mathfrak{S}_n$. As a set, it consists of all maximal chains

$$\varnothing \subset \{w_1\} \subset \{w_1, w_2\} \subset \cdots \subset \{w_1, w_2, \ldots, w_{n-1}, w_n\} = \{1, 2, \ldots, n\} \qquad (5)$$

of elements in $B(n,0)$. Such chains biject with the linear orders $w = (w_1, \ldots, w_n)$ in which the elements are added, which can be read as permutations w in $\mathfrak{S}_n$.

To order $B(n,1)$, recall the *(left-)inversion set* $\mathrm{Inv}_2(w)$ of w is the collection of pairs $i < j$ (which we will call *2-packets*) for which j appears before i in the order w. Define $w \leq w'$ in $B_{\subseteq}(n,1)$ if $\mathrm{Inv}_2(w) \subseteq \mathrm{Inv}_2(w)$. On the other hand, one can define the *single-step* ordering $B(n,1)$ as the transitive closure of the relation $w < w'$ if

$\text{Inv}_2(w) \subset \text{Inv}_2(w')$ and $|\text{Inv}_2(w')| = |\text{Inv}_2(w)| + 1$. It is a classical result that these two orders on $\mathfrak{S}_n$ are the same, and define the *weak Bruhat order*.

Note that a linear order w can be recovered from its inversion set $\text{Inv}_2(w)$. Also note that inversion sets $\text{Inv}_2(w)$ of permutations w are not arbitrary subsets of all 2-packets $\left(\binom{\{1,2,\dots,n\}}{2}\right)$: given any 3-packet $i < j < k$, transitivity forces that if ij, jk lie in $\text{Inv}_2(w)$ or if both are absent from $\text{Inv}_2(w)$, then the same must be true for ik. Said differently, for each 3-packet $i < j < k$, an inversion set $\text{Inv}_2(w)$ must intersect the lexicographic order on the 2-packets

$$ij, \quad ik, \quad jk \tag{6}$$

in either an initial segment, or a final segment.

Example 8.3. Things become more interesting when $k = 2$. Again $B(n,2)$ can be derived from consideration of maximal chains c in $B(n,1)(= B_{\subseteq}(n,2))$. Such chains are sequences of permutations

$$e = w^{(0)} < w^{(1)} < \cdots < w^{(\binom{n}{2})} = w_0 \tag{7}$$

from the identity e to the *longest element* $w_0 = (n, n-1, \dots, 2, 1)$, in which one adds one element to $\text{Inv}_2(w^{(i)})$ at each stage. These maximal chains correspond to *reduced decompositions* for w_0 in terms of the adjacent transposition *Coxeter generators* $s_i = (i, i+1)$ for $\mathfrak{S}_n$.

One can derive from such a chain c a 3-packet inversion set $\text{Inv}_3(c)$ by considering for each 3-packet $i < j < k$ the two possible orders in which its 2-packet subsets ij, ik, jk are added to the sets $\text{Inv}_2(w^{(i)})$ for $i = 1, 2, \dots, \binom{n}{2}$: either they are added in the lexicographic order from (6) and one decrees $ijk \notin \text{Inv}_3(c)$, or they are added in the reverse of this order and one decrees $ijk \in \text{Inv}_3(c)$.

As a set $B(n,2)$ is defined to be all equivalence classes $\bar{c}$ of such chains c, where c, c' are equivalent if $\text{Inv}_3(c) = \text{Inv}_3(c')$. In terms of reduced words for w_0, this is the same as equivalence under *commuting braid moves* $s_i s_j = s_j s_i$ for $|i - j| \geq 2$.

Similarly to $B_{\subseteq}(n,1)$, define $\bar{c} \leq \bar{c}'$ in $B_{\subseteq}(n,2)$ if $\text{Inv}_3(\bar{c}) \subseteq \text{Inv}_3(\bar{c}')$. On the other hand, one can define the *single-step* ordering $B(n,2)$ as the transitive closure of the covering relation $\bar{c} \lessdot \bar{c}'$ if their 3-packet inversion sets are nested and differ in cardinality by one.

It is no longer obvious that these two orders $B(n,2)$ and $B_{\subseteq}(n,2)$ on the set are the same; this nontrivial fact was proven by Felsner and Weil [18].

And it is again true that inversions sets $\text{Inv}_3(\bar{c})$ are not arbitrary subsets of all 3-packets $\left(\binom{\{1,2,\dots,n\}}{3}\right)$: given any 4-packet $i_1 < i_2 < i_3 < i_4$, they are characterized by a nontrivial *biconvexity condition* [46, Lemma 2.4] asserting that $\text{Inv}_3(\bar{c})$ must intersect the lexicographic order on the 3-packets within this 4-packet

$$i_1 i_2 i_3, \quad i_1 i_2 i_4, \quad i_1 i_3 i_4, \quad i_2 i_3 i_4$$

in either an initial segment or a final segment.

This picture continues, allowing one to derive $B(n,k)$ from considering maximal chains c of elements in $B_{ss}(n,k-1)$. Each such maximal chain can be considered as the sequence of k-packets added to the k-inversion sets of the elements in the chain. For each fixed $(k+1)$-packet S, its subset k-packets are added either in lexicographic, or reverse lexicographic order, as one proceeds up the chain, and one uses this to decree whether the $(k+1)$-packet does not or does lie in $\text{Inv}_{k+1}(c)$.

One defines $B(n,k)$ as a set to be the equivalence classes $\bar{c}$ of these chains c in $B_{ss}(n,k-1)$ having the same $(k+1)$-packet inversion sets $\text{Inv}_{k+1}(c)$. One orders the set $B(n,k)$ either as $B_{\subseteq}(n,k)$ via *inclusion* of the sets $\text{Inv}_{k+1}(c)$, or denotes the analogous poset $B(n,k)$ defined via *single-step inclusion*.

Although these definitions are recursive, one can also make them nonrecursive; see Ziegler [46]. On the other hand, the recursive description builds in a classical result by Manin and Schechtman on how the structures of the higher Bruhat orders $B(n,k)$ are intertwined for different k; Theorem 6.6 was actually inspired by the following older result.

Theorem 8.4 ([25, Thm. 3]). *There is a one-to-one correspondence between certain equivalence classes of maximal chains in $B(n,k)$ and elements of $B(n,k+1)$.*

On the other hand, when $k \geq 3$, a subtlety appears, in that Ziegler shows [46, Theorem 4.5] that $B_{ss}(n,3)$ and $B_{\subseteq}(n,3)$ do *not* coincide for $n \geq 8$. In light of this fact, and the existence of the map $B(n,k) \to \text{HST}_1(n+2,k+1)$, to be discussed in Section 8.3 below, the resolution of Problem 3.3 becomes even more interesting.

8.2 Some geometry of higher Bruhat orders

The geometry that originally motivated Manin and Schechtman to define $B(n,d)$ comes from generalizations of the classical braid arrangement known as *discriminantal arrangements*. We will not discuss this here, but instead focus on the interpretation of $B(n,d)$ via tilings of the d-dimensional *cyclic zonotope* $\mathbf{Z}(n,d)$, defined to be the *Minkowski sum*

$$\mathbf{Z}(n,d) = \left\{ \sum_{i=1}^{n} c_i v_d(t_i) : -1 \leq c_i \leq +1 \right\}$$

consisting of all $[-1,+1]$ combinations of n distinct vectors $v_d(t_1),\ldots,v_d(t_n)$ lying on the moment curve in $\mathbb{R}^k$, with $t_1 < \cdots < t_n$, as usual.

This interpretation is essentially stated without proof in the paper of Kapranov and Voevodsky [24, Theorem 4.9]. It was carefully justified and explained later by Thomas in [44, Thm. 2.1], as well as by Ziegler's discussion following [46, Theorem 4.1] when one takes into account the equivalence between zonotopal tilings and oriented matroid single-element liftings given by the Bohne-Dress Theorem [9, Theorem 2.2.13].

The story begins with the observation that any zonotope which is a Minkowski sum $\sum_{i=1}^{n}[-1,+1]v_i$ generated by n vectors $v_1,\ldots,v_n$ in $\mathbb{R}^d$ is simply the projection

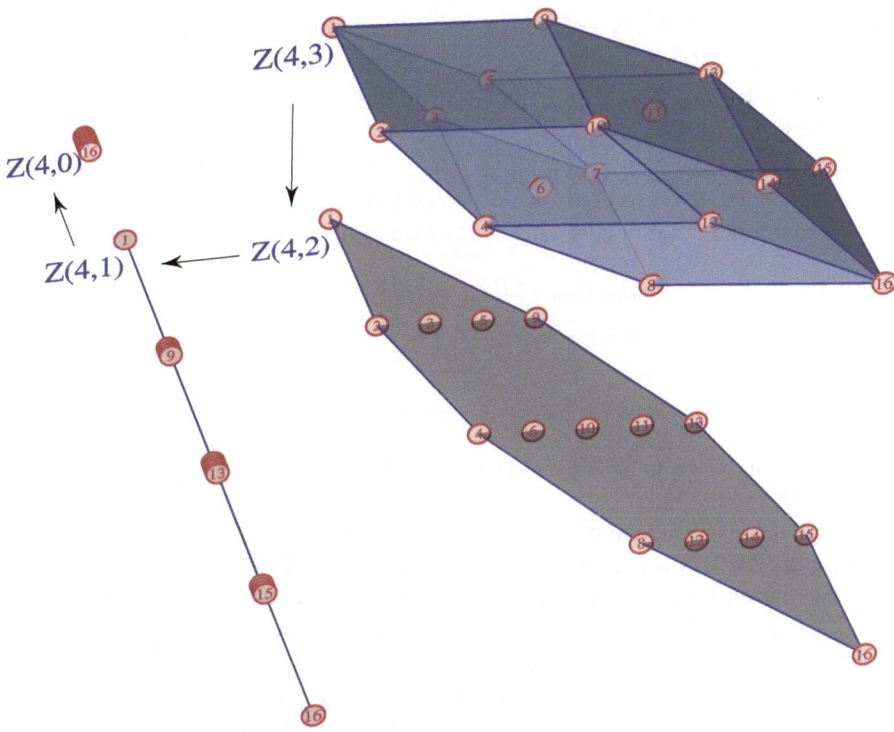

Fig. 18 Cyclic zonotopes $\mathbf{Z}(4,3)$, $\mathbf{Z}(4,2)$, $\mathbf{Z}(4,1)$, and $\mathbf{Z}(4,0)$ ($2^4 = 16$ repeated points at the origin) together with the canonical projections forgetting the last coordinate. Note that already $\mathbf{Z}(4,3)$ has interior points, namely 6 and 11, and already $\mathbf{Z}(4,2)$ has repeated points, namely 7 and 10. $\mathbf{Z}(4,4)$ is simply a 4-cube.

of the n-cube $[+1,-1]^n$ in $\mathbb{R}^n$ generated by the standard basis vectors $e_1,\ldots,e_n$, under the linear projection $\mathbb{R}^n \to \mathbb{R}^d$ that sends e_i to v_i. Because of this, one has a natural tower of projections $\mathbf{Z}(n,d') \to \mathbf{Z}(n,d)$, depicted for $n = 4$ in Figure 18, analogous to the tower of projections $\mathbf{C}(n,d') \to \mathbf{C}(n,d)$ discussed in Section 2 and depicted in Figure 1.

To explain the interpretation of $B(n,d)$ in terms of *tight zonotopal tilings* of $\mathbf{Z}(n,d)$, that is, tilings by subzonotopes which cannot be further refined, we will work our way up from the low-dimensional cases, where the geometry is simpler. For a careful discussion of the definitions of zonotopal tilings, we refer the reader to Richter-Gebert and Ziegler [37], Billera and Sturmfels [7, §4], or DeLoera, Rambau and Santos [11, §9.1.2].

Example 8.5. When $d = 0$, each vector $v_0(t_i)$ lies at the origin which is the unique point in $\mathbb{R}^0$, and equals the zero-dimensional zonotope $\mathbf{Z}(n,0)$. However, we regard the point $\mathbf{Z}(n,0)$ as having 2^n different labels by subsets S of $\{1,2,\ldots,n\}$, each corresponding to the vertex of the n-cube $Z(n,n)$ that projects to it. Thus, an element

S of $B(n,0) = 2^{\{1,2,\ldots,n\}}$, is a choice of such a label, and is considered a zonotopal tiling of $\mathbf{Z}(n,0)$. Alternatively, it gives a section of the map $\mathbf{Z}(n,n) \overset{\pi}{\to} \mathbf{Z}(n,0)$.

Note also that the covering relation between subsets $S \lessdot S'$ in $B(n,0)$ corresponds to two vertices lying along an edge of the n-cube.

Example 8.6. When $d = 1$, each vector $v_1(t_i) = t_i$ points along the (x_1-)axis of $\mathbb{R}^1$, and $Z(n,1)$ is the line segment whose two endpoints $v_{\min}, v_{\max}$ are $\pm(t_1 + \cdots + t_n)$. A tight zonotopal tiling of $Z(n,1)$ is a sequence of intervals

$$[v_{\min}, v_{\min} + 2t_{w_1}],$$
$$[v_{\min} + 2t_{w_1}, v_{\min} + 2t_{w_1} + 2t_{w_2}],$$
$$\ldots,$$
$$[v_{\min} + 2t_{w_1} + 2t_{w_2} + \cdots + 2t_{w_{n-1}}, v_{\max}]$$

corresponding to a permutation $w = (w_1, \ldots, w_n)$ in $\mathfrak{S}_n$, or an element of $B(n,1)$; see Example 8.2.

On the other hand, such permutations or elements of $B(n,1)$ correspond to maximal chains in $B(n,0)$, that is, sequences of nested subsets as in (5), and hence by our observation for $d = 0$, to edge-paths in the cube $\mathbf{Z}(n,n)$ which proceed in a monotone fashion from the vertex labelled by the empty set $\varnothing$ to the vertex labelled by $\{1,2,\ldots,n\}$. In other words, they give sections of the map $\mathbf{Z}(n,n) \overset{\pi}{\to} \mathbf{Z}(n,1)$. See Figure 21 and following for some examples of such edges paths with $n = 3$.

Note also that covering relation between two permutations $w \lessdot w'$ in $B(n,1)$ corresponds to two monotone edge paths in the cube $\mathbf{Z}(n,n)$ that differ only in two adjacent steps that proceed in opposite ways around a quadrilateral face of the cube

Example 8.7. Again, things become interesting when $d = 2$. Now the vectors $v_2(t_i)$ in $\mathbb{R}^2$ generate a zonotopal polygon $\mathbf{Z}(n,2)$, that is, a centrally symmetric $2n$-gon.

An element of $B(n,2)$ can be thought of as a maximal chain of permutations in $B(n,1)$ as in (7), up to a certain equivalence relation. It is possible to model this equivalence relation in at least two ways. One way considers the associated *pseudoline arrangement* or *wiring diagram*, as in Figure 19, whose vertical slices record the permutations in the chain as the ordering of the strands. These diagrams are considered only up to the equivalence relation of isotopies in the plane that never allow one strand to slide over the crossing of two other strands.

The other way considers each permutation w_i in the chain as a monotone edge path in the cube, and each covering relation $w_i \lessdot w_{i+1}$ in the chain as giving a quadrilateral face of the cube on which the two paths take two adjacent steps that disagree. The union of all such quadrilateral faces is a 2-dimensional surface inside the cube $Z(n,n)$, which is a section of the map $\mathbf{Z}(n,n) \to \mathbf{Z}(n,2)$.

The concordance between these two models is that the quadrilateral faces in this 2-dimensional surface map under π to a tight zonotopal tiling of the $2n$-gon $\mathbf{Z}(n,2)$. This tiling can be recovered as the planar dual graph to the graph given by the pseudoline arrangement, considered as having vertices only at the strand crossings; see Figure 19.

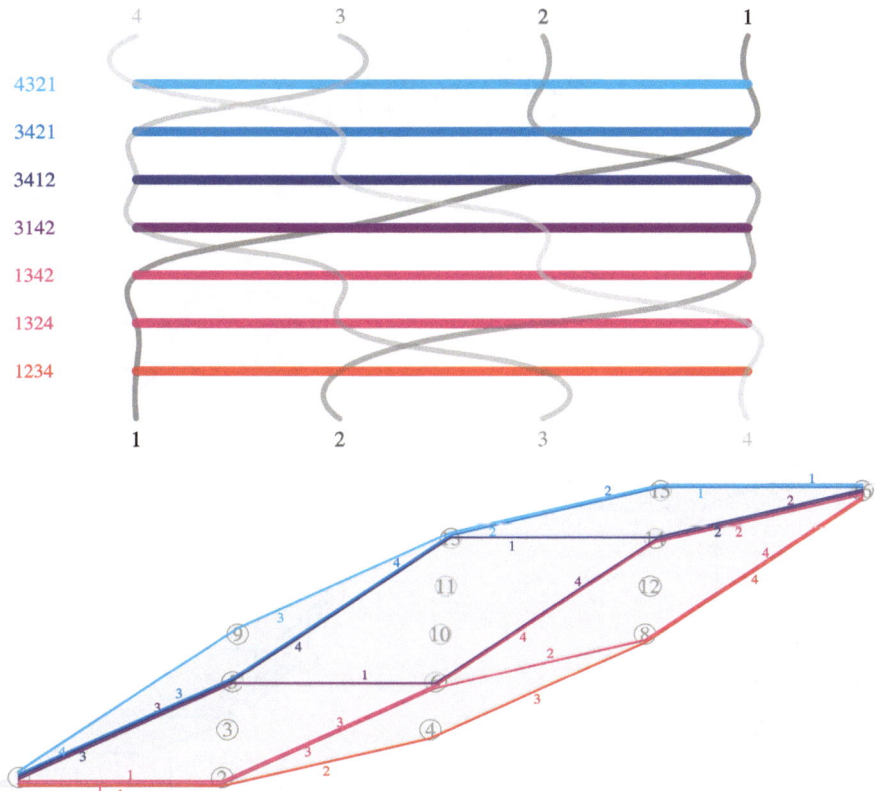

Fig. 19 An element of $B(4,2)$ derived from a maximal chain of permutations in $B(4,1)$, the weak Bruhat order on $\mathfrak{S}_4$. The chain of permutations (colored from red to cyan) leads to an arrangement of pseudolines, also called a wiring diagram: horizontal slices have the strands ordered as in the permutations in the chain. The planar dual of the pseudoline graph can be drawn as a tight subdivision of the zonotope $\mathbf{Z}(4,2)$, in which the pseudoline strand i for $i = 1,2,3,4$ is dual to the edges of the tiles in the parallelism class labelled by i. Moreover, the chain of permutations can be recovered in the zonotopal tiling as a sequence of monotone paths (colored from red to cyan) with covering relations coming from "flipping" the paths "upwards" through a quadrilateral.

This picture continues. The work of Thomas [44, Prop. 2.1], Ziegler [46, Theorem 4.1] shows that an element of $B(n,d)$ can be thought of as unions of d-dimensional faces inside the cube $\mathbf{Z}(n,n)$, corresponding to the image of a section of the map $\mathbf{Z}(n,n) \xrightarrow{\pi} \mathbf{Z}(n,d)$, projecting to a tight zonotopal subdivision of $\mathbf{Z}(n,d)$.

One can furthermore show that if one instead associates to these tight zonotopal subdivisions $\mathscr{S}$ of $\mathbf{Z}(n,d)$ a section $s_{\mathscr{S}}$ of the map $\mathbf{Z}(n,d+1) \xrightarrow{\pi} \mathbf{Z}(n,d)$, then one has $\mathscr{S} \leq \mathscr{S}'$ in the higher Bruhat order $B_{\subseteq}(n,d)$ exactly when $s_{\mathscr{S}}(x)_{d+1} \leq s_{\mathscr{S}'}(x)_{d+1}$ for all x in $\mathbf{Z}(n,d)$; see Figure 20 for this picture of $B_{\subseteq}(4,2)$.

Analogously to the situation for cyclic polytopes $\mathbf{C}(n,d)$, these tight zonotopal subdivisions and the edges between them in the Hasse diagram for $B(n,d)$ are special cases of the more general notion of a *zonotopal subdivision* of $\mathbf{Z}(n,d)$, which is

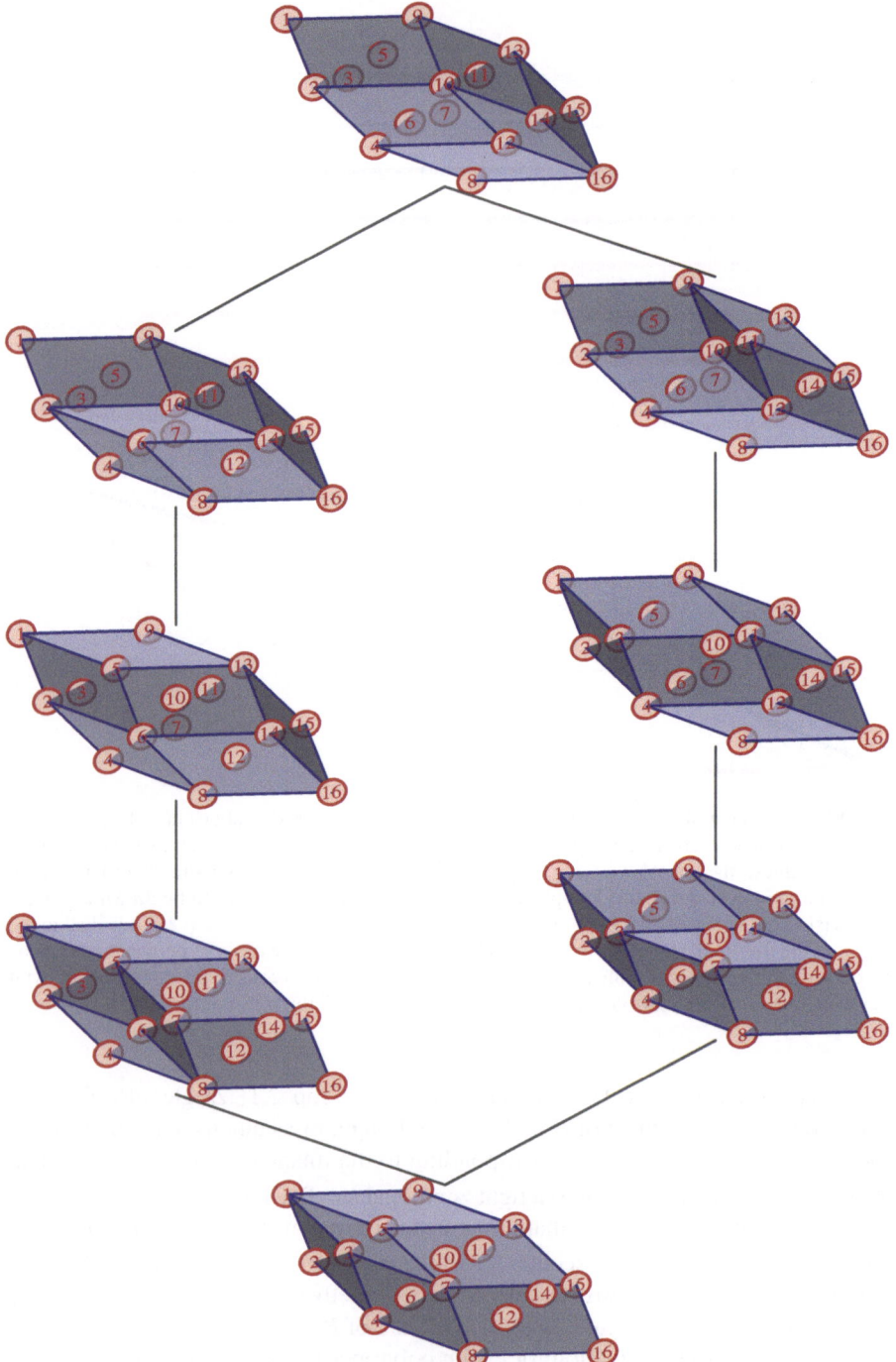

Fig. 20 A picture of $B(4,2)$ with elements drawn as the sections of zonotopal tilings of $\mathbf{Z}(4,2)$ in $\mathbf{Z}(4,3)$, partially ordered by height; it can be seen how the sections, on their way to the top, submerge more and more points. Each chain can be built by stacking cubes, and the cubes corresponding to a chain form a zonotopal tiling of $\mathbf{Z}(4,3)$, which represents an element of $B(4,3)$.

a π-compatible subdivision for the projection $\mathbf{Z}(n,n) \xrightarrow{\pi} \mathbf{Z}(n,d)$. There is again a Baues poset of all such subdivisions, ordered by refinement, and the Baues problem asks for its homotopy type. Athanasiadis [2] investigated the Baues problem for all of the canonical projections $\mathbf{Z}(n,d') \xrightarrow{\pi} \mathbf{Z}(n,d)$, as in Figure 18.

Theorem 8.8 ([2, Thm. 1.1]). *For all $d' > d$, the generalized Baues poset of the canonical projection from $\mathbf{Z}(n,d')$ to $\mathbf{Z}(n,d)$ has the homotopy type of a $d' - d - 1$-sphere.*

8.3 The map from higher Bruhat to higher Stasheff-Tamari orders

The similarity of the description between the higher Bruhat orders $B(n,k)$ in the last section should make their analogy to the higher Stasheff-Tamari orders $\mathrm{HST}_1(n,d)$ apparent.

Tightening the connection, Kapranov and Voevodsky [24] claimed, and later Rambau [31] proved, that there actually is a poset map between them. Later, Thomas shed more light on this connection in [44, §4] (see Figures 21 through 24 for an illustration).

Theorem 8.9 ([31, Cor. 8.16]). *There is an order-preserving map*

$$B(n,k) \xrightarrow{f} \mathrm{HST}_1(n+2,k+1).$$

In low dimensions, the map f is familiar. Example 8.1 noted the isomorphisms

$$B(n,0) = 2^{\{1,2,\dots,n\}} \cong \mathrm{HST}_1(n+2,1).$$

In the next dimension up, the map $B(n,1) \xrightarrow{f} \mathrm{HST}_1(n+2,2)$ is the same as the map from the weak Bruhat order on $\mathfrak{S}_n$ to the Tamari order on triangulations of $\mathbf{C}(n+2,2)$ discussed in the survey by Reading [35, §1] in this volume[5]. To describe it in our geometric setting, one must assign a triangulation of $\mathbf{C}(n+2,2)$ to each permutation w in $\mathfrak{S}_n$, or to each monotone edge path in the n-cube. To this end, think of $\mathbf{C}(n+2,2)$ as labeled by $0,1,2,\dots,n+1$, with $\{0,n+1\}$ its only upper edge. In the order of the permutation w, cut off any remaining vertex i of $\mathbf{C}(n+2,2)$ by inserting the diagonal from its left to its right neighbor. Once all vertices $0 < i < n+1$ have been cut off, the set of inserted diagonals forms a triangulation. Note that two distinct permutations can map to the same triangulation because i,j that are adjacent in the permutation but not adjacent during the cut-off procedure can be cut off in an arbitrary order. Compare this with the description of this map in the survey by Reading [35, §1], and in particular, compare [35, Figure 3], with Figures 22 through 24 below.

This f extends (modulo technical details) to a map $B(n,d) \xrightarrow{f} \mathrm{HST}_1(n+2,d+1)$ via induction on d. Elements of $B(n,d)$ are equivalence classes of maximal chains

---

[5] This map also appears implicitly in the survey by Hohlweg [21], where it is explained how to embed the associahedron in such a way that its normal fan coarsens that of the permutohedron.

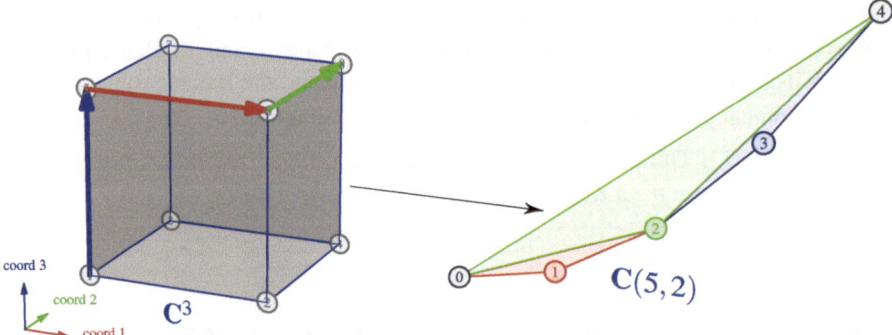

Fig. 21 Each permutation in $B(3,1)$ corresponds to a monotone path in the 3-cube, which induces a triangulation of $\mathbf{C}(5,2)$ by using the order in which the coordinates change as the order in which the vertices $1,2,3$ are cut-off by the triangulation. Note that this can be interpreted as an upflip sequence in $\mathrm{HST}_1(5,1)$. Thus, what we see here is the flip map $\mathscr{T}_{\mathrm{flip}}$ from [31].

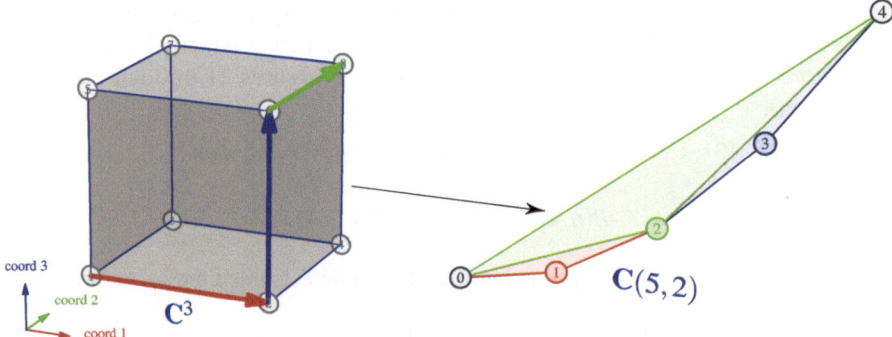

Fig. 22 A different monotone path can lead to an identical triangulation.

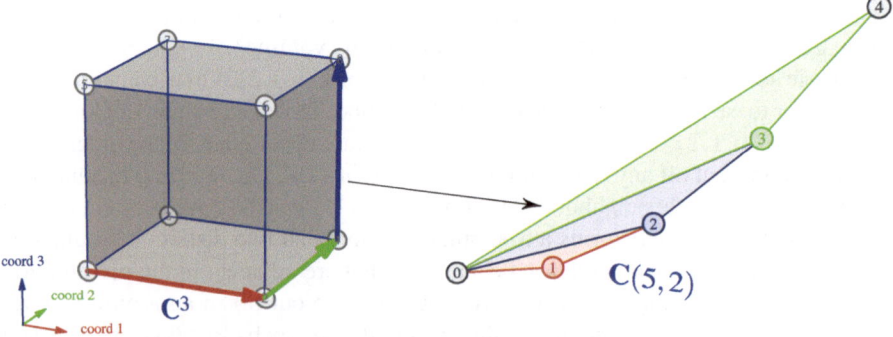

Fig. 23 A different monotone path can also lead to a different triangulation.

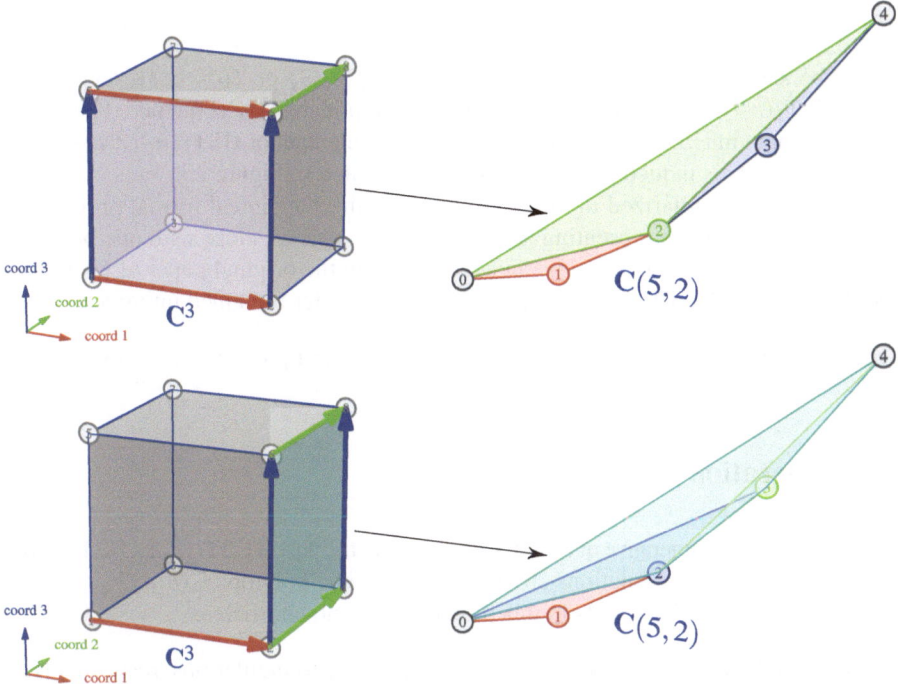

Fig. 24 Monotone paths that differ by a "face flip" (that is, the corresponding permutations are connected by an inversion) lead to triangulations that are either identical or are connected by a bistellar flip.

Fig. 25 Illustration of the inductive structure of the map from higher Bruhat orders to higher Stasheff-Tamari orders: A zonotopal tiling of $\mathbf{Z}(4,2)$ (the one from Figure 19) can be traversed upwards by monotone paths (colored from red to cyan), which map to triangulations of $\mathbf{C}(6,2)$ that form a chain (from red to cyan) inducing a triangulation of $\mathbf{C}(6,3)$ consisting of the flip simplices in the chain – determining an element of $\mathrm{HST}_1(6,3)$.

$$c = c_1 \lessdot c_2 \lessdot \cdots$$

of elements c_i in $B(n, d-1)$. Each $f(c_i)$ in $\mathrm{HST}_1(n+2, d)$ is already defined by induction, and thereby gives a sequence of triangulations of $\mathbf{C}(n+2, d)$

$$f(c_1) \le f(c_2) \le \cdots \tag{8}$$

It can be shown that for each i, either $f(c_i) = f(c_{i+1})$ or $f(c_i) \lessdot f(c_{i+1})$ in the order $\text{HST}_1(n+2,d)$. Hence, after eliminating duplicates, the sequence (8) gives a maximal chain in $\text{HST}_1(n+2,d)$, and therefore an element of $\text{HST}_1(n+2,d+1)$ by Theorem 6.6. This inductive construction is illustrated in Figure 25.

The results summarized in this section all required technical formal proofs, for which we refrain from presenting any details. However, we close with one problem on the above map f, suggested by an assertion from the original paper of Kapranov and Voevodsky [24, Theorem 4.10], but which has so far remained unproven.

Open Problem 8.10. *Prove that the map* $B(n,d) \xrightarrow{f} \text{HST}_1(n+2,d+1)$ *is surjective.*

9 Enumeration

We close with an enumerative question: How large are the posets $\text{HST}_1(n,d), \text{HST}_2(n,d)$, that is, how many triangulations are there of the cyclic polytope $\mathbf{C}(n,d)$?

A few mostly trivial results in this direction are known, such as

- $\mathbf{C}(n,0), \mathbf{C}(n,1), \mathbf{C}(n,2)$ have $n, 2^{n-2}, \frac{1}{n-1}\binom{2(n-2)}{n-2}$ triangulations, respectively,
- $\mathbf{C}(n,n-1), \mathbf{C}(n,n-2), \mathbf{C}(n,n-3)$, have $1, 2, n$ triangulations, respectively.

The following nontrivial result was proven by Azaola and Santos [4].

Theorem 9.1 ([4]). *The number of triangulations of* $\mathbf{C}(n,n-4)$ *is*

$$\begin{cases} (n+4) \cdot 2^{\frac{n-4}{2}} - n & \text{for } n \text{ even, and} \\ \frac{3n+11}{2} \cdot 2^{\frac{n-5}{2}} - n & \text{for } n \text{ odd.} \end{cases}$$

Another interesting unsolved problem is the following.

Open Problem 9.2. *Count the triangulations of* $\mathbf{C}(n,3)$.

How about computer-based enumeration? Table 1 below compiles a few results achieved by the general purpose enumeration program for triangulations TOPCOM [32]. With special purpose codes it should be possible to generate more numbers that can be used to check conjectural enumeration formulas.

References

1. J.F. Adams, "On the cobar construction", *Proceedings of the National Academy of Science* **42** (1956) 409–412.
2. C. Athanasiadis, "Zonotopal subdivisions of cyclic zonotopes", *Geometriae Dedicata* **86** (2001) 37–57.

$c \backslash d{:}$ 0	1	2	3	4	5	6	7	8	9	10	11	12	13
1 1	1	1	1	1	1	1	1	1	1	1	1	1	1
2 2	2	2	2	2	2	2	2	2	2	2	2	2	2
3 3	4	5	6	7	8	9	10	11	12	13	14	15	16
4 4	8	14	25	40	67	102	165	244	387	562	881	1264	1967
5 5	16	42	138	357	1233	3278	12589	35789	159613	499900	2677865	9421400	62226044
6 6	32	132	972	4824	51676	340560	6429428						
7 7	64	429	8477	96426	5049932	132943239							
8 8	128	1430	89405	2800212									
9 9	256	4862	1119280	116447760									
10 10	516	16796	16384508										
11 11	1028	58786	276961252										

Table 1 Some computations done with TOPCOM [32] for some dimensions d and some codimensions $c := n - d$; the computation of the largest numbers in the table for $C(13, 6)$ and $C(14, 3)$ needed around 40 GB of main memory.

3. C. Athanasiadis, J. Rambau, and F. Santos, "The Generalized Baues Problem for cyclic polytopes II", *Publications De l'Institut Mathématique, Belgrade* **66** (1999) 3–15.
4. M. Azaola and F. Santos, "The number of triangulations of the cyclic polytope $C(n, n - 4)$", *Discrete Comput. Geom.* **27** (2002) 29–48, Geometric combinatorics (San Francisco, CA/Davis, CA, 2000).
5. H.J. Baues, "Geometry of loop spaces and the cobar construction", *Memoirs of the American Mathematical Society* **25** (1980) 1–171.
6. L.J. Billera, M.M. Kapranov, and B. Sturmfels, "Cellular strings on polytopes", *Proceedings of the American Mathematical Society* **122** (1994) 549–555.
7. L.J. Billera and B. Sturmfels, "Fiber polytopes", *Annals of Mathematics* **135** (1992) 527–549.
8. A. Björner, "Topological methods", in *Handbook of Combinatorics*, R.L. Graham, M. Grötschel, and L. Lovász, eds., North-Holland, Amsterdam, 1995, 1819–1872.
9. A. Björner, M. Las Vergnas, B. Sturmfels, N. White, and G.M. Ziegler, *Oriented matroids*, second ed., Encyclopedia of Mathematics and its Applications, vol. 46, Cambridge University Press, Cambridge, 1999.
10. M. de Berg, O. Cheong, M. van Kreveld, and M. Overmars, *Computational Geometry: Algorithms and Applications*, 3rd ed., Springer, 2008.
11. J. De Loera, J. Rambau, and F. Santos, *Triangulations – Structures for Applications and Algorithms*, Algorithms and Computation in Mathematics, vol. 25, Springer, 2010.
12. P. Dehornoy, "Tamari lattices and the symmetric Thompson monoid", in *this volume*.
13. T.K. Dey, "On counting triangulations in d dimensions", *Comput. Geom.* **3** (1993) 315–325.
14. P. Edelman, V. Reiner, and J. Rambau, "On subdivision posets of cyclic polytopes", *European Journal of Combinatorics* **21** (2000) 85–101.
15. P.H. Edelman and V. Reiner, "The higher Stasheff-Tamari posets", *Mathematika* **43** (1996) 127–154.
16. H. Edelsbrunner, *Geometry and topology for mesh generation*, Cambridge Monographs on Applied and Computational Mathematics, vol. 7, Cambridge University Press, Cambridge, 2001.
17. H. Edelsbrunner and N.R. Shah, "Incremental topological flipping works for regular triangulations", in *Proceedings of the 8th annual ACM Symposium on Computational Geometry*, ACM press, 1992, 43–52.
18. S. Felsner and H. Weil, "A theorem on higher Bruhat orders", *Discrete & Computational Geometry* **23** (2000) 121–127.
19. I.M. Gelfand, M.M. Kapranov, and A.V. Zelevinsky, "Discriminants of polynomials in several variables and triangulations of Newton polyhedra", *Leningrad Mathematical Journal* **2** (1991) 449–505.

20. _____, *Discriminants, Resultants, and Multidimensional Determinants*, Mathematics: Theory & Applications, Birkhäuser, Boston, 1994.

21. C. Hohlweg, "Permutahedra and associahedra: Generalized associahedra from the geometry of finite reflection groups", in *this volume*.

22. S. Huang and D. Tamari, "Problems of associativity: A simple proof for the lattice property of systems ordered by a semi-associative law", *J. Combinatorial Theory Ser. A* **13** (1972) 7–13.

23. D. Huguet and D. Tamari, "La structure polyédrale des complexes de parenthésages", *J. Combin. Inform. System Sci.* **3** (1978) 69–81.

24. M.M. Kapranov and V.A. Voevodsky, "Combinatorial-geometric aspects of polycategory theory: pasting schemes and higher Bruhat orders (list of results)", *Cahiers de Topologie et Géométrie différentielle catégoriques* **32** (1991) 11–27.

25. Y.I. Manin and V.V. Schechtman, "Arrangements of hyperplanes, higher braid groups and higher Bruhat orders", *Advanced Studies in Pure Mathematics* **17** (1989) 289–308.

26. S. Oppermann and H. Thomas, "Higher dimensional cluster combinatorics and representation theory", *arxiv.org/abs/1001.5437*.

27. U. Pachner, "P.L. homeomorphic manifolds are equivalent by elementary shellings", *European J. Combin.* **12** (1991) 129–145.

28. J.M. Pallo, "An algorithm to compute the Möbius function of the rotation lattice of binary trees", *RAIRO Inform. Théor. Appl.* **27** (1993) 341–348.

29. J. Rambau, *Projections of Polytopes and Polyhedral Subdivisions*, Berichte aus der Mathematik, Shaker, Aachen, 1996, Dissertation, TU Berlin.

30. _____, "A suspension lemma for bounded posets", *J. Combin. Theory Ser. A* **80** (1997) 374–379.

31. _____, "Triangulations of cyclic polytopes and higher Bruhat orders", *Mathematika* **44** (1997) 162–194.

32. _____, "TOPCOM: Triangulations of point configurations and oriented matroids", in *Mathematical Software – ICMS 2002*, A.M. Cohen, X.-S. Gao, and N. Takayama, eds., World Scientific, 2002, 330–340.

33. J. Rambau and F. Santos, "The Generalized Baues Problem for cyclic polytopes I", *European Journal of Combinatorics* **21** (2000) 65–83.

34. J. Rambau and G.M. Ziegler, "Projections of polytopes and the Generalized Baues Conjecture", *Discrete & Computational Geometry* **16** (1996) 215–237.

35. N. Reading, "From the Tamari lattice to Cambrian lattices and beyond", in *this volume*.

36. V. Reiner, "The generalized Baues problem", in *New Perspectives in Algebraic Combinatorics (Berkeley, CA, 1996–97)*, Math. Sci. Res. Inst. Publ., vol. 38, Cambridge Univ. Press, Cambridge, 1999, 293–336.

37. J. Richter-Gebert and G.M. Ziegler, "Zonotopal tilings and the Bohne-Dress theorem", in *Proceedings "Jerusalem Combinatorics '93"*, H. Barcelo and G. Kalai, eds., Contemporary Mathematics, vol. 178, American Mathematical Society, 1994, 211–232.

38. F. Santos, "A point configuration whose space of triangulations is disconnected", *Journal of the American Mathematical Society* **13** (2000) 611–637.

39. _____, "Non-connected toric Hilbert schemes", *Mathematische Annalen* **332** (2005) 645–665.

40. D.D. Sleator, R.E. Tarjan, and W.P. Thurston, "Rotation distance, triangulations, and hyperbolic geometry", *Journal of the American Mathematical Society* **1** (1988) 647–681.

41. J.D. Stasheff, "How I 'met' Dov Tamari", in *this volume*.

42. J.R. Stembridge, "A Maple package for posets", Free software, available online, 2008.

43. H. Thomas, "New combinatorial descriptions of the triangulations of cyclic polytopes and the second higher Stasheff-Tamari posets", *Order* **19** (2002) 327–342.

44. _____, "Maps between higher Bruhat orders and higher Stasheff-Tamari posets", in *Formal Power Series and Algebraic Combinatorics Conference – Linköping, Sweden*, 2003.

45. _____, "The Tamari lattice as it arises in quiver representations", in *this volume*.

46. G.M. Ziegler, "Higher Bruhat orders and cyclic hyperplane arrangements", *Topology* **32** (1993) 259–279.

47. G.M. Ziegler, *Lectures on polytopes*, Graduate Texts in Mathematics, vol. 152, Springer-Verlag, New York, 1995.

KP Solitons, Higher Bruhat and Tamari Orders

Aristophanes Dimakis and Folkert Müller-Hoissen

Abstract In a tropical approximation, any tree-shaped line soliton solution, a member of the simplest class of soliton solutions of the Kadomtsev-Petviashvili (KP-II) equation, determines a chain of planar rooted binary trees, connected by right rotation. More precisely, it determines a maximal chain of a Tamari lattice. We show that an analysis of these solutions naturally involves higher Bruhat and higher Tamari orders.

1 Introduction

Waves on a fluid surface show a very complex behavior in general. Only under special circumstances can we expect to observe a more regular pattern. For shallow water waves, the Kadomtsev-Petviashvili (KP) equation

$$(-4u_t + u_{xxx} + 6uu_x)_x + 3u_{yy} = 0$$

(where, e.g., $u_t = \partial u / \partial t$) provides an approximation under the conditions that the wave dominantly travels in the x-direction, the wave length is long as compared with the water depth, and the effect of the nonlinearity is about the same order as that of dispersion.[1] More precisely, this is the KP-II equation, but we will write KP, for short. It generalizes the famous Korteweg-deVries (KdV) equation, which describes waves moving in only one spatial dimension. Although the KdV equation is much

Aristophanes Dimakis
Department of Financial and Management Engineering, University of the Aegean, 41, Kountourioti Str., GR-82100 Chios, Greece, e-mail: *dimakis@aegean.gr*

Folkert Müller-Hoissen
Max-Planck-Institute for Dynamics and Self-Organization, Bunsenstrasse 10, D-37073 Göttingen, Germany, e-mail: *folkert.mueller-hoissen@ds.mpg.de*

[1] The physical form of this equation is obtained by suitable rescalings of x, y, t and u, involving physical parameters.

better established as an approximation of the more general water wave equations, recent studies also confirm the physical relevance of KP [9].

In [2] we studied the soliton solutions of the KP equation in a *tropical* approximation, which reduces them to networks formed by line segments in the *xy*-plane, evolving in time *t*. A subclass corresponds to evolutions (in time *t*) in the set of (planar) rooted binary trees. At transition events, the binary tree type changes (through a tree that is not binary). It turned out that the time evolution is simply given by right rotation in a tree, and the solution evolves according to a maximal chain of a *Tamari lattice* [24, 25], see Figure 1.

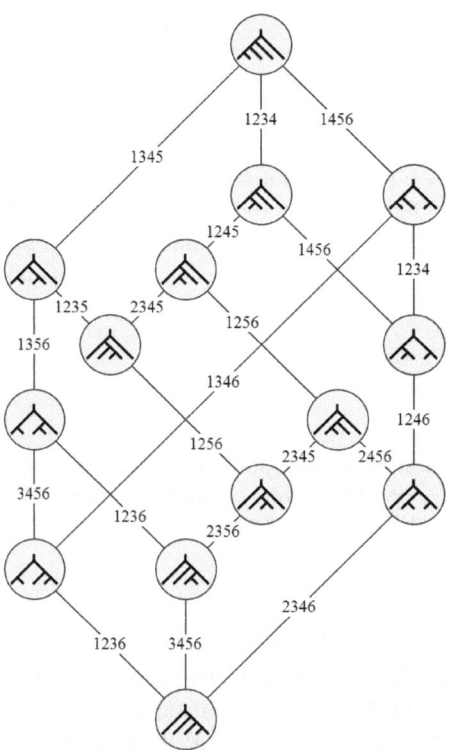

Fig. 1 The Tamari lattice $\mathbb{T}_4$ in terms of rooted binary trees. The top node shows a left comb tree that represents the structure of a certain KP line soliton family (with six asymptotic branches in the *xy*-plane) as $t \to -\infty$. A label $ijkl$ assigned to an edge indicates the transition time t_{ijkl} at which the soliton graph changes its tree type via a 'rotation' (see Section 3). The values of the parameters, on which the solutions depend, determine the linear order of the 'critical times' t_{ijkl}, and thus decide which chain is realized. The bottom node shows a right comb tree, which represents the tree type of the soliton as $t \to \infty$. The special family of solutions thus splits into classes corresponding to the maximal chains of $\mathbb{T}_4$. For each Tamari lattice, there is a family of KP line solitons that realizes its maximal chains in this way.

In this realization of Tamari lattices, the underlying set consists of states of a physical system, here the tree-types of a soliton configuration in the xy-plane. The Tamari poset (partially ordered set) structure describes the possible ways in which these states are allowed to evolve in time, starting from an initial state (the top node) and ending in a final state (the bottom node).

Figure 2 displays a solution evolving via a tree rotation, and further provides an idea how this can be understood in terms of an arrangement of planes in three-dimensional space-time (after idealizing line soliton branches to lines in the xy-plane, see Section 3).

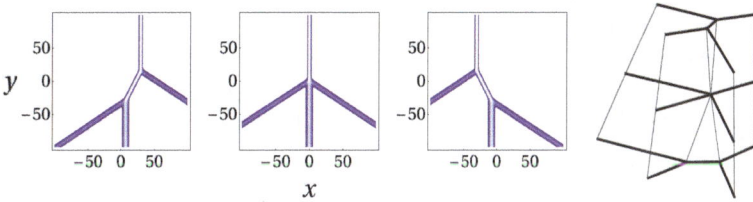

Fig. 2 Density plots of a line soliton solution at three successive times, exhibiting a tree rotation. To the right is a corresponding space-time view in terms of intersecting planes (here time flows upward).

In this work we show that the classification of possible evolutions of tree-shaped KP line solitons involves *higher Bruhat orders* [17, 18, 19, 30]. Moreover, we are led to associate with each higher Bruhat order a *higher Tamari order* via a surjection, in a way different from what has been considered previously. There is some evidence that our higher Tamari orders coincide with 'higher Stasheff-Tamari posets' introduced by Kapranov and Voevodsky [8] (also see [3]), but a closer comparison will not be undertaken in this work.

In Section 2, we briefly describe the general class of KP soliton solutions. In Section 3, we concentrate on the above-mentioned subclass of tree-shaped solutions, in the tropical approximation, and somewhat improve results in [2]. Section 4 recalls results about higher Bruhat orders and extracts from the analysis of tree-shaped KP line solitons a reduction to higher Tamari orders. Section 7 proposes a hierarchy of monoids that expresses the hierarchical structure present in the KP soliton problem. This makes contact with *simplex equations* [29, 1, 6, 13] and provides us with an algebraic method to construct higher Bruhat and higher Tamari orders. Section 8 contains some additional remarks. Throughout this work, Hasse diagrams of posets will be displayed upside down (i.e., with the lowest element(s) at the top).

2 KP solitons

The M-soliton solutions of the KP-II equation are parametrized by the totally non-negative Grassmannians $\mathrm{Gr}_{n,M+1}^{\geq}$ [9], which is easily recognized in the Wronskian form of the solutions. Translating the KP equation via

$$u = 2\log(\tau)_{xx},$$

into a bilinear equation in the variable τ, these solutions are given by

$$\tau = f_1 \wedge f_2 \wedge \cdots \wedge f_n,$$

where

$$f_i = \sum_{j=1}^{M+1} a_{ij}\, e_j, \qquad e_j = e^{\theta_j}, \qquad \theta_j = \sum_{r=1}^{M} p_j^r\, t^{(r)} + c_j.$$

Here $t^{(r)}$, $r = 1,\ldots,M$, are independent real variables that may be regarded as coordinates on $\mathbb{R}^M$, and we set $t^{(1)} = x$, $t^{(2)} = y$, $t^{(3)} = t$. The variables $t^{(r)}$, $r > 3$, are the additional evolution variables that appear in the *KP hierarchy*, which extends the KP equation to an infinite set of compatible PDEs. Furthermore, p_j, c_j, a_{ij} are real constants, and without restriction of generality we can and will assume that

$$p_1 < p_2 < \cdots < p_{M+1}.$$

The exterior product on the space of functions generated by the exponential functions e_j, $j = 1,\ldots,M+1$, is defined by

$$e_{i_1} \wedge \cdots \wedge e_{i_m} = \Delta(p_{i_1},\ldots,p_{i_m})\, e_{i_1}\cdots e_{i_m},$$

with the Vandermonde determinant

$$\Delta(p_{i_1},\ldots,p_{i_m}) = \begin{vmatrix} 1 & p_{i_1} & \cdots & p_{i_1}^{m-1} \\ 1 & p_{i_2} & \cdots & p_{i_2}^{m-1} \\ \vdots & \vdots & \cdots & \vdots \\ 1 & p_{i_m} & \cdots & p_{i_m}^{m-1} \end{vmatrix} = \prod_{1 \leq r < s \leq m} (p_{i_s} - p_{i_r}).$$

Now we can express τ as

$$\tau = \sum_{I \in \binom{[M+1]}{n}} A_I\, \Delta(p_I)\, e_I,$$

where $[m] = \{1, 2, \ldots, m\}$, and $\binom{[m]}{n}$ denotes the set of n-element subsets of $[m]$. Numbering the elements of a subset $I = \{i_1, \ldots, i_n\}$ such that $i_1 < \cdots < i_n$, we set $\Delta(p_I) = \Delta(p_{i_1}, \ldots, p_{i_n})$. Finally, A_I denotes the maximal minor with columns $i_1, \ldots, i_n$ of the matrix

$$A = \begin{pmatrix} a_{1,1} & \cdots & a_{1,M+1} \\ \vdots & \ddots & \vdots \\ a_{n,1} & \cdots & a_{n,M+1} \end{pmatrix}.$$

For regular (soliton) solutions, the Plücker coordinates A_I have to be *non-negative* real numbers (and at least one has to be different from zero). In the following we concentrate on the subclass of solutions parametrized by $\mathrm{Gr}^{\geq}_{1,M+1}$, i.e.,

$$\tau = e_1 + \cdots + e_{M+1}, \tag{1}$$

where we absorbed the positive constants $a_{1,j}$ into the constants c_j. To good approximation, a *general* line soliton solution can be understood as a *superimposition* of solutions from the subclass (see [2]).

3 Tropical approximation of a subclass of KP line solitons

Let us fix $M \in \mathbb{N}$, constants $p_1 < p_2 < \cdots < p_{M+1}$ and c_i, $i = 1, \ldots, M+1$. The behavior of $\tau : \mathbb{R}^M \to \mathbb{R}$, given by (1), is best understood in a tropical approximation of $\log(\tau)$. In a region where some phase, say θ_i, *dominates* all others, i.e., $\theta_i > \theta_j$ for all $j \neq i$, we have

$$\log(\tau) = \theta_i + \log\left(1 + \sum_{\substack{j=1 \\ j \neq i}}^{M+1} e^{-(\theta_i - \theta_j)}\right) \simeq \theta_i \,.$$

As a consequence,

$$\log(\tau) \simeq \max\{\theta_1, \ldots, \theta_{M+1}\},$$

where the right-hand side can be regarded as a *tropical* version of $\log(\tau)$. Sufficiently away from the boundary of a dominating-phase region, $\log(\tau)$ is linear in x, so that u vanishes. A crucial observation is that a line soliton branch in the xy-plane, for fixed $t^{(r)}$, $r > 2$, corresponds to a boundary line between two dominating-phase regions. Viewing it in space-time, by regarding t as a coordinate of an additional dimension, or more generally in the extended space $\mathbb{R}^M$ by adding dimensions corresponding to the evolution variables $t^{(r)}$, $r = 3, \ldots, M$, the boundary consists piecewise of affine hyperplanes. Let $\mathcal{U}_i$ denote the region where θ_i is not dominated by any other phase, i.e.,

$$\mathcal{U}_i = \{\mathbf{t} \in \mathbb{R}^M \mid \max\{\theta_1, \ldots, \theta_{M+1}\} = \theta_i\} = \bigcap_{k \neq i}\{\mathbf{t} \in \mathbb{R}^M \mid \theta_k \leq \theta_i\}. \tag{2}$$

In the tropical approximation, a description of KP line solitons amounts to an analysis of intersections of such regions, i.e.,

$$\mathcal{U}_I = \mathcal{U}_{i_1} \cap \cdots \cap \mathcal{U}_{i_n}, \quad I = \{i_1, \ldots, i_n\} \in \binom{\Omega}{n}, \quad \Omega := [M+1] = \{1, \ldots, M+1\}.$$

This is a subset of the affine space

$$\mathscr{P}_I = \{\mathbf{t} \in \mathbb{R}^M \,|\, \theta_{i_1} = \cdots = \theta_{i_n}\} \qquad n > 1,$$

which is easy to deal with (see below). It is more difficult to determine which parts of $\mathscr{P}_I$ are *visible*, i.e., belong to $\mathscr{U}_I$. Fixing the values of $t^{(r)}$, $r > 2$, determines a line soliton segment in the xy-plane if $n = 2$, and a meeting point of n such segments if $n > 2$.[2] In order to decide about visibility, i.e., whether a point of $\mathscr{P}_I$ lies in $\mathscr{U}_I$, a formula is needed to compare the values of all phases at this point, see (6) below.

Let us first look at $\mathscr{P}_I$ in more detail. Introducing a real *auxiliary variable* $t^{(0)}$, the equation $\theta_{i_1} = \cdots = \theta_{i_n} = -t^{(0)}$ results in the linear system[3]

$$t^{(0)} + p_{i_j} t^{(1)} + p_{i_j}^2 t^{(2)} + \cdots + p_{i_j}^{n-1} t^{(n-1)} = -\tilde{c}_{i_j}, \qquad j = 1, \ldots, n,$$

where

$$\tilde{c}_{i_j} = c_{i_j} + p_{i_j}^n t^{(n)} + \cdots + p_{i_j}^M t^{(M)}.$$

This fixes the first $n - 1$ coordinates as linear functions of the remaining coordinates, $t_I^{(k)} = t_I^{(k)}(t^{(n)}, \ldots, t^{(M)})$, $k = 1, \ldots, n - 1$. In particular, we obtain (also see [2], Appendix A)

$$
\begin{aligned}
t_I^{(n-1)} &= -\sum_{r=1}^{M+1-n} h_r(p_I) t^{(n+r-1)} - c_I \\
&= -(p_{i_1} + \cdots + p_{i_n}) t^{(n)} - \tilde{c}_I,
\end{aligned}
\tag{3}
$$

where $h_r(p_I) = h_r(p_{i_1}, \ldots, p_{i_n})$ is the r-th *complete symmetric polynomial* [15] in the variables $p_{i_1}, \ldots, p_{i_n}$, and

$$c_I = \frac{1}{\Delta(p_I)} \sum_{s=1}^n (-1)^{n-s} c_{i_s} \Delta(p_{I \setminus \{i_s\}}), \qquad \tilde{c}_I = \sum_{r=2}^{M+1-n} h_r(p_I) t^{(n+r-1)} + c_I.$$

We note that $\tilde{c}_I$ depends on $t^{(n+1)}, \ldots, t^{(M)}$. Here are some immediate consequences:

- Since obviously $\mathscr{P}_I \subset \mathscr{P}_J$ for $J \subset I$, on $\mathscr{P}_I$ we have

$$t_I^{(n-2)} = t_{I \setminus \{i_n\}}^{(n-2)}(t_I^{(n-1)}, t^{(n)}, \ldots, t^{(M)}),$$

and corresponding expressions for $t_I^{(r)}$, $r = 1, \ldots, n - 3$.

---

[2] For generic values of $t^{(r)}$, $r > 2$, we see line soliton segments and meeting points of *three* segments in the xy-plane. Meeting points of more than three segments only occur for special values.

[3] We note that the full set of equations $t^{(0)} + p_i t^{(1)} + p_i^2 t^{(2)} + \cdots + p_i^M t^{(M)} = -c_i$, $i = 1, \ldots, M+1$, defines a *cyclic hyperplane arrangement* [30] in $\mathbb{R}^{M+1}$ with coordinates $t^{(0)}, \ldots, t^{(M)}$.

- $\mathscr{P}_\Omega$ is a common point of all $\mathscr{P}_I$, $I \subset \Omega$. According to (3), on $\mathscr{P}_\Omega$ we have

$$t_\Omega^{(M)} = -c_\Omega .$$

Clearly, $\mathscr{P}_\Omega = \mathscr{U}_\Omega$, and is thus visible.
- The hyperplane $\mathscr{P}_{\{i_1,i_2\}}$ is given by

$$t_{\{i_1,i_2\}}^{(1)} = -(p_{i_1} + p_{i_2})t^{(2)} - \tilde{c}_{\{i_1,i_2\}} ,$$

and we have

$$\theta_{i_2} - \theta_{i_1} = (p_{i_2} - p_{i_1})(t^{(1)} - t_{\{i_1,i_2\}}^{(1)}), \qquad (4)$$

so that, for $i_1 < i_2$,

$$t^{(1)} \lessgtr t_{\{i_1,i_2\}}^{(1)} \quad \Longleftrightarrow \quad \theta_{i_2} \lessgtr \theta_{i_1} .$$

Together with (2), this implies in particular that each $\mathscr{U}_i$ is the intersection of half-spaces, and thus a closed convex set. None of these sets is empty since they all contain $\mathscr{U}_\Omega$. It follows in turn that each set $\mathscr{U}_I$ (with non-empty I) is non-empty, closed and convex (and thus in particular connected).

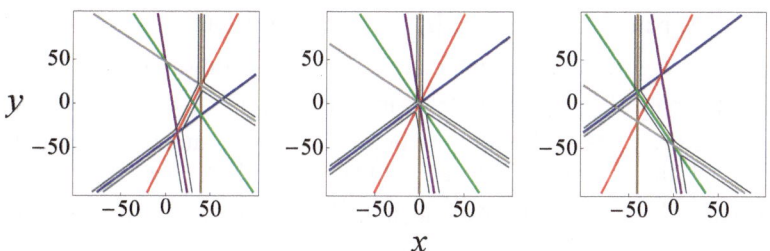

Fig. 3 A soliton solution with $M = 3$, hence $\Omega = \{1,2,3,4\}$, at times $t < t_\Omega$, $t = t_\Omega$ and $t > t_\Omega$. A thin line is the coincidence of two phases. It corresponds to some $\mathscr{P}_{\{i,j\}}$ restricted to the respective value of t. Only a thick part of such a line is visible at the respective value of time. It corresponds to some $\mathscr{U}_{\{i,j\}}$, restricted to that value of t. The left and also the right plot shows two visible coincidences of three phases, corresponding to points in some $\mathscr{U}_{\{i,j,k\}}$. They coincide in the middle plot to form a visible four phase coincidence, the point $\mathscr{U}_\Omega$.

Lemma 3.1 ([2, Proposition A.3]). *For $K = \{k_1, \ldots, k_{n+1}\}$, $n \in [M]$, we have*[4]

$$t_{K\setminus\{k_j\}}^{(n-1)} - t_{K\setminus\{k_l\}}^{(n-1)} = (p_{k_j} - p_{k_l})(t^{(n)} - t_K^{(n)}) \qquad j,l \in \{1,\ldots,n+1\} .$$

---

[4] For $n = 1$, this is (4), since $t_i^{(0)} = -\theta_i$.

With the help of this important lemma, we obtain the following result.

Proposition 3.2. *If $K = \{k_1, \ldots, k_{n+1}\}$ is in linear order, i.e., $k_1 < k_2 < \cdots < k_{n+1}$, then*

$$
\begin{array}{ll}
t^{(n-1)}_{K\setminus\{k_{n+1}\}} < t^{(n-1)}_{K\setminus\{k_n\}} < \cdots < t^{(n-1)}_{K\setminus\{k_1\}} & t^{(n)} < t^{(n)}_K \\[2mm]
t^{(n-1)}_{K\setminus\{k_1\}} < t^{(n-1)}_{K\setminus\{k_2\}} < \cdots < t^{(n-1)}_{K\setminus\{k_{n+1}\}} & \quad for \quad \quad t^{(n)} > t^{(n)}_K .
\end{array}
\tag{5}
$$

The first chain in (5) is in *lexicographic order*, the second in *reverse lexicographic order*, with respect to the index sets. This makes contact with *higher Bruhat orders*, see Section 4. The following result is crucial for determining (non-)visible events.

Proposition 3.3 ([2, Corollary A.6]). *Let $I = \{i_1, \ldots, i_n\}$ and $k \in \Omega \setminus I$. On $\mathscr{P}_I$ we have*

$$
\theta_k - \theta_{i_1} = (p_k - p_{i_1}) \cdots (p_k - p_{i_n})(t^{(n)} - t^{(n)}_{I\cup\{k\}}) .
\tag{6}
$$

Example 3.4. Let $n = 2$ and thus $I = \{i, i'\}$ with $i < i'$. On $\mathscr{P}_I$, (6) reads

$$
\theta_k - \theta_i = (p_k - p_i)(p_k - p_{i'})(t^{(2)} - t^{(2)}_{\{i,i',k\}}) .
$$

If I is an interval, i.e., $i' = i+1$, we have either $k < i$ or $k > i+1$, and thus $\theta_k \lessgtr \theta_i$ iff $t^{(2)} \lessgtr t^{(2)}_{\{i,i+1,k\}}$. As a consequence, the part of $\mathscr{P}_{\{i,i+1\}}$ with $t^{(2)} \leq \min_k\{t^{(2)}_{\{i,i+1,k\}}\}$ is visible and the part with $t^{(2)} > \min_k\{t^{(2)}_{\{i,i+1,k\}}\}$ is non-visible. If I is not an interval, then there is a $k \in \{2, \ldots, M\}$ such that $i < k < i'$. It follows that $\theta_k > \theta_i$ if $t^{(2)} < t^{(2)}_{\{i,i',k\}}$, so the part of $\mathscr{P}_{\{i,i'\}}$ with $t^{(2)} < \max_k\{t^{(2)}_{\{i,i',k\}} \mid i < k < i'\}$ is non-visible. If there is a $k \in \{1, \ldots, M+1\}$ with $k < i$ or $k > i'$, then the situation is as in the case of an interval. In the remaining case $I = \{1\} \cup \{M+1\}$, the part of $\mathscr{P}_{\{1,M+1\}}$ with $t^{(2)} \geq \max_k\{t^{(2)}_{\{i,i',k\}}\}$ is visible. We conclude that $\mathscr{P}_I$, with I of the form $\{i,i+1\}$, $i \in \{1, \ldots, M\}$, has a visible part extending to arbitrary negative values of $y = t^{(2)}$, and only $\mathscr{P}_{\{1,M+1\}}$ has a visible part extending to arbitrary positive values of y. Any visible part of another $\mathscr{P}_{\{i,i'\}}$ has to be bounded in the xy-plane. Furthermore, (5) shows that

$$
t^{(1)}_{\{1,2\}} < t^{(1)}_{\{2,3\}} < \cdots < t^{(1)}_{\{M,M+1\}} \qquad for \qquad t^{(2)} < \min_{i,k}\{t^{(2)}_{\{i,i+1,k\}}\} .
$$

All this information determines the asymptotic line soliton structure in the xy-plane depicted in Figure 4.

For $k \leq l$, let $[k, l]$ denote the *interval* $\{k, k+1, \ldots, l\}$. We call an interval *even* (respectively *odd*) if its cardinality is even (respectively odd). Now we formulate a generalization of results in Example 3.4. The proof is by inspection of (6), which depends on the structure of I. We note that an even interval cannot influence the sign of the right-hand side of (6).

Fig. 4 Asymptotic structure in the xy-plane ($x = t^{(1)}$ horizontal, $y = t^{(2)}$ vertical coordinate) for the line soliton solutions given by (1). Outside a large enough disk, the xy-plane is divided into regions as shown in the figure, where one of the phases θ_i dominates all others. This structure is independent of the values of $t^{(r)}$, $r > 2$.

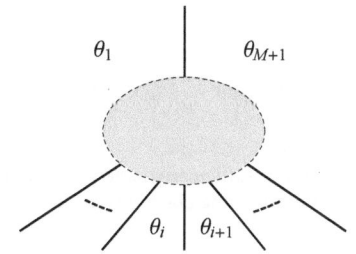

Proposition 3.5. *Let I be an n-subset of $\Omega = [M+1]$, $1 < n < M+1$.*

(1) *Let I be the disjoint union of even intervals, and also $\{1\}$ if n is odd. Then*

$$\{\mathbf{t} \in \mathscr{P}_I \,|\, t^{(n)} \leq \min\{t_K^{(n)} \,|\, K \in \binom{\Omega}{n+1}, I \subset K\}\}$$

is visible, but its complement

$$\{\mathbf{t} \in \mathscr{P}_I \,|\, t^{(n)} > \min\{t_K^{(n)} \,|\, K \in \binom{\Omega}{n+1}, I \subset K\}\}$$

not.

(2) *Let I be the disjoint union of $\{M+1\}$ and any number of even intervals, and also $\{1\}$ if n is even. Then*

$$\{\mathbf{t} \in \mathscr{P}_I \,|\, t^{(n)} \geq \max\{t_K^{(n)} \,|\, K \in \binom{\Omega}{n+1}, I \subset K\}\}$$

is visible, but its complement

$$\{\mathbf{t} \in \mathscr{P}_I \,|\, t^{(n)} < \max\{t_K^{(n)} \,|\, K \in \binom{\Omega}{n+1}, I \subset K\}\}$$

not.

(3) *If I is not of the form specified in (1) or (2), a visible part of $\mathscr{P}_I$ can only appear for $t^{(n)}$ between $\min\{t_K^{(n)} \,|\, K \in \binom{\Omega}{n+1}, I \subset K\}$ and $\max\{t_K^{(n)} \,|\, K \in \binom{\Omega}{n+1}, I \subset K\}$.*

Example 3.6. Let $n = 3$. According to Proposition 3.5, for $t^{(3)} \leq \min\{t_K^{(3)} \,|\, K \in \binom{\Omega}{4}\}$ only (the corresponding parts of) $\mathscr{P}_{\{1,i,i+1\}}$, $i = 2, \ldots, M$, are visible, and for $t^{(3)} \geq \max\{t_K^{(3)} \,|\, K \in \binom{\Omega}{4}\}$ only (the corresponding parts of) $\mathscr{P}_{\{i,i+1,M+1\}}$, $i = 1, \ldots, M-1$, are visible. The order of the values $t^{(2)}_{\{1,i,i+1\}}$, respectively $t^{(2)}_{\{i,i+1,M+1\}}$, follows from (5). All this leads to the structure of a line soliton solution for large negative time (left comb), respectively large positive time (right comb), shown in Figure 5.

Example 3.7. Let $M = 5$ and $n = 4$. For $t^{(4)} \leq \min\{t_K^{(4)} \,|\, K \in \binom{[6]}{5}\}$, the corresponding part of $\mathscr{P}_I$ is only visible if I is one of the sets $\{1,2,3,4\}$, $\{1,2,4,5\}$, $\{1,2,5,6\}$,

Fig. 5 For sufficiently large negative (respectively positive) values of time $t = t^{(3)}$, any line soliton solution from the class (1) has the tree shape in the xy-plane shown by the left (right) graph.

$\{2,3,4,5\}, \{2,3,5,6\}, \{3,4,5,6\}$. For $t^{(4)} \geq \max\{t_K^{(4)} \mid K \in \binom{[6]}{5}\}$, the corresponding part of $\mathscr{P}_I$ is only visible if I is one of $\{1,2,3,6\}, \{1,3,4,6\}, \{1,4,5,6\}$. Also see Figure 16.

In the following, K denotes the set $\{k_1, \ldots, k_{n+1}\}$, and will be assumed to be in linear order, so that $k_1 < \cdots < k_{n+1}$. We split such a set into[5]

$$K_< = \{k_{n-2r} \mid r = 0, \ldots, \lceil n/2 \rceil - 1\} = \{k_{2-(n \bmod 2)}, \ldots, k_n\},$$
$$K_> = \{k_{n-2r+1} \mid r = 0, \ldots, \lfloor n/2 \rfloor\} = \{k_{1+(n \bmod 2)}, \ldots, k_{n+1}\}.$$

Choosing a point $\mathbf{t}_0 \in \mathscr{P}_K$ means fixing the free coordinates on $\mathscr{P}_K$ to values $t_0^{(n+1)}, \ldots, t_0^{(M)}$. For each $k \in K$, it determines a line

$$\{\mathbf{t}_{K \setminus \{k\}}(\lambda, t_0^{(n+1)}, \ldots, t_0^{(M)}) \mid \lambda \in \mathbb{R}\} \subset \mathscr{P}_{K \setminus \{k\}}.$$

Proposition 3.8 ([2, Proposition A.7]). *The following half-lines are non-visible:*

$$\{\mathbf{t}_{K \setminus \{k\}}(\lambda, t_0^{(n+1)}, \ldots, t_0^{(M)}) \mid \lambda < t_K^{(n)}(t_0^{(n+1)}, \ldots, t_0^{(M)})\} \text{ for } k \in K_<,$$
$$\{\mathbf{t}_{K \setminus \{k\}}(\lambda, t_0^{(n+1)}, \ldots, t_0^{(M)}) \mid \lambda > t_K^{(n)}(t_0^{(n+1)}, \ldots, t_0^{(M)})\} \text{ for } k \in K_>.$$

Proposition 3.9. *If* $\mathbf{t}_0 = \mathbf{t}_I(t_0^{(n)}, \ldots, t_0^{(M)})$ *is non-visible, then there is a* $k \in \Omega \setminus I$ *such that* $\mathbf{t}_0$ *lies on the non-visible side of the point* $\mathbf{t}_{I \cup \{k\}}(t_0^{(n+1)}, \ldots, t_0^{(M)})$ *on the line* $\{\mathbf{t}_I(\lambda, t_0^{(n+1)}, \ldots, t_0^{(M)}) \mid \lambda \in \mathbb{R}\}.$

Proof. Since $\mathbf{t}_0$ is non-visible, it lies in some $\mathscr{U}_k$, $k \notin I$. Let $K = I \cup \{k\}$. Then $\mathbf{t}_1 := \mathbf{t}_K(t_0^{(n+1)}, \ldots, t_0^{(M)})$ lies on the above line. Let us consider the case $t_0^{(n)} < t_K^{(n)}(t_0^{(n+1)}, \ldots, t_0^{(M)})$. If $k \in K_>$, then (6) shows that $\theta_k < \theta_{i_1}$, which contradicts $\mathbf{t}_0 \in \mathscr{U}_k$. Hence $k \in K_<$ and $\mathbf{t}_0$ lies on the non-visible side of $\mathbf{t}_1$ according to Proposition 3.8. A similar argument applies in the case $t_0^{(n)} > t_K^{(n)}(t_0^{(n+1)}, \ldots, t_0^{(M)})$. $\qquad\square$

According to Proposition 3.9, Proposition 3.8 provides us with a method to determine *all* non-visible points, and thus also all visible points.

A point $\mathbf{t}_0 \in \mathscr{P}_I$ is called *generic* if $\mathbf{t}_0 \notin \mathscr{P}_J$ for every $J \subset \Omega$, $J \not\subset I$.

---

[5] $\lceil n/2 \rceil$ denotes the smallest integer greater than or equal to $n/2$, and $\lfloor n/2 \rfloor$ the largest integer smaller than or equal to $n/2$.

Proposition 3.10. *Let* $\mathbf{t}_0 \in \mathscr{P}_I$ *be generic and visible, and* U *any convex neigh-borhood of* $\mathbf{t}_0$ *in* $\mathscr{P}_I$ *that does not intersect any* $\mathscr{P}_K$ *with* $I \subset K$, $I \neq K$. *Then* U *is visible.*

Proof. Let U be a neighborhood as specified in the assumptions. Suppose a non-visible point $\mathbf{t}_1 \in U$ exists. Then there is some $k \in \Omega \setminus I$ such that, at $\mathbf{t}_1$, θ_k dominates all phases associated with elements of I. Let $\mathbf{t}'$ be the point where the line segment connecting $\mathbf{t}_0$ and $\mathbf{t}_1$ intersects the boundary of $\mathscr{U}_k$. Then $\mathbf{t}' \in U \cap \mathscr{U}_k \subset \mathscr{P}_{I \cup \{k\}}$ contradicts one of our assumptions. □

Proposition 3.11. *If* $\mathbf{t}_0 \in \mathscr{P}_K$ *is generic and visible, then points sufficiently close to* $\mathbf{t}_0$ *on the complementary half-line of any of the half-lines in Proposition 3.8 are also visible.*

Proof. We have $\mathbf{t}_0 \in \mathscr{U}_{k_r}$, $r = 1, \ldots, n+1$. The lines

$$\mathscr{L}_r = \{\mathbf{t}_{K \setminus \{k_r\}}(\lambda, t_0^{(n+1)}, \ldots, t_0^{(M)}) \,|\, \lambda \in \mathbb{R}\} \qquad r - 1, \ldots, n+1$$

all lie in the n-dimensional space $\mathscr{E}$ defined by $t^{(s)} = t_0^{(s)}$, $s = n+1, \ldots, M$. Since $\mathbf{t}_0$ is assumed to be visible and generic, there is a neighborhood of $\mathbf{t}_0$ covered by the sets $\mathscr{U}_{k_r} \cap \mathscr{E}$. Since each line contains the visible point $\mathbf{t}_0$, its visible part extends on the complementary side of that in Proposition 3.8, either indefinitely or until it meets some $\mathscr{U}_m$ with $m \notin K$. □

Proposition 3.12. *Let* $I \in \binom{\Omega}{n}$, $n \in \{1, \ldots, M\}$, *and* $t_0^{(n+1)}, \ldots, t_0^{(M)} \in \mathbb{R}$. *If all points* $\mathbf{t}_{I \cup \{k\}}(t_0^{(n+1)}, \ldots, t_0^{(M)})$, $k \in \Omega \setminus I$, *are non-visible, then the whole line*

$$\{\mathbf{t}_I(\lambda, t_0^{(n+1)}, \ldots, t_0^{(M)}) \,|\, \lambda \in \mathbb{R}\}$$

is non-visible.

Proof. Suppose there is a visible point $\mathbf{t}_0 = \mathbf{t}_I(\lambda_0, t_0^{(n+1)}, \ldots, t_0^{(M)})$ on the above line, which we denote as $\mathscr{L}$. Let $\mathbf{t}_1 := \mathbf{t}_{I \cup \{m\}}(t_0^{(n+1)}, \ldots, t_0^{(M)})$ be the nearest of the non-visible points specified in the assumption. Then there is some $m' \in \Omega \setminus I$, $m' \neq m$, such that $\mathbf{t}_1 \in \mathscr{U}_{m'} \cap \mathscr{L}$. The line segment between $\mathbf{t}_0$ and $\mathbf{t}_1$ meets the boundary of the convex set $\mathscr{U}_{m'}$ at the point $\mathbf{t}_{I \cup \{m'\}}(t_0^{(n+1)}, \ldots, t_0^{(M)})$. Since the latter would then be visible, we have a contradiction. □

Proposition 3.13. *For each* $I \in \binom{\Omega}{n}$, $n \in \{1, \ldots, M\}$, *there are* $t_0^{(n+1)}, \ldots, t_0^{(M)} \in \mathbb{R}$ *such that the line* $\{\mathbf{t}_I(\lambda, t_0^{(n+1)}, \ldots, t_0^{(M)}) \,|\, \lambda \in \mathbb{R}\}$ *has a visible part.*

Proof. Since $\mathscr{P}_\Omega$ is a visible point, we can use Proposition 3.11 iteratively. □

Now we arrived at the following situation. From Proposition 3.2, we recall that

$$
\begin{array}{cc}
\begin{aligned}
t^{(n-1)}_{K\backslash\{k_{n+1}\}} < t^{(n-1)}_{K\backslash\{k_{n-1}\}} < \cdots < t^{(n-1)}_{K\backslash\{k_{1+(n\bmod 2)}\}} \\
t^{(n-1)}_{K\backslash\{k_{2-(n\bmod 2)}\}} < \cdots < t^{(n-1)}_{K\backslash\{k_{n-2}\}} < t^{(n-1)}_{K\backslash\{k_n\}}
\end{aligned}
& \text{for}
& \begin{aligned}
t^{(n)} < t^{(n)}_K \\
t^{(n)} > t^{(n)}_K .
\end{aligned}
\end{array}
\tag{7}
$$

For all $t^{(n-1)}_{K\backslash\{k_i\}}$ that are absent in the respective chain, there is no visible event in the respective half-space ($t^{(n)} < t^{(n)}_K$, respectively $t^{(n)} > t^{(n)}_K$). Let $\mathbf{t}_0 \in \mathscr{P}_K$ be given by $t^{(n)} = t^{(n)}_K$ and fixing the higher variables to $t_0^{(n+1)}, \ldots, t_0^{(M)}$. Let $\mathbf{t}_0$ be visible and generic, and $t^{(n-1)}_{K\backslash\{k_i\}}$ in one of the chains in (7). For $t^{(n)}$ close enough to $t^{(n)}_K$, and on the respective side according to (7), every event in $\mathscr{P}_{K\backslash\{k_i\}}$ with remaining coordinates $t^{(n)}, t_0^{(n+1)}, \ldots, t_0^{(M)}$ is visible.

Example 3.14. For $n = 3$ and $k_1 < k_2 < k_3 < k_4$, (7) takes the form

$$
\begin{aligned}
y_{\{k_1,k_2,k_3\}} < y_{\{k_1,k_3,k_4\}} \\
y_{\{k_2,k_3,k_4\}} < y_{\{k_1,k_2,k_4\}}
\end{aligned}
\quad \text{if} \quad t \lesseqgtr t_{\{k_1,k_2,k_3,k_4\}} .
$$

Fixing all variables $t^{(n)}$, $n > 3$, for $t < t_{\{k_1,k_2,k_3,k_4\}}$ close enough to $t_{\{k_1,k_2,k_3,k_4\}}$, the corresponding points of $\mathscr{P}_{\{k_1,k_2,k_3\}}$ and $\mathscr{P}_{\{k_1,k_3,k_4\}}$ are visible, but not the corresponding points of $\mathscr{P}_{\{k_2,k_3,k_4\}}$ and $\mathscr{P}_{\{k_1,k_2,k_4\}}$, whereas for $t > t_{\{k_1,k_2,k_3,k_4\}}$ it is the other way around. All this describes a *tree rotation*, see the plots in Figure 2.

We described line solitons as objects moving in the xy-plane (where $x = t^{(1)}$ and $y = t^{(2)}$). They evolve according to the KP equation (with evolution parameter $t = t^{(3)}$), the first equation of the KP hierarchy. A higher KP hierarchy equation has one of the parameters $t^{(r)}$, $r > 3$, as its evolution parameter. We do not consider the corresponding evolutions in this work, but it turned out that these parameters are important in order to classify the various evolutions. We showed that solitons from a subclass have the form of rooted binary trees with leaves extending to infinity in the xy-plane and evolving by right rotation as time t proceeds. They all start with the same asymptotic form as $t \sim -\infty$ and end with the same asymptotic form as $t \sim +\infty$. These are the maximal and minimal element, respectively, of a Tamari lattice, and any generic evolution thus corresponds to a maximal chain. For a soliton configuration with $M + 1$ leaves, which chain is realized depends on the values of the parameters $t^{(r)}$, $r = 1, \ldots, M$.

4 Higher Bruhat and higher Tamari orders

According to Proposition 3.2, a substantial role in the combinatorics underlying the tree-shaped line solitons is played by the order relations (5). They are at the roots of the generalization by Manin and Schechtman of the weak Bruhat order on the set of permutations of $[m]$ to 'higher Bruhat orders' [17, 18, 19], also see [30]. In the following subsection we recall some definitions and results mainly from [30]. In section 4.2 we introduce 'higher Tamari orders'.

4.1 Higher Bruhat orders

Let $n,N \in \mathbb{N}$ with $1 \leq n \leq N-1$. An element $K \in \binom{[N]}{n+1}$ will be written as $K = \{k_1,\ldots,k_{n+1}\}$ with $k_1 < \cdots < k_{n+1}$. $P(K)$ denotes the *packet* of K, i.e., the set of n-subsets of K. The *lexicographic order* on $P(K)$ is given by $K \setminus \{k_{n+1}\}, K \setminus \{k_n\},\ldots,K \setminus \{k_1\}$. A *beginning segment* of $P(K)$ has the form $\{K \setminus \{k_{n+1}\}, K \setminus \{k_n\},\ldots,K \setminus \{k_j\}\}$ for some j. An *ending segment* is of the form $\{K \setminus \{k_j\}, K \setminus \{k_{j-1}\},\ldots,K \setminus \{k_1\}\}$.[6]

A subset $U \subset \binom{[N]}{n+1}$ is called *consistent* if its intersection with any $(n+1)$-packet[7] is either a beginning or an ending segment. The *higher Bruhat order* $B(N,n)$ is the set of consistent subsets of $\binom{[N]}{n+1}$, ordered by single-step inclusion.[8]

Example 4.1. The consistent subsets of $\binom{[3]}{2}$ are $\emptyset$, $\{\{1,2\}\}$, $\{\{2,3\}\}$, $\{\{1,2\},\{1,3\}\}$, $\{\{1,3\},\{2,3\}\}$, $\{\{1,2\},\{1,3\},\{2,3\}\}$. Single-step inclusion leads to $B(3,1)$, which has a hexagonal Hasse diagram (also see Figure 7 in Section 4.2).

A linear order ρ on $\binom{[N]}{n}$ (which may be regarded as a permutation of $\binom{[N]}{n}$) is called *admissible* if, for every $K \in \binom{[N]}{n+1}$, the packet of K appears in it either in lexicographic or in reverse lexicographic order. Let $A(N,n)$ be the set of admissible linear orders of $\binom{[N]}{n}$. Two elements ρ,ρ' of $A(N,n)$ are *elementarily equivalent*, if they differ only by the exchange of two neighboring elements that are not contained in a common packet. The resulting equivalence relation will be denoted by $\sim$. For each $\rho \in A(N,n)$, the *inversion set* $\mathrm{inv}(\rho)$ is the set of all $K \in \binom{[N]}{n+1}$ for which $P(K)$ appears in reverse lexicographic order in ρ. We have $\rho \sim \rho'$ iff $\mathrm{inv}(\rho) = \mathrm{inv}(\rho')$, so that the inversion set only depends on the equivalence class of ρ. All this results in a poset isomorphism $U = \mathrm{inv}(\rho) \mapsto [\rho]$ between $B(N,n)$ and $A(N,n)/\sim$.[9]

For $[\rho] \in A(N,n)/\sim$, let $Q[\rho]$ be the intersection of all *linear* orders in $[\rho]$, i.e., the partial order on $\binom{[N]}{n}$ given by $I' < I$ iff $I' <_\sigma I$ for all $\sigma \in [\rho]$. The set of linear extensions of $Q[\rho]$ coincides with $[\rho]$. We set $Q(U) := Q[\rho]$ where $U = \mathrm{inv}(\rho)$.

Of great help for the construction of higher Bruhat orders is the existence [17, 18, 30] of a natural bijection between $A(N,n)$ and the set of maximal chains of $B(N,n-1)$, which we describe next. With $\rho = (I_1,\ldots,I_s) \in A(N,n)$ (where $I_i \in \binom{[N]}{n}$ and $s = \binom{N}{n}$) we associate the chain of consistent sets $\emptyset \to \{I_1\} \to \{I_1,I_2\} \to \cdots \to$

---

[6] $\emptyset$ and $P(K)$ are considered as being both, beginning and ending.

[7] An $(n+1)$-packet is the packet of some element of $\binom{[N]}{n+2}$.

[8] By a theorem of Ziegler [30], this definition is equivalent to the original one of Manin and Schechtman [17, 18], also see [4]. For finite sets U,U', *single-step inclusion* is defined by $U \subset U'$ and $|U'| = |U|+1$.

[9] The difficult part is to show that the set of inversion sets coincides with the set of consistent sets, see [30].

$\{I_1,\ldots,I_s\} = \binom{[N]}{n}$ in $B(N,n-1)$. Conversely, given a maximal chain $\emptyset \to U_1 \to U_2 \to \cdots \to \binom{[N]}{n}$ of consistent sets in $B(N,n-1)$, these are ordered by single step inclusion, hence $U_{r+1} \setminus U_r = \{I_r\}$ with some $I_r \in \binom{[N]}{n}$. Thus we obtain an admissible linear order $(I_1,\ldots,I_s) \in A(N,n)$. If, for some $K \in \binom{[N]}{n+1}$, $P(K)$ appears in (reverse) lexicographic order in ρ, then $U_r \cap P(K)$ is a beginning (ending) segment, for $r = 0,1,\ldots,s$. Conversely, if U is an element of a maximal chain of $B(N,n-1)$, and if $U \cap P(K)$ is a beginning (ending) segment, then this holds for all elements of this chain. Thus $P(K)$ appears in the corresponding ρ in (reverse) lexicographic order. Furthermore, if two maximal chains have a common edge $U \xrightarrow{I} U \cup \{I\}$, then, obviously, both contain all packets having I as a member in the same way (i.e., lexicographically, respectively reverse lexicographically).

With the help of these results, all higher Bruhat orders can be constructed iteratively, starting from the highest, i.e., $B(N,N-1)$. The latter consists of only two elements, $\emptyset$ and $\{[N]\}$.

Suppose we have constructed $B(N,n)$. Its elements are consistent sets of the form $\{K_1,\ldots,K_r\}$, where $K_i \in \binom{[N]}{n+1}$ and $r \in \{0,1,\ldots,\binom{N}{n+1}\}$ ($\emptyset$ if $r = 0$). Associated with each such consistent set U is an equivalence class $[\rho] \in A(N,n)/\sim$ such that $U = \text{inv}(\rho)$. For each $\rho \in [\rho]$ we construct the corresponding maximal chain of $B(N,n-1)$, as explained above. The collection of all such maximal chains constitutes $B(N,n-1)$. Its elements are consistent sets $\{I_1,\ldots,I_r\}$, $r \in \{0,1,\ldots,s\}$. Now we can continue to construct $B(N,n-2)$, and so forth.

Having arrived at $B(N,1)$, the weak Bruhat order on the permutation group S_N, we can even proceed once more. We consider a permutation $\pi = (\pi_1,\pi_2,\ldots,\pi_N) \in S_N$ as an order $\{\pi_1\} < \{\pi_2\} < \cdots < \{\pi_N\}$, which in turn determines the chain $\emptyset \to \{\{\pi_1\}\} \to \{\{\pi_1\},\{\pi_2\}\} \to \cdots \to \{\{\pi_1\},\ldots,\{\pi_N\}\} = \binom{[N]}{1}$. These are the maximal chains of $B(N,0)$, which is isomorphic to the Boolean lattice of subsets of $[N]$ and forms an N-cube.[10]

In the following, we represent a higher Bruhat order $B(N,n)$ by a diagram, suppressing the labels of the vertices and expressing a maximal chain $\emptyset \to \{K_1\} \to \{K_1,K_2\} \to \cdots \to \{K_1,\ldots,K_q\} = \binom{[N]}{n+1}$, $q = \binom{N}{n+1}$, graphically as

$$\bullet \xrightarrow{K_1} \bullet \xrightarrow{K_2} \bullet \cdots \bullet \xrightarrow{K_q} \bullet \,.$$

The edges are thus labelled by the sets K_i, which are sequentially added to the preceding vertex, starting with the empty set. The vertices are thus given by sets of the form $\{K_1,\ldots,K_r\}$, $r \in \{0,1,\ldots,q\}$. They are ordered by single step inclusion, and we have $(K_1,\ldots,K_q) \in A(N,n+1)$.

Remark 4.2. The weak Bruhat order $B(N,1)$ is a lattice and can be visualized as a polytope in $N-1$ Euclidean dimensions, called *permutohedron*. Not all higher Bruhat orders are lattices [30] and not all can be realized as polytopes [5].

---

[10] More generally, the elements of $B(N,n)$ can be represented as sets of n-faces of the N-cube [27, 26].

In the context of KP line solitons, the relevance of higher Bruhat orders is evident from Proposition 3.2, as already mentioned there. For fixed $M \in \mathbb{N}$ and parameters $p_i, c_i, i = 1, \ldots, M+1$, the order $p_1 < \cdots < p_{M+1}$ induces an order on the 'critical values' $t_J^{(n-1)}$, $J \in \binom{\Omega}{n}$, according to the following rule. For $t^{(n)} < t_K^{(n)}$, the values $t_J^{(n-1)}$, $J \in P(K)$, are ordered lexicographically, and for $t^{(n)} > t_K^{(n)}$ they are ordered reverse lexicographically. Via the bijection $I \mapsto t_I^{(|I|-1)}$ of subsets of $\Omega = [M+1]$ and the set of critical values, this corresponds to admissible permutations on $\binom{\Omega}{n}$. Without further restriction of the parameters, the resulting partial order is $B(M+1, n)$.

4.1.1 How to obtain the poset $Q(U)$ for a consistent set U: an example

This subsection explains in an elementary way the construction of the poset $Q(U)$ for a consistent set U in the case $N = 4$. There is only one 3-packet, namely

$$\binom{[4]}{3} = \{\{1,2,3\}, \{1,2,4\}, \{1,3,4\}, \{2,3,4\}\},$$

and thus the following 8 consistent subsets of $\binom{[4]}{3}$,

$$\emptyset, \ \{\{1,2,3\}\}, \ \{\{1,2,3\},\{1,2,4\}\}, \ \{\{1,2,3\},\{1,2,4\},\{1,3,4\}\},$$
$$\{\{1,2,3\},\{1,2,4\},\{1,3,4\},\{2,3,4\}\},$$
$$\{\{2,3,4\}\}, \ \{\{1,3,4\},\{2,3,4\}\}, \ \{\{1,2,4\},\{1,3,4\},\{2,3,4\}\} \, .$$

Single-step inclusion results in $B(4,2)$, which has an octagonal Hasse diagram (see Figure 6). The packets of the elements of $\binom{[4]}{3}$ are given by

$$P(\{1,2,3\}) = \{\{1,2\},\{1,3\},\{2,3\}\}, \quad P(\{1,2,4\}) = \{\{1,2\},\{1,4\},\{2,4\}\},$$
$$P(\{1,3,4\}) = \{\{1,3\},\{1,4\},\{3,4\}\}, \quad P(\{2,3,4\}) = \{\{2,3\},\{2,4\},\{3,4\}\} \, .$$

For each consistent subset U of $\binom{[4]}{3}$, we consider a table, which displays the packet of each element of U downwards in reverse lexicographic order (and the remaining packets in lexicographic order). The left one of the following two tables describes the case $U = \emptyset$, hence all packets are in lexicographic order.

1 2 3	1 2 4	1 3 4	2 3 4
1 2	1 2	1 3	2 3
1 3	1 4	1 4	2 4
2 3	2 4	3 4	3 4

1 2 3	1 2 4	1 3 4	2 3 4
2 3	2 4	1 3	2 3
1 3	1 4	1 4	2 4
1 2	1 2	3 4	3 4

From the table, where, e.g., 12 stands for $\{1,2\}$, we read off[11] cover relations and deduce a poset, drawn as a Hasse diagram. This leads to the very first poset (of

---

[11] For example, the top node can only be an entry from the first row of the table. But the first column only leaves us with 12. It is also obvious that 34 is the bottom node. Furthermore, we see that from 13 to 24 there are two ways, via 14 respectively 23.

both horizontal chains) in Figure 6. It has two linear extensions, which are elements $\rho, \rho' \in A(4,2)$, one with $14, 23$, the other one with $23, 14$ instead. Since 14 and 23 belong to different packets, $\rho \sim \rho'$. Since no packet is in reverse lexicographic order, $[\rho] \in A(4,2)/\sim$ corresponds to the empty (consistent) set. The poset is $Q(\emptyset)$.

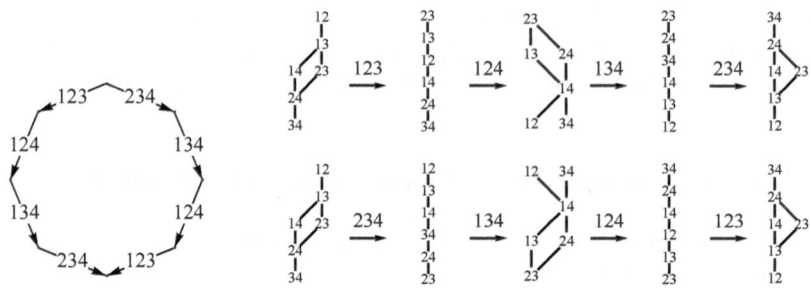

Fig. 6 $B(4,2)$ and its two maximal chains in terms of the Q-posets, which can be constructed in the way explained in section 4.1.1.

The second table above decribes the case $U = \{\{1,2,3\},\{1,2,4\}\}$. Its evaluation leads to $Q(U)$, the third poset in the first horizontal chain in Figure 6. It has four linear extensions. Since 13 and 24, as well as 12 and 34, belong to different packets, all these extensions are equivalent, hence U determines a single element of $A(4,2)/\sim$. Evaluating the remaining tables, we finally obtain the two chains of posets in Figure 6.

4.2 Higher Tamari orders

Let $K \in \binom{[N]}{n+1}$. In the KP line soliton context in Section 3, where $N = M + 1$, we found the following rules concerning non-visible critical events. A point $\mathbf{t}_{K\setminus\{k\}}$ is non-visible if either (1) $t^{(n)} < t_K^{(n)}$ and $k \in K_<$, or (2) $t^{(n)} > t_K^{(n)}$ and $k \in K_>$. In the first case, the critical values $t_{K\setminus\{k\}}^{(n)}$ are ordered lexicographically, in the second case reverse lexicographically. This induces a corresponding combinatorial rule on the sets that enumerate the critical values, and we can resolve the notion of 'non-visibility' from the special KP line soliton context as follows.

Definition 4.3. Let $\rho \in A(N,n)$. $I \in \rho$ is called *non-visible* in ρ if there is a $K \in \binom{[N]}{n+1}$ and a $k \in K$ such that $I = K \setminus \{k\}$, and if either $K \notin \text{inv}(\rho)$ and $k \in K_<$, or $K \in \text{inv}(\rho)$ and $k \in K_>$ (with $K_<$ and $K_>$ defined in Section 3). I is called *visible* in ρ if it is *not* non-visible in ρ.

Since $\text{inv}(\rho)$ only depends on the equivalence class of ρ, this definition induces a notion of non-visibility (visibility) in $[\rho] \in A(N,n)/\sim$. Moreover, since an element of an admissible linear order $\rho \in A(N,n)$ corresponds to an edge $U \xrightarrow{I} U'$ of a maximal chain of $B(N,n-1)$ (cf. Section 4.1), the above definition induces a notion of non-visibility of such an edge: $U \xrightarrow{I} U'$ is *non-visible* in a maximal chain of $B(N,n-1)$ if $I = K \setminus \{k\}$ for some $K \in \binom{[N]}{n+1}$, and $k \in K_<$ if $U' \cap P(K)$ is a beginning segment, $k \in K_>$ if $U' \cap P(K)$ is an ending segment. If $U \xrightarrow{I} U'$ is non-visible in one maximal chain of $B(N,n-1)$, then it is non-visible in *every* maximal chain that contains it. Therefore we can drop the reference to a maximal chain.

We have seen in Section 4.1 that any $U \in B(N,n)$ determines a poset $Q(U)$. Eliminating all non-visible elements (in admissible linear orders) of $Q(U)$, by application of the rules in the preceding definition, results in a subposet that we denote by $R(U)$.

Proposition 4.4. *An edge $U \xrightarrow{K} U'$ in $B(N,n)$ is non-visible iff $R(U) = R(U')$.*

Proof. We will show that if $U \xrightarrow{K} U'$ is non-visible, then all elements of $P(K)$ are non-visible in (any linear extension of) $Q(U)$ and $Q(U')$, and if $U \xrightarrow{K} U'$ is visible, then $R(U)$ and $R(U')$ differ by elements of $P(K)$. Let $I \in P(K)$, i.e., $K = I \cup \{k\}$ with some $k \in [N] \setminus I$, and $l \in [N] \setminus K$. We set $L = K \cup \{l\} \in \binom{[N]}{n+2}$ and $K' = L \setminus \{k\} \in \binom{[N]}{n+1}$. Let $l \in L_<$. We recall that in this case

(a) $U' \cap P(L)$ is a beginning segment if $U \xrightarrow{K} U'$ is non-visible,

(b) $U' \cap P(L)$ is an ending segment if $U \xrightarrow{K} U'$ is visible.

If $k > l$, then $l \in K'_>$. In case (a) we have $K' \in U, U'$, hence $I = K' \setminus \{l\}$ is non-visible. In case (b) we have $K' \notin U, U'$, hence $I = K' \setminus \{l\}$ is *not* non-visible with respect to K'. If $k < l$, then $l \in K'_<$. In case (a) we have $K' \notin U, U'$ and I is again non-visible. In case (b) we have $K' \in U, U'$ and I is again *not* non-visible with respect to K'.

For $l \in L_>$, we have to exchange 'beginning' and 'ending' in the above conditions (a) and (b). If $k > l$, we have $l \in K'_<$. Then I is non-visible in case (a) and not non-visible with respect to K' in case (b). If $k < l$, then $l \in K'_>$. Again, I is non-visible in case (a) and not non-visible with respect to K' in case (b).

We conclude from case (a) that non-visibility of $U \xrightarrow{K} U'$ implies that all elements of $P(K)$ are non-visible in $Q(U)$, as well as in $Q(U')$. Since K and all the K' exhaust the elements of $\binom{[N]}{n+1}$ whose packets contain I, it follows from case (b) that $I \in R(U)$ and $I \notin R(U')$ if $k \in K_>$, whereas $I \notin R(U)$ and $I \in R(U')$ if $k \in K_<$.   □

Example 4.5. Let $N = 6$, $n = 3$ and $K = 1346$, which stands for $\{1,3,4,6\}$. There are only two elements of $\binom{[6]}{5}$ such that their packet contains K. These are $L_1 = 12346$ and $L_2 = 13456$, hence $l_1 = 2 \in (L_1)_<$ and $l_2 = 5 \in (L_2)_<$.

If $U \xrightarrow{K} U'$ is non-visible, then $U' \cap P(L_1)$ is a beginning segment: $U' \cap P(L_1) = \{1234, 1236, 1246, 1346\}$. The left table below displays the corresponding information for the construction of $Q(U')$ and $R(U')$, obtained via Definition 4.3. Non-visible elements are marked in red. We see that the whole packet of K (marked in the table in light red) is already non-visible as a consequence of the information obtained from the other elements of $\binom{[6]}{4}$. Hence $R(U') = R(U)$.

If $U \xrightarrow{K} U'$ is visible, then $U' \cap P(L_1)$ is an ending segment: $U' \cap P(L_1) = \{1346, 2346\}$. In this case we obtain the right table. $R(U)$ contains the elements 134 and 146 of

$P(K)$. They are not present in $R(U')$, in which the new complementary elements 136 and 346 of $P(K)$ show up. For L_2 an analogous discussion applies. Finally we can conclude that $R(U') \neq R(U)$.

1234	1236	1246	1346	2346
234	236	246	346	234
134	136	146	146	236
124	126	126	136	246
123	123	124	134	346

1234	1236	1246	1346	2346
123	123	124	346	346
124	126	126	146	246
134	136	146	136	236
234	236	246	134	234

Corollary 4.6. *If $U_0 \overset{K_1}{\to} U_1 \overset{K_2}{\to} \cdots \overset{K_r}{\to} U_r$ is any (not necessarily maximal) chain in $B(N,n)$, containing at least one visible edge, then $R(U_0) \neq R(U_r)$.*

Proof. Without restriction of generality we can assume that $U_0 \overset{K_1}{\to} U_1$ is visible and $K_1 = I \cup \{k\}$ with $k \in (K_1)_>$, so that $I \in R(U_0)$ and $I \notin R(U_1)$, see the proof of Proposition 4.4. Since $K_1 \in U_s$ for $s = 1, \ldots, r$, and $k \in (K_1)_>$, I does not appear in any $R(U_s)$, and in particular not in $R(U_r)$. $\qquad\square$

Definition 4.7. The *higher Tamari order* $T(N,n)$ is the poset with set of vertices $\{R(U) \mid U \in B(N,n)\}$ and the order given by $R(U) \leq R(U')$ if $U \leq U'$ in $B(N,n)$.

Remark 4.8. The map $B(N,n) \to T(N,n)$, given by $U \mapsto R(U)$, is surjective and order preserving. The Tamari orders also inherit the following property from the Bruhat orders. There is a bijection between the maximal chains of $T(N,n)$ and the linear extensions of the R-posets that form the vertices of $T(N,n+1)$.

In order to construct the Tamari order $T(N,n)$, in each of the maximal chains $\bullet \overset{K_1}{\longrightarrow} \bullet \overset{K_2}{\longrightarrow} \bullet \cdots \bullet \overset{K_q}{\longrightarrow} \bullet$ of $B(N,n)$ we locate the edges associated with non-visible K's and eliminate them. This results in a reduced chain $\bullet \overset{K_{i_1}}{\longrightarrow} \bullet \overset{K_{i_2}}{\longrightarrow} \bullet \cdots \bullet \overset{K_{i_w}}{\longrightarrow} \bullet$.[12] Such an elimination involves identifying the two vertices that are connected by this edge in the Bruhat order. The consistency of this identification is guaranteed by Proposition 4.4. As a consequence of Corollary 4.6, the elimination process cannot lead to cycles, so indeed defines a partial order. Figure 7 shows a simple example.

Fig. 7 $B(3,1)$ (weak Bruhat order on S_3), and the corresponding Tamari order $T(3,1)$. The latter is obtained from the former by eliminating in the left maximal chain the non-visible 13 and in the right maximal chain the non-visible 12 and 23.

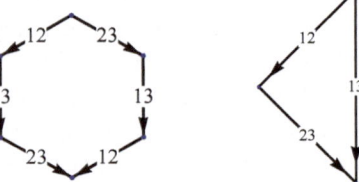

---

[12] The information that resides in the edge labels of the Tamari (i.e., reduced Bruhat) chains is not sufficient to construct the R-posets, which are the vertices of $T(N,n)$. The latter have to be constructed from the Q-posets via elimination of non-visible elements.

Since $T(N, N-1) \cong B(N, N-1)$, both are represented by $\bullet \xrightarrow{[N]} \bullet$. For $N = 3$ and $N = 4$, the vertices are represented in Figure 8, respectively Figure 9, in terms of the Q-posets, respectively R posets, and the information contained in the latter is translated into a soliton graph.

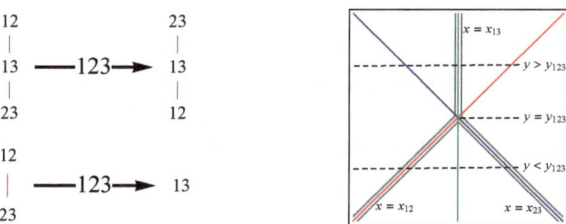

Fig. 8 To the left is the poset $B(3,2)$ and, below it, its visible part $T(3,2)$. Here the vertices are labelled by $Q(\emptyset)$ and $Q(\{[3]\})$, respectively $R(\emptyset)$ and $R(\{[3]\})$. The latter data translate into the soliton solution with $M = 2$, as indicated in the plot on the right-hand side (also see [2]). At a thin line, two phases coincide. Only the thickened parts are visible. We note that $Q(\emptyset)$ and $Q(\{[3]\})$ are linear orders in this particular example, hence elements of $A(3,2)$ and, by a general result, maximal chains of $B(3,1)$, see Figure 7.

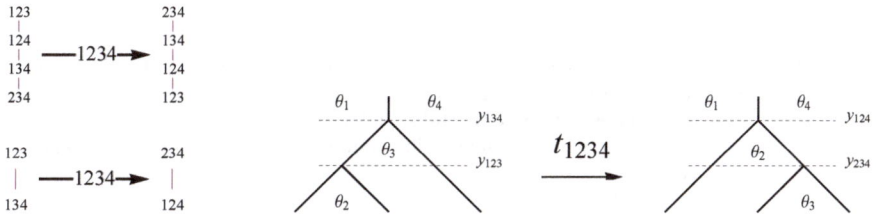

Fig. 9 $B(4,3)$ with vertices the posets $Q(\emptyset)$ and $Q(\{[4]\})$. The diagram below it is $T(4,3)$, with vertices $R(\emptyset)$ and $R(\{[4]\})$. To the right is the translation of $T(4,3)$ into a chain of two rooted binary trees related by a tree rotation. This chain is the Tamari lattice $\mathbb{T}_2$ (in terms of rooted binary trees).

The Bruhat order $B(N, N-2)$ consists of two maximal chains,

$$\bullet \xrightarrow{[N-1]} \bullet \xrightarrow{[N]\backslash\{N-1\}} \bullet \cdots \bullet \xrightarrow{[N]\backslash\{1\}} \bullet \, , \qquad \bullet \xrightarrow{[N]\backslash\{1\}} \bullet \xrightarrow{[N]\backslash\{2\}} \bullet \cdots \bullet \xrightarrow{[N-1]} \bullet \, ,$$

in which $P([N])$ appears in lexicographic, respectively reverse lexicographic order. In the first chain we have to eliminate the edges corresponding to all sets $[N] \backslash \{k\}$ with $k \in [N]_< = \{N-1, N-3, \ldots, 1 + (N \bmod 2)\}$, in the second chain those corresponding to such sets with $k \in [N]_> = \{N, N-2, \ldots, 2 - (N \bmod 2)\}$. This results in the two reduced chains

$$\bullet \xrightarrow{[N-1]} \bullet \xrightarrow{[N]\backslash\{N-2\}} \bullet \cdots \bullet \xrightarrow{[N]\backslash\{2-(N\bmod 2)\}} \bullet$$
$$\bullet \xrightarrow{[N]\backslash\{1+(N\bmod 2)\}} \bullet \cdots \bullet \xrightarrow{[N]\backslash\{N-3\}} \bullet \xrightarrow{[N]\backslash\{N-1\}} \bullet \, ,$$

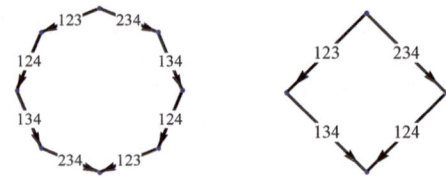

Fig. 10 $B(4,2)$ and $T(4,2)$.

which form $T(N, N-2)$. See Figure 10 for the case $N = 4$. The two maximal chains of $B(4,2)$ with the Q-posets as vertices have already been displayed in Figure 6. Elimination of non-visible elements yields the chains of $T(4,2)$ in Figure 11.

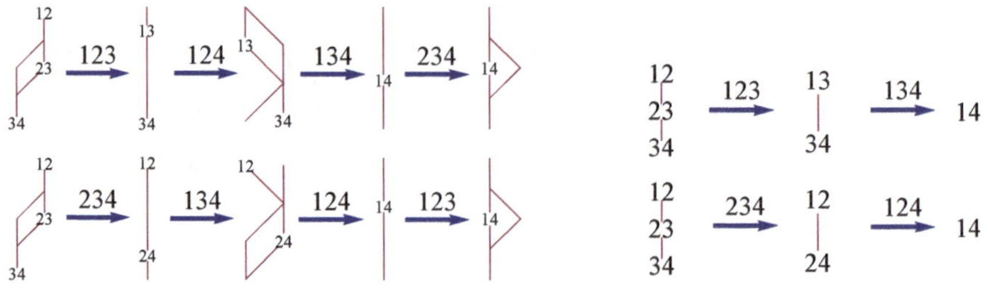

Fig. 11 The two maximal chains of $T(4,2)$ are displayed on the right-hand side. On the left-hand side, in order to illustrate Proposition 4.4, we show an intermediate elimination step applied to the $B(4,2)$-chains of Q-posets in Figure 6. Here we still kept the edges 124 and 234 in the upper chain, and 134, 123 in the lower chain, which are non-visible in the Tamari order. We observe that they indeed connect identical posets.

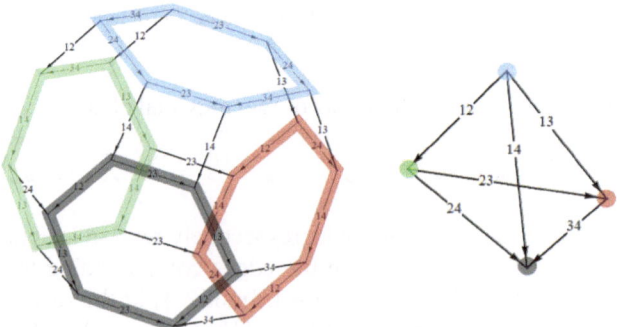

Fig. 12 The left figure shows $B(4,1)$. Non-visible edges connecting vertices that are mapped to the same vertex of $T(4,1)$ (tetrahedral poset) are marked with the same color.

By determining the linear extensions $K_1 \to K_2 \to \cdots$ of the posets in Figure 6, we obtain maximal chains $\bullet \xrightarrow{K_1} \bullet \xrightarrow{K_2} \cdots$ of $B(4,1)$. In this way we construct $B(4,1)$ and then obtain $T(4,1)$ from it by elimination of non-visible edges, see Figure 12. Figure 13 shows the corresponding construction of $T(5,2)$ from $B(5,2)$.

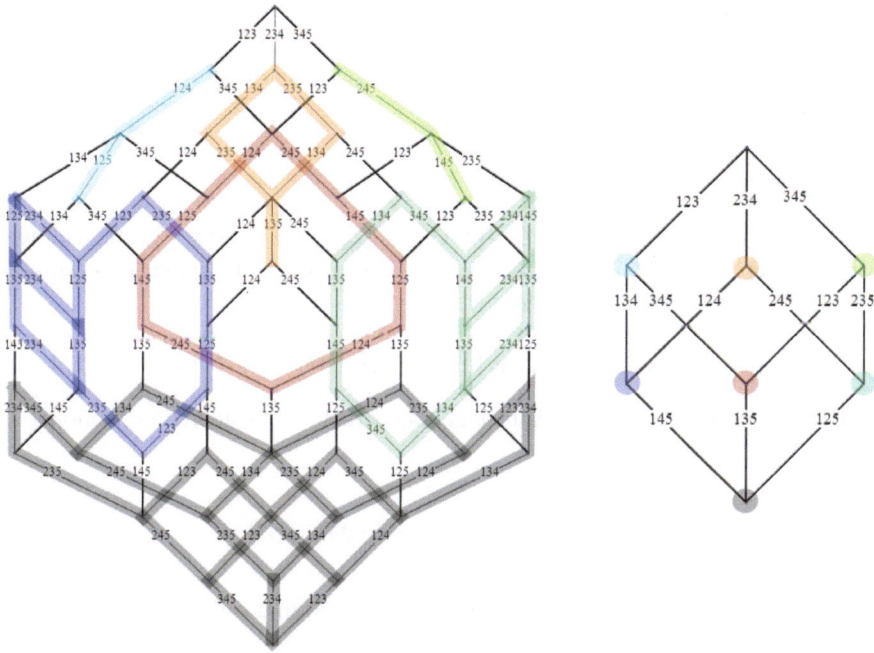

Fig. 13 $B(5,2)$ and $T(5,2)$. Vertices of $B(5,2)$ connected by edges marked with the same color are mapped to the same vertex of $T(5,2)$.

4.2.1 An $(n+1)$-gonal equivalence relation

The vertices of $B(N,n)$ can be described by the Q-posets, and on this level we defined the transition to $T(N,n)$. But the vertices of $B(N,n)$ are equivalently given by consistent sets (which was in fact our original definition). Along a maximal chain, moving from a vertex to the next means increasing the consistent set associated with the first vertex by inclusion of a new set which we use to label the edge between the two vertices. The transition from a Bruhat to a Tamari order means elimination of some of these sets. Whereas every maximal chain of $B(N,n)$ ends in the same set, this is not so for different Tamari chains because of different eliminations along different chains, see Figure 14. If U is the set that corresponds to the first vertex from which the two chains in Figure 14 descend, the first ends in $U \cup (K \setminus \{k_{n+1}\}) \cup \cdots \cup$

Fig. 14 For $K \in \binom{[N]}{n+1}$, the two chains of the diagram start in the same vertex and end in the same vertex, which includes the union of all sets associated with the edges of the left chain, but also the union of all sets associated with the edges of the right chain. The two resulting sets have to be identified, which leads to an $(n+1)$-gonal equivalence relation. If $K = \{[N]\}$, i.e., $n = N - 1$, the chains are the two maximal chains of $T(N, N-2)$.

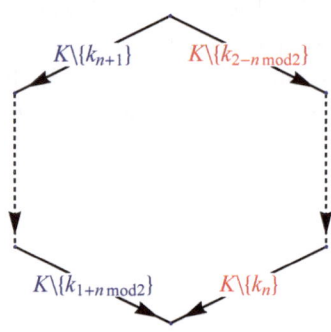

$(K \setminus \{k_{1+n \bmod 2}\})$, the second in $U \cup (K \setminus \{k_{2-n \bmod 2}\}) \cup \cdots \cup (K \setminus \{k_n\})$. The two resulting sets have to be identified, since the final vertex is the same. This requires the $(n+1)$-*gonal equivalence relation*

$$U \cup \{K \setminus \{k_{n+1}\}, \dots, K \setminus \{k_{1+n \bmod 2}\}\} \sim U \cup \{K \setminus \{k_n\}, \dots, K \setminus \{k_{2-n \bmod 2}\}\},$$

for a Tamari order, and motivates the algebraic structure considered in Section 7.

5 KP line soliton evolutions in the case $M = 5$

For $M = 5$ (i.e., $N = 6$) the τ-function of a tree-shaped line soliton solution is given by

$$\tau = e^{\theta_1} + \cdots + e^{\theta_6}, \qquad \theta_i = p_i x + p_i^2 y + p_i^3 t + p_i^4 t^{(4)} + p_i^5 t^{(5)} + c_i,$$

with real constants $p_1 < \cdots < p_6$ and c_i. We have $\Omega = [6] = \{1, 2, 3, 4, 5, 6\}$, and $\bullet \xrightarrow{[6]} \bullet$ represents $B(6, 5)$ and also $T(6, 5)$. The evolution of the soliton corresponds to the left vertex if $t^{(5)} < t^{(5)}_{123456}$, and to the right vertex if $t^{(5)} > t^{(5)}_{123456}$. Associated with the two vertices are maximal chains of $B(6, 4)$, from which $T(6, 4)$ is obtained by elimination of non-visible events, see Figure 15.

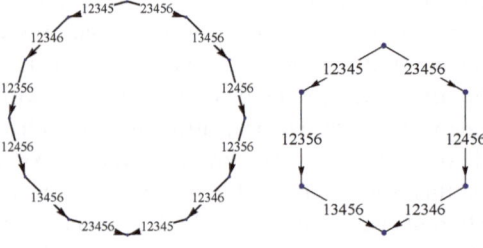

Fig. 15 $B(6,4)$ and $T(6,4)$. Here, e.g., 12356 (which stands for $\{1, 2, 3, 5, 6\}$) translates to the value $t^{(4)}_{12356}$ of the parameter $t^{(4)}$.

If $t^{(5)} < t^{(5)}_{123456}$, the left chain of $B(6,4)$ applies, which means

$$t^{(4)}_{12345} < t^{(4)}_{12346} < t^{(4)}_{12356} < t^{(4)}_{12456} < t^{(4)}_{13456} < t^{(4)}_{23456},$$

where the second, fourth and sixth value corresponds to a non-visible event. If $t^{(5)} > t^{(5)}_{123456}$, the right chain of $B(6,4)$ applies, hence

$$t^{(4)}_{23456} < t^{(4)}_{13456} < t^{(4)}_{12456} < t^{(4)}_{12356} < t^{(4)}_{12346} < t^{(4)}_{12345},$$

where the second, fourth and sixth value is non-visible. If $t^{(5)} = t^{(5)}_{123456}$, all the values $t^{(4)}_{ijklm}$ coincide.

Fig. 16 The two maximal chains of $T(6,4)$. The vertices are the R-posets obtained from the Q-posets associated with the vertices of $B(6,4)$. The upper chain applies if $t^{(5)} < t^{(5)}_{123456}$, the lower if $t^{(5)} > t^{(5)}_{123456}$.

The vertices of $T(6,4)$ are the R-posets in Figure 16. The linear extensions of a poset determine the possible orders of critical values of time $t = t^{(3)}$. For example, if $t^{(5)} > t^{(5)}_{123456}$, the lower horizontal chain in Figure 16 applies. If furthermore $t^{(4)}_{12456} < t^{(4)} < t^{(4)}_{12346}$ (third poset), then we have either $t_{1234} < t_{1456} < t_{1246} < t_{2346}$ or $t_{1456} < t_{1234} < t_{1246} < t_{2346}$ (since the poset has two linear extensions). In order to decide which of these orders is realized by the soliton, further conditions on the parameters p_i (not the c_i) are required (see [2]). From the linear extensions of the R-posets, we obtain (the maximal chains of) $T(6,3)$, which is the Tamari lattice $\mathbb{T}_4$. Figure 17 displays it as a (Tamari-Stasheff) polytope. In order to identify the vertices of $T(6,3)$, we take the pentagonal equivalence

$$\{\{i,j,k,l\}, \{i,j,l,m\}, \{j,k,l,m\}\} \sim \{\{i,j,k,m\}, \{i,k,l,m\}\}$$

(where $i < j < k < l < m$) into account (see Section 4.2.1). The binary trees labelling its vertices in Figure 1 are obtained from the R-posets, listed in Figure 18.

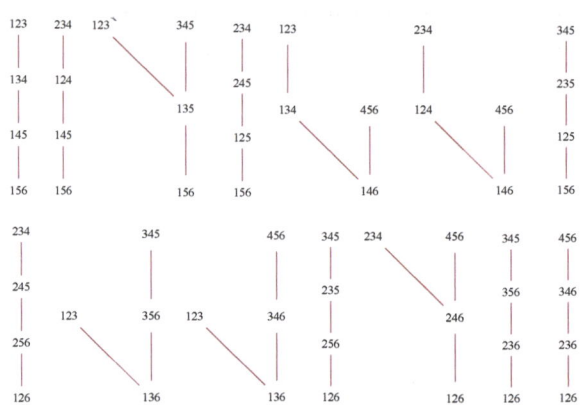

Fig. 17 The Tamari lattice $\mathbb{T}_4 = T(6,3)$ as a Tamari-Stasheff polytope. Such a representation first appeared in Tamari's thesis in 1951 [24].

Fig. 18 The fourteen R-posets that label the vertices of $T(6,3) = \mathbb{T}_4$. They translate into the trees labelling the vertices of the Tamari lattice $\mathbb{T}_4$ in Figure 1. Each poset determines more directly a triangulation of a hexagon, the vertices of which are numbered (anticlockwise) by $1,2,\ldots,6$. A triple ijk then specifies a triangle.

Figure 19 shows an example of a line soliton evolution and the caption identifies the corresponding maximal chain of $\mathbb{T}_4$.

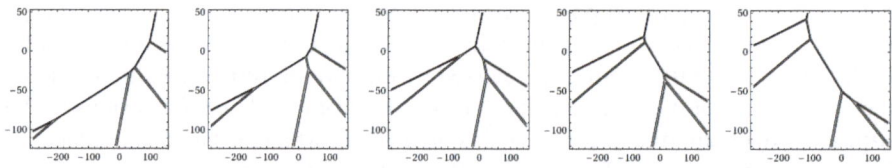

Fig. 19 Plots of an $M = 5$ soliton in the xy-plane at successive values of time. Here we have chosen the parameters such that $t^{(5)} < t^{(5)}_{123456}$ and $t^{(4)}_{12356} < t^{(4)} < t^{(4)}_{13456}$. The evolution corresponds to $\bullet \xrightarrow{1345} \bullet \xrightarrow{1356} \bullet \xrightarrow{1236} \bullet \xrightarrow{3456} \bullet$ on the Tamari lattice $T(6,3) = \mathbb{T}_4$ in Figure 1. It is a linear extension of the third poset of the first horizontal chain in Figure 16.

In the next steps, we obtain $T(6,2)$ and $T(6,1)$, see Figure 20. Finally, $T(6,0)$ is given by the Hasse diagram with two vertices, the upper connected with the lower by six edges (labelled by $1,2,\ldots,6$).

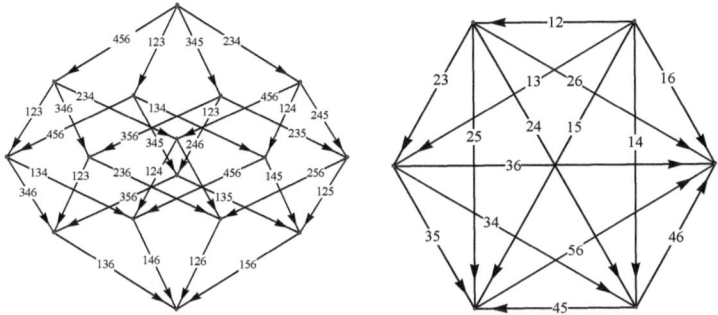

Fig. 20 The Tamari order $T(6,2)$, which forms a cube in four dimensions, and $T(6,1)$, which is a 5-simplex.

6 Some insights into the case $M > 5$

The subclass of tree-shaped line soliton solutions with $M = 6$, hence $N = 7$ and $\Omega = [N] = \{1,2,3,4,5,6,7\}$, is given by

$$\tau = e^{\theta_1} + \cdots + e^{\theta_7}, \qquad \theta_i = p_i x + p_i^2 y + p_i^3 t + p_i^4 t^{(4)} + p_i^5 t^{(5)} + p_i^6 t^{(6)} + c_i .$$

$T(7,5)$ is the heptagon in Figure 21. The left chain is realized if $t^{(6)} < t_\Omega^{(6)}$, the right chain if $t^{(6)} > t_\Omega^{(6)}$. Each element (vertex) is an R-poset, they are displayed in Figure 22.

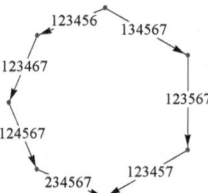

Fig. 21 The Tamari order $T(7,5)$.

From these R-posets, we obtain in turn the maximal chains of $T(7,4)$, hence we can construct $T(7,4)$ by putting all these chains together (joining a minimal and a maximal element). In this way we recover a poset that first appeared in [3] (Figure 4

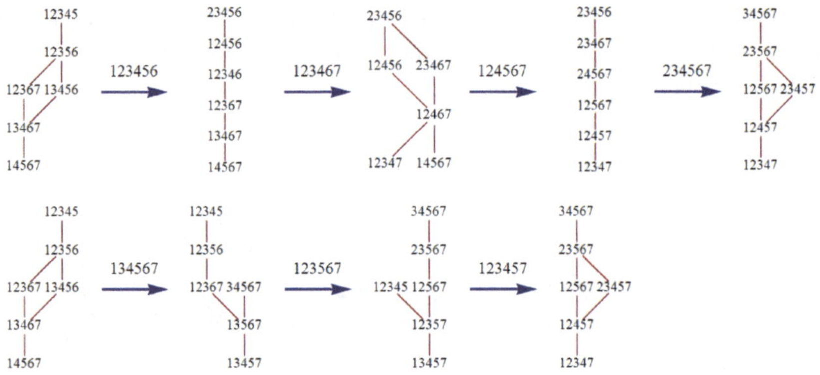

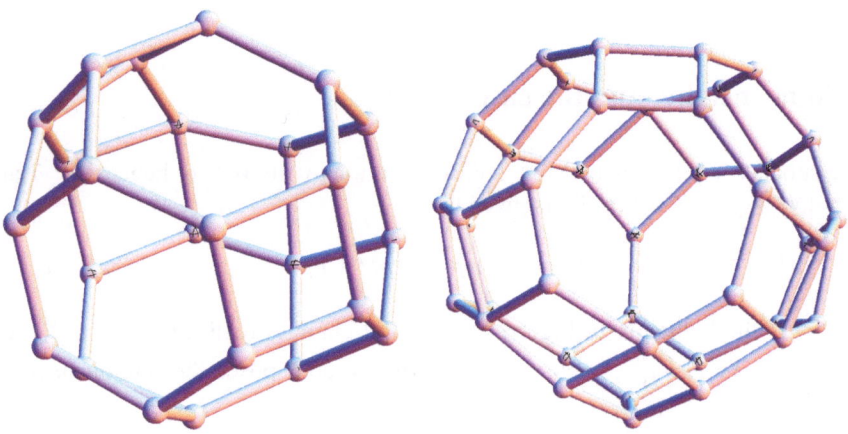

Fig. 22 The two maximal chains of $T(7,5)$, with vertices resolved into R-posets.

therein). Using MATHEMATICA [28], we obtained a pseudo-realization as a polytope, see Figure 23.

Fig. 23 Polytopes on which the higher Tamari orders $T(7,4)$ and $T(8,5)$ live. It should be noticed, however, that not all faces are *regular* or *flat* quadrangles or hexagons, respectively heptagons (as also in Figure 17 with pentagons, cf. [14]).

In the next step, we obtain the Tamari lattice $\mathbb{T}_5 = T(7,3)$, which can be realized as the 4-dimensional associahedron. Its maximal chains classify the possible evolutions of a tree-shaped line soliton with seven phases.

Figure 23 also shows that $T(8,5)$ is polytopal. Figure 24 displays polytope-like representations of some other higher Tamari orders. These are *not* polytopes, however.

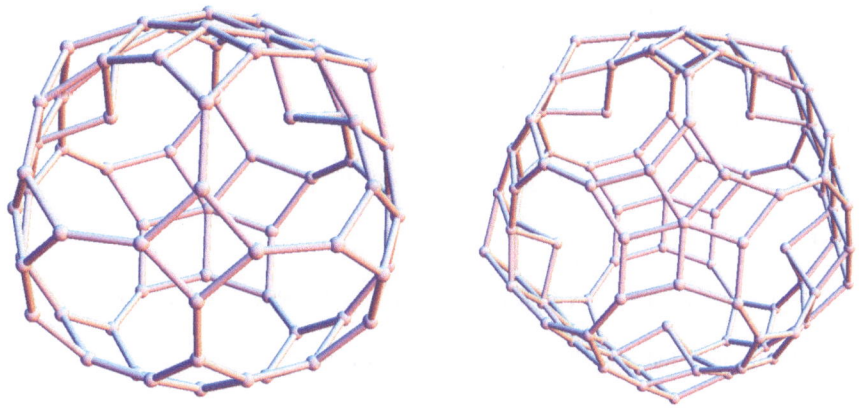

Fig. 24 Polytope-like structures on which the higher Tamari orders $T(9,6)$ and $T(10,7)$ live. The existence of 'small cubes' (cf. [5], proof of Observation 5.3) indicates that these are *not* polytopes. There are two of them in the left figure, and five in the right.

7 An algebraic construction of higher Bruhat and Tamari orders

7.1 Higher Bruhat orders and simplex equations

Let n, N be integers with $0 < n < N$. Let $\mathcal{B}_{N,n}$ be the monoid generated by symbols R_I, $I \in \binom{[N]}{n}$, and R_K, $K \in \binom{[N]}{n+1}$, subject to the following relations.

(a) $R_J R_{J'} = R_{J'} R_J$ if $J, J' \in \binom{[N]}{k}$, $k \in \{n, n+1\}$, are such that $|J \cup J'| > k+1$.

(b) $R_I R_K = R_K R_I$ if $I \not\subset K$.

(c)   For $K = \{k_1, \ldots, k_{n+1}\}$, $k_1 < k_2 < \cdots < k_{n+1}$,

$$R_{K \setminus \{k_{n+1}\}} R_{K \setminus \{k_n\}} \cdots R_{K \setminus \{k_1\}} = R_K \, R_{K \setminus \{k_1\}} R_{K \setminus \{k_2\}} \cdots R_{K \setminus \{k_{n+1}\}} \, . \tag{8}$$

Recall that, if two neighbors $I_j, I_{j+1} \in \binom{[N]}{n}$ in a linear order $\rho = (I_1, \ldots, I_s) \in A(N,n)$ (where $s = \binom{N}{n}$) are not contained in a common packet (or, equivalently, if $|I_j \cup I_{j+1}| > n+1$), exchanging them leads to an elementarily equivalent linear order ρ', i.e., $\rho \sim \rho'$. As a consequence of the above relations, the map that sends ρ to the monomial $R_{I_1} R_{I_2} \cdots R_{I_s}$ induces a correspondence between equivalence classes $[\rho]$, and thus elements of the higher Bruhat order $B(N,n)$, and such monomials. Relation (c) encodes the order relations of $B(N,n)$. It relates a lexicographically ordered product to the reverse lexicographically ordered product. In the following, we write

$$R_{i_1 \ldots i_n} := R_{\{i_1, \ldots, i_n\}} \qquad i_1 < i_2 < \cdots < i_n \, .$$

For $n = 1$, we have

$$R_i R_j = R_{ij} R_j R_i \qquad i < j,$$

and, for $i < j < k$,

$$(R_i R_j) R_k = R_{ij} R_j (R_i R_k) = R_{ij} R_j R_{ik} R_k R_i = R_{ij} R_{ik} (R_j R_k) R_i = R_{ij} R_{ik} R_{jk} R_k R_j R_i,$$

and also

$$R_i (R_j R_k) = R_i R_{jk} R_k R_j = R_{jk} (R_i R_k) R_j = R_{jk} R_{ik} R_k (R_i R_j) = R_{jk} R_{ik} R_{ij} R_k R_j R_i.$$

Associativity now leads to the *consistency condition*

$$R_{ij} R_{ik} R_{jk} = R_{jk} R_{ik} R_{ij} \qquad i < j < k. \tag{9}$$

For $N = 3$, this is $R_{12} R_{13} R_{23} = R_{23} R_{13} R_{12}$, which has the form of the *Yang-Baxter equation* [21]. (9) is a special case of (8) for $n = 2$,

$$R_{ij} R_{ik} R_{jk} = R_{ijk} R_{jk} R_{ik} R_{ij} \qquad i < j < k.$$

This generalization of (9) has been considered in particular in [16, 10, 11, 7].

For example, in $\mathscr{B}_{4,1}$, we have

$$\begin{aligned}(R_1 R_2) R_3 R_4 &= R_{12} R_2 (R_1 R_3) R_4 = R_{12} R_{13} (R_2 R_3) R_1 R_4 = R_{12} R_{13} R_{23} R_3 R_2 (R_1 R_4) \\ &= R_{12} R_{13} R_{23} R_{14} R_3 (R_2 R_4) R_1 = R_{12} R_{13} R_{23} R_{14} R_{24} (R_3 R_4) R_2 R_1 \\ &= R_{12} R_{13} R_{23} R_{14} R_{24} R_{34} R_4 R_3 R_2 R_1.\end{aligned}$$

The monomial with which we started corresponds to the minimal element of $B(4,1)$, i.e., $(\{1\}, \{2\}, \{3\}, \{4\})$. After a sequence of inversions, leading to new elements of $B(4,1)$, the resulting monomial $R_4 R_3 R_2 R_1$ corresponds to the maximal element $(\{4\}, \{3\}, \{2\}, \{1\})$. The final sequence of R_{ij}'s represents an element of $A(4,2)$. In $\mathscr{B}_{4,2}$, we have

$$\begin{aligned}&(R_{12} R_{13} R_{23}) R_{14} R_{24} R_{34} = R_{123} R_{23} R_{13} (R_{12} R_{14} R_{24}) R_{34} \\ &= R_{123} R_{124} R_{23} [R_{13} R_{24}] R_{14} [R_{12} R_{34}] = R_{123} R_{124} R_{23} R_{24} (R_{13} R_{14} R_{34}) R_{12} \\ &= R_{123} R_{124} R_{134} (R_{23} R_{24} R_{34}) R_{14} R_{13} R_{12} = R_{123} R_{124} R_{134} R_{234} R_{34} R_{24} R_{23} R_{14} R_{13} R_{12},\end{aligned}$$

where square brackets indicate an application of a commutativity relation, and

$$R_{12} R_{13} [R_{23} R_{14}] R_{24} R_{34} = \ldots = R_{234} R_{134} R_{124} R_{123} R_{34} R_{24} R_{23} R_{14} R_{13} R_{12}.$$

These calculations reproduce by stepwise inversions the two maximal chains of $B(4,2)$, starting with the minimal element $(\{1,2\}, \{1,3\}, \{2,3\}, \{1,4\}, \{3,4\})$, and ending with the maximal element $(\{3,4\}, \{1,4\}, \{2,3\}, \{1,3\}, \{1,2\})$ (cf. Figure 6). As a consequence of associativity, we obtain the consistency condition

$$R_{123}R_{124}R_{134}R_{234} = R_{234}R_{134}R_{124}R_{123} .$$

The two sides of this equation represent the two elements of $A(4,3)$. In $\mathscr{B}_{4,3}$, it generalizes to

$$R_{123}R_{124}R_{134}R_{234} = R_{1234}\, R_{234}R_{134}R_{124}R_{123} .$$

We expect that, more generally, the consistency conditions of (8) are given by

$$R_{L\backslash\{l_{n+2}\}} R_{L\backslash\{l_{n+1}\}} \cdots R_{L\backslash\{l_1\}} = R_{L\backslash\{l_1\}} R_{L\backslash\{l_2\}} \cdots R_{L\backslash\{l_{n+2}\}} \tag{10}$$

for $L = \{l_1,\dots,l_{n+2}\} \in \binom{[N]}{n+2}$ in linear order. These consistency conditions have the form of generalized Yang-Baxter equations, the so-called *simplex equations* [29, 1, 6, 12, 13], see the following remark. A derivation of these equations as consistency conditions, in the way described above, apparently first appeared in [16] (called 'obstruction method' in [20, 7]).

Remark 7.1. For given positive integers $n < N$, we choose two finite-dimensional vector spaces V, W, and $S_I \in V \otimes \mathrm{End}(W)$, $I \in \binom{[N]}{n}$, with the property

$$S_I \tilde{\otimes} S_{I'} = P S_{I'} \tilde{\otimes} S_I \qquad \text{if} \quad |I \cup I'| > n+1 ,$$

where $P(w \tilde{\otimes} w') = w' \tilde{\otimes} w$, and $\tilde{\otimes}$ denotes the tensor product over $\mathrm{End}(W)$. For $K = \{k_1,\dots,k_{n+1}\} \in \binom{[N]}{n+1}$, $k_1 < \dots < k_{n+1}$, we write (8) in the form

$$S_{K\backslash\{k_{n+1}\}} \tilde{\otimes} S_{K\backslash\{k_n\}} \tilde{\otimes} \cdots \tilde{\otimes} S_{K\backslash\{k_1\}} = R\, S_{K\backslash\{k_1\}} \tilde{\otimes} S_{K\backslash\{k_2\}} \tilde{\otimes} \cdots \tilde{\otimes} S_{K\backslash\{k_{n+1}\}} ,$$

with some $R \in \mathrm{End}(\otimes^{n+1} W)$. In components, this takes the form

$$S_{K\backslash\{k_{n+1}\}}^{a_{n+1}} S_{K\backslash\{k_n\}}^{a_n} \cdots S_{K\backslash\{k_1\}}^{a_1} = \sum_{b_1,\dots,b_{n+1}} R_{b_{n+1}\dots b_1}^{a_{n+1}\dots a_1}\, S_{K\backslash\{k_1\}}^{b_1} S_{K\backslash\{k_2\}}^{b_2} \cdots S_{K\backslash\{k_{n+1}\}}^{a_{n+1}} .$$

The consistency conditions are the $(n+1)$-simplex equations

$$R_{L\backslash\{l_{n+2}\}} R_{L\backslash\{l_{n+1}\}} \cdots R_{L\backslash\{l_1\}} = R_{L\backslash\{l_1\}} R_{L\backslash\{l_2\}} \cdots R_{L\backslash\{l_{n+2}\}} ,$$

for all $L = \{l_1,\dots,l_{n+2}\} \in \binom{[N]}{n+2}$, $l_1 < l_2 < \cdots < l_{n+2}$. Here $R_K \in \mathrm{End}(\otimes^N V)$ is given by R acting non-trivially only on factors of the N-fold tensor product of V at those positions that are given by the numbers $k_1,\dots,k_{n+1}$.

Remark 7.2. Assuming an extension of the monoid $\mathscr{B}_{N,n}$ to a unital ring of formal power series in an indeterminate ε, and an expansion

$$R_K = 1 + \varepsilon\, \mathfrak{r}_K + \cdots$$

in powers of ε, from (10) we obtain to first nontrivial order (which is ε^2)

$$\sum_{1 \leq a < b \leq n+2} [\mathfrak{r}_{L\backslash\{l_b\}}, \mathfrak{r}_{L\backslash\{l_a\}}] = 0 ,$$

which has the form of a *classical simplex equation* [6]. For $N = 3$, this is the *classical Yang-Baxter equation*

$$[\mathfrak{r}_{12}, \mathfrak{r}_{13}] + [\mathfrak{r}_{12}, \mathfrak{r}_{23}] + [\mathfrak{r}_{13}, \mathfrak{r}_{23}] = 0 .$$

The 'classical limit' does not work, however, on the level of the 'obstruction equations' (8).

7.2 Equations associated with higher Tamari orders

Let $\mathscr{T}_{N,n}$ be the monoid obtained from $\mathscr{B}_{N,n}$ by replacing (8) with the following $(n+1)$-*gonal relations*,

$$T_{K\setminus\{k_{n+1}\}} T_{K\setminus\{k_{n-1}\}} \cdots T_{K\setminus\{k_{1+(n\bmod 2)}\}} = T_K \, T_{K\setminus\{k_{2-(n\bmod 2)}\}} \cdots T_{K\setminus\{k_{n-2}\}} T_{K\setminus\{k_n\}} . \tag{11}$$

This means that we omit all factors in (8) with a non-visible index set.

Assuming that the corresponding statement for $\mathscr{B}_{N,n}$ holds, it follows that these relations imply the *consistency conditions*

$$T_{L\setminus\{l_{n+2}\}} T_{L\setminus\{l_n\}} \cdots T_{L\setminus\{l_{2+(n\bmod 2)}\}} = T_{L\setminus\{l_{1+(n\bmod 2)}\}} \cdots T_{L\setminus\{l_{n-1}\}} T_{L\setminus\{l_{n+1}\}}$$

for all $L = \{l_1, \ldots, l_{n+2}\} \in \binom{[N]}{n+2}$ in linear order. These are special $(n+2)$-gonal relations. For $n = 1, 2, 3, 4$, they are listed in the following table.

n	$(n+1)$-gonal relations	consistency conditions
1	$\Theta_i = X_{ij}\Theta_j$	$X_{ij}X_{jk} = X_{ik}$
2	$X_{ij}X_{jk} = Y_{ijk}X_{ik}$	$Y_{ijk}Y_{ikl} = Y_{jkl}Y_{ijl}$
3	$Y_{ijk}Y_{ikl} = T_{ijkl}\,Y_{jkl}Y_{ijl}$	$T_{ijkl}T_{ijlm}T_{jklm} = T_{iklm}T_{ijkm}$
4	$T_{ijkl}T_{ijlm}T_{jklm} = S_{ijklm}\,T_{iklm}T_{ijkm}$	$S_{ijklm}S_{ijkmq}S_{iklmq} = S_{jklmq}S_{ijlmq}S_{ijklq}$

Here we demand that the indices are ordered such that, e.g., $i < j$ for $n = 1$ and $i < j < k < l < m < q$ for $n = 4$, and we set

$$\Theta_i = T_{\{i\}} , \quad X_{ij} = T_{\{i,j\}} , \quad Y_{ijk} = T_{\{i,j,k\}} , \quad T_{ijkl} = T_{\{i,j,k,l\}} , \quad S_{ijklm} = T_{\{i,j,k,l,m\}} .$$

For example, for $n = 3$, the tetragonal relations imply

$$
\begin{aligned}
(Y_{ijk}Y_{ikl})Y_{ilm} &= T_{ijkl}Y_{jkl}Y_{ijl}Y_{ilm} = T_{ijkl}Y_{jkl}(Y_{ijl}Y_{ilm}) = T_{ijkl}Y_{jkl}T_{ijlm}Y_{jlm}Y_{ijm} \\
&= T_{ijkl}T_{ijlm}(Y_{jkl}Y_{jlm})Y_{ijm} = T_{ijkl}T_{ijlm}T_{jklm}Y_{klm}Y_{jkm}Y_{ijm} \\
Y_{ijk}(Y_{ikl}Y_{ilm}) &= Y_{ijk}T_{iklm}Y_{klm}Y_{ikm} = T_{iklm}Y_{ijk}Y_{klm}Y_{ikm} = T_{iklm}Y_{klm}(Y_{ijk}Y_{ikm}) \\
&= T_{iklm}Y_{klm}T_{ijkm}Y_{jkm}Y_{ijm} = T_{iklm}T_{ijkm}Y_{klm}Y_{jkm}Y_{ijm} ,
\end{aligned}
$$

where $i < j < k < l < m$, and we obtain the consistency conditions for the T's.

For $N = 5$, the indices of Y_{ijk} determine the three vertices of a triangle inside a pentagon, whose vertices are numbered consecutively counterclockwise from 1 to 5. A product of three Y's of the kind that appears in the above computations corresponds to a triangulation of the pentagon, see Figure 25. The above computation starts with $Y_{ijk}Y_{ikl}Y_{ilm}$, i.e., the triangulation of the top vertex, and the computation proceeds step by step along the left, respectively right maximal chain.

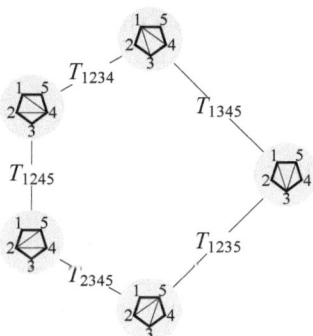

Fig. 25 The two different ways of computing a product of three Y's correspond to the maximal chains of the pentagonal lattice, which is the Tamari lattice $\mathbb{T}_3 = T(5,3)$.

A particularly interesting aspect is that we can construct any of the Tamari orders $T(N,n)$ by starting from $X_{12}X_{23}\cdots X_{N-1,N}$, which corresponds to the minimal element of the triangulations of the cyclic polytope $C(N,1)$ (see [22]). Applying the trigonal relations, we obtain from it

$$(\cdots(X_{12}X_{23})\cdots)X_{N-2,N-1})X_{N-1,N} = Y_{123}Y_{134}Y_{145}\cdots Y_{1,N-1,N}\,X_{1,N}\,.$$

The chain of Y's corresponds to the minimal element of the Tamari lattice $\mathbb{T}_{N-2} = T(N,3)$. Elaborating all possible proper bracketings of this chain, by applications of the tetragonal relations, one obtains the maximal chains of $T(N,3)$. For $N = 6$,

$$(Y_{123}Y_{134})Y_{145}Y_{156} = \cdots = T_{1234}T_{1245}T_{2345}T_{1256}T_{2356}T_{3456}\,Y_{456}Y_{346}Y_{236}Y_{126}$$

produces the coefficient $T_{1234}T_{1245}T_{2345}T_{1256}T_{2356}T_{3456}$, which we recognize as one of the longest maximal chains of the Tamari lattice $\mathbb{T}_4$ in Figure 1. This computation moves step by step along that chain, from one tree to the next, since each product of four Y's appearing in the computation determines the tree (which is dual to a triangulation of the 6-gon) assigned to the respective vertex. In conclusion, we have an algebraic method to construct Tamari lattices and, more generally, higher Tamari orders.

Remark 7.1 can be translated to the present setting. The significance of the equations obtained in this way has still to be explored, but an interesting observation is made in the following remark.

Remark 7.3. Assuming an extension of the monoid $\mathscr{T}_{N,n}$ to a (unital) ring of formal power series in an indeterminate ε, and an expansion

$$T_I = 1 + \varepsilon\, t_I + \cdots,$$

to first order in ε, we obtain from (11) the (simplicial) coboundary conditions

$$t_K = (\delta t)_K \quad \text{where} \quad (\delta t)_K := t_{K\setminus\{k_{n+1}\}} - t_{K\setminus\{k_n\}} + t_{K\setminus\{k_{n-1}\}} - \cdots + (-1)^n t_{K\setminus\{k_1\}}.$$

The consistency conditions are then a consequence of $\delta^2 = 0$. In contrast to the 'Bruhat case' considered in section 7.1, see Remark 7.2, in the present 'Tamari case' all equations possess a (formal) 'classical limit', which moreover turns out to be 'cohomological'.

8 Further remarks

The KP equation (and its hierarchy) has connections with various areas of mathematics and (mathematical) physics. The relation with higher Bruhat and higher Tamari orders established in [2] and in the present work provides another example.

We did not explore the precise relation of the higher Tamari orders introduced in this work and the 'higher Stasheff-Tamari posets' of Kapranov and Voevodsky [8] (also see [3, 22]). It may well be that these orders coincide, which is indeed so for low-order examples.

The higher Bruhat orders and the induced Tamari orders are supplied with a higher-dimensional category structure via the evolution variables of the KP hierarchy. It has obvious relations to Ross Street's algebra of oriented simplexes [23] (also see [8]).

Acknowledgement We would like to thank Jim Stasheff for very valuable comments and questions that led to several improvements of this work.

References

1. V. Bazhanov and Y. Stroganov, "Conditions of commutativity of transfer matrices on a multidimensional lattice", *Theor. Math. Phys.* **52** (1982) 685–691.
2. A. Dimakis and F. Müller-Hoissen, "KP line solitons and Tamari lattices", *J. Phys. A: Math. Theor.* **44** (2011) 025203.
3. P. Edelman and V. Reiner, "The higher Tamari posets", *Mathematika* **43** (1996) 127–154.
4. S. Felsner and H. Weil, "A theorem on higher Bruhat orders", *Discr. Comput. Geom.* **23** (2000) 121–127.
5. S. Felsner and G. Ziegler, "Zonotopes associated with higher Bruhat orders", *Discr. Math.* **241** (2001) 301–312.
6. I. Frenkel and G. Moore, "Simplex equations and their solutions", *Commun. Math. Phys.* **138** (1991) 259–271.
7. J. Hietarinta and F. Nijhoff, "The eight tetrahedron equations", *J. Math. Phys.* **38** (1997) 3603–3615.

8. M. Kapranov and V. Voevodsky, "Combinatorial-geometric aspects of polycategory theory: pasting schemes and higher bruhat orders (list of results)", *Cahiers de topologie et géométrie différentielle catégoriques* **32** (1991) 11–27.

9. Y. Kodama, "KP solitons in shallow water", *J. Phys. A: Math. Theor.* **43** (2010) 434004.

10. I. Korepanov, "Tetrahedral Zamolodchikov algebras corresponding to Baxter's L-operators", *Comm. Math. Phys.* **154** (1993) 85–97.

11. _____, "The tetrahedron equation and algebraic geometry", *J. Math. Sci.* **83** (1997) 85–92.

12. R. Lawrence, "Algebras and triangle relations", *J. Pure Appl. Alg.* **100** (1995) 43–72.

13. _____, "Yang-Baxter type equations and posets of maximal chains", *J. Comb. Theory, Ser. A* **79** (1997) 68–104.

14. J.-L. Loday, "Dichotomy of the addition of natural numbers", in *this volume*.

15. I. Macdonald, *Symmetric functions and Hall polynomials*, 2nd ed., Oxford University Press, Oxford, 1995.

16. J. Maillet and F. Nijhoff, "Multidimensional latttice integrability and the simplex equations", in *Nonlinear Evolution Equations: Integrability and Spectral Methods*, A. Degasperis, A. Fordy, and M. Lakshmanan, eds., Manchester University Press, 1990, 537–548.

17. Y. Manin and V. Schechtman, "Arrangements of real hyperplanes and Zamolodchikov equations", in *Group Theoretical Methods in Physics*, M. Markov, V. Man'ko, and V. Dodonov, eds., Adv. Stud. Pure Math., vol. 17, VNU Science Press, 1986, 151–165.

18. _____, "Higher Bruhat orders, related to the symmetric group", *Funct. Anal. Appl.* **20** (1986) 148–150.

19. _____, "Arrangements of hyperplanes, higher braid groups and higher Bruhat orders", in *Algebraic Number Theory – in honor of K. Iwasawa*, J. Coates, R. Greenberg, B. Mazur, and I. Satake, eds., Adv. Stud. Pure Math., vol. 17, Academic Press, 1989, 289–308.

20. F. Michielsen and F. Nijhoff, "D-algebras, the D-simplex equations, and multidimensional integrability", in *Quantum Topology*, L. Kauffman and R. Baadhio, eds., Series on Knots and Everything, vol. 3, World Scientific, 1993, 230–243.

21. J. Perk and H. Au-Yang, "Yang-Baxter equations", in *Encyclopedia of Mathematical Physics*, J.-P. Françoise, G. Naber, and S. Tsou, eds., vol. 5, Elsevier Science, 2006, 465–473.

22. J. Rambau and V. Reiner, "A survey of the higher Stasheff-Tamari orders", in *this volume*.

23. R. Street, "The algebra of oriented simplexes", *J. Pure Appl. Alg.* **49** (1987) 283–335.

24. D. Tamari, "Monoides préordonnés et chaînes de Malcev", Doctorat ès-Sciences Mathématiques Thèse de Mathématique, Paris, 1951.

25. _____, "The algebra of bracketings and their enumeration", *Nieuw Arch. Wisk.* **10** (1962) 131–146.

26. H. Thomas, "Maps between higher Bruhat orders and higher Stasheff-Tamari posets", in *FPSAC 15th Anniversary International Conference on Formal Power Series and Algebraic Combinatorics*, K. Eriksson, A. Björner, and S. Linusson, eds., 2003.

27. M. Voevodski and M. Kapranov, "Free n-categories generated by a cube, oriented matroids, and higher Bruhat orders", *Funct. Anal. Appl.* **25** (1990) 50–52.

28. Wolfram Research, Inc., *Mathematica 8.0*, Wolfram Research, Inc., Champaign, Illinois, 2010.

29. A. Zamolodchikov, "Tetrahedron equations and the relativistic S-matrix of straight-strings in 2+1-dimensions", *Commun. Math. Phys.* **79** (1981) 489–505.

30. G. Ziegler, "Higher Bruhat orders and cyclic hyperplane arrangements", *Topology* **32** (1993) 259–297.

Appendix
Dov Tamari's Publications

1. "On a certain classification of rings and semigroups", *Bulletin of the American Mathematical Society* **54** (1948) 153–158.
2. "Caractérisation des semi-groupes à un paramètre", *Comptes rendus hebdomadaires des séances de l'Académie des sciences* **228** (1949) 1092–1094.
3. "Groupoïdes reliés et demi-groupes ordonnés", *Comptes rendus hebdomadaires des séances de l'Académie des sciences* **228** (1949) 1184–1186.
4. "Groupoïdes ordonnés, l'ordre lexicographique pondéré", *Comptes rendus hebdomadaires des séances de l'Académie des sciences* **228** (1949) 1909–1911.
5. "Ordres pondérés. Caractérisation de l'ordre naturel comme l'ordre du semigroupe multiplicatif des nombres naturels", *Comptes rendus hebdomadaires des séances de l'Académie des sciences* **229** (1949) 98–100.
6. "Les images homomorphes des groupoides de Brandt et l'immersion des semigroupes", *Comptes rendus hebdomadaires des séances de l'Académie des sciences* **229** (1949) 1291–1293.
7. "Sur l'immersion d'un semi-groupe topologique dans un groupe topologique", *Algèbre et Théorie des Nombres, Colloques Internationaux du Centre National de la Recherche Scientifique*, no. 24 (1950) 217–221.
8. "Représentations isomorphes par des systèmes de relations –. Systèmes associatifs", *Comptes rendus hebdomadaires des séances de l'Académie des sciences* **232** (1951) 1332–1334.
9. "Monoides préordonnés et chaînes de Malcev", Doctorat ès-Sciences Mathématiques Thèse de Mathématique, Paris (1951).
10. "Machines logiques et problèmes de mots. I: les machines de Turing (T.M.)", *Séminaire N. Bourbaki*, Exp. No. 55 (1951) 47–58.
11. "Machines logiques et problèmes de mots. II: Problèmes des mots indecidable", *Séminaire N. Bourbaki*, Exp. No. 61 (1952) 109–119.
12. "On the embedding of Birkhoff-Witt rings in quotient fields", *Proceedings of the American Mathematical Society* **4** (1953) 197–202.
13. "Some mutual applications of logic and mathematics", in: *Applications Scientifiques de la Logique Mathématique, Actes du 2ᵉ Colloque International de*

Logique Mathématique, Paris – 25–30 Aout 1952, Gauthier-Villars, Paris, 1954, 89–90.

14. "Monoïdes préordonnés et chaînes de Malcev", *Bulletin de la Société Mathématique de France* **82** (1954) 53–96.

15. "On a generalization of uniform structures and spaces", *Bulletin of the Research Council of Israel* **3** (1954) 417–428.

16. "Une contribution aux théories modernes de communication: Machines de Turing et problèmes de mot", *Synthèse* **9** (1955) 205–227.

17. "A refined classification of semi-groups leading to generalized polynomial rings with a generalized degree concept", in: *Proceedings of the International Congress of Mathematicians, Amsterdam* 1954, North-Holland, Amsterdam, 1957, 439–440.

18. "Nonassociative systems satisfying the Malcev conditions", *Notices of the American Mathematical Society* (1959) **6** 765.

19. "Generalization of a theorem of Rees", *Notices of the American Mathematical Society* **6** (1959) 812.

20. "Interpretations en mathématiques", *Synthese* **11** (1959) 167–176.

21. " 'Near-groups' as generalized normal multiplication tables", *Notices of the American Mathematical Society* **7** (1960) 77.

22. "Imbeddings of partial (incomplete) multiplicative systems (monoids), associativity and word problem", *Notices of the American Mathematical Society* **7** (1960) 760.

23. "Families of binary relations and analytical functions", *Notices of the American Mathematical Society* **7** (1960) 982.

24. "Representations of multiplicative systems by families of binary relations (I)", with Abraham Ginzburg, *Journal of the London Mathematical Society* **37** (1962) 410–423.

25. "Representação de sistemas multiplicativos por famílias de relações binárias e suas conexões com a teoria de funções analíticas e da continuação analítica", *Summa Brasiliensis Mathematicae* (Atas do terceiro Colóquio Brasileiro de Matemática) **5** (1962) 217–258.

26. "Sur l'associativité partielle des symétrisations de semi-groupes", with João Bosco de Carvalho, *Portugaliae Mathematica* **21** (1962) 157–169.

27. "The algebra of bracketings and their enumeration", *Nieuw Archief voor Wiskunde (3)* **10** (1962) 131–146.

28. "Problèmes d'associativité des monoïdes et problèmes des mots pour les groupes", *Séminaire Dubreil-Pisot (Algèbre et Théorie des nombres)* **16**(7) (1962/63) 1–29.

29. "Sur quelques problèmes d'associativité", *Annales scientifiques de l'Université de Clermont-Ferrand 2, Série Mathématiques* **24** (1964) 91–107.

30. "The unsolvability of the associativity problem for finite monoids and its equivalence to the word problem for groups", abstract on page 159 of R.M. Smullyan, Meeting of the Association for Symbolic Logic, *The Journal of Symbolic Logic* **29** (1964) 150–162.

31. "Problèmes d' associativité: une structure de treillis finis induite par une loi demi-associative", with Haya Friedman, *Journal of Combinatorial Theory* **2** (1967) 215–242.

32. "Representation of binary systems by families of binary relations", with Abraham Ginzburg, *Israel Journal of Mathematics* **7** (1969) 21–32.

33. "Representation of generalized groups by families of binary relations", with Abraham Ginzburg, *Israel Journal of Mathematics* **7** (1969) 33–45.

34. "The equivalence of associativity and word-problems", *Queen's Papers in Pure and Appl. Math.* **25** (1970) 171–189 (Proceedings of the Conference on Universal Algebra, October 1969, Queen's University, Kingston, Ontario, Canada).

35. "Les problèmes d'associativité des monoïdes et le problème des mots pour les demi-groupes; algèbres partielles et chaînes élémentaires", *Séminaire Dubreil-Pisot (Algèbre et Théorie des nombres)* **24**(8) (1970/71) 1–15.

36. "Formulae for well formed formulae", *Notices of the American Mathematical Society* **18** (1971) 928–929.

37. "Polyhedral structure of bracket complexes", with D. de Fougères, *Notices of the American Mathematical Society* **19** (1972) A-31.

38. "Problems of associativity: a simple proof for the lattice property of systems ordered by a semi-associative law", with Samuel Huang, *Journal of Combinatorial Theory (A)* **13** (1972) 7–13.

39. "The associativity problem for monoids and the word problem for semigroups and groups", in: *Word Problems – Decision Problems and the Burnside Problem in Group Theory*, Studies in Logic and the Foundations of Mathematics, vol. 71, W. Boone, F. Cannonito, R. Lyndon, eds., North-Holland, Amsterdam, 1973, 591–607.

40. "Formulae for well formed formulae and their enumeration", *Journal of the Australian Mathematical Society* **17** (1974) 154–162.

41. "Embedding a semigroup in a group", with Kevin E. Osondu, *Notices of the American Mathematical Society* (1975) A-5.

42. "Deciding associativity for partial multiplication tables of order 3", with Paul W. Bunting and Jan van Leeuwen, *Mathematics of Computation* **32** (1978) 593–605.

43. "La structure polyédrale des complexes de parenthésages", with Danièle Huguet, *Journal of Combinatorics, Information & System Sciences* **3** (1978) 69–81.

44. "ALGEBRA – its place in Mathematics and its role, past and present, for mankind. A historical and critical essay.", ΕΛΕΥΘΕΡΙΑ (*Eleutheria*), *Mathematical Journal of the Seminar P. Zervos* (September 1978) 187–201.

45. "A graphic theory of associativity and wordchain patterns", *Lecture Notes in Mathematics* **969** (1982) 302–320.

46. "Une théorie constructive de l'associativité", ΕΛΕΥΘΕΡΙΑ (*Eleutheria*), *Mathematical Journal of the Seminar P. Zervos*, Volume in honor of M. Krasner (1986) 93–127.

47. "Associativity theory and the theory of lists. Their applications from abstract algebra to the four-colour map problem", *Congressus Numerantium* **54** (1986) 39–53.

48. "Dual representation of standard maps as double triangulations. Preliminary Report", with Douglas Bowman, *Abstracts of papers presented to the AMS* **9** (1988) 415.
49. "Coordinatization and anatomy of standard maps (polyhedra) with Whitney cycles. Preliminary Report", with Douglas Bowman, *Abstracts of papers presented to the AMS* **9** (1988) 502.
50. "Directed 3-colored (3c) Triangulation (T.) Part II. Compositions and Congruences. Conjecture $W = \overline{\overline{A^*}}$. Preliminary Report", *Abstracts of papers presented to the AMS* **10** (1989) 276.
51. "Directed 3-colored (3c) double Triangulation (T.) Part I. Definitions. Associativity (A.). Preliminary Report", *Abstracts of papers presented to the AMS* **10** (1989) 304.
52. *Moritz Pasch* (1843–1930). *Vater der modernen Axiomatik. Seine Zeit mit Klein und Hilbert und seine Nachwelt. Eine Richtigstellung*, Shaker Verlag, Aachen, 2007.

This list may not be complete. We included brief announcements of results (abstracts), like the publications in the *Notices* and *Abstracts AMS*. Tamari wrote many reviews for *Zentralblatt für Mathematik*, some of quite unusual length. Extreme examples are his review of the book *Automata Studies* by Claude E. Shannon and John McCarthy (Ann. Math. Studies **34**, Princeton Univ. Press, 1956), see *Zentralblatt für Mathematik* **74** (1960) 112–116, and his review on "The word problem" by William W. Boone, *Ann. Math.* **70** (1959) 207–265, see *Zentralblatt für Mathematik* **102** (1963) 9–14. His *unpublished* work includes the following manuscripts.

1. "Foundations of graphic associativity theory (G.A.T.) and 4CM" (1990).
2. "Special graphic associativity theory (G.A.T.1 or Contraction A.), Part I: Parsing and lists" (later than 1990).
3. "Elementary theory of special lists and their poset-lattices" (1996).
4. "200 years theorem of Gauß 1797–1997" (1997).
5. "The Euclidean way. Euclid, the man and the mathematician. Euclid and modern civilization. From Euclid to Pasch and Hilbert" (1997).
6. "The story: On proofs and editorial policies" (1997).

Four further papers are cited in those listed above. Either the stated publication data could not be verified or we were unable to obtain a corresponding manuscript.

Index

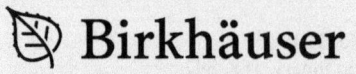
Birkhäuser · birkhauser-science.com

Progress in Mathematics (PM)

Edited by
Hyman Bass, University of Michigan, USA
Joseph Oesterlé, Institut Henri Poincaré, Université Paris VI, France
Alan Weinstein, University of California, Berkeley, USA
Yuri Tschinkel, Courant Institute of Mathematical Sciences, New York, USA

Progress in Mathematics is a series of books intended for professional mathematicians and scientists, encompassing all areas of pure mathematics. This distinguished series, which began in 1979, includes research level monographs, polished notes arising from seminars or lecture series, graduate level textbooks, and proceedings of focused and refereed conferences. It is designed as a vehicle for reporting ongoing research as well as expositions of particular subject areas.